Neurotransmitter Interactions and Cognitive Function

COVER: Cluster of cholinergic neurons in the basal nuclear complex of a rat demonstrated by choline acetyltransferase immunohistochemistry. The overlapping dendrites and logic symbols suggest the rich interactions and complex computational operations that could occur in this region of the brain in relation to cognitive function. Photomicrograph from the laboratory of Larry L. Butcher.

Neurotransmitter Interactions and Cognitive Function

Edward D. Levin
Michael W. Decker
Larry L. Butcher
Editors

James L. McGaugh
Foreword

73 Illustrations with some color

Birkhäuser
Boston • Basel • Berlin

Edward D. Levin
Department of Psychiatry
Duke University Medical Center
Durham, NC 27710
USA

Michael W. Decker
Abbott Laboratories
Abbott Park, IL 60064
USA

Larry L. Butcher
Laboratory of Chemical Neuroanatomy
Department of Psychology
University of California
Los Angeles, CA 90024-1563
USA

Library of Congress Cataloging-in-Publication Data

Neurotransmitter interactions and cognitive function / Edward D.
Levin, Michael Decker, Larry L. Butcher, editors.
p. cm.
Includes bibliographical references and index.

1. Neurotransmitters. 2. Cognition. 3. Cognitive neuroscience.
I. Levin, Edward D. II. Decker, Michael, 1951– . III. Butcher,
Larry L.
[DNLM: 1. Cognition—physiology—congresses. 2. Neuroregulators-
-physiology—congresses. QV 126 N49385]
QP364.7.N468 1992
612.8′2—dc20
DNLM/DLC
for Library of Congress 92-17848
CIP

Printed on acid-free paper.

Softcover reprint of the hardcover 1st edition 1992

ISBN-13: 978-1-4615-9845-9 e-ISBN-13: 978-1-4615-9843-5
DOI: 10.1007/978-1-4615-9843-5

Typeset by Atlis Graphics and Design, Mechanicsburg, PA

9 8 7 6 5 4 3 2 1

Contents

Foreword

JAMES L. MCGAUGH

Understanding of the nature and functions of neurotransmitter systems in the brain has increased enormously in recent decades. Lack of knowledge required us, not too long ago, to use the adjective "putative" when discussing transmitters. Such caution is no longer essential (at least for a number of transmitters). Impressive progress has been achieved in understanding the pharmacology, biochemistry and anatomy of transmitter systems. There has, however, been relatively less progress in understanding the functioning of brain transmitters in regulating and mediating behavior. A simple and certainly correct explanation for this is, of course, that understanding of neurotransmitter functions requires prior detailed knowledge of basic pharmacology, biochemistry and anatomy. Beyond that, it now seems likely that progress in understanding the functions of brain neurotransmitters will proceed only as we examine the *interactions* of neurotransmitter systems in regulating behavioral functions. This premise is, of course, suggested by the findings of studies of the chemical neuroanatomy of the brain: Neurotransmitter systems are influenced by other neurotransmitter systems and, in turn, influence the same as well as other systems. No system works alone. The chapters in this book explicitly examine the interactions of neurotransmitter systems involved in the regulation of cognitive processes. The facts and interpretations offered provide compelling support for the premise that cognitive processes are orchestrated by interactions among neurotransmitter systems. And, they offer promise that understanding of such interactions will be of critical importance in the development of treatments for brain diseases affecting cognitive functioning.

James L. McGaugh
Center for the Neurobiology of
Learning and Memory and
Department of Psychobiology
University of California, Irvine

May 1992

Preface

Cholinergic systems have been extensively studied over a period of decades as neural substrates for cognitive function. However, acetylcholine is not monolithic in its importance for learning and memory. A variety of other transmitter systems, including dopamine, norepinephrine, serotonin, and GABA, have also been found to be critical for cognitive processing. Recently, the study of transmitter system involvement in cognitive function has evolved to the point where interactions between transmitter systems has become a focus of investigation. This involves a transition in thinking from a reductionistic approach to determine the critical involvement of specific anatomic and physiological processes in cognitive function to an integrative approach to determine how neural systems interact to form the complex nature of cognition.

The purpose of this volume is to convey the importance of the study of neural integration as well as to elucidate the specific details of the anatomical and physiological interations of transmitter systems with regard to cognitive function. This project began with a Satellite Symposium to the Society for Neuroscience meeting in New Orleans on November 10, 1991. Participants of that symposium, as well as additional experts in the field, have contributed chapters for this book. We hope that this volume is not only useful in the transmission of new information about the importance of neural integration for cognitive function, but that it also serves to encourage the interaction of scientists studying a variety of systems to further advance our understanding of neurobehavioral relationships.

We would like to thank the following companies for their support of this endeavor: Abbott Laboratories; Roerig, a division of Pfizer Pharmaceuticals; Hoechst-Roussel Pharmaceuticals; and Schering-Plough Research.

Edward D. Levin
Michael W. Decker
Larry L. Butcher

March 1992

List of Contributors

Harvey J. Altman, Department of Psychiatry, Wayne State University School of Medicine, Lafayette Clinic, Detroit, Michigan 48207, USA

Marcus L. Brandão, Laboratory of Psychobiology, FFCLRP, University of São Paulo, 14.049 Ribeirão Preto, SP, Brazil

Gilberto N. O. Brito, Setor de Neurociencias, Universidade Federal Fluminense, Niteroi, RJ 24001, Brazil

John P. Bruno, Department of Psychology and Neuroscience Program, Ohio State University, Columbus, Ohio 43210, USA

Larry L. Butcher, Laboratory of Chemical Neuroanatomy, Department of Psychology, University of California, Los Angeles, California 90024-1563, USA

James J. Chrobak, Center for Molecular and Behavioral Neuroscience, Rutgers University, Newark, New Jersey 07102, USA

Kenneth L. Davis, Department of Psychiatry, The Mount Sinai School of Medicine, New York, New York 10029-6574, USA

Michael W. Decker, Abbott Laboratories, Abbott Park, Illinois 60064, USA

Paul Dudchenko, Department of Psychology and Neuroscience Program, Ohio State University, Columbus, Ohio 43210, USA

Norberto Garcia-Cairasco, Neurophysiology and Neuroethology Laboratory, FMRP, University of São Paulo, SP, Brazil

Bennet Givens, Department of Psychology, Ohio State University, Columbus, Ohio 43210, USA

Vahram Haroutunian, Department of Psychiatry, The Mount Sinai School of Medicine, New York, New York 10029-6574, USA

Lee Ann Holley, Department of Psychology and Neuroscience Program, Ohio State University, Columbus, Ohio 43210, USA

Robert Jaffard, Laboratoire de Neurosciences Comportementales et Cognitives, URA CNRS, Université de Bordeaux I, Institut de Biologie Animale, 33405 Talence Cedex, France

Pekka Jäkälä, Department of Neurology, University of Kuopio, SF-70211 Kuopio, Finland

Esa Koivisto, Department of Neurology, University of Kuopio, SF-70211 Kuopio, Finland

Risto Lammintausta, Orion Corporation FARMOS, R&D Pharmaceuticals, SF-20101 Turku, Finland

Edward D. Levin, Department of Psychiatry, Duke University Medical Center, Behavioral Neuroscience Laboratory, Box 3557, Durham, North Carolina 27710, USA

Aline Marighetto, Laboratoire de Neurosciences Comportementales et Cognitives, URA CNRS, Université de Bordeaux I, Institut de Biologie Animale, 33405 Talence Cedex, France

Jacques Micheau, Laboratoire de Neurosciences Comportementales et Cognitives, URA CNRS, Université de Bordeaux I, Institut de Biologie Animale, 33405 Talence Cedex, France

Holly Moore, Department of Psychology and Neuroscience Program, Ohio State University, Columbus, Ohio 43210, USA

T. Celeste Napier, Department of Pharmacology, and the Neuroscience Program, Loyola University of Chicago, Stritch School of Medicine, Maywood, Illinois 60153, USA

Howard J. Normile, Department of Psychiatry, Wayne State University School of Medicine, Lafayette Clinic, Detroit, Michigan 48207, USA

David S. Olton, Department of Psychology, The Johns Hopkins University, Baltimore, Maryland 21218, USA

Kevin Pang, Department of Psychology, The Johns Hopkins University, Baltimore, Maryland 21218, USA

D. Penava, Department of Psychology, University of Western Ontario, London, Ontario, Canada N6A 5C2

Gal Richter-Levin, The Department of Neurobiology, The Weizmann Institute of Science, Rehovot 76100, Israel

Paavo J. Riekkinen, Department of Neurology, University of Kuopio, SF-70211 Kuopio, Finland

Paavo Riekkinen, Jr., Department of Neurology, University of Kuopio, SF-70211 Kuopio, Finland

Jed E. Rose, Department of Psychiatry, Duke University Medical Center, Nicotine Research Laboratory, VA Medical Center, Durham, North Carolina 27705, USA

Roger W. Russell, Center for the Neurobiology of Learning and Memory and the Department of Psychobiology, University of California, Irvine, California 92717, USA

A. C. Santucci, Department of Psychology, Manhattanville College, Purchase, New York 10577, USA

Martin Sarter, Department of Psychology and Neuroscience Program, Ohio State University, Columbus, Ohio 43210, USA

Menahem Segal, The Department of Neurobiology, The Weizmann Institute of Science, Rehovot 76100, Israel

Jouni Sirviö, Department of Neurology, University of Kuopio, SF-70211 Kuopio, Finland

Robert W. Stackman, Department of Psychology, Rutgers University, New Brunswick, New Jersey 08903, USA

Carlos Tomaz, Laboratory of Psychobiology, FFCLRP, University of São Paulo, SP, Brazil

Antti Valjakka, Department of Neurology, University of Kuopio, SF-70211 Kuopio, Finland

C. H. Vanderwolf, Department of Psychology, University of Western Ontario, London, Ontario, Canada N6A 5C2

Thomas J. Walsh, Department of Psychology, Rutgers University, New Brunswick, New Jersey 08903, USA

Gary L. Wenk, Division of Neural Systems, Memory, and Aging, University of Arizona, Tucson, Arizona 85724, USA

László Záborszky, Departments of Neurology, Neurosurgery, and Otolaryngology, University of Virginia Health Sciences Center, Charlottesville, Virginia 22908, USA

1

Interactions among Neurotransmitters: Their Importance to the "Integrated Organism"

ROGER W. RUSSELL

Introduction

Knowledge develops by quantal steps that depend upon the invention of new concepts and new techniques. Our area of the biobehavioral sciences is now in a period of rapid development, which provides exciting challenges for all who care to give it thought. It was almost a half-century ago when I participated in a symposium on "Physiological Mechanisms in Animal Behavior" organized by the Society for Experimental Biology and held at Cambridge University in England. The most dramatic moment of the meeting was provided by Karl Lashley in his presentation entitled, "In Search of the Engram." In ending a summary of 30 years of his research using surgical ablation techniques, he commented, "I sometimes feel, in reviewing the evidence on the localization of the memory trace, that the necessary conclusion is that learning just is not possible" (Lashley, 1950). Riding back to London on the train I asked him: "What next?" He expounded at some length on the necessity of seeking the nature of dynamic events occurring within neuronal sites and of looking for interactions between them and the behavior of the "integrated organism." Some 15 years later two professors—this time eminent biochemists—anticipated an agenda, an approach to uncovering the nature of that integration. They published an article entitled, "Introduction to Molecular Psychology" (Moore and Mahler, 1965), in which they wrote: "We are today close to a critical point in history at which the mind of man by taking thought (and making experiments) will be able to understand the chemical and physical mechanisms responsible for thought itself. Just as molecular biology has provided a chemical basis for heredity and evolution, molecular psychology will elucidate the chemistry of memory, learning, sensation, emotion and finally human consciousness." The agenda sounded familiar: Search for the basic building blocks and then recreate the integrated organism. It was familiar because it was—and is—the approach of the physical and biological sciences generally that, it was being proposed, applied to the neurobehavioral sciences.

Thus our branch of the biobehavioral sciences has proceeded. Has this been "bad" in some sense of that term? Unfortunate? Certainly not. However, it has given us the long-range "responsibility of putting it all together," as my friends in psychology and psychiatry occasionally remind me. After they are discovered, ways in which neurotransmitter systems interact must be studied and effects on even the most complex functions—the "cognitive" upon which organisms depend for adjustments to their physical and social environments—eventually must be understood. What should be remembered is that, from the beginning, the term *neurotransmitter interactions* included more than interactions between different transmitter systems. It included molecular events within neurotransmitter systems that eventually influence interactions between them. My comments will be couched in terms of this broader definition.

Neurotransmitters in Theoretical Models of Behavior

Now, there is a considerable distance between a transmitter molecule and its effects on behavior. It is not possible to forget the modulatory influences of events intervening between the two, even when we are concentrating specifically on their interactions per se. What interactions? There are a large number of possibilities.

Levels of biological organization at or between which interactions affecting cognitive behaviors may occur are shown schematically in Figure 1. They range from the chemical conformation of molecules to the involvement of different organ systems, such as modulation of events in the nervous system by actions of the endocrine system. Within this very broad range the search for interactions may focus on variations in which anatomical, biochemical, and/or electrophysiological properties serve as the independent variables. For our present purposes today, cognitive behaviors constitute the dependent variables. Examination of the research literature provides multiple examples of ways in which, as knowledge progresses, it becomes possible to investigate combinations of those independent variables.

Some Basic Postulates

To put my comments into proper perspective, it will be well to state briefly certain basic postulates from the theoretical models of neurobehavioral functions with which we all work, postulates that are consistent with empirical findings. First, the primary stimulus by which endogenous or exogenous chemicals affect subsequent physiological and behavioral functions consists of a physiochemical binding between a chemical and a specific receptor site on a biologically active macromolecule in the body. A second postulate is that there exist a finite number of such receptor sites. Thirdly, the magnitudes of events that follow the occurrence of a chemical-receptor complex are related by some function to the number of sites occupied. Fourthly, the nervous system is capable of synthesizing receptors *de novo*. Fifthly, different neurotransmitters are localized in different anatomical sites within the central nervous system (CNS). These postulates will appear frequently in what follows.

What kind of conceptual framework is suggested by them? The most obvious are biostatistical models (Ashford, 1981; Rothman et al., 1980) based on the

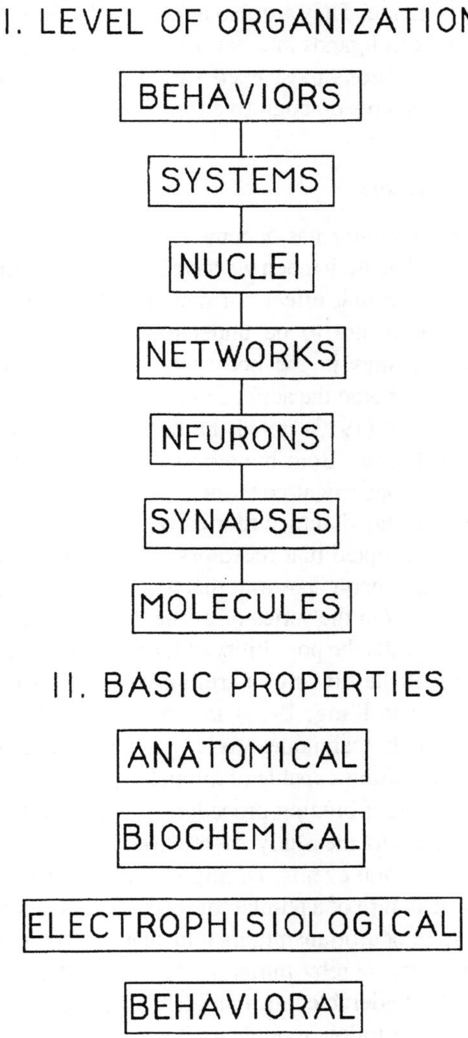

FIGURE 1. Levels of biological organization range from the chemical conformation of molecules to the behaviors of living organisms as they adjust to their physical and psychosocial environments. The search for interactions among them focuses on one or more of the organism's four basic properties: anatomical, biochemical, electrophysiological, and behavioral.

general biological premise that each individual in a population has a definite threshold for responses to stimulation from both internal and external environments. This implies a probability distribution. According to the postulates I have described, receptors play primary roles in the physicochemical events that may eventually influence cognitive behaviors. These behaviors show threshold effects as receptor densities change. They are also influenced by the efficiency with which endogenous or exogenous ligands interact with them. I plan, first, to look at ways in which receptor populations have been varied experimentally and the consequences such variations have on cognitive functions.

Primary Roles of Receptors

Today when the term *receptor* has become almost a household expression, it is often difficult to appreciate how much we have learned about them since the 1970s began. In fact the notion that effects of natural substances and of compounds introduced into the body are to be understood in terms of physicochemical interactions at specific sites in the body was introduced early in the present century. Langley (1906) noted the actions of nicotine and curare on nerve-muscle preparations, and Ehrlich (1913) noted the high degree of antibody specificity against antigenic substances. Both hypothesized the existence of receptors and receptor substances. A theoretical construct of "receptor" had come into being. How was it to be translated into empirically observable terms? As research progressed it became accepted that receptors consist basically of recognition or binding sites and of some second portions that translate the recognition into second messengers, the next step in the series of events leading eventually to effects on behavior. By the mid-1960s the possibility of labelling neurotransmitter receptors was tested in studies of the binding of tritium-labelled cholinergic antagonists, both reversible (Paton and Rang, 1965) and irreversible (Gill and Rang, 1966). The importance of such techniques to an understanding of the actions of neurotransmitters and of drugs capable of mimicking or antagonizing their actions led to the standardization of binding procedures (described by Yamamura et al., 1978) and this, in turn, to the study of relations between receptors and both physiological and behavioral events. During recent years much has been learned about the molecular structure of a number of receptors and about how the receptor translates the binding of neurotransmitters into changes in ion conductance across postsynaptic membranes, thereby initiating communication between cells (McCarthy et al., 1986). Considerable information has also begun to accumulate about ways in which events that follow modulate effects on behavior, but there is much more to be learned.

Multiple Receptors in Behavior

It would seem a most unlikely hypothesis that would propose that "behavior"—particularly as complex cognitive functions—depends upon but one neurotransmitter system. Even such simple actions as the stretch reflex, which involves only an afferent and an efferent neuron, interfaces with other systems by way of receptors on its cell surfaces. Clearly, in the normally functioning organism interactions between neurotransmitter systems are paramount. The several approaches to searching for the nature of such interactions will be illustrated in today's discussions. I wish to comment on approaches of an experimental nature, while realizing that evidence for interactions may also come from epidemiological and clinical sources.

Manipulating Receptors

Accepting this theoretical role for neurotransmitter receptors, the question arises as to how they may be manipulated empirically to test hypotheses about their roles within a system and about how they may influence interactions between systems. Manipulation has been based primarily on techniques for altering receptor populations. Results of some of the earliest studies in several laboratories provided evidence that normal behavior is dependent upon the densities of receptor populations. Within limits a homeostatic condition prevailed whereby hyperstimulation by agonists led to downregulation and hypostimulation, and to upregulation of receptor populations. Beyond such threshold limits abnormal behavioral effects were observed. Since these findings several approaches to manipulating receptor populations have been put to use experimentally.

Selective Breeding

One approach has been to determine whether rodent families could be selectively bred for differing receptor populations. Results of extensive research using this experimental approach have been reported by several investigators. In one extensive series (Overstreet et al., 1984) using selective mating procedures, a line or rats (FSL) has been bred that has elevated numbers of muscarinic receptors in the striatum and hippocampus, is supersensitive to cholinergic stimulation, and shows behavioral and physiological characteristics analogous to depressive states in humans, including cognitive functions. The mating procedure has also produced a resistant line (FRL). Biometric genetic analyses have suggested that muscarinic sensitivity in the FSL animals is under genetic control, with the greatest contribution arising from additive genetic factors (Overstreet et al., 1992). The availability of the two lines has stimulated interest in behavioral effects of challenges with agonists and antagonists specific to other neurotransmitter systems. Both inbred lines of animals have now been challenged with a wide variety of chemical agents. Behavioral and physiological functions in FSL animals have been shown to be more sensitive to cholinergic agonists and to other agents tested, including a serotonergic agonist. An exception is challenges by dopaminergic agents, which produced sub- or supersensitivity depending upon the function measured. This effect occurred despite the failure to find differences in concentrations of D_2 receptors in the two lines inbred for very significant differences in concentrations of cholinergic receptors (Crocker and Overstreet, 1991). The results have been interpreted to ". . . indicate that the changes in dopamine sensitivity which accompany cholinergic supersensitivity are function-dependent, but are not associated with parallel changes in dopamine receptor concentration." In other words, interactions between neurotransmitter systems, although affecting some behaviors, are not inevitably involved directly in all behavioral functions.

Recovery from Irreversible Binding

Another experimental approach to manipulating a receptor population is by the administration of irreversible agonists (e.g., Russell et al., 1986, 1989) or antagonists. Such compounds interact covalently, sharing electrons with their binding site. Replacement of a normal population of functional sites depends upon *de novo* synthesis. Results of a series of experiments designed to study effects of such a compound (BM123) will serve as an example of interactions between neurotransmitter functions and behavioral outputs (Russell et al., 1986). The central objective was to observe interactions between muscarinic and nicotinic transmitter systems when the former was varied and the latter was held at its normal level. BM123 was chosen as the pharmacological tool because of its selective action in reducing the binding of [^{3}H]NMS and [^{3}H](−)QNB, while being almost inactive at nicotine receptors. BM123 was injected into the tail vein of Sprague-Dawley rats, reducing the population of muscarinic receptor to approximately 10% of normal. Recovery followed an exponential time course, with return to control levels in approximately 8 days. A variety of behavioral and physiological end points were measured during the recovery. It is a reasonable hypothesis that different end points are differentially receptor dependent, that they require different densities of receptor occupancy to function normally. The probability of a threshold being reached is related to the density of the receptor population active at the time. Results of these experiments demonstrated a hierarchy of recovery times in which thresholds for normal functioning of the various end points measured were in the following order from low to high: (a) physiological variables, (b) sensory-perceptual, (c) motoric, and (d) cognitive. Learned responses and those requiring temporal discrimination were the only variables paralleling the return of the muscarinic receptors to their normal population levels. Remember that throughout the recovery period nicotinic receptors remained unchanged. Such experiments make it possible to study hypothesized interactions between neurotransmitter systems by holding one essentially constant while varying the other (e.g., Crocker and Russell, 1990).

Aging

Other research has demonstrated that differential effects may be related to aging, as well as genetic factors. For example, two strains of mice born with the same number of cholinergic neurons developed clear differences by 45 days of age, one strain characterized ". . . by a lower cell density in the *nucleus basalis Meynert*" (Castellano et al., 1988). The strains also differed in sensitivity to stimulation by muscarinic (oxotremorine) and nicotinic (nicotine) agonists. As aging processes progress, declines occur both functionally and structurally in the cholinergic system (Decker, 1987). Such changes have been seen in other neurotransmitter systems as well and in interactions between them: ". . . one of the striking findings in the aging brain appears to be the progressive imbalance between dopaminergic and cholinergic tone. . ." (Pradhan, 1986). With aging there also occurs serum protein leakage in the human brain, accompanied by interference of ligand binding at α_2-adrenergic and cholinergic binding sites (Pappolla and Andorn, 1987). We should note that such a condition could give rise to erroneous interpretations of neuronal loss and could also alter the relative toxicities of exogenous compounds affecting the neurotransmitter systems involved (Benke and Murphy, 1975). Aging may also affect the normal process involving increases and decreases of receptor numbers in response to hypo- and hyperfunctioning of the cholinergic system, which enable the CNS to compensate for changes of chemical neurotransmission, including those that are pathological. Research using animal models has shown that such compensatory regulation occurred in young but not in old animals (Pilch and Müller, 1988). Studies of over 140 receptor systems have reported that only a few change affinity during aging: "Essentially all the correlations established between age changes in receptors and responsiveness (behavior) have been in cases of reductions with increasing age" (Roth and Hess, 1982).

Regional Variations

Variations in receptor density or in the efficiency of coupling between receptor occupancy and cellular response in different regions of the CNS may also affect interactions between neurotransmitters (Kenakin, 1986). There is experimental evidence that compounds with high affinity and low intrinsic efficacy are the most likely to exert selectivity by this mechanism: "In particular, responses that appear to be associated with a small receptor reserve for muscarinic agonists . . . may be separated from those that have a large receptor reserve . . ." (Ringdahl et al., 1987). This suggests that selectivity of the effects of such agonists may occur even in the absence of distinct receptor subtypes. A corollary is that subtle changes in molecular structure may influence significantly the effects of interactions between neurotransmitters and their receptors, as well as between transmitters.

Transmitter–Receptor Interactions

"False" Transmitter

The very important effects on cognitive functions of even very small changes in the chemical structure of a neurotransmitter are clearly apparent in reports of a series of experiments designed to test a hypothesis about effects on cognitive functions of competition for available choline (Ch) between biochemical pathways involved in phospholipid metabolism and acetylcholine (ACh) synthesis. A result could be failure of synaptic transmission, of mechanisms for cell membrane renewal, or both. The experimental approach was to replace Ch in the animals' diet with a less efficient analog, N-aminodeanol (NADe) (Newton and Jenden, 1986). The first task was to validate the neurochemical features of the NADe model. After an extensive series of experiments it was established that NADe does, in fact, progress through exactly the same metabolic pathway as Ch, replacing ACh with AcNADe, which is hydrolyzed by acetylcholinesterase (AChE) and interacts with both muscarinic and nicotinic receptors. However, the potency of this false transmitter at those receptors is only 4% to 17%, respectively, of that true transmitter. The second major phase of the project was to examine the progressive effects of replacing dietary Ch with NADe on a broad spectrum of behavioral and physiological functions (Russell et al, 1990). Among these functions cognitive processes were by far the most sensitive, showing significant deficits that increased progressively as time on the diet increased. They also showed the least adaptability, even when animals were returned to normal Ch diets. Looking at the chemical structures of Ch and NADe, one can see that the only difference is the replacement of one methyl group of the Ch molecule by one amine group. A very small change in a transmitter-receptor interaction may be reflected in a catastrophic change in cognitive function.

Partial Agonists

Having looked at a way in which transmitter-receptor interactions may disrupt cognitive functions, I wish now to give an example in which such interactions may be used to ameliorate cognitive impairment. Simultaneous stimulation of pre- and postsynaptic receptors provides striking examples of the behavioral significance of interactions between two receptor subtypes within the same neurotransmitter system. In order to maximize cholinergic transmission when hypocholinergics interfere with cognitive processes, it would be advantageous to stimulate simultaneously M_2 postsynaptic and M_1 presynaptic receptors. Activity of the former facilitates cellular excitation; activity of the latter inhibits ACh release. Compounds that act both as postsynaptic agonists and as presynaptic antagonists already exist as partial agonists (Ariens, 1983; Ringdahl et al., 1987). Among these are tertiary amines that combine low intrinsic efficacy with high affinity for muscarinic receptors. Research on one of these, compound BM-5 (Dahlbom, 1982), indicated that it met the requirements of a partial agonist. During these experiments several measures of behavior were recorded. Measures of both learning and memory showed ". . . a tendency toward improved performance, reminiscent of muscarinic receptor agonists. . ." (Nordström et al., 1986). Our own experiments with BM-5 have not been as successful as we had hoped in light of this report. Nevertheless, the potential values in treating several neurological and psychiatric conditions of such agents, with their differential actions on pre- and postsynaptic muscarinic receptors, warrant continuing investigation. The search should be extended to include partial agonists acting on receptors of other neurotransmitter systems, since such compounds have already been identified for dopamine receptors (Meller et al., 1986).

Toxicological Effects of Multiple Chemical Interactions

It would be amiss not to include some reference to what Calabrese (1990) in his very recent volume, *Multiple Chemical Interactions*, has termed ". . . the reality of exposures to the complex mixtures within the sea of environmental toxins." In view of their large numbers, the toxicological effects on cognitive functions of interaction among such chemicals, many of which act on neurotransmitter systems, are matters of great practical concern—and of enormous technological difficulties. It will not be possible within the objectives of the present symposium to consider these difficulties in any satisfactory manner. However, the importance of the subject should be recorded.

Localizing Sites of Interaction

Tracing Transmitter Pathways

I turn next to the search for differential sites of neurotransmitter activity. During the second century AD, Galen had pronounced the dictum that "the brain is the organ of the mind" and had begun to study experimentally the localization of behavioral functions in the brain. However, it was not until Broca's reports some 16 centuries later of a "center" for speech, that credence was given to the question of just how specific such functions may be. Even in 1949 Lashley felt compelled to comment about his 30 years of experiments to which I referred earlier: "This series of experiments has yielded a good bit of information about what and where the memory trace is not" (Lashley, 1950). The invention of new techniques to trace the localization of neurotransmitter activities is one of the striking examples of ways in which knowledge develops by quantal steps (Butcher and Woolf, 1982). From fuller understanding of neurotransmitters and of their receptors developed radioactive labelling techniques that could trace neurotransmitter pathways more precisely. Further advances have come from applications of methods of histochemistry and immunocytochemistry, and the refinement of procedures based on *in situ* hybridization of the mRNA for enzymes involved in the syntheses of neurotransmitters awaits in the wings. Applications of these techniques are providing much clearer pictures of neurotransmitter pathways than we have had in the past. They are also making the story of neurotransmitter interactions more complex by showing that central neurons may possess receptors for more than one transmitter system and, thus, provide a basis for interactions among them.

Neurotrophic Factors

Having localized sites of action for various transmitter systems, it becomes possible to use lesioning techniques with considerable precision. The discovery of the roles of neurotrophic factors in the developing CNS and in the "repair" of tissue damage has added to interest in manipulating the anatomical features of neurotransmitter systems to study interactions with their dynamic biochemical properties. The major design features of such studies begin with a lesion and conclude with assays of the activity of the transmitter involved and of the enzymes associated with it. A project currently underway in our laboratory serves as an example. In phase 1, effects of bilateral lesions in the rat insular cortex on conditional taste avoidance (CTA) are measured. This is followed in phase 2 by implants of fetal brain tissue, impregnated with nerve growth factor, into the sites of the lesions. In phase 3 measurements of CTA are repeated after an interval that has been shown to be adequate for implants to become viable, and performance in an inhibited avoidance task is observed. This is followed in phase 4 by measurements of ACh turnover in the area of the implants. Suitable control and experimental groups make it possible to test hypotheses about behavioral effects of the lesions and of the implants. In our present experiments of primary importance for us is the ways in which changes in the turnover of the neurotransmitter *in vivo* are related to these effects.

Continuing Challenges

Despite such advances there still remain many important issues to be clarified. For example, how do neural networks retain their essential characteristics and continue to function within their characteristic limits in spite of the enormous diversity of modulatory actions to which they are exposed (Harris-Warrick and Marder, 1991)? Recently questions have arisen about the apparent loss of neurons in disorders such as Alzheimer's disease, characterized by very severe dysfunctions of memory. Could the apparent loss be a methodological artifact resulting from a reduction in deductibility of the neurotransmitter-related enzyme used as a cell marker (Hagg et al., 1988)? Such matters continue to challenge those interested in *where* the dynamic events of neurotransmission are located. We will be hearing from several of our colleagues in the symposium who are experts at inventing and applying new techniques to answer questions about sites of neurotransmitter actions and interactions, so I will not expand these comments further.

Nonsynaptic Mechanisms

There is, however, one other potential mechanism to which reference should be made briefly. Some 30 years ago the possibility that interactions could arise from nonsynaptic mechanisms formed the basis for hypotheses that began with observations on hippocampal neurotransmission. Investigators were studying the electrophysiological and anatomical characteristics of electrical activity in that site (Green and Maxwell, 1961). The close apposition of neuronal membranes and observed seizure activity when neurochemical synaptic activity was blocked experimentally led to suggestions that interactions might be dependent upon electronic coupling through gap junctions, the spread of current from an electrically active neuron through extracellular space (ephaptic interactions), and as shifts in the concentration of extracellular ions. Research continues to examine the potential importance of such nonsynaptic, electrophysiological mechanisms (Pumain et al., 1985; Traynelis and Dingledine, 1989).

In the End: An Integrated Organism

I hope that these comments will serve as a useful introduction to a much more detailed discussion of interactions between neurotransmitter systems per se.

FIGURE 2. There is a potential analogy with the ancient fable of the three wise, blind men describing an elephant: one by feeling the trunk; another, the limb; and the third, the tail. Biochemists examine body fluids; anatomists, bodily cells, tissues, and organs; electrophysiologists, biopotentials wherever they appear. To this trio has been added behavioral biologists, who have opportunities to meld these properties into an "integrated organism" capable of adjusting to the demands of an ever-changing environment. (This figure was drawn specifically for the present manuscript by my good friend and former colleague, Dr. Grant D. Schiller of The Flinders University of South Australia.)

Interest in interactions at this level has been growing rapidly during recent years, as evidenced, for example, by some 250 "selected" references included in a 1991 review by two of our fellow participants (Decker and McGaugh, 1991). We are still far away from understanding the full chain of events leading from neurotransmitter-receptor binding to cognitive functions. The fact that we can devote this symposium to discussions of effects of neurotransmitter interactions on such functions indicates that progress toward characterizing the "integrated organism" is being made.

There is an ancient fable about three wise, blind men describing an elephant. One felt its trunk; another, its leg; the third, its tail. There is an analogy in the biosciences today. Biochemists examine body fluids; anatomists, the structures of the body from cells to tissues to organs to organ systems; electrophysiologists, biopotentials wherever they appear; and behavioral biologists, the capabilities of integrated organisms to adjust to their physical and psychosocial environments. It is stimulating to participate in a symposium that seeks to consider interactions among all these basic biological properties. Remember the comment by the great clinical neurologist, Hulings Jackson: "The highest nervous processes are potentially the whole organism."

I leave you with an "Ode To The Symposium":

We may live without conscience; we may live without heart.
We may live without poetry, music and art.
We may live without politics or nuclear fission.
But no one can live without neurotransmission.

References

Ariens EJ (1983): Intrinsic activity: Partial agonists and partial antagonists. *J Cardiovasc Pharmacol* 5 (Suppl 1):S8–S15

Ashford JR (1981): General models for the joint action of mixtures of drugs. *Biometrics* 37:457–474

Benke GM, Murphy SD (1975): The influence of age on the toxicology and metabolism of methyl parathion and parathion in male and female rats. *Toxicol Appl Pharmacol* 31:254–269

Butcher LL, Woolf NJ (1982): Cholinergic and serotonergic systems in the brain and spinal cord: Anatomic organization, role in inter-cellular communication processes, and interactive mechanisms. In: *Chemical Transmission in the Brain*, Buigs RM, Pevet P, Swaab DF, eds. Amsterdam: Elsevier Biomedical Press, pp 3–40

Calabrese EJ (1990): *Multiple Chemical Interactions*. Boca Raton, FL: Lewis Publishers, p 1200

Castellano C, Oliverio A, Schwab C, Brueckner C, Biesold D (1988): Age dependent differences in cholinergic drug response in two strains of mice. *Neurosci Lett* 84:335–338

Crocker AD, Overstreet DH (1991): Dopamine sensitivity in rats selectively bred for increases in cholinergic function. *Pharmacol Biochem Behav* 38:105–108

Crocker AD, Russell RW (1990): Pretreatment with an irreversible muscarinic agonist affects responses to apomorphine. *Pharmacol Biochem Behav* 35:511–516

Dahlbom R (1982): Structure and stearic aspects of compounds related to oxotremorine. In: *Cholinergic Mechanisms: Phylogenetic Aspects, Central and Peripheral Synapses, and Clinical Significance*, Pepell G, Ladinsky H, eds. New York: Plenum Press

Decker MW (1987): The effects of aging on hippocampal and cortical projections of the forebrain cholinergic system. Brain Res 434:423–438

Decker MW, McGaugh JL (1991): The role of interactions between the cholinergic system and other neuromodulatory systems in learning and memory. *Synapse* 7:151–168

Ehrlich P (1913): Chemotherapeutics: Scientific principles, methods and results. *Lancet* 2:445–451

Gill EW, Rang HP (1966): An alkylating relative to benzilye-choline with specific and long-lasting parasympatholytic activity. *Mol Pharmacol* 2:284–297

Green JD, Maxwell DS (1961): Hippocampal electrical activity. 1. Morphological aspects. *Electroencephalogr Clin Neurophysiol* 13:837–846

Hagg T, Manthrope M, Vahlsing HL, Varon S (1988): Delayed treatment with nerve growth factor reverses the apparent loss of cholinergic neurons after acute brain damage. *Exper Neurol* 101:303–312

Harris-Warrick RM, Marder E (1991): Modulation of neural networks for behavior. *Ann Rev Neurosci* 14:39–57

Kenakin TP (1986): Tissue and receptor selectivity: Similarities and differences. *Adv Drug Res* 15:71–109

Langley, JN (1906): Croonian Lecture: On nerve endings and on special excitable substances in cells. *Proc R Soc, B* 78:170–194

Lashley KS (1950): In search of the engram. In: *Physiological Mechanisms in Animal Behavior*, Danielli JF, Brown R, eds. Cambridge: University Press

McCarthy MP, Earnest JP, Young EF, Choe S, Stroud RM (1986): The molecular neurobiology of the acetylcholine receptor. *Ann Rev Neurosci* 9:383–413

Meller E, Helmer-Matyjek E, Bohmaker K, Adler CH, Friedhoff AJ, Goldstein M (1986): Receptor reserve at striatal dopamine autoreceptors: Implications for selectivity of dopamine agonists. *Eur J Pharmacol* 123:311–314

Moore WJ, Mahler H (1965): Introduction to molecular psychology. *J Chem Educ* 42:49–60

Newton MW, Jenden DJ (1986): False transmitters as presynaptic probes for cholinergic mechanisms and functions. *Trends Pharmacol Sci* 7:316–320

Nordström Ö, Undén A, Grimm V, Frieder B, Ladinsky H, Bartfai T (1986): *In vivo* and *in vitro* studies on a muscarinic presynaptic antagonist and postsynaptic agonist: BM-5. In: *Biology of Cholinergic Function*, Hanin I, ed. New York: Plenum Press

Overstreet DH, Russell RW, Crocker AD, Schiller GD (1984): Selective breeding for differences in cholinergic function: Pre- and post-synaptic mechanisms involved in sensitivity to the anticholinesterase, DFP. *Brain Res* 294:327–332

Overstreet DH, Russell RW, Hay DA, Crocker AD (1992): Selective breeding for increased cholinergic function: Biometrical genetic analysis of muscarinic responses. *Neuropsychopharmacology* 00:00–00

Pappolla MA, Andorn AC (1987): Serum protein leakage in aged human brain and inhibition of ligand binding at alpha$_2$-adrenergic and cholinergic binding sites. *Synapse* 1:82–89

Paton WDM, Rang HP (1965): The uptake of atropine and related drugs by intestinal smooth muscle of the guinea-pig in relation to acetylcholine receptors. *Proc R Soc, B* 163:1–44

Pilch H, Müller WE (1988): Chronic treatment with choline or scopolamine indicates the presence of muscarinic cholinergic receptor plasticity in the frontal cortex of young but not of aged mice. *J Neural Transm* 71:39–43

Pradhan SN (1986): Central neurotransmitters and aging. *Life Sci* 26:1643–1656

Pumain R, Menini C, Heinemann U, Louvel J, Silva-Barrat C (1985): Chemical synaptic transmission is not necessary for epileptic seizures to persist in the baboon *Papio papio*. *Exp Neurol* 89:250–258

Ringdahl B, Roch M, Jenden UJ (1987): Regional differences in receptor reserve for analogs of oxotremorine *in vivo:* Implications for development of selective muscarinic agonists. *J Pharm Exp Therap* 242:464–471

Roth GS, Hess GD (1982): Changes in the mechanisms of hormone and neurotransmitter action during aging: Current status of the roles of receptor and post-receptor alterations. *Mech Aging Dev* 20:175–194

Rothman KJ, Greenland S, Walker AM (1980): Concepts of interaction. *Am J Epidemiol* 112:467–470

Russell RW, Smith CA, Booth RA, Jenden DJ, Waite JJ (1986): Behavioral and physiological effects associated with changes in muscarinic receptors following administration of an irreversible cholinergic agonist (BM123). *Psychopharmacology* 90:308–315

Russell RW, Booth RA, Smith CA, Jenden DJ, Roch M, Rice KM, Lauretz SD (1989): Roles of neurotransmitter receptors in behavior: Recovery of function following decreases in muscarinic sensitivity induced by cholinesterase inhibition. *Behav Neurosci* 103:881–892

Russell RW, Jenden DJ, Booth RA, Lauretz SD, Roch M, Rice KM (1990): Global *in vivo* replacement of choline by N-aminodeanal. Testing a hypothesis about progressive degenerative dementia. II. Physiological and behavioral effects. *Pharm Biochem Behav* 37:811–820

Traynelis SF, Dingledine R (1989): Role of extracellular space in hyper-osmotic suppression of potassium-induced electrographic seizures. *J Neurophysiol* 61:927–938

Yamamura HI, Enna SJ, Kuhar MJ (1978): *Neurotransmitter Receptor Binding*. New York: Raven Press

2

The Cholinergic Basal Forebrain and its Telencephalic Targets: Interrelations and Implications for Cognitive Function

LARRY L. BUTCHER

Introduction

The brain, the organ of cognition, possesses the remarkable ability to process, integrate, and transform information from diverse knowledge domains. The mechanisms by which it performs those operations have remained elusive, however, not only because of the intricacies of the neural systems involved but also because of the inherent complexities at the behavioral level of cognitive processes themselves.

In this chapter, I will develop the thesis that the basal nuclear complex (for nomenclature, see Butcher and Semba, 1989), particularly its cholinergic elements, possesses a number of structural and organizational properties that make it particularly well suited to play an important role in cognition. These properties fall into two broad, but not necessarily mutually exclusive, categories, which, for purposes of the present discourse, will be called *point* features and *system* features. The former term refers to characteristics of the individual neurons themselves; the latter designation denotes attributes that can be revealed only by analyzing collections of neurons and their networks. Before considering these properties in greater detail, I will summarize first the basic plan of the cholinergic basal forebrain.

Organization of the Cholinergic Basal Forebrain

Cholinergic somata in the basal nuclear complex form a contiguous, if not continuous, constellation of large, isodendritic neurons that range from the medial septal nucleus rostrally to rostral portions of the lateral hypothalamus and the so-called nucleus of the ansa lenticularis caudally (Butcher and Woolf, 1986; Oh et al. 1992). Insofar as is known, the vast majority, if not all, of these cholinergic cells are projection neurons that form essentially three different hodologic entities in the telencephalon: (a) an obliquely arching hippocampal pathway, commonly but incorrectly called the septohippocampal projection, deriving mostly from the medial septal nucleus and the vertical limb of the diagonal band; (b) a medial fiber bundle, originating largely in the vertical and horizontal limbs of the diagonal band, and, to a somewhat lesser extent the magnocellular preoptic area and the so-called substantia innominata, which projects to medial cortical regions, including saliently the cingulate, retrosplenial, and medial occipital cortices; and (c) a collection of laterally coursing pathways emanating prominently from the magnocellular preoptic area, substantia innominata, nucleus basalis, and the so-called nucleus of the ansa lenticularis, which provides afferents to the remaining allocortex and lateral isocortex, as well as to the amygdala (Fig. 1).

A major target of the cholinergic basal forebrain neurons appears to be pyramidal cells in the hippocampus and cortex (Fig. 2; Luiten et al., 1989). Pyramidal cells in deeper layers of certain cortical regions also reciprocally innervate the cholinergic ventral forebrain, and afferents from a number of those neurons have been suggested to terminate directly on cells of the basal nuclear complex (Irle and Markowitsch, 1986). Although evidence is currently most compelling for direct reciprocal innervation between basal forebrain neurons and their targets in the limbic telencephalon, information conveyed from other cortical regions could influence basal nuclear cells in a reciprocating fashion as well, conceivably by interneuronal nonsynaptic mechanisms (Woolf and Butcher, 1991) and/or processes involving glial intermediates (Milner, 1991), if not by synaptic means.

Point Features of Cholinergic Neurons in the Basal Nuclear Complex: Relevance to Cognition

The Dendritic Arbors of Individual Basal Forebrain Cholinergic Neurons Are Extensive and Highly Overlapping

Many cholinergic neurons in the basal forebrain possess long, radiating dendrites that intersect other neuronal processes and tissue elements (Figs. 3–5). This anatomic feature affords the possibility of a single cholinergic neuron sampling information from numerous sources and/or transmitting it to multiple tissue elements, a structural feature that could have important implications for the diversity and multidimensionality of cognition.

The Axonal Terminal Fields of Individual Basal Forebrain Cholinergic Neurons Are Highly Circumscribed

It is not inconsequential that the slab of cortex innervated by a single cholinergic neuron in the basal nuclear complex, approximately 1–2 mm^2 (Bigl et al., 1982; Price and Stern, 1983; see Fig. 4), is of the same approximate dimensions as a cortical module. This means that a single cholinergic neurons could influence, by mechanisms not fully established at present, a cognitively important unit of cortical organization. One possibility is that activation of the cholinergic input to a given cortical module or hippocampal domain enables or modulates the ability of that region to engage in the function(s) it performs (Woolf, 1991).

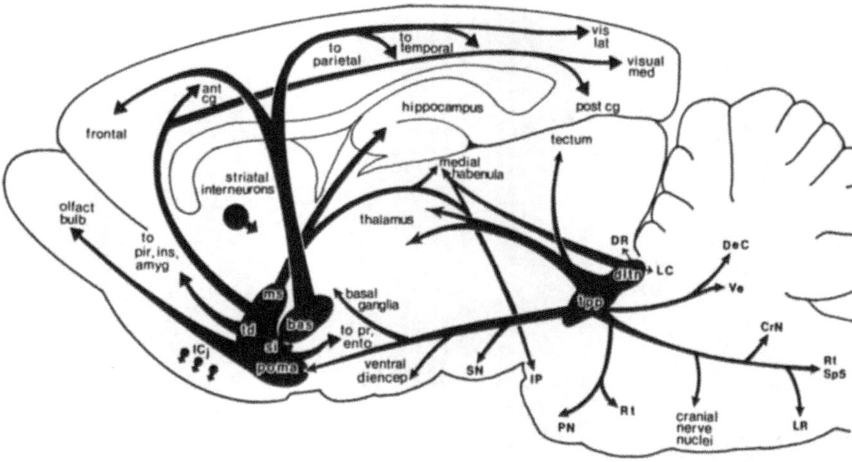

FIGURE 1. Schematic representation of the major cholinergic systems in the mammalian brain. The depiction is modified from Butcher and Woolf (1986) and Woolf and Butcher (1989), and is based on data derived from the use of choline acetyltransferase immunohistochemistry in combination with tract-tracing histology as described first in Woolf et al. (1983). As illustrated, central cholinergic neurons evince two basic organizational schemata: (a) local circuit cells (i.e., those morphologically arrayed wholly within the neural structure in which they are found), exemplified by the interneurons of the caudate-putamen nucleus, nucleus accumbens, olfactory tubercle, and islands of Calleja complex (ICj) and (b) projection neurons (i.e., those that connect two or more different regions). Of the cholinergic projection neurons that interconnect central structures, two major subconstellations have been identified: (a) the basal nuclear complex, composed of cholinergic neurons in the medial septal nucleus (ms), diagonal band nuclei (td), substantia innominata (si), magnocellular preoptic field (poma), and nucleus basalis (bas) and projecting prominently to the entire nonstriatal telencephalon, and (b) the pontomesencephalotegmental cholinergic complex, composed of cholinergic cells in the penduncolopontine (tpp) and laterodorsal (dltn) tegmental nuclei and projecting ascendingly to the thalamus and other diencephalic loci and descendingly to the pontine and medullary reticular formations (Rt), deep cerebellar (DeC) and vestibular (Ve) nuclei, and cranial nerve nuclei. Not shown are the somatic and parasympathetic cholinergic neurons of cranial nerves III–VII and IX-XII, the projections of the vestibular and cochlear efferent nuclei, and the cholinergic α- and γ-motor and autonomic neurons of the spinal cord. amyg = amygdala; ant cg = anterior cingulate cortex; CrN = dorsal cranial nerve nuclei; diencep = diencephalon; DR = dorsal raphe nucleus; ento = entorhinal cortex; frontal = frontal cortex; IP = interpeduncular nucleus; ins = insular cortex; LC = locus ceruleus; LR = lateral reticular nucleus; olfact = olfactory; pir = piriform cortex; PN = pontine nuclei; pr = perirhinal cortex; parietal = parietal cortex; post cg = posterior cingulate cortex; SN = substantia nigra; Sp5 = spinal nucleus of cranial nerve five; temporal = temporal cortex; vis lat = lateral visual cortex; visual med = medial visual cortex.

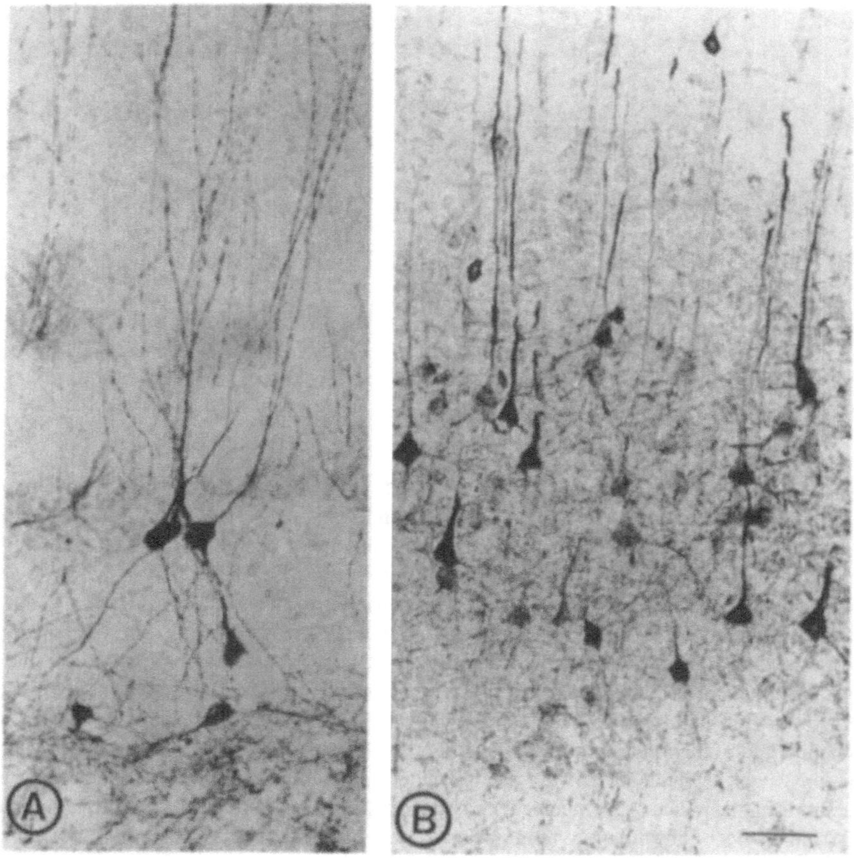

FIGURE 2. Muscarinic receptors associated with cells in the CA2 region of the hippocampus (A) and with neurons of the parietal cortex (B) in the rat. An immunohistochemical procedure (Woolf et al., 1989) utilizing the M35 antibody against muscarinic receptors was used (Luiten et al., 1989). Scale bar equals 40 μm.

Basal Forebrain Cholinergic Neurons Exhibit Plasticity

The cells of the cholinergic basal forebrain evince considerable morphologic and physiologic plasticity under a variety of conditions, a property desirable, if not essential, for neurons involved in cognitive processes. Axons of the septohippocampal pathway, for example, approximately 60% of which are cholinergic (Woolf et al., 1984), sprout to invade target regions vacated by degenerated axon terminals of ablated entorhinal neurons projecting to the hippocampus (Lynch et al., 1972). Indeed, the axonal arbors of cholinergic neurons in the basal forebrain normally display morphologic profiles reminiscent of fiber restructuring following injury to the cortex (Butcher and Woolf, 1987; Woolf and Butcher,

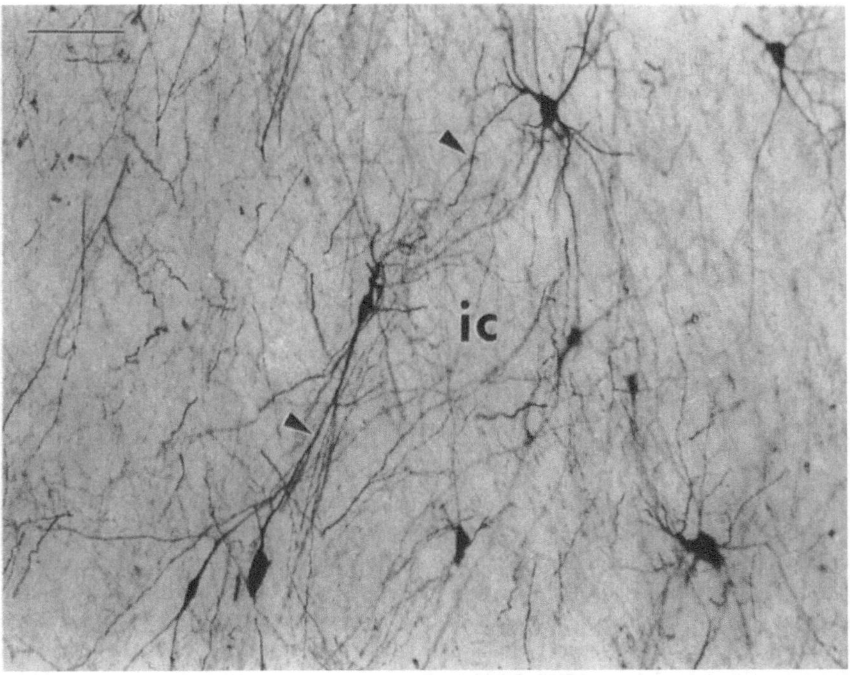

FIGURE 3. Neurons demonstrating choline acetyltransferase-like immunoreactivity in the nucleus basalis of the cat. Immunohistochemical procedures were performed as described in Woolf et al. (1989). Observe that cholinergic somata and their processes are embedded within the myelinated fibers of the posterior limb of the internal capsule (ic) and that many of those processes intersect or are confluent (arrowheads). Scale bar equals 100 μm.

1990), as well as of anatomic remodelling occurring under essentially physiologic conditions in mouse superior cervical ganglion cells (Purves and Voyvodic, 1987).

Further evidence of morphologic plasticity in forebrain cholinergic neurons is suggested by the observations that fetal neuronal cell suspensions from the nucleus basalis and septum-diagonal band region can be grafted viably into the denervated hippocampal formation and frontoparietal cortex and can provide cholinergic fibers ostensibly reinnervating those targets (Björklund and Gage, 1988). Whether or not this "reinervation" restores normal physiology or merely represents neuronal growth run amok, however, remains to be established (for discussion in relation to Alzheimer's disease, see Butcher and Woolf, 1989; Woolf and Butcher, 1990).

Recently, transverse knife cuts of the medial cholinergic pathway in adult rats have been found to produce an initial accumulation of choline acetyltransferase and acetylcholinesterase in severed axons proximal to the transection, and a subsequent reduction in those enzymes and the number of fiber elements

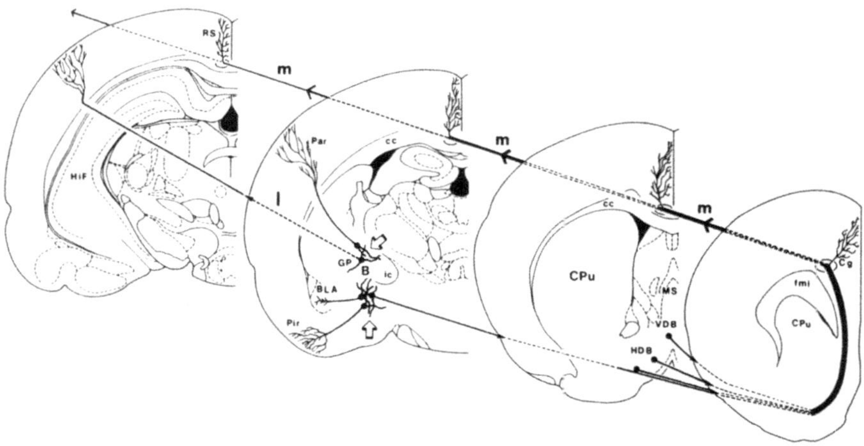

FIGURE 4. Schematic representation of the somata of origin and trajectories of the medial (m) and lateral (l and remaining unlabeled fibers) cholinergic pathways. The septohippocampal projection, cholinergic fibers to the olfactory bulb, and descending projections are not shown. The axonal terminal fields are represented by circumscribed branch-like entities in the cortex and basolateral amygdala (BLA). Arrows point to overlapping dendrites of different neurons within the nucleus basalis (B). cc = corpus callosum; Cg = cingulate cortex; CPu = caudate-putamen nucleus; fmi = forceps minor; GP = globus pallidus; HDB = horizontal limb of the diagonal band; HiF = hippocampal formation; ic = internal capsule; MS = medial septal nucleus; Par = parietal cortex; Pir = piriform cortex; RS = retrosplenial cortex; SI = substantia innominata; VDB = vertical limb of the diagonal band.

containing them distal to the surgical insult (Butcher and Woolf, 1987; Woolf and Butcher, 1990). This pathologic ensemble is followed by reappearance of cholinergic fibers in distal cortical regions approximately 30 days later (Butcher and Woolf, 1987; Woolf and Butcher, 1990). One mechanism accounting for such fiber reappearance, in part at least, is sprouting and regrowth of transected cholinergic axons, some neuritic extensions of which penetrate the glial scar, presumably *en route* to denervated cortical areas (Butcher and Woolf, 1987; Woolf and Butcher, 1990). Indeed, many of the severed neuronal processes demonstrate bulbous, and in some cases "growth cone-like," profiles shortly after surgical procedures, which is consistent with a sprouting hypothesis (Butcher and Woolf, 1987; Woolf and Butcher, 1990; see also Wendt and Ayyad, 1985). Comparable neuronal restructuring following injury has not yet been established for neurons of the pontomesencephalotegmental cholinergic complex (Farris and Butcher, 1991). Furthermore, chronic administration during development of triiodothyronine, also called T3, produces filopodia-like protuberances from the cell bodies and proximal processes of basal forebrain neurons similar to those seen in Alzheimer's disease (Butcher and Woolf, 1987; Woolf and Butcher, 1990).

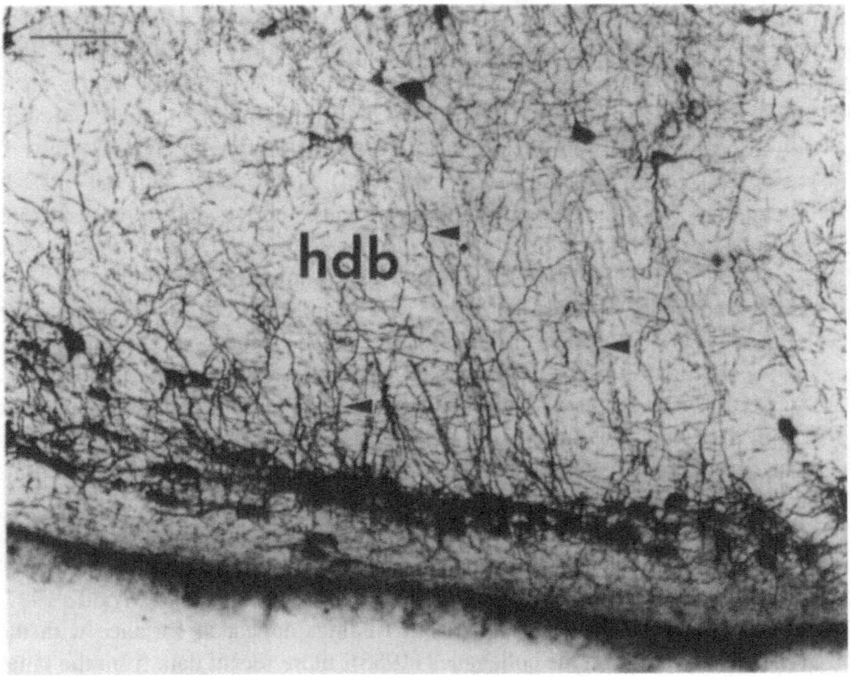

FIGURE 5. Neurons demonstrating choline acetyltransferase-like immunoreactivity in the horizontal limb of the diagonal band (hdb) of the cat. Immunohistochemical procedures were performed as described in Woolf et al. (1989). Individual cholinergic neurons possess dendritic processes (arrowheads) that trasverse almost perpendicularly the fibers of the diagonal band. Scale bar equals 100 μm.

Such hormone treatment does not give rise to morphologic alterations in cholinergic cells of the pontomesencephalotegmental complex (Gould and Butcher, 1989).

If cholinergic neurons demonstrate morphologic modifiability as an important aspect of their adult physiology, a suggestion that does not appear unwarranted in view of the foregoing discussion and the findings of Purves and his associates (Purves and Voyvodic, 1987), what factors are instrumental in enabling that structural variability? Because substances with neurotrophic properties have been identified in extracts of the central nervous system and in wound fluids (Manthorpe et al., 1983), one possibility is that cytoarchitectonic remodeling is chemically mediated by endogenous growth-promoting and/or growth-inhibiting agents. Indeed, various lines of experimental evidence indicate that nerve growth factor (NGF) may be a specific, but probably not the only trophic factor for cholinergic systems in the basal forebrain. First, NGF infused into the hippocampus or neocortex is taken up, presumably by a receptor-mediated process, and transported retrogradely to the cholinergic cells in the medial septal nucleus, vertical limb of the diagonal band, and nucleus basalis, but not to other neurons projecting to those targets (Seiler and Schwab, 1984). Second, intraventricular injection of NGF in neonatal rats elevates choline acetyltransferase activity in the basal forebrain (Gnahn et al., 1983). Third, NGF administration increases choline acetyltransferase activity in the hippocampus of adult rats with partial lesions of the septohippocampal pathway (Hefti et al., 1984). Fourth, NGF and NGF-mRNA have been detected in normal adult rats, both in the targets and in most regions containing the cell bodies of origin of the basal forebrain cholinergic system (Korsching et al., 1985; Woolf et al., 1989). Fifth, cholinergic neurons in the basal forebrain bind NGF (Richardson et al., 1986) and demonstrate NGF receptor (Hefti et al., 1986; Woolf et al., 1989), but those in the pontomesencephalotegmental cholinergic complex do not (Richardson et al., 1986; Woolf et al., 1989). Sixth, NGF increases choline acetyltransferase but not survival or fiber growth in pure neuronal cultures of septal cells (Hefti et al., 1985), but it does enhance survival of presumed septal cholinergic cells after fimbrial transections (Hefti, 1986).

Recently, we have found that NGF promotes fiber reappearance, due in part to enhanced neuritic outgrowth from cut axons, following transections of the medial pathway of the basal forebrain cholinergic system (Butcher and Woolf, 1987; Woolf and Butcher, 1990). Although our findings appear at variance with the observations of Hefti and his colleagues (1985), more recent data from the latter investigators' laboratory indicate that NGF does promote fiber outgrowth from septal cholinergic neurons in culture but only in mixed glial-neuronal preparations (Hartikka and Hefti, 1986). Coupling this latter observation with the absence of NGF's effect on cholinergic fiber growth in pure neuronal cultures suggests that in *in vivo* preparations NGF may require additional conditions (e.g., glia-mediated enabling of NGF receptors) and/or factors to effect neurite formation. Although it has not been established what these factors are, it is known that other substances can affect cholinergic neurons in ways similar to NGF. These include gangliosides (Pedata et al., 1984), estrogens (Luine, 1985), triiodothyronine (Hefti et al., 1986; Gould and Butcher, 1989), thyrotropin-releasing hormone (Yarbrough, 1983), and possibly laminin (Varon and Manthorpe, 1984).

System Features of Cholinergic Neurons in the Basal Nuclear Complex: Relevance to Cognition

Diffuse Projections

Although each cholinergic neuron projects to a relatively restricted region of cortex or hippocampus, the system as a whole innervates the entire nonstriatal telencephalon, a global presence that may be expressed physiologically. Cortical cup studies, for example, have indicated that electrical stimulation of the brainstem elicits widespread release of acetylcholine from virtually all cortical areas (e.g., see De Feudis, 1974), although not necessarily uniformly, even though brainstem input to the basal forebrain appears relatively sparse. How could this be accomplished? One possibility is that neurons of the basal forebrain communicate interdendritically and that activation of one such neuron could result in the spread of activity to numerous other cells. Because neurons projecting to widely disparate cortical areas lie next to one another (Fig. 4), this would result in acetylcholine being released in a number of different regions. "Diffuseness" as it applies to the cholinergic basal forebrain, therefore, is clearly a property of the system as a whole and not of individual cholinergic neurons.

Such a system property could have important implications for cognitive function in that it would allow the cholinergic modulation of a particular restricted cortical area to also be expressed in other regions, thereby allowing a number of different attributes to be associated with the primary target module(s) (Woolf, 1991).

Cells Embedded in Fiber Tracts

Many cholinergic basal forebrain neurons and/or their processes are embedded in or traverse the corticofugal fiber tracts from the cortical region to which they project (see data and commentary in Saper, 1984; Woolf and Butcher, 1991; Woolf, 1991; see Figs. 4 and 5). Neurons projecting to the visual cortex, for example, lie in the path of the occipitogeniculate projections (see Saper, 1984). As illustrated in Figures 3 and 5, cholinergic neurons of the basal forebrain possess dense plexuses of dendrites extending into fiber tracts, such as the posterior internal capsule (Fig. 3), which contains corticofugal fibers deriving from the visual and parietal cortices, and the diagonal band (Fig. 5), which contains corticofugal fibers arising from the limbic cortex. The existence of cholinergic dendrites highly interdigitated with fiber bundles deriving from the cortex provides seemingly ample opportunities for cortically derived axons to communicate with basal forebrain neurons. Even in the absence of synaptic linkages, transneuronal field effects, which do not require specialized junctions (Korn and Faber, 1979), could mediate the transfer of information from the cerebral cortex to cells in the basal forebrain.

Cholinergic Neurons Projecting to Different Telencephalic Areas Lie Close to One Another

It is known that rich interconnections exist between the neurons of the cholinergic basal forebrain and those of the cerebral mantle. Cholinergic basal forebrain cells project to every region of the cerebral cortex, including the hippocampus and olfactory bulb, as well as to the amygdala (Bigl et al., 1982; Butcher and Woolf, 1986; see Fig. 1). Furthermore, basal forebrain cells projecting to totally different regions of the cerebral mantle can lie in close proximity (Bigl et al., 1982; see Fig. 4), allowing for relationships to be established between widely disparate cortical areas having ostensibly different functions. Activation of a cortical module(s) relevant to the primary visualization of a lion, for example, could also allow for the activation of auditory cortex where memories of the lion's roar could be stored.

Clusters of Cholinergic Neurons Are Larger in Animals Higher on the Phylogenetic Scale

Clusters of cholinergic neurons in the basal forebrain were described first by Bigl et al. in 1982, wherein it was noted that the dendrites of many individual cholinergic basal forebrain neurons were concatenated, giving rise to the speculation that the cells of a given cluster could operate as a unit or "module" within the basal nuclear complex itself (Woolf and Butcher, 1991). In fact, the presence and size of such cellular aggregates appears to be species related in that the tendency for cholinergic cells of the nucleus basalis to aggregate in soma clusters of three or more neurons is much less common in the rat and cat than in the monkey (Woolf and Butcher, 1991). Furthermore, the number of cells (>15) comprising a single such ensemble in the nonhuman primate is greater than in the rat or cat, both latter species demonstrating maximum cluster sizes of three to five neurons. If the number of basal forebrain neurons in a cluster and their interconnections reflect a capacity for cognitive complexity, then a structural basis would appear to exist for the increased proficiency in this regard that primates appear to enjoy.

Conspectus

Cholinergic bashing, I regret to observe, has become increasingly fashionable in recent years as investigators attempt to dissect the contributions of other transmitter systems to various cognitive processes, both normal and pathologic. Nonetheless, cholinergic systems, particularly in the forebrain, remain intriguing and possess a number of anatomic features that make them particularly attractive in relation to neural mechanisms of cognition. Emphasis on how different transmitters interact to mediate and modulate cognition might be a better strategy than focusing on a single chemically specific system to the exclusion or detriment of others.

References

Bigl V, Woolf NJ, Butcher LL (1982): Cholinergic projections from the basal forebrain to frontal, parietal, temporal, occipital, and cingulate cortices: A combined fluorescent tracer and acetylcholinesterase analysis. *Brain Res Bull* 8:727–749

Björklund A, Gage FH (1988): Grafts of fetal cholinergic neurons in rat models of aging and dementia. In: *Aging and the Brain*, Terry RD, ed. New York: Raven Press, pp 243–257

Butcher LL, Semba K (1989): Reassessing the cholinergic basal forebrain: Nomenclature schemata and concepts. *Trends Neurosci* 12:483–485

Butcher LL, Woolf NJ (1986): Central cholinergic systems: Synopsis of anatomy and overview of physiology and pathology. In: *The Biological Substrates of Alzheimer's Disease*, Scheibel AB, Wechsler AF, eds. New York: Academic Press, pp 73–86

Butcher LL, Woolf NJ (1987): Cholinergic neuronal regeneration can be modified by growth factors. In: *Cellular and Molecular Basis for Cholinergic Function*, Dowdall MJ, Hawthorne JN, eds. Chichester, UK: Ellis Horwood Press, pp 395–402

Butcher LL, Woolf NJ (1989): Neurotrophic agents exacerbate the pathologic cascade of Alzheimer's disease. *Neurobiol Aging* 10:557–570

DeFeudis FV (1974): *Central Cholinergic Systems and Behaviour*. London, UK: Academic Press

Farris TW, Butcher LL (1991): Sparse regeneration after axotomy of cholinergic projections from pontine tegmentum to anterior thalamus. *Soc Neurosci Abstr* 17:942

Gnahn H, Hefti F, Heumann R, Schwab ME, Thoenen H. (1983): NGF-mediated increase of choline acetyltransferase (ChAT) in the neonatal rat forebrain: Evidence for a physiological role of NGF in the brain? *Dev Brain Res* 9:45–52

Gould E, Butcher LL (1989): Developing cholinergic basal forebrain neurons are sensitive to thyroid hormone. *J Neurosci* 9:3347–3358

Hartikka J, Hefti F (1986): Effect of nerve growth factor on septal cholinergic neurons in mixed glial-neuronal cultures. *Soc Neurosci Abstr* 12:1099

Hefti F (1986): Nerve growth factor promotes survival of septal cholinergic neurons after fimbrial transections. *J Neurosci* 6:2155–2162

Hefti F, Dravid A, Hartikka J. (1984): Chronic intraventricular injections of nerve growth factor elevate hippocampal choline acetyltransferase activity in adult rats with partial septo-hippocampal lesions. *Brain Res* 293:305–311

Hefti F, Hartikka J, Eckenstein F, Gnahn H, Heumann R, Schwab M (1985): Nerve growth factor increases choline acetyltransferase but not survival or fiber outgrowth of cultured fetal septal cholinergic neurons. *Neuroscience* 14:55–68

Hefti F, Hartikka J, Salvatierra A, Weiner WJ, Mash DC (1986): Localization of nerve growth factor receptors in cholinergic neurons of the human basal forebrain. *Neurosci Lett* 69:37–41

Irle E, Markowitsch HJ (1986): Afferent connections of the substantia innominata/basal nucleus of Meynert in carnivores and primates. *J Hirnforsch* 27:343–367

Korn H, Faber DS (1979): Electrical interactions between vertebrate neurons: Field effects and electrotonic coupling. In: *The Neurosciences Fourth Study Program*, Schmitt FO, Worden FG, eds. Cambridge, MA: MIT Press, pp 333–358

Korsching S, Auburger G, Heumann R, Scott J, Thoenen H (1985): Levels of nerve growth factor and its mRNA in the central nervous system of the rat correlate with cholinergic innervation. *EMBO J* 4:1389–1393

Luine VN (1985): Estradiol increases choline acetyltransferase activity in specific basal forebrain nuclei and projection areas of female rats. *Exp Neurol* 89:484–490

Luiten PGM, Gaal G, Gaykema RPA, Strosberg AD, Schroeder H (1989): Immunohistochemical demonstration of muscarinic acetylcholine receptor proteins in rat and human neocortex. Interactions with cortical projections from the magnocellular basal nucleus. *Soc Neurosci Abstr* 15:811

Lynch G, Matthews DA, Mosko S, Parks T, Cotman C (1972): Induced acetylcholinesterase-rich layer in rat dentate gyrus following entorhinal lesions. *Brain Res* 42:311–318

Manthorpe M, Nieto-Sampedro M, Skaper SD, Lewis ER, Barbin G, Longo FM, Cotman CW, Varon S (1983): Neuronotrophic activity in brain wounds of the developing rat. Correlation with implant survival in the wound cavity. *Brain Res* 267:47–56

Milner TA (1991): Cholinergic neurons in the rat septal complex: Ultrastructural characterization and synaptic relations with catecholaminergic terminals. *J Comp Neurol*, 314:37–54

Oh JD, Woolf NJ, Roghani A, Edwards RH, Butcher LL (1992): Cholinergic neurons in the rat central nervous system demonstrated by *in situ* hybridization of choline acetyltransferase mRNA. *Neuroscience*, 47:807–822

Pedata F, Giovannelli L, Pepeu G (1984): GM1 ganglioside facilitates the recovery of high-affinity choline uptake in the cerebral cortex of rats with a lesion of the nucleus basalis magnocellularis. *J Neurosci Res* 12:421–427

Price JL, Stern R (1983): Individual cells in the nucleus basalis-diagonal band complex have restricted axonal projections to the cerebral cortex in the rat. *Brain Res* 269:352–356

Purves D, Voyvodic JT (1987): Imaging mammalian nerve cells and their connections over time in living animals. *Trends Neurosci* 10:398–404

Richardson PM, Verge Issa VMK, Riopelle RJ (1986): Distribution of neuronal receptors for nerve growth factor in the rat. *J Neurosci* 6:2312–2321

Saper CB (1984): Organization of cerebral cortical afferent systems in the rat. I. Magnocellular basal nucleus. *J Comp Neurol* 222:313–342

Seiler M, Schwab ME (1984): Specific retrograde transport of nerve growth factor (NGF) from neocortex to nucleus basalis in the rat. *Brain Res* 300:33–39

Varon S, Manthorpe M (1984): Trophic and neurite-promoting factors for cholinergic neurons. In: *Cellular and Molecular Biology of Neuronal Development*, Black IB, ed. New York: Plenum Press, pp 251–275

Wendt JS, Ayyad KA (1985): AChE-positive axonal sprouting and regeneration across scar tissue in adult rat brain. *Soc Neurosci Abstr* 11:975

Woolf NJ (1991): Cholinergic systems in mammalian brain and spinal cord. *Progr Neurobiol* 37:475–524

Woolf NJ, Butcher LL (1989): Cholinergic systems in the rat brain: IV. Descending projections of the pontomesencephalic tegmentum. *Brain Res Bull* 23:519–540

Woolf NJ, Butcher LL (1990): Dysdifferentiation of structurally plastic neurons initiates the pathologic cascade of Alzheimer's disease: Toward a unifying hypothesis. In: *Brain Cholinergic Systems*, Steriade M, Biesold D, eds. Oxford, UK: Oxford University Press, pp 387–438

Woolf NJ, Butcher LL (1991): The cholinergic basal forebrain as a cognitive machine. In: *Activation to Acquisition: Functional Aspects of the Basal Forebrain Cholinergic System*, Richardson RT, ed. Boston: Birkhäuser Press, pp 347–380

Woolf NJ, Eckenstein F, Butcher LL (1983): Cholinergic projections from the basal forebrain to the frontal cortex: A combined fluorescent tracer and immunohistochemical analysis. *Neurosci Lett* 40:93–98

Woolf NJ, Eckenstein F, Butcher LL (1984): Cholinergic systems in the rat brain: I. Projections to the limbic telencephalon. *Brain Res Bull* 13:751–784

Woolf NJ, Gould E, Butcher LL (1989): Nerve growth factor receptor is associated with cholinergic neurons of the basal forebrain but not the pontomesencephalon. *Neuroscience* 30:143–152

Yarbrough GG (1983): Thyrotropin releasing hormone and CNS cholinergic neurons. *Life Sci* 33:111–118

3

Synaptic Organization of Basal Forebrain Cholinergic Projection Neurons

László Záborszky

Introduction

Since the discovery in the late 1970s (Bowen et al., 1976; Davies and Maloney, 1976; Perry et al., 1977) that Alzheimer's disease is characterized by severe losses of cortical presynaptic cholinergic markers with concomitant cell loss in the basal forebrain (for review see Butcher and Woolf, 1986; Whitehouse, 1991), considerable attention has been devoted to the basal forebrain cholinergic projection (BFC) system.

Cholinergic projection neurons are distributed across a number of classically defined regions of the basal forebrain and collectively project to the entire cortical mantle, including allocortical areas, such as the hippocampus, the amygdala, and the olfactory bulb. While this apparently diffuse organization makes the BFC an ideal candidate to serve general cortical activation (Buzsaki et al., 1988; Steriade et al., 1990), the manner in which BFC neurons participate in more restricted activation of specific portions of the cortex (for example, in sensory processing, selective attention, or learning) is less well understood (Richardson and DeLong, 1991,b; Wilson, 1991). This is largely due to our rudimentary knowledge about the synaptic organization of BFC neurons. Cholinergic neurons are intermingled among numerous noncholinergic cells (Záborszky et al., 1986a; Walker et al., 1989) and are distributed in close proximity to several major ascending and descending fiber systems (for review see Záborszky, 1989a). These circumstance make the BFC less accessible to morphological or physiological studies at the level of single cell or population (system) analysis.

The present review focuses on synaptic input to basal forebrain corticopetal (basalo-cortical) cholinergic neurons as revealed using strict criteria by cross-correlating anatomical (topographical) and transmitter properties of individual neurons and synapses. Data on septohippocampal neurons are only partially covered by this review. Despite many morphological similarities, both electrophysiological (Dutar et al., 1986) and pharmacological studies (Miller and Chmielewski, 1990) suggest that the two systems have many different properties.

Efferent Projections of BFC Neurons

BFC neurons show a rough ventrolateral-dorsomedial and rostro-caudal topography towards their target areas, although some projections, possibly determined by developmental factors (see Bayer, 1985), do not fit into a simple topographical scheme. In the rat, neurons within the medial septum and vertical limb of the diagonal band innervate the hippocampus and the entorhinal cortex. Cholinergic cells within the horizontal limb of the diagonal band, together with the caudolateral and angular part of the vertical limb of the diagonal band, provide efferents to the olfactory bulb, and to the entorhinal, piriform, medial prefrontal, cingulate, and occipital cortices. Cholinergic neurons in the ventral pallidum, globus pallidus, sublenticular substantia innominata (SI), internal capsule, and nucleus ansae lenticularis project to the basolateral amygdala and the lateral frontal, parietal, insular, and perirhinal cortices (for reviews see Koliatsos et al., 1990; Gaykema et al., 1991; Koliatsos and Price, 1991; Price and Carnes, 1991; nomenclature: Butcher and Semba, 1989). Approximately 30% to 50% of the hippocampopetal, 50% to 75% of the amygdalopetal, and 80% to 90% of the corticopetal neurons are cholinergic, while the rest are GABAergic and/or peptidergic (Rye et al., 1984; Carlsen et al., 1985; Záborszky et al., 1986a; Lamour et al., 1989; Wainer and Mesulam, 1990; Koliatsos and Price, 1991). Although the majority of cholinergic neurons projecting to a given target area show some clustering in the basal forebrain, cholinergic neurons projecting to different target areas are often intermingled, especially at the periphery of such clusters.

Both morphological (Price and Stern, 1983; Koliatsos et al., 1988; Price and Carnes, 1991) and electrophysiological studies (Aston-Jones et al., 1985) suggest that individual cholinergic axons have restricted cortical arborization fields. Cholinergic terminals show specific distribution patterns in different cortical areas (Lysakowski et al., 1989) and possess distinct synaptic arrangements in the various layers (Houser et al., 1985; Beaulieu and Somogyi, 1991), which are in agreement with the differential actions of acetylcholine (ACh) across individual cortical layers (Sillito and Kemp, 1983; Sato et al., 1987; Lamour et al., 1988).

Dendritic Architecture: Compartmentalization of BFC Neurons

Cholinergic cell bodies are multipolar, round to oval, or fusiform; the two to five primary dendrites are long and rectilinear, with sparse branching. The primary dendrites are smooth; however, secondary or tertiary dendrites often showed a beaded appearance. In a few cases, fine varicose dendrites originate directly from the cell body. Spine-like protrusions are only occasionally observed. The long dendrites form a continuum of overlapping dendritic fields. The dendrites have a tendency to follow the orientation of the myelinated fiber bundles within which they are embedded. Whenever the dendrites are markedly oriented, the cell body follows the same trend. Thus, the dendritic organization of BFC neurons resembles that of the isodendritic type of neurons (Ramón-Moliner and Nauta, 1966) or the interstitial neurons characterized by Das and Kreutzberg (1968). Interestingly, cholinergic neurons often form clusters consisting of 5–10 tightly packed cell bodies. The dendrites of such clusters seem to radiate to all directions in more ventral and medial parts of the horizontal limb of the diagonal band; however, they often form bundles in the more dorsal part of the SI in which the predominant orientation of the dendrites is in an oblique dorso-ventral direction (Fig. 1). The axon originates from the cell body or proximal dendrite, and after an initial short distance becomes beaded and often produces collaterals.

Although several studies have dealt with the morphology of basal forebrain neurons, including the cholinergic cells, no systematic study is available on the dendritic architecture of BFC neurons, especially in quantitative terms (Semba et al., 1987; Brauer et al., 1988; Dinopoulos et al., 1988). Cholinergic neurons show differences in size and shape in different subdivisions of the basal forebrain (Sofroniew et al., 1987; Brauer et al., 1991). The smallest neurons are found in the medial septum, while progressively more caudally the neuronal size gradually changes in favor of larger neurons. Electron microscopic (Milner, 1991) studies suggest that cholinergic neurons, even in one region such as the medial septum-diagonal band complex, display significant morphological variations.

Immuno- and histochemical studies support the notion that BFC neurons are neurochemically compartmentalized. For example, galanin (Melander et al., 1986), GABA (Brashear et al., 1986; Fisher et al., 1988; Freund and Antal, 1988; Kosaka et al., 1988; Fisher and Levine, 1989; Lamour et al., 1989), N-acetyl-aspartyl-glutamate (Forloni et al., 1987), and tyrosine hydroxylase (Henderson, 1987) have been colocalized in basal forebrain cholinergic projection neurons. Interestingly, the occurrence and proportion of these double-labeled neurons show species differences. For example, in the septohippocampal system of rats, a proportion of cholinergic neurons are colocalized with galanin, but not within corticopetal cholinergic neurons. In contrast, galanin is colocalized throughout the basal forebrain in monkey (Melander and Staines, 1986). Again, in human the majority of BFC neurons do not contain galanin (for references see Kordower and Mufson, 1990). Interestingly, in almost all of the cholinergic neurons in the medial septum-diagonal band complex that contain galanin, NADPH-diaphorase, another histochemical marker, is also coexpressed (Brauer et al., 1991; Pasqualotto and Vincent, 1991). Most cholinergic neurons in the monkey and in the human nucleus basalis are immunoreactive for calbindin-D-28 (CaBP), a vitamin D-dependent calcium-binding protein; on the other hand, none of the rat BFC neurons express CaBP immunoreactivity (Celio and Norman, 1985; Ichimiya et al., 1989; Chang and Kuo, 1991).

Another aspect of the neurochemical heterogeneity of BFC neurons is indicated by the fact that estrogen and thyroid hormones differentially regulate subpopulations of cholinergic neurons in males as compared to females (Luine, 1985; Loy and Sheldon, 1987; Patel et al., 1988; McEwen et al., 1991; Westlind-Danielsson et al., 1991), which may be due to regional differences in the density of the corresponding receptors (Simasko and Horita, 1984; Toran-Allerand et al., 1991).

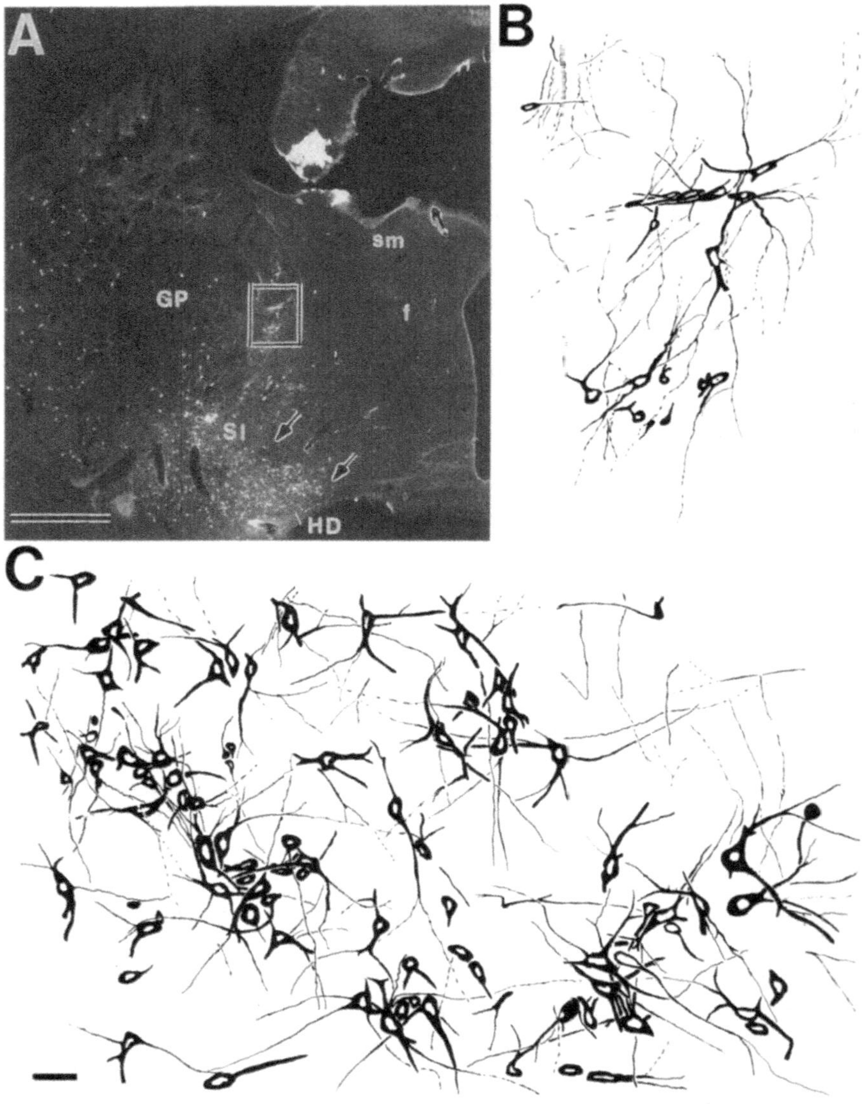

FIGURE 1. Distribution of cholinergic neurons in a frontal section of the rat brain approximately 1 mm behind the bregma, depicted by a low-magnification dark-field photomicrograph of this area A: Individual cholinergic cell bodies appear as white dots at this magnification. The section was immunostained with an antibody against choline acetyltransferase (ChAT). GP = globus pallidus; HD = horizontal limb of the diagonal band; f = fornix; sm = stria medullaris. B: Camera lucida drawing of the area indicated by box in A. Note that the main axis of the neurons is oriented either horizontally or dorso-ventrally. C: Camera lucida drawing from the dorsal part of the HD from an adjacent section. The approximate location of this area corresponds to the region indicated by arrows in A. Note that cholinergic cell bodies often form clusters. Scale: A, 1 mm; C, 50 μm.

It is not known, at present, whether morphological and/or neurochemical heterogeneity corresponds to physiologically defined heterogeneity in BFC neurons (Aston-Jones et al., 1985; Reiner et al., 1987).

Pattern of Afferent Terminals in the Basal Forebrain in Relation to BFC Neurons

Before describing the synaptic organization of BFC neurons, it is necessary to provide a general summary of afferents to the forebrain areas containing corticopetal cholinergic neurons. Several recent reviews dealt in great detail with this topic (Alheid and Heimer, 1988; Záborszky, 1989a; Heimer et al., 1991; Price and Carnes, 1991; Záborszky et al., 1991). In general, cholinergic neurons are located in the way station for several ascending and descending pathways: Fibers from the prefrontal cortex, brainstem, hypothalamic, and thalamic fibers all pass through the areas containing BFC neurons. As a result, only the application of correlated light and electron microscopic studies, along with a rigorous reconstruction of the spatial relationships between forebrain afferent terminals and cholinergic projection neurons, make it possible to extract the principles governing afferents to BFC neurons, which may be summarized as follows: (a) Inputs to BFC neurons are in one sense nonspecific, i.e., cholinergic neurons do not maintain afferent connections distinct from neighboring noncholinergic cells but receive the same type of input. In fact, the frequency of close contacts with cholinergic neurons appears to be proportional to the density of terminals present in a given area. The majority of afferent terminals in the forebrain, however, are related to noncholinergic neurons. (b) The distribution patterns of various terminals on BFC neurons correspond to the general topographical arrangement of basal forebrain fibers (Fig. 2). On the other hand, not all fiber systems that traverse the forebrain provide terminals to cholinergic cells. In many cases, fibers completely devoid of terminal varicosities course through areas rich in cholinergic cells. (c) Afferents to BFC neurons may be restricted or relatively diffuse. The majority of afferents appear to have a preferential distribution on subsets of BFC neurons. For example, fibers from different divisions of the prefrontal cortex seem to innervate subpopulations of cholinergic neurons in the horizontal limb of the diagonal band and ventral SI, according to a medio-lateral topography (Sesack et al., 1989; Gaykema et al., 1991b; Hurley et al., 1991; Záborszky et al., 1991). No other cortical area seems to directly innervate BFC neurons. Another striking example of preferential innervation is that ventral pallidal cholinergic neurons receive massive terminations from afferents originating in the ventral striatum (Záborszky and Cullinan, 1992). In addition, the distribution of several peptides in the basal forebrain suggests that they might contact subpopulations of cholinergic neurons (Beach et al., 1987; Záborszky, 1989b; Fig. 3a). Thus the emerging view is that different subsets of BFC neurons receive different combinations of afferents according to their location in the basal forebrain (Semba and Fibiger, 1989). For example, anterior and medial cortical areas, which are innervated by cholinergic neurons

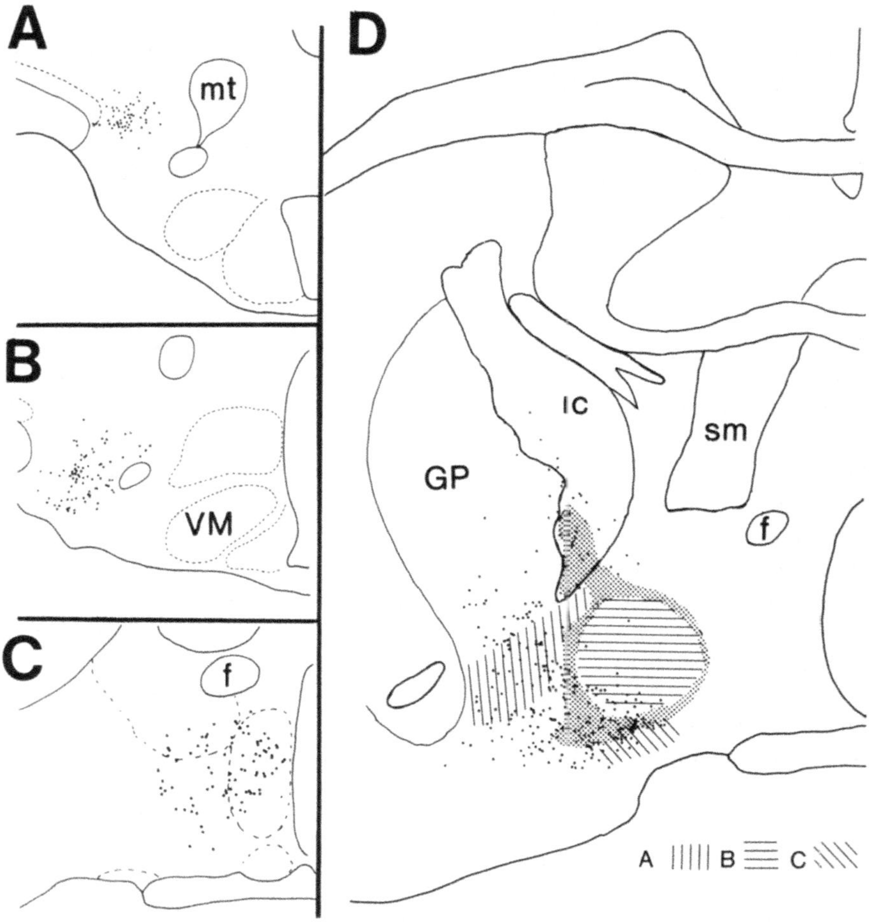

FIGURE 2. Topographical relationship between ascending hypothalamic axons/terminals and BFC neurons. The anterograde tracer PHA-L was injected into the far-lateral (A), mid-lateral (B), and medial hypothalamus (C). Dots in A–C represent the location of the labeled neurons at the injection sites. D: After uptake and anterograde transport of the tracer, areas of terminal arborizations from cases A–C were plotted onto a coronal section, approximately at the same level as Figure 1A. Dots symbolize cholinergic projection neurons. Hatching represent areas of dense fiber/terminal labeling, whereas stippling denotes areas of less dense labeling in case B. GP = globus pallidus; VM = ventromedial hypothalamic nucleus; f = fornix; ic = internal capsule, mt = mammillothalamic tract; sm = stria medullaris.

located in the more rostral and medial SI, may be preferentially innervated by hypothalamic neurons located more medially within the lateral hypothalamus (Záborszky and Cullinan, 1989). Posterior and lateral cortical areas, which appear to innervated by cholinergic neurons situated more caudally and laterally within the SI, may receive afferents preferentially from cells located more laterally within

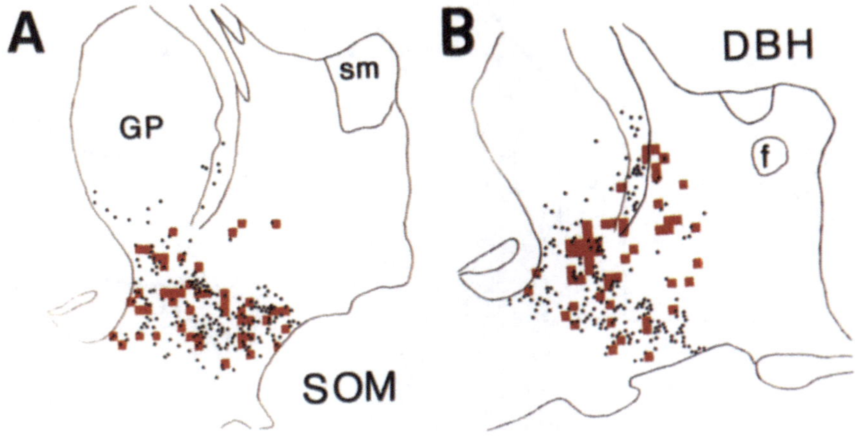

FIGURE 3 Schematic drawings to illustrate the relation between somatostatinergic (SS) and dopamine-β-hydroxylase (DBH) containing fibers/terminals and BFC neurons at the same level as that of Figure 1. Double immunostained sections (SS/ChAT or DBH/ChAT) were mapped at high-magnification light microscopy for the presence of putative contacts between immunostained axonterminals and cholinergic profiles. Cholinergic neurons are represented by dots. Zone of putative contacts between cholinergic neurons and immunostained terminals are depicted as red squares (corresponding 80 × 80 μm in the section). GP = globus pallidus; f = fornix; sm = stria medullaris.

the lateral hypothalamus (Cullinan and Záborszky, 1991; Fig. 2). Hippocampopetal cholinergic neurons are located in the medial septum-diagonal band complex (Amaral and Kurz, 1985) and are likely to receive input from the anterior hypothalamic and medial preoptic areas. Midlateral hypothalamic cells may also influence the hippocampus through connections with cholinergic neurons in the dorsomedial portion of the septum. On the other hand, noradrenergic afferents, particularly those from the locus coeruleus, apparently contact extensive portions of the BFC system (Záborszky et al., 1991; Záborszky et al., 1993; Fig. 3b).

Synaptic Pattern of BFC Neurons

Overview

The generally held view is that synaptic input to the cell bodies and proximal dendrites of BFC neurons is sparse, but it increases on more distal dendritic segments (Armstrong, 1986; Dinopoulos et al., 1986; Bialowas and Frotscher, 1987; Ingham et al., 1985, 1988; Martinez-Murillo et al., 1990; Milner, 1991). A more systematic analysis, however, suggests that the synaptic density on the cell body shows considerable regional variation. Cholinergic cell bodies in the internal capsule or in the area of the globus pallidus are surrounded by myelinated fibers,

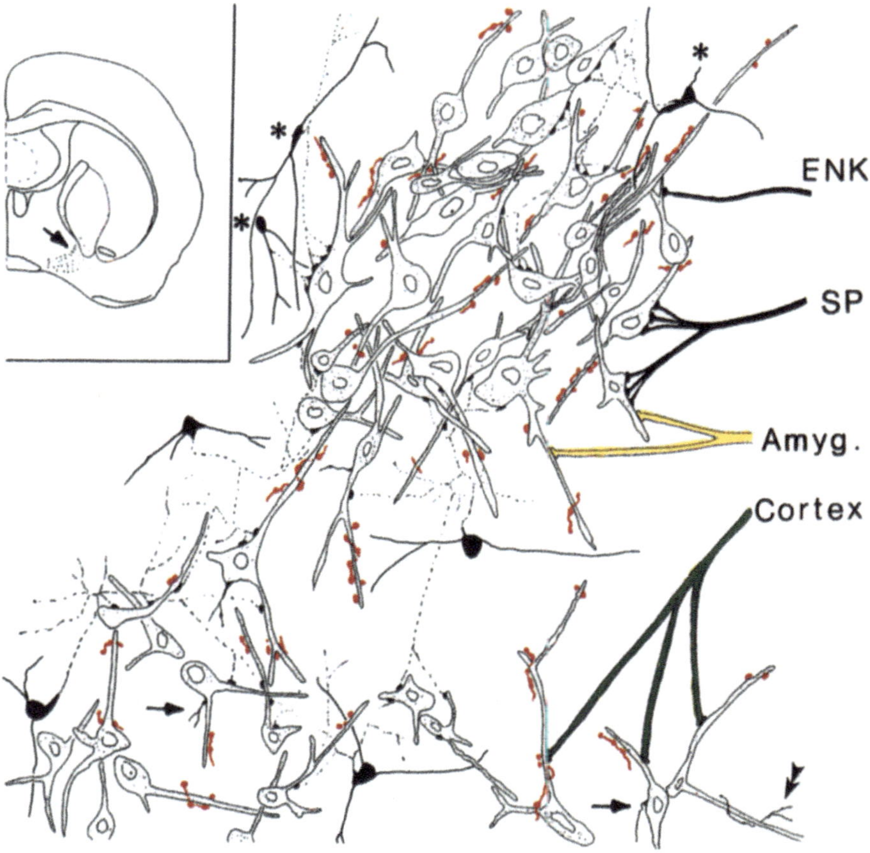

FIGURE 9. Schematic diagram summarizing the structural features of BFC neurons in the sublenticular substantia innominata, their relationship to local interneurons, and the different incoming afferents. Cholinergic neurons were drawn by the aid of a camera lucida from the area indicated by the arrow in the inset. The initial segment of the axon of some cholinergic neurons is indicated by arrows. The double arrowhead at the lower right points to a cholinergic axon climbing around a distal cholinergic dendrite. Neuropeptide Y (asterisk) and other interneurons (somatostatin) are in full black, and their axonal arborizations are reconstructed from corresponding double-labeled sections. Enkephalin-containing fibers terminate on distal dendrites, while substance P-containing axons establish symmetric synapses with the perikaryon and proximal dendrites (both axons in black). Amygdalofugal fibers (yellow) and axons from the prefrontal cortex (green) contact proximal dendrites. Noradrenergic (DBH-positive) axons (red) terminate primarily on dendrites. For ultrastructural features of the different synapses see Figure 8.

and the majority of corticopetal cholinergic cell bodies in these areas receive synaptic boutons only extremely rarely, as seen in Figure 4. Figure 5 depicts another cholinergic neuron in the medial part of the horizontal limb of the diagonal band, which receives some synapses, although a large proportion of the plasmalemmal surface is surrounded by astrocytic processes or is apposed by myelinated

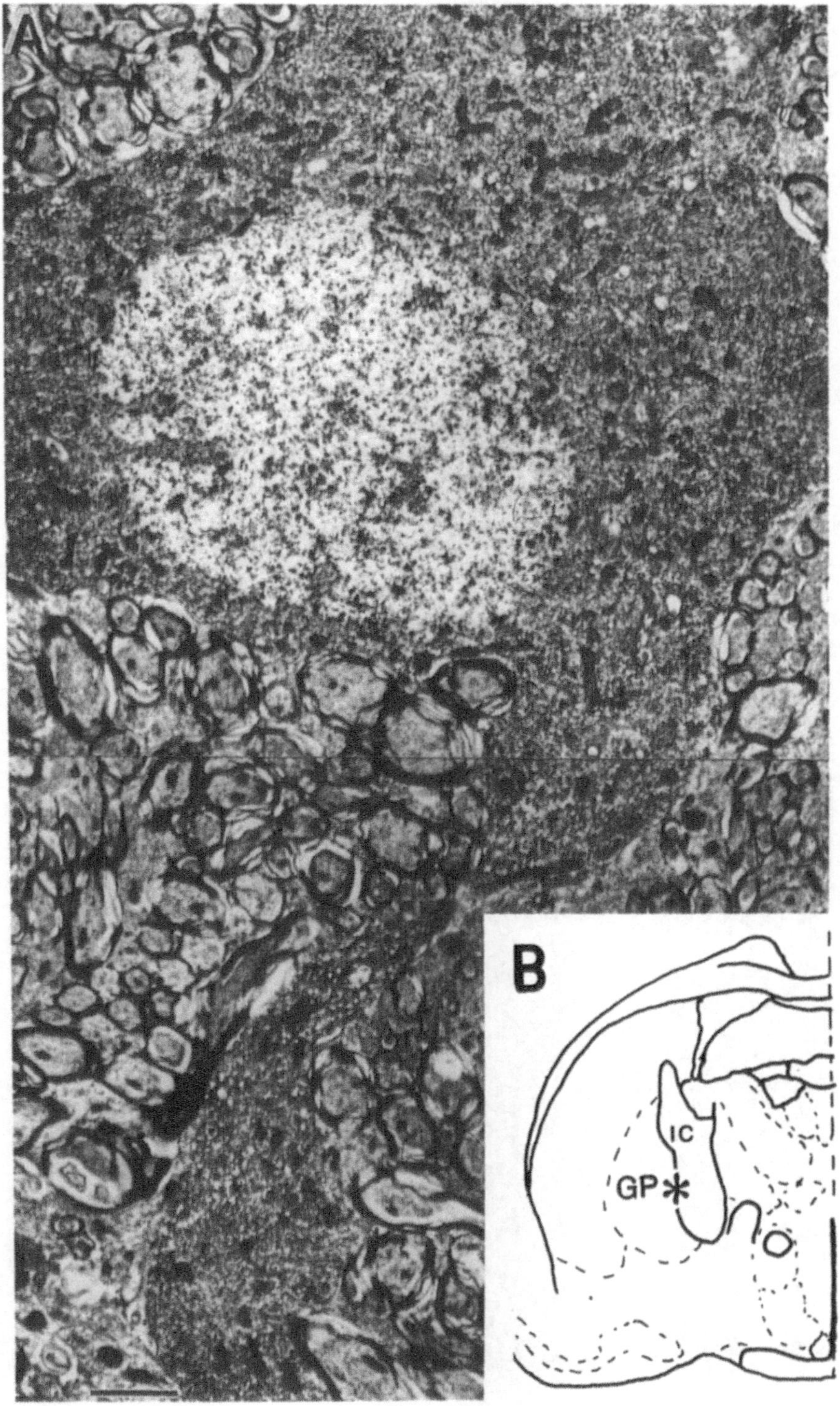

FIGURE 4 A: Electronmicrograph of a cholinergic neuron located at the border between the globus pallidus (GP) and internal capsule (ic). The majority of the perikaryal plasmalemmal surface is surrounded by myelinated fibers. B: Asterisk denotes the location of the cholinergic neuron shown in A. Scale: 2 μm.

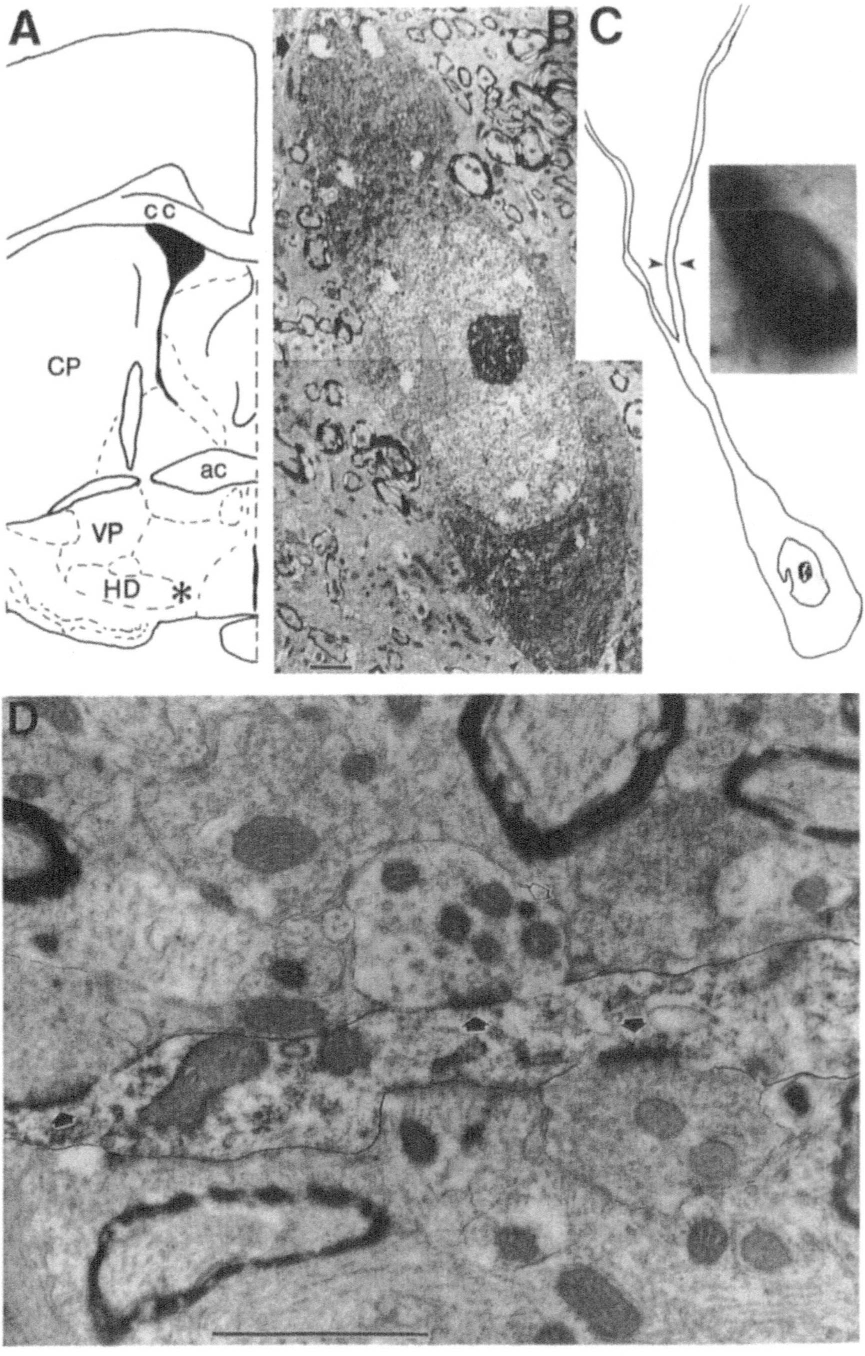
A
B
C
D
cc
CP
ac
VP
HD

axons, as described by Armstrong (1986) and Dinopoulos et al. (1986). A similar situation exists for cholinergic neurons in the medial septal nucleus, as described recently by Milner (1991). Finally Figure 6 shows a cholinergic cell body from the ventral pallidum, near the substriatal gray, which receives a large number of synaptic inputs to the perikaryon. Indeed, in serial reconstruction this cell body received 27 synaptic boutons. Most terminals on the perikaryon usually formed symmetric (Gray type II) synapses, which are believed to mediate inhibitory transmission. The cell body depicted in Figure 6 received only two asymmetric synapses.

Synaptic contacts are more frequent on progressively more distal dendritic segments of cholinergic neurons and here asymmetric (Gray type I) synapses predominate. In one study (Chang et al., 1987), 81% of synapses found on cholinergic dendrites were of the asymmetric type. Some of the asymmetrical contacts displayed thick postsynaptic specializations with prominent subsynaptic dense bodies (Fig. 6G). Asymmetric synapses are believed to mediate excitatory synaptic transmission. Often an individual bouton contacted both a ChAT-positive cell body or dendritic shaft, and an adjacent unlabeled dendritic shaft (Fig. 7). In a few cases the initial segment of the axon was identified, which received several symmetric contacts from boutons containing clear round to ovoid synaptic vesicles (Fig. 6F). In general, boutons synapsing with cholinergic profiles contained clear, round, or pleomorphic vessels, with occasionally small or large, dense-core vesicles (Figs. 5D, 6E–G); however, in material processed for single immunocytochemistry, the less then optimal preservation of the tissue frequently does not allow a further categorization of boutons based upon vesicle morphology. Although double immunolabeling techniques can cause further deterioration of the ultrastructure, they are capable of identifying the transmitter content of both pre- and postsynaptic profiles, thereby allowing tentative conclusions to be made about the nature of transmitter interactions. A brief summary is given below for transmitters identified in boutons contacting BFC neurons. Figure 8 schematically illustrates the morphological characteristics of the different transmitter-specific synapses.

FIGURE 5. Electronmicrographs showing a cholinergic cell body (B) and a portion of its dendrite (D). A: Asterisk indicates the location of the neuron, medial to the horizontal limb of the diagonal band (HD). CP = nucleus caudate putamen; VP = ventral pallidum; ac = anterior commissure; cc = corpus callosum. B: Electron micrograph of the cell body. A large part of the plasmalemmal surface is covered by glial processes (arrowheads at lower right). At this plane only one synapse (arrow at upper left) is recognizable. C: Schematic drawing of the neuron. Arrowheads denote the portion of the dendrite shown in D. The insert is a photomicrograph from the cell body. D: Electron micrograph showing several boutons in close association (synapses are indicated by arrows) with this dendrite. The open arrow points to a large, dense core vesicle reminiscent of a neurophysin-containing neurosecretory granule. Scale: B, 2 μm; D, 1 μm.

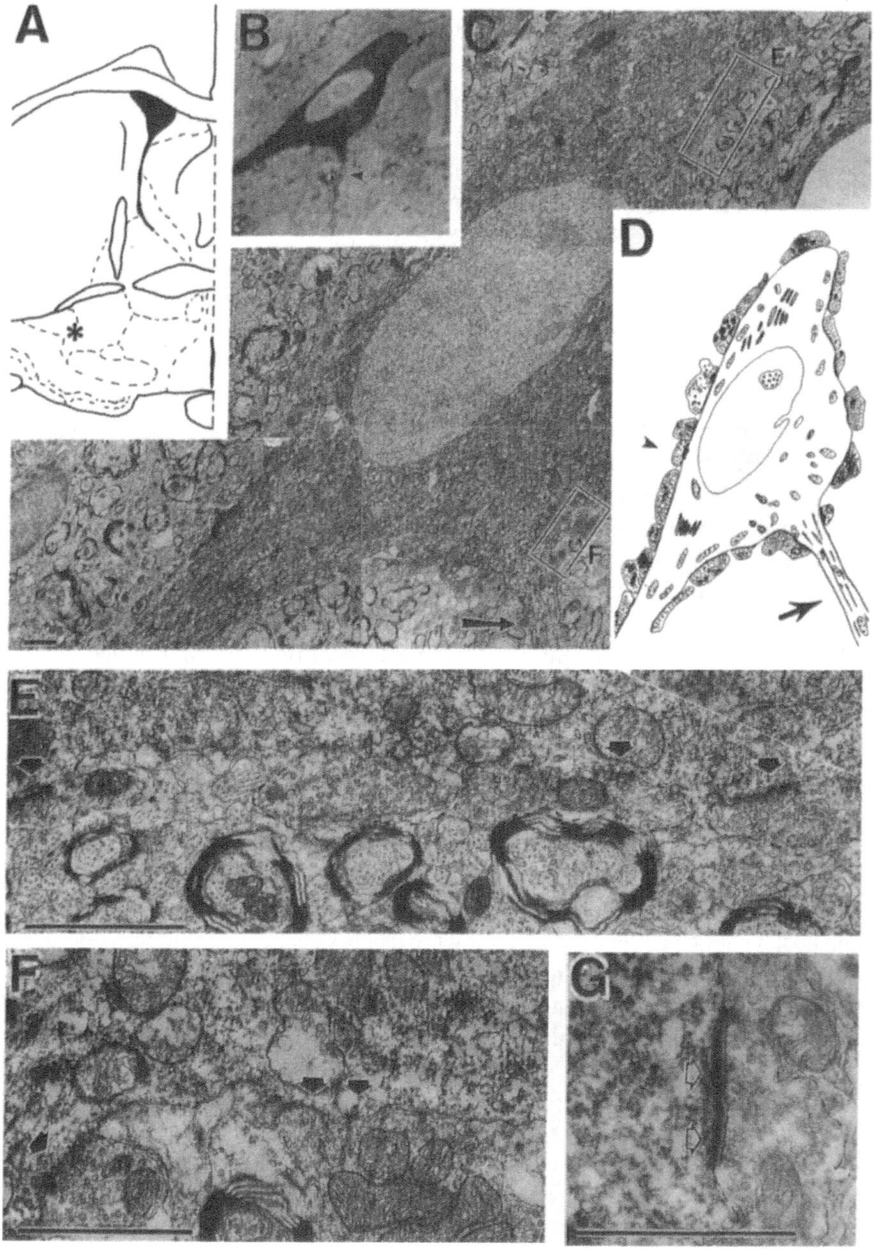

FIGURE 6. Electron micrographs showing several synapses on the cell body of a cholinergic neuron located near the substriatal gray (asterisk in A). B: Appearance of the cholinergic neuron in a semithin section. Arrows in B–D point to the initial segment of the axon. C: Low-magnification electron micrograph of the perikaryon. Boxed areas are enlarged in E and F. D: Schematic diagram showing the distribution of boutons around the cell body. All of the boutons except one terminal (arrowhead), in this or adjacent sections have symmetric synapses. E: Enlarged view of the upper box in C. Four boutons are seen in this field; three of them are in synaptic contact (arrows) with the perikaryon. F: Enlarged view of the lower box in C showing the axon hillock area. Both terminals enter symmetric synaptic contacts with this neuron. G: Asymmetric axosomatic synapse with prominent subjunctional bodies (arrowheads) from a different plane of the section. Scale: C, 2 μm; E–G, 1 μm.

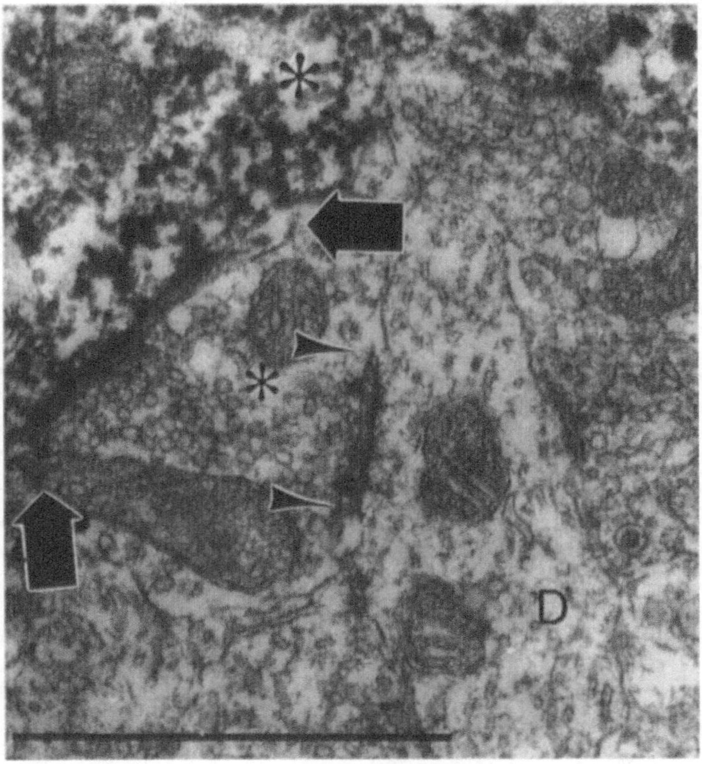

FIGURE 7. Electron micrograph showing an axon terminal (asterisk) in direct contact with both a cholinergic cell body (heavy immunostained profile on the upper left) and an unlabeled dendritic profile (D). Arrows and arrowheads denote the synaptic thickening. Scale: 1 μm.

Transmitter-Specific Synapses

Amino Acids

GABA. GABAergic synapses on cholinergic projection neurons were first described in the ventral pallidum (Záborszky et al., 1986b), and have since been identified on corticopetal, presumably cholinergic, neurons in the globus pallidus (Ingham et al., 1988), on septal cholinergic neurons (Leranth and Frotscher, 1989), and on cholinergic cells in the anterior amygdaloid area (Nitecka and Frotscher, 1989). GABAergic boutons containing pleomorphic vesicles form symmetric membrane specializations, which in some areas comprise over half of the input to the perikarya and proximal dendrites (Ingham et al., 1988). Due to the differential distribution of GABAergic terminals in the basal forebrain, it is likely that subpopulations of cholinergic neurons are exposed to varying degrees of GABAergic inhibition. In the ventral pallidum, a significant proportion of

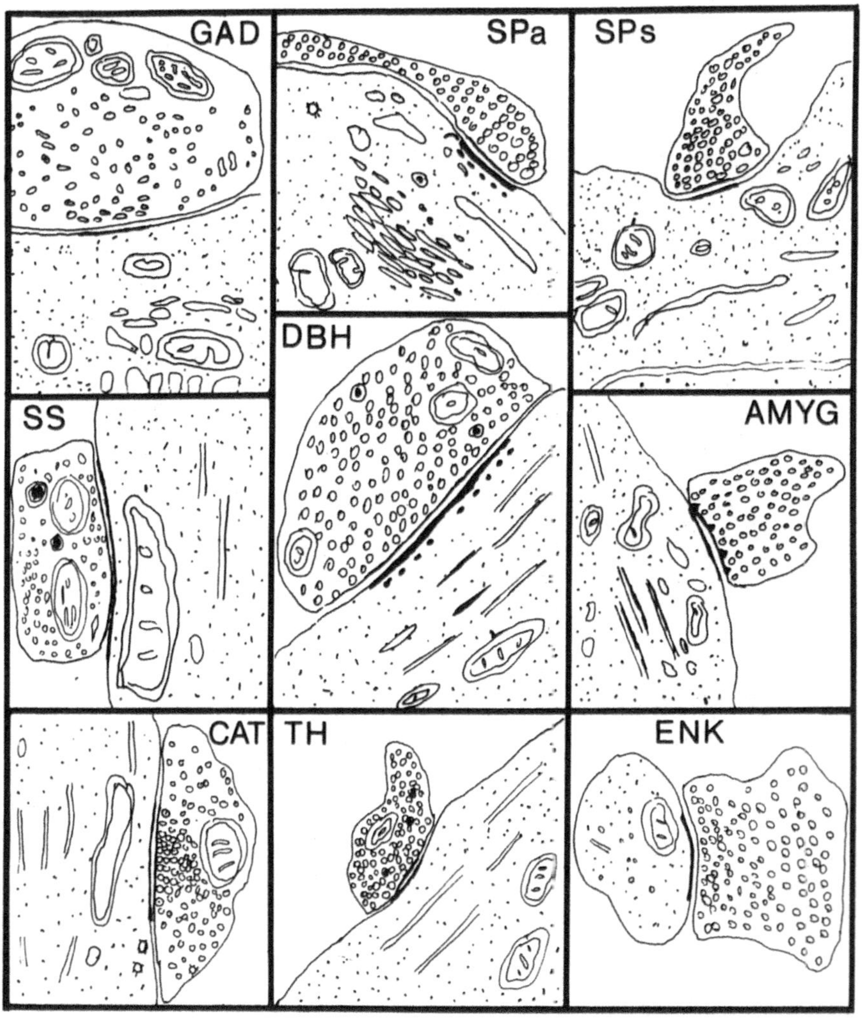

FIGURE 8. Schematic figure illustrating the different transmitter-specific synapses on BFC neurons compiled from our data as well as that of Bolam et al. (1986; substance P) and Chang et al. (1987; enkephalin). Drawings were made from the normalized original photomicrographs, so the sizes of the boutons are comparable. GAD = glutamic acid decarboxylase-containing bouton in synaptic connection with a cholinergic cell body; SPa = substance P-containing terminal in asymmetric contact with the proximal portion of the dendrite; SPs = substance P containing terminal establishes symmetric synapse with the cell body; SS = somatostatin-positive bouton in symmetric synaptic contact with a proximal portion of a cholinergic dendrite; DBH = dopamine-β-hydroxylase containing terminal makes asymmetric synapse with a proximal dendrite. AMYG = amygdalofugal terminal in synaptic contact with the proximal dendrite of a cholinergic neuron; CAT = cholinergic terminal in symmetric synaptic contact with a proximal dendrite. TH = tyrosine hydroxylase-positive terminal in symmetric synaptic contact with a distal dendrite. ENK = enkephalin-positive bouton establishes synaptic contact with a small dendrite. Cholinergic profiles are indicated by light stippling.

GABAergic terminals comes from the nucleus accumbens, since lesions of this structure result in a significant decrease of GAD in the ventral pallidum (Walaas and Fonnum, 1979; Záborszky et al., 1982). Experiments using PHA-L tracing and choline acetyltransferase (ChAT) immunocytochemistry at the ultrastructural level (Záborszky and Cullinan, 1992) confirm our preliminary electron microscopic degeneration data (Záborszky et al., 1984a) that suggest nucleus accumbens axons establish symmetric synaptic contacts with ventral pallidal cholinergic neurons. Another source of GABAergic terminals may be local GABAergic neurons, although no direct data exist for such a possibility. GABAergic terminals may costore enkephalin, substance P, and/or CaBP, since medium-sized striatal projection neurons contain CaBP (Gerfen et al., 1985), GABA (Mugnaini and Oertel, 1985), and either substance P or enkephalin (Penny et al., 1986).

Iontophoretically applied GABA strongly inhibits identified basal forebrain corticopetal neurons in anesthetized rats (Lamour et al., 1986). GABA elicits chloride-dependent inhibitory postsynaptic currents in cultured nucleus basalis neurons, which are mediated through $GABA_A$ receptors located on the cholinergic neurons (Akaike et al., 1992). It is further suggested (Akaike et al., 1992) that $GABA_B$ receptors located on presynaptic GABAergic axons regulate the synaptic transmission between these axons and their cholinergic cell targets via negative feedback (negative autoreceptor).

ACETYLCHOLINE. It has been generally noted that cholinergic neurons do not receive significant cholinergic input (Ingham et al., 1985; Armstrong, 1986; Dinopoulos et al., 1986; Záborszky et al., 1986b). We and others (Bialowas and Frotscher, 1987; Martinez-Murillo et al., 1990) have occasionally found ChAT-immunoreactive boutons in contact with cholinergic cell bodies or proximal dendrites in the SI, and these synapses were usually of the symmetric type (Záborszky et al., 1991). On the other hand, Martinez-Murillo et al. (1990) observed very rarely cholinergic synapses with distal dendrites, which were asymmetric and they were frequently associated with subjunctional dense bodies. Milner (1991) found that in the septal complex only 7% of the contacts on cholinergic perikarya and dendrites were from ChAT-containing terminals. Such terminals established symmetric synapses with cell bodies, but cholinergic synapses on dendrites were either symmetric, asymmetric, or without clear-cut membrane specializations.

The origin of cholinergic axons in the basal forebrain remains unclear. One possible source is the ascending cholinergic projection from the pedunculopontine and laterodorsal tegmental nuclei (Satoh and Fibiger, 1986), although a PHA-L study did not find evidence for terminations on BFC neurons, but rather, on noncholinergic cells (Hallanger et al., 1988). Another possibility is that forebrain cholinergic terminals are of intrinsic origin. Indeed, physiologically identified corticopetal neurons (Semba et al., 1987) and neurons stained with ChAT in the SI (personal observation) are readily observed to have local axon collaterals; however, no direct evidence exists that these collaterals innervate cholinergic neurons locally.

The majority of identified basal forebrain corticopetal neurons were excited by iontophoretic application of ACh or cholinergic agonists (Lamour et al., 1986). A study using intracellular recording from septal, presumably cholinergic, neurons in slices of guinea pig brain showed that muscarinic agonists have a variable effect on passive membrane properties (depolarize or hyperpolarize the membrane), but they consistently inhibited a long afterhyperpolarization (AHP) in these neurons (Sim and Griffith, 1991). Block of the long AHP, however, was not associated with a dramatic change in the spike discharge, as is the case in the hippocampus (Brown et al., 1990), and cannot explain the excitation produced by ACh in a large number of putative BFC neurons from *in vivo* studies (Lamour et al., 1986). How far these findings may be extrapolated to *in vivo* basalo-cortical cholinergic neurons is not clear, but these observations, together with the presence of mixed population of cholinergic synapses on cholinergic neurons, suggest that ACh may have diverse action on basal forebrain cholinergic neurons.

GLUTAMATE. It has been suggested that amygdalofugal axons use glutamate as a transmitter, as evidenced by decreases in glutamate uptake in the SI following amygdaloid lesions (Francis et al., 1987), as well as by the presence of ^{3}H-D-Asp-containing cell bodies in the amygdala after injection of this tracer into the ventral pallidum (Fuller et al., 1987). We have found that asymmetric contacts are formed on BFC neurons by amygdalofugal axons (Záborszky et al., 1984), although the transmitter of these terminals was not identified. A recent retrograde tracing study injecting ^{3}H-D-Asp into different basal forebrain areas containing BFC neurons suggests a more widespread origin of glutamatergic projection to basal forebrain areas, including the prefrontal cortex, intralaminar thalamic nuclei, lateral septum, habenula, and several hypothalamic and brainstem sites (Carnes et al., 1990; Clements et al., 1991; Price and Carnes, 1991). The identification of glutamate in synapses on BFC neurons awaits ultrastructural double-labeling studies.

Peptides

SUBSTANCE P. Substance P-containing terminals have been found to contact BFC neurons in the ventral pallidum and ventromedial globus pallidus (Bolam et al., 1986). Such synapses were detected on the proximal dendrites and cell bodies of cholinergic neurons. It has been estimated that as many as one third of the observed contacts contained substance P. The majority of immunoreactive terminals formed symmetrical synaptic specializations. However, a few boutons formed asymmetric contacts with the proximal dendrites of cholinergic neurons. Individual cholinergic neurons apparently receive multiple substance P-containing terminals. One potential site of origin of substance P-containing terminals to BFC neurons in the ventral pallidum or ventromedial globus pallidus is the striatum (nucleus accumbens), since lesions of this region cause a reduction of substance P immunostaining in the ventral pallidum (Záborszky et al., 1982; Haber and Nauta, 1983). However, the different morphological types of substance P synapses, as well as the fact that several other structures that project to the areas containing BFC neurons also contain substance P-positive cells, suggests other potential input sources, including the amygdala, bed nucleus of the stria terminalis, hypothalamus, and pontine tegmentum (Beach et al., 1987).

Iontophoretic application of substance P has been shown to produce increased excitability of identified cholinergic neurons from cultures of the basal forebrain, an effect caused by reducing the inward-rectifying K currents (Nakajima et al., 1985, 1988, 1991). BFC neurons express substance P receptors (Gerfen, 1991), and the postsynaptic action of substance P through its receptor is mediated by a pertussis toxin-insensitive G protein (Nakajima et al., 1991).

ENKEPHALIN. An immunocytochemical double-labeling study has shown that enkephalin-positive terminals contact cholinergic dendrites in the globus pallidus (Chang et al., 1987), although these were reported to be few. In fact, in this study, out of a total of 156 immunopositive terminals, only two contacted with cholinergic dendrites. Such boutons contained large pleomorphic vesicles and established symmetric synapses with small cholinergic dendrites. Another study (Martinez-Murillo et al., 1988a) found somewhat more enkephalin-positive terminals, mainly in contact with distal dendrites of putative cholinergic neurons. Due to the use of AChE, a less specific marker for cholinergic neurons, this study leaves some margin of doubt about the postsynaptic neurons; however, the presence of enkephalinergic afferents on cholinergic cells appears likely. No electrophysiological study has investigated the effect of enkephalin on identified BFC neurons, although some pharmacological studies (Wenk, 1984; Baud et al., 1988) suggest an inhibitory role.

SOMATOSTATIN. Somatostatin is found in axon terminals in contact with BFC neurons in the SI (Záborszky, 1989a). The synapses observed are of the symmetric type and were found primarily on proximal dendrites or on cell bodies. The results of a high-magnification light microscopic analysis of the distribution of putative contact sites suggest that BFC neurons in the SI and horizontal limb of the diagonal band may receive a rather diffuse somatostatinergic innervation (Záborszky and Braun, 1988: Fig. 3), and at least a proportion of such terminals are likely to originate from local interneurons.

Somatostatin has been reported to inhibit ACh release from cholinergic neurons of the myenteric plexus (Yau et al., 1983). The functional impact of such somatostatinergic/cholinergic interactions in the basal forebrain remains to be elucidated.

NEUROPEPTIDE Y. Neuropeptide Y (NPY) has been found in axon terminals in synaptic contact with BFC neurons in the SI (Záborszky and Braun, 1988) and in the horizontal limb of the diagonal band (Tamiya et al., 1991). In many cases, multiple NPY-containing boutons were found to encompass cholinergic cell bodies. The synapses were always of the symmetric type. Similar to somatostatin, at least a proportion of NPY terminals are likely to be derived from local interneurons (Záborszky, 1989a,b). No electrophysiological study has examined the putative NPY/cholinergic interaction in the basal forebrain.

Other Peptides. Various peptide-containing fiber systems are differentially distributed in the basal forebrain (for references see Palkovits, 1984). As a result, different populations of BFC neurons are likely to be contacted by different afferent peptidergic fibers (Záborszky, 1989b). For example, the majority of BFC neurons in the rostral globus pallidus and ventral pallidum may receive a rich neurotensin innervation (Záborszky et al., 1991), but appear to be contacted only occasionally by other peptidergic afferents. BFC neurons in the SI appear to receive a substantial input from a number of different peptidergic systems, including NPY, somatostatin, and neurotensin.

The possibility of a rich innervation of a subpopulation of BFC neurons by *neurotensin* fibers is supported by a study reporting localization of radiolabeled neurotensin-binding sites on AChE-positive neurons in the area of the nucleus basalis (Szigethy et al., 1990). The significance of neurotensin in the regulation of basal forebrain cholinergic function is discussed by Wenk in this volume.

Vasopressin axons emanating from the hypothalamic paraventricular nucleus *en route* to the posterior pituitary cross cholinergic dendrites of the SI or horizontal limb of the diagonal band, and boutons containing large neurosecretory vesicles are occasionally found in synaptic contact with distal cholinergic dendrites (Záborszky et al., 1991; Fig. 5D). The functional significance of such connections is presently unclear.

Galanin has been shown to be present in virtually all cholinergic neurons within the monkey nucleus basalis (Melander and Staines, 1986). In contrast to this, few magnocellular neurons in the human substantia innominata contained galanin (Kordower and Mufson, 1990). However, in human brains galanin was localized in parvicellular interneurons, and in fibers that seem to innervate cholinergic cell bodies and proximal dendrites in extensive basal forebrain areas (Chan-Palay, 1988; Gentleman et al., 1989; Kowall and Beal, 1989; Kordower and Mufson, 1990). Lesion studies in rats suggest that most medial septal neurons that contain both galanin and acetylcholine project to the ventral hippocampus (Melander et al., 1986). In the ventral hippocampus exogenously administered galanin appears to attenuate scopolamine-stimulated ACh release (Fisone et al., 1987) and to inhibit the muscarinic stimulation of phosphoinositide turnover (Palazzi et al., 1988). Galanin also has been shown as an inhibitory peptide affecting myenteric cholinergic neurons (Yau et al., 1986), although no data are yet available for their central action at the level of the cholinergic neurons in the basal forebrain (see also Crawley and Wenk, 1989).

Monoamines

Noradrenaline/adrenaline. With the exception of those neurons in the dorsal part of the globus pallidus and internal capsule, cholinergic neurons are approximated by dopamine-β-hydroxylase (DBH)-positive varicosities in most portions of the BFC system. DBH is the enzyme synthesizing noradrenaline, and therefore boutons containing DBH-positive immunoreactivity are likely to contain noradrenaline and/or adrenaline. Due to the low level of adrenaline in the basal forebrain, it is suggested that the predominant catecholamine in DBH-positive terminals is noradrenaline (Chang, 1989). In some cases, particularly within the caudal SI, distal segments of cholinergic dendrites appeared to receive repetitive contacts in the form of a "climbing" arrangement. The distribution of DBH-positive fibers/terminals in relation to BFC neurons caudal to the level of the anterior commissure is shown in Figure 3B. Parallel experiments at the electron microscopic level confirmed synaptic contact between DBH-positive terminals and distal cholinergic dendrites (Záborszky et al., 1992). DBH-positive boutons are usually large, containing round clear and dense core vesicles, and the synapses are always of the asymmetric type with prominent postsynaptic subjunctional dense bodies (Fig. 8). Preliminary experiments following PHA-L injection in the locus coeruleus suggest that at least part of the noradrenaline innervation of the basal forebrain originates in the locus coeruleus. The results of these studies suggest that the overall noradrenergic innervation of the BFC system is rather diffuse, although the input is apparently not entirely uniform. Regional variations in the type and distribution of labeled fibers further suggests contributions to the innervation of the BFC from noradrenergic cell groups, such as the A1 and A2 groups (Záborszky et al., 1991).

DOPAMINE. Several lines of evidence suggest a dopaminergic innervation of forebrain regions containing BFC neurons (Brownstein et al., 1974; Lindvall and Björklund, 1979; Haring and Wang, 1986; Martinez-Murillo et al., 1988b; Semba et al., 1988; Jones and Cuello, 1989; Napier and Potter, 1989, Napier et al., 1991). Tyrosine hydroxylase (TH) is the rate-limiting enzyme for dopamine and other catecholamines; therefore, its presence in boutons contacting BFC neurons cannot be taken as evidence for its dopaminergic nature. However, the differential distribution of TH versus DBH varicosities in relation to cholinergic neurons (Záborszky et al., 1992), as well as the different morphological characteristics of noradrenergic and dopaminergic synapses (see below) suggest that some TH-positive terminals in the septum (Milner, 1991), and especially in the ventromedial globus pallidus and internal capsule, may represent dopaminergic input to BFC neurons. TH axons in the latter two regions establish symmetric synapses with BFC cells (Fig. 8). While a few TH-positive boutons contacted the cell body, the majority of such synapses were found on distal dendrites (Záborszky et al., 1992). In the medial septum-diagonal band complex, according to Milner (1991) 25% of all boutons on cholinergic neurons were TH positive. Such boutons formed both asymmetric and symmetric synapses with cholinergic dendrites, and it is suggested that the symmetric synapses are dopaminergic and the asymmetric ones contain noradrenaline. A definitive answer to the question of direct dopaminergic/cholinergic interactions in the basal forebrain awaits double-labeling studies using antisera for ChAT and dopamine.

SEROTONIN. Our preliminary light microscopic morphologic (Záborszky et al., 1991) and biochemical data (Záborszky and Luine, 1987) suggest serotoninergic innervation of basal forebrain cholinergic neurons, although electron microscopic studies are lacking so far.

Input–Output Organization of BFC Neurons: Basic Circuits

Participating Elements

Figure 9 illustrates some basic properties of forebrain circuits involved in the transmission of information through the basal forebrain cholinergic system to the cortex. In the upper right part of the figure it is especially clear that a large number of cholinergic cell bodies are very closely spaced, while other neurons are located some distance from this "cluster" (such clusters were originally described by Bigl et al., 1982; see Woolf and Butcher, 1991). The arborization pattern of local somatostatin and NPY axons (Záborszky and Braun, 1988) suggest that they indiscriminately innervate cholinergic and noncholinergic neurons. Usually one somatostatin neuron can innervate as many as 5–10 cholinergic neurons, supplying mainly the cell body and the proximal dendrite. Similarly, NPY-containing axons innervate several cholinergic neurons. Although the axonal arborization pattern of local GABAergic neurons is not known, it is likely that one GABAergic neuron innervates a group of cholinergic neurons. On the other hand, the same cholinergic neuron may be innervated by as many as three to four somatostatin, NPY, and perhaps GABA neurons.

Local Processing of Information

Because the axonal arborization of somatostatin and NPY neurons is often confined within their dendritic fields, it is likely that local events will influence the activity of cholinergic neurons through these interneurons and their strategically located synapses around the cell body and proximal dendrites of cholinergic neurons. The long-range putative excitatory afferents (cortical, amygdaloid, some hypothalamic) to cholinergic cells may also project to these local interneurons (for simplicity this is not shown in Fig. 9; but see Fig. 7 in Záborszky, 1989a), so that the interneurons could participate in feedforward inhibition, by "shaping" excitatory events in the cholinergic neurons. The axons of cholinergic neurons give off local axon collaterals, and since GABAergic neurons receive cholinergic input (Záborszky et al., 1986b), presumably these inhibitory interneurons receive input from proximal cholinergic neurons, thus contributing to feedback inhibition. Since both the long-range excitatory and inhibitory inputs to basal forebrain areas show a specific regional pattern (see Fig. 2) and the number of inputs to the cell body varies between different regions, it is likely that cholinergic neurons are subject in varying degree to local influence. For example, cholinergic neurons receiving a relatively large number of inhibitory inputs on their cell body, such as shown in Figure 6, are more likely to reflect local circuit activity than those that receive only very few such contacts. Noradrenergic input, on the other hand, is distributed to more distal dendrites of cholinergic neurons in extensive regions of the basal forebrain, thus involving most of the BFC neurons in a general pattern of excitation, in accord with the presumed function of the noradrenergic cells of the locus coeruleus in alerting, attention, or arousal (Foote et al., 1983). Noradrenergic input, in other words, can change the responsiveness of BFC neurons to more "specific" (topographically organized) information arriving through interneurons or other topographically organized excitatory afferents.

Input–Output Features

NETWORK PROPERTIES. Although it is very likely that the same type of neuronal elements participate in information processing for different regions of the basal forebrain, the number of such neurons and their inputs may vary between regions. While cholinergic neurons in the sublenticular substantia innominata, in particular, cluster together, where a few dozen neurons may receive the same input BFC neurons located ventrolaterally in the horizontal limb of the diagonal band are much more loosely arranged, and thus the number of BFC neurons with shared afferents may be less. As suggested by retrograde tracing experiments (Pearson et al., 1983; Koliatsos and Price, 1991), cholinergic cells projecting to the same or interconnected cortical areas show clustering. It is unclear what determines the size of these "efferent clusters." Limbic and paralimbic cortical areas are more heavily innervated than sensory or association regions (Mesulam et al., 1986; Koliatsos and Price, 1991). It is not known if more neurons project to these areas or whether the axons of cholinergic neurons simply arborize more extensively in these cortical territories. The proportion of cholinergic neurons that participate in individual projections appears to vary not only among individual targets, but also among different sectors of the basal forebrain cholinergic system (Koliatsos and Price, 1991).

An understanding of the three-dimensional relationship between projection neurons and afferents is necessary to appreciate how clusters of cholinergic neurons may function as information-processing "modules" or units (Woolf and Butcher, 1991) and whether the information transmitted through restricted afferents to BFC cells is maintained through the efferent projection of these neurons. It is clear from simple observations of normal material that intermingling of information could occur between neighboring clusters, since it is often the case that dendrites of remote neurons extend into the cell body area of neighboring BFC groups. It is unclear, however, whether these "outlying" cholinergic neurons project to the same target as the majority of cholinergic neurons in such clusters. Also because interneurons often innervate cholinergic neurons belonging to different clusters, the activation of basal forebrain afferents could inhibit a varying number of neighboring cholinergic cell clusters. The role of the dendritic bundles or intermingling dendrites also remains to be determined. Do they help to synchronize the electrical activity of individual cholinergic neurons (Woolf and Butcher, 1991) or are they separated by glial sheets, such as the dendrites of preganglionic cholinergic neurons (Markham and Vaughn, 1990), and therefore remain capable of responding individually to activation from a particular afferent?

SPECIFIC VERSUS DIFFUSE ORGANIZATION. In spite of apparent diffuse characteristics, the BFC system has a spatial organization that seems to allow restricted afferents to influence information processing in selected areas of the cortex, as evidenced by the early observation of Szerb and his colleagues (Kanai and Szerb, 1965) that peripheral sensory stimulation elicits maximum ACh release over specific cortical areas, while other nonrelated cortical areas show only minor increases in ACh release. The basis of this phenomenon may be that specific sensory systems reach selected portions of the BFC system (Záborszky et al., 1991).

It is interesting to note that in primates cholinergic neurons projecting to a given target area show stronger separation than in rodents (Mesulam et al., 1983a,b; Mesulam and Geula, 1988, see also Butcher et al., this volume). It can be speculated that the phylogenetic differentiation of the BFC system (Gorry, 1963; Woolf and Butcher, 1991) would make this system in primates more suited to selective modulation of information processing in specific cortical areas underlying cognitive functions. No matter how attractive this hypothesis (Bigl and Arendt, 1991), electrophysiological studies in behaving monkeys so far lend little support to the proposition that differential responsiveness of basal forebrain units relates to their spatial distribution (see more comprehensive discussion of this topic in Richardson and DeLong, 1991b; Wilson, 1991; Woolf, 1991; Woolf and Butcher, 1991).

Transynaptic Regulation of Cortical Acetylcholine Release

Basal forebrain corticopetal, presumably cholinergic, neurons show rhythmic, spontaneous firing pattern (Aston-Jones et al., 1985; Dutar et al., 1986; Reiner et al., 1987), and the discharge rate of these neurons is tightly coupled with cortical electrical activity. Increased discharge frequency of basal forebrain neurons during waking and REM sleep is consistently associated with EEG desynchronization, while lower firing of BFC neurons is paralleled with EEG synchronization (Detari and Vanderwolf, 1987; Richardson and Delong, 1991a,b). As opposed to this tonic changes in the firing rate, identified basal forebrain corticopetal neurons show phasic changes in firing in response to noxious stimuli (Lamour et al., 1986). It has been additionally shown that basal forebrain neurons are particularly sensitive to appetitive stimuli or those sensory stimuli that signal the availability of reinforcement, irrespective of their sensory modality (Richardson and DeLong, 1991a,b; Wilson and Rolls, 1990; Wilson, 1991). These phasic changes in the response characteristics of BFC neurons may induce a transient increase of activity in functionally specific cortical areas and thus may optimize their particular functions.

Since the mid-1960s (Kanai and Szerb, 1965; Celesia and Jasper, 1966; Collier and Mitchell, 1967; Szerb, 1967; Jasper and Tessier, 1971) it has been known that stimulation of the reticular formation, similar to peripheral sensory stimulation, elicits increased ACh release from the cortex that parallels EEG activation.

Several indirect pieces of evidence (for review see Buzsaki et al., 1988; Steriade et al., 1990) suggest that the tonic or phasic activation of the BFC system mediates this response. Which pathways or transmitters activate the BFC is still only partially understood (Durkin, 1989; Rolls, 1989; Wilson, 1991; Záborszky et al., 1991).

It has been established that cortical or hippocampal sodium-dependent high-affinity choline uptake, ACh turnover, and ACh release are dynamic and sensitive indices of the state of activity of septal or basal forebrain corticopetal cholinergic neurons (for review see Durkin, 1989; Pepeu et al., 1990; Fibiger et al., 1991; Olton et al., 1991). Early pharmacological data, using a cortical cup technique coupled to a bioassay (Mitchell, 1963; Mullin and Phillis, 1975) have been confirmed using the more sophisticated method of *in vivo* dialysis coupled with HPLC and using electrochemical detection or radioenzymatic methods. These experiments demonstrated that sensory stimulation, which is paralleled by the activation of septohippocampal neurons, elicits increased ACh overflow from the hippocampus (Nilsson et al., 1990).

A brief summary of pharmacological studies is given below in which different neurotransmitter agonists or antagonists were infused into the basal forebrain and indices of presynaptic cholinergic functions were measured in the cortex or hippocampus. A detailed discussion of the concomitant behavioral changes can be found in the reviews by Durkin (1989), Decker and McGaugh (1991), Fibiger et al. (1991), Levin et al. (1990), and Olton et al. (1991) and in several chapters of this volume. Considering the circuit diagram of Figure 9, it is obvious that local microinfusion of drugs will cause diverse effects; the distribution of various receptors and the diffusional distance of the ejected drugs are but some of the factors that might confound the interpretation of the resulting changes in presynaptic cortical function.

GABA

There is ample evidence that ACh release and turnover in the cortex is controlled by highly efficient inhibitory GABAergic mechanisms in the basal forebrain (Wood and Cheney, 1979; Wood and Richard, 1982; Wenk, 1984; Blaker, 1985; Casamenti et al., 1986; Wood and McQuade, 1986; Sarter et al., 1990; Dudchenko and Sarter, 1991). When the $GABA_A$ receptor agonist muscimol is infused into the basal forebrain, cortical acetylcholine release, turnover, and high-affinity choline uptake are reduced, while picrotoxin or bicuculin, two $GABA_A$ antagonists, stimulate high-affinity choline uptake in cortical synaptosomes. In addition, microinjection of GABA into the substantia innominata inhibits cue-elicited unit responses in the frontal cortex (Pirch et al., 1991). Other behavioral studies suggest that $GABA_A$ antagonists enhance and muscimol downregulates memory consolidation. The $GABA_A/Cl^-$ channel receptor, which is located on or near cholinergic cell bodies, is associated with the benzodiazepine receptor (BNZO). Recent studies (for review see Izquierdo and Medina, 1991) suggest that during learning experiments, depending on the level of anxiety and/or associated stress, endogeneous benzodiazepines are released in basal forebrain, which bind the

$BNZ/GABA/Cl^-$ channel receptor complex and allosterically modulate the effect of GABA on the Cl^- channel. The cognition-enhancing role of benzodiazepine receptor inverse agonists is associated with upregulation of ACh release in the cortex and may be related to inhibition of GABAergic/cholinergic synaptic transmission in the basal forebrain (Miller and Chmielewski, 1990; Sarter et al., 1990; Sarter, 1991; Moran et al., 1992).

Acetylcholine

The observations that *muscarinic* agonists injected into the SI reduce cortical ACh turnover (Robinson et al., 1988) or ACh efflux (Bertorelli et al., 1991), that the muscarinic antagonist scopolamine increases cortical ACh output, and furthermore, that the GABAergic antagonist picrotoxin is able to block the scopolamine effect on ACh efflux, prompted the suggestion that ACh released in the basal forebrain from axons originating in the pedunculopontine tegmental nucleus (Consolo et al., 1990) or arising from local collaterals acts at muscarinic receptors on GABAergic neurons, which in turn inhibit the basal forebrain corticopetal cholinergic neurons. This is consistent with the data that GABAergic neurons in the ventral pallidum (Záborszky et al., 1986b) are known to receive cholinergic input. Until the cellular identification of cholinergic receptors in the basal forebrain is resolved, these data are subject to the alternative interpretation that BFC neurons and/or presynaptic cholinergic axons possess impulse-modulating (M_2 muscarinic) autoreceptors, similar to pedunculopontine cholinergic neurons (Leonard and Llinas, 1990). The localization of M_2 receptors in the basal forebrain (Spencer et al., 1986; Levey et al., 1991) is compatible with either explanation. In addition, it was suggested that ACh acts also through *nicotinic* receptors in the basal forebrain (Akaike et al., 1991; Khateb et al., 1991), although their cellular localization is now known.

Glutamate

Depolarizing effects of glutamate have been reported in dissociated basal forebrain cultures (Nakajima et al., 1985) and identified basalocortical neurons were readily excited by iontophoretically applied glutamate (Lamour et al., 1986). Injections of glutamate in the SI significantly increased cortical high-affinity choline uptake (Wenk, 1984), and electrophysiological studies (Akaike et al., 1991) established the presence of NMDA receptors on neurons dissociated from basal forebrain neurons; however, the transmitter identity of the neurons was not determined. Until more detailed morphological and electrophysiological investigations are done in the basal forebrain, it is unclear whether glutamate acts directly or indirectly on cholinergic neurons.

Substance P and Enkephalin

Substance P (Malthe-Sorenssen et al., 1978) has been found to cause a significant decrease in hippocampal ACh turnover when injected locally in the septum and, similarly, injections of enkephalin or its derivatives to the SI have been shown to reduce neocortical high-affinity choline uptake by approximately 50% (Wenk, 1984) and to produce naloxone-reversible locomotor hyperactivity (Baud et al., 1988). In another study, however, local delivery of enkephalin derivatives into the SI failed to induce changes in cortical ACh turnover, although a significant decrease resulted after parenteral administration (Wood and McQuade, 1986).

Apparently no study has investigated directly the effect of basal forebrain-injected substance P on cortical presynaptic cholinergic function. However, it is likely to be complex, due to the presence of different types of substance P synapses (see above), the occurrence of different tachykinin receptors, the differential distribution of substance P in the basal forebrain (Beach et al., 1987; Záborszky et al., 1989a), as well as the possible involvement of noncholinergic neurons. In addition, both substance P and enkephalin may be colocalized with GABA, although their release and time course of action may be different. Iontophoretically applied morphine in the ventral pallidum/SI area resulted in current-dependent suppression (40%) and excitation (30%) in the targeted cells, although the transmitter content of these neurons was not identified (Chrobak and Napier, 1991). These data underline the need for more sophisticated physiological and morphological studies to elucidate the action of substance P and opiates on basal forebrain corticopetal cholinergic function.

Noradrenaline

Parenteral administration of amphetamine increases ACh turnover in the hippocampus and cortex. This response is prevented by intraseptal injection of phenoxybenzamine, an irreversible α-adrenergic blocker (Costa et al., 1983), or by 6-OHDA lesions of the ventral or dorsal noradrenergic bundles (Robinson, 1986a,b), indicating that this response may be mediated through noradrenergic terminals on septal or corticopetal cholinergic neurons. A preliminary electrophysiological study showed that basal forebrain neurons in guinea pig slices were strongly depolarized and excited by noradrenaline, although the chemical identity of the neurons was not determined (Khateb et al., 1991). Our study clearly indicates the presence of a massive noradrenergic input to BFC neurons; however, the possible involvement of a GABAergic mechanisms has to be investigated, as it is suggested to exist in the septum (Durkin, 1992).

Dopamine

In *in vitro* and *in vivo* experiments (Maslowski et al., 1991; Napier et al., 1991; Napier, this volume) dopamine induced depolarization and an increased firing rate of ventral pallidal neurons. On the other hand, nigral stimulation resulted in inhibition of ventral pallidum/SI neuronal activity. While these studies established the presence of dopaminergic transmission, they did not disclose the cellular site of action. In earlier biochemical studies it was suggested that dopamine modulates cortical ACh turnover indirectly through an influence on GABAergic interneurons in the basal forebrain (Robinson et al., 1979; Costa et al., 1983; Casamenti et al.,

1986). On the other hand, morphological data cited in this review suggest that dopaminergic terminals may impinge directly upon septal (Milner, 1991) or basaloycortical (Záborszky et al., 1992) cholinergic neurons.

The high proportion of inhibitory terminals on BFC neurons is compatible with the notion that the GABAergic link acts as a "final common pathway" (Chrobak et al., 1989) for other afferents in the regulation of BFC neurons. The confirmation of this proposal requires further sophisticated morphological studies, since it is unknown whether GABAergic neurons, which may receive dopaminergic or other afferents, project, in turn, to corticopetal cholinergic neurons and/or project directly to the cortex and cause disinhibition there through their action on local GABAergic neurons, as suggested by Freund and Antal (1988) in the hippocampus.

Chronic Deprivation of Catecholaminergic Afferents

Ten to 14 days after electrolytic lesion in the mesopontine tegmentum (Consolo et al., 1990) or 6-hydroxydopamine lesions of the ascending catecholaminergic pathways (Záborszky and Luine, 1987; Záborszky et al., 1992), ChAT activity is significantly reduced in basal forebrain areas rich in cholinergic cell bodies. This finding, together with the ultrastructural evidence of catecholaminergic/cholinergic interaction, suggests that catecholamine afferents transynaptically regulate ChAT levels in the postsynaptic neuron. ChAT is not the rate-limiting factor in the ACh biosynthetic pathway *in vitro* at saturating concentrations of its substrates (Pepeu et al., 1990). However, it cannot be excluded that *in vivo* it is involved in the regulation of ACh synthesis (Tucek, 1985). It is possible that the chronic deprivation of catecholaminergic afferents to the basal forebrain that may occur during aging, in Alzheimer's disease or in Parkinson's disease (for references see Jellinger, 1986; Mann, 1988; McGeer et al., 1990; Gallagher et al., 1990; Fisher et al., 1991; Záborszky et al., 1991) contribute to a downregulation of ChAT expression and/or the degeneration of BFC neurons. Reduced ChAT activity would ultimately result in a reduction of the synthesized ACh available for release at presynaptic cortical sites. Although several mechanisms might lead to dementias with cholinergic deficits, such an anterograde transsynaptic mechanism is consistent with recent observations about the afferent regulation of gene expression of transmitters or their associated enzymes (Black et al., 1984; Adler and Black, 1986; Joh and Baker, 1988; Salvaterra and Vaughn, 1989), as well as with the proposition of transynaptic system degeneration in neurological diseases (Saper et al., 1987; Arendash et al., 1989). Our data, together with the pharmacological-behavioral studies of Durkin and his colleagues (Durkin, 1989; Marighetto et al., 1989; Lebrun et al., 1990), suggest that, in addition to GABA (Záborszky et al., 1986b; Sarter et al., 1990) noradrenaline is also a major transmitter controlling the activity of BFC neurons.

Concluding Remarks

The complicated anatomy of the basal forebrain, the uncontrolled diffusion of injected drugs, and their occasional nonspecific effects often conspire against firm conclusions with regard to their cellular site of action. In fact, changes in cortical cholinergic function after local injections of different putative transmitters into the basal forebrain are often compatible with modulation of the cholinergic cells in the SI either directly, via interneurons, or even through presynaptic receptors on afferents to cholinergic or noncholinergic projection neurons or interneurons. In addition, cortical ACh release is also under local influence through presynaptic receptors located on cholinergic axons (Vizi, 1979; Illes, 1986; Pepeu et al., 1990). It is also important to recognize that synaptically released transmitters in the forebrain, with concomitant changes in ACh release in the cortex, may vary in their behavioral effects as compared to "flooding" a local circuit with exogenous drugs. The morphological data discussed in this review suggest that the situation is even more complex; interactions between other peptides or transmitters and the cholinergic system can also be postulated. Additional morphological data are necessary to understand the whole complexity of organization of the BFC system. At present the way in which synaptic conductances determinate the firing pattern of BFC neurons is poorly understood (Griffith et al., 1991). It is expected that identification of the presence and distribution of specific receptors and monitoring ionic currents on cholinergic and other basal forebrain neurons *in vitro* will provide a foundation for exploring *in vivo* the modulation of BFC neurons by synaptically released transmitters. These studies will help to understand the basic mechanisms of how cellular activity of BFC neurons relate to cortical ACh release in learning, memory, sensory processing, and other behaviors in which the BFC neurons have been implicated. Defining the precise anatomy and regulatory mechanisms in specific basal forebrain circuits will be essential for understanding the deficiencies in information processing within these systems that characterize aging and disease states, such as Alzheimer's and Parkinson's diseases.

Acknowledgments. I would like to thank Drs. H.T. Chang and P. Bolam for allowing me to use some of their published electron micrographs to illustrate the morphology of transmitter-specific inputs to the basal forebrain. I am grateful to Drs. V.E. Koliatsos and G. Alheid for comments and critical reading of this manuscript. Ms. Vickie Loeser has provided expert secretarial assistance. This review contains paraphrases from Záborszky et al. (1991). The original research summarized in this review is supported by USPHS grant nos. 23945, 17743, and P01 30024.

References

Adler JE, Black IB (1986): Membrane contact regulates transmitter phenotypic expression. *Dev Brain Res* 30:237–241

Akaike N, Harata N, Tateishi N (1991): Modulatory action of cholinergic drugs on N-methyl-D-aspartate response in dissociated rat nucleus basalis of Meynert neurons. *Neurosci Lett* 130:243–247

Akaike N, Harata N, Ueno S, Tateishi N (1992): GABAergic synaptic current in dissociated nucleus basalis of Meynert neurons of the rat. *Brain Res* 570:102–108

Alheid GF, Heimer L (1988): New perspectives in basal forebrain organization of special relevance for neuropsychiatric disorders: The striatopallidal, amygdaloid, and corticopetal components of substantia innominata. *Neuroscience* 27:1–39

Amaral DG, Kurz J (1985): An analysis of the origins of the cholinergic and noncholinergic septal projections to the hippocampal formation in the rat. *J Comp Neurol* 240:37–59

Arendash GW, Sengstock GJ, Shaw G, Millard WJ (1989): Transneuronal neurochemical and neuropathological changes induced by nucleus basalis lesions: A possible degenerative mechanism in Alzheimer's disease. *Adv Behav Biol* 36:235–254

Armstrong DM (1986): Ultrastructural characterization of choline acetyltransferase-containing neurons in the basal forebrain: Evidence for a cholinergic innervation of intracerebral blood vessels. *J Comp Neurol* 250:81–92

Aston-Jones G, Shaver R, Dinan TG (1985): Nucleus basalis neurons exhibit axonal branching with decreased impulse conduction velocity in rat cerebrocortex. *Brain Res* 325:271–285

Baud P, Mayo W, LeMoal M, Simon H (1988): Locomotor hyperactivity in the rat after infusion of muscimol and [D-Ala^2]Met-enkephalin into the nucleus basalis magnocellularis: Possible interaction with cortical cholinergic projections. *Brain Res* 452:203–211

Bayer SA (1985): Neurogenesis of the magnocellular basal telencephalic nuclei in the rat. *Int J Dev Neurosci* 3:229–243

Beach TG, Tago H, McGeer EG (1987): Light microscopic evidence for a substance P-containing innervation of the human nucleus basalis of Meynert. *Brain Res* 408:251–257

Beaulieu C, Somogyi P (1991) Enrichment of cholinergic synaptic terminals on GABAergic neurons and coexistence of immunoreactive GABA and choline acetyltransferase in the same synaptic terminals in the striate cortex of the cat. *J Comp Neurol* 304:666–680

Bertorelli R, Forloni G, Consolo S (1991) Modulation of cortical in vivo acetylcholine release by the basal nuclear complex: Role of the pontomesencephalic tegmental area. *Brain Res* 563:353–356

Bialowas J, Frotscher M (1987) Choline acetyltransferase-immunoreactive neurons and terminals in the rat septal complex: A combined light and electron microscopic study. *J Comp Neurol* 259:298–307

Bigl V, Arendt T (1991): Cholinergic neurons of the central nervous system: Morphofunctional aspects. *Acta Psychiatr Scand Suppl* 366:7–13

Bigl V, Woolf NJ, Butcher LL (1982): Cholinergic projections from the basal forebrain to frontal, parietal, temporal, occipital, and cingulate cortices: A combined fluorescent tracer and acetylcholinesterase analysis. *Brain Res Bull* 8:727–749

Black IB, Adler JE, Dreyfus CF, Jonakait GM, Katz DM, LaGamma EF, Markey KM (1984): Neurotransmitter plasticity at the molecular level. *Science* 225:1266–1270

Blaker WD (1985): GABAergic control of the cholinergic projections to the frontal cortex is not tonic. *Brain Res* 325:389–390

Bolam JP, Ingham CA, Izzo PN, Levey AI, Rye DB, Smith AD, Wainer BH (1986): Substance P-containing terminals in synaptic contact with cholinergic neurons in the neostriatum and basal forebrain: A double immunocytochemical study in the rat. *Brain Res* 397:279–289

Bowen DM, Smith CB, White P, Davison AN (1976): Neurotransmitter-related enzymes and indices of hypoxia in senile dementia and other abiotrophies. *Brain Res* 99:459–495

Brashear HR, Záborszky L, Heimer L (1986): Distribution of GABAergic and cholinergic neurons in the rat diagonal band. *Neuroscience* 17:439–451

Brauer K, Schober W, Werner L, Winkelman E, Lungwitz W, Hajdu F (1988): Neurons in the basal forebrain complex of the rat: A Golgi study. *J Hirnforsch* 29:43–71

Brauer K, Schober A, Wolff JR, Winkelmann E, Luppa H, Lüth H-J, Böttcher H (1991): Morphology of neurons in the rat basal forebrain nuclei: Comparison between NADPH-diaphorase histochemistry and immunohistochemistry of glutamic acid decarboxylase, choline acetyltransferase, somatostatin and parvalbumin. *J Hirnforsch* 32:1–17

Brown DA, Gähwiler BH, Griffith WH, Halliwell JV (1990): Membrane currents in hippocampal neurons. *Prog Brain Res* 83:141–160

Brownstein M, Saavedra JM, Palkovits M (1974): Norepinephrine and dopamine in the limbic system of the rat. *Brain Res* 79:431–436

Butcher LL, and Semba K (1989): Reassessing the cholinergic basal forebrain: Nomenclature schemata and concepts. *Trends in Neurosci* 12:483–485

Butcher LL, Woolf NJ (1986): Cholinergic systems: Synopsis of anatomy and overview of physiology and pathology. In: Biological Substrates of Alzheimer's Disease, Scheibel AB, Wechsler AF, eds. New York: Academic Press, pp 73–83

Buzsaki G, Bickford RG, Ponomareff G, Thal LJ, Mandel R, Gage FH (1988): Nucleus basalis and thalamic control of neocortical activity in the freely moving rat. *J Neurosci* 8:4007–4026

Carlsen J, Záborszky L, Heimer L (1985): Cholinergic projections from the basal forebrain to the basolateral amygdaloid complex: A combined retrograde fluorescent and immunohistochemical study. *J Comp Neurol* 234:155–167

Carnes KM, Fuller TA, Price JL (1990): Sources of presumptive glutamatergic/aspartatergic afferents to the magnocellular basal forebrain in the rat. *J Comp Neurol* 302:824–852

Casamenti F, Deffenu G, Abbamondi AL, Pepeu G (1986): Changes in cortical acetylcholine output induced by modulation of the nucleus basalis. *Brain Res Bull* 16:689–695

Celesia GG, Jasper HH (1966): Acetylcholine released from cerebral cortex in relation to state of activation. *Neurology* 16:1053–1064

Celio MR, Norman AW (1985): Nucleus basalis of Meynert neurons contain the vitamin D-induced calcium-binding protein (Calbindin-D-28K). *Anat Embryol* 173:143–148

Chang HT (1989): Noradrenergic innervation of the substantia innominata: A light and electron microscopic analysis of dopamine-β-hydroxylase immunoreactive elements in the rat. *Exp Neurol* 104:101–112

Chang HT, Kuo H (1991): Calcium-binding protein (Calbindin D-28K) immunoreactive neurons in the basal forebrain of the monkey and the rat: Relationship with the cholinergic neurons. In: *The Basal Forebrain: Anatomy to Function,* Napier TC, Kalivas PW, Hanin I, eds. New York: Plenum Press, pp 119–126

Chang HT, Penny GR, Kitai ST (1987): Enkephalinergic-cholinergic interaction in the rat globus pallidus: A pre-embedding double-labeling immunocytochemistry study. *Brain Res* 426:197–203

Chan-Palay V (1988): Neurons with galanin innervate cholinergic cells in the human basal forebrain and galanin and acetylcholine coexist. *Brain Res Bull* 21:465–472

Chrobak JJ, Napier TC (1991): Ventral pallidal and substantia innominata neurons respond to systemic and microiontophoretic morphine treatment. *Soc Neurosci Abst* 17:266

Chrobak JJ, Stackman RW, Walsh TJ (1989): Intraseptal administration of muscimol produces dose-dependent memory impairments in the rat. *Behav Neural Biol* 52:357–369

Clements JR, Toth DD, Highfield DA, Grant SJ (1991): Glutamate-like immunoreactivity is present within cholinergic neurons of the laterodorsal tegmental and pedunculopontine nuclei. In: *The Basal Forebrain: Anatomy to Function*, Napier TC, Kalivas PW, Hanin I, eds. New York: Plenum Press, pp 127–142

Collier B, Mitchell JF (1967): The central release of acetylcholine during consciousness and after brain lesions. *J Physiol* 188:83–98

Consolo S, Bertorelli R, Forloni GL, Butcher LL (1990): Cholinergic neurons of the pontomesencephalic tegmentum release acetylcholine in the basal nuclear complex of freely moving rats. *Neuroscience* 37:717–723

Costa E, Panula P, Thompson HK, Cheney DL (1983): The transsynaptic regulation of the septal-hippocampal cholinergic neurons. *Life Sci* 32:165–179

Crawley JN, Wenk GL (1989): Co-existence of galanin and acetylcholine: Is galanin involved in memory processes and dementia? *Trends in Neurosci* 12:278–282

Cullinan WE, Záborszky L (1991): Organization of ascending hypothalamic projections to the rostral forebrain with special reference to the innervation of cholinergic projection neurons. *J Comp Neurol* 306:631–667

Das GD, Kreutzberg GW (1968): Evaluation of interstitial nerve cells in the central nervous system: A correlative study using acetylcholinesterase and Golgi techniques. *Adv Anat Embryol* 41:1–58

Davies P, Maloney AJR (1976): Selective loss of central cholinergic neurons in Alzheimer's disease. *Lancet* 2:1403

Decker MW, McGaugh JL (1991): The role of interactions between the cholinergic system and other neuromodulatory systems in learning and memory. *Synapse* 7:151–168

Détári L, Vanderwolf CH (1987): Activity of identified cortically projecting and other basal forebrain neurons during large slow waves and cortical activation in anaesthetized rats. *Brain Res* 437:1–8

Dinopoulous A, Parnavelas JG, Eckenstein F (1986): Morphological characterization of cholinergic neurons in the horizontal limb of the diagonal band of Broca in the basal forebrain of the rat. *J Neurocytol* 15:619–628

Dinopoulous A, Parnavelas JG, Uylings HBM, Van Eden CG (1988): Morphology of neurons in the basal forebrain nuclei of the rat: A Golgi study. *J Comp Neurol* 272:461–474

Dudchenko P, Sarter M (1991): GABAergic control of basal forebrain cholinergic neurons and memory. *Behav Brain Res* 42:33–71

Durkin T (1989): Central cholinergic pathways and learning and memory processes: Presynaptic aspects. *Comp Biochem Physiol* 93A:273–280

Durkin TP (1992): GABA-ergic mediation of indirect trans-synaptic control over basal and spatial memory testing-induced activation of septo-hippocampal cholinergic activity in mice. *Behav Brain Res*. (In press).

Dutar P, Rascol O, Jobert A, Lamour Y (1986): Comparison of septo-hippocampal with basalo-cortical projection neurons in the rat: An electrophysiological approach. *Neurosci Lett* 63:86–90

Fibiger HC, Damsma G, Day JC (1991): Behavioral pharmacology and biochemistry of central cholinergic neurotransmission. In: *The Basal Forebrain: Anatomy to Function*, Napier TC, Kalivas PW, Hanin I, eds. New York: Plenum Press, pp 399–414

Fischer W, Chen KS, Gage FH, Björklund A (1991): Progressive decline in spatial learning and integrity of forebrain cholinergic neurons in rats during aging. *Neurobiol Aging* 13:9–23

Fisher RS, Levine MS (1989): Transmitter cosynthesis by corticopetal basal forebrain neurons. *Brain Res* 491:163–168

Fisher RS, Buchwald NA, Hull CD, Levine MS (1988): GABAergic basal forebrain neurons project to the neocortex: The localization of glutamic acid decarboxylase and choline acetyltransferase in feline corticopetal neurons. *J Comp Neurol* 272:489–502

Fisone G, Wu CF, Consolo S, Nordström O, Brynne N, Bartfai T, Melander T, Hökfelt T (1987): Galanin inhibits acetylcholine release in the ventral hippocampus of the rat: Histochemical, autoradiographic in vivo, and in vitro studies. *Proc Natl Acad Sci USA* 84:7339–7343

Foote SL, Bloom FE, Aston-Jones G (1983): Nucleus locus coerlueus: New evidence of anatomical and physiological specificity. *Physiol Rev* 63:844–914

Forloni G, Grzanna R, Blakely RD, Coyle JT (1987): Co-localization of N-acetyl-aspartyl-glutamate in central cholinergic, noradrenergic, and serotonergic neurons. *Synapse* 1:455–460

Francis PT, Carl R, Pearson A, Lowe SL, Neal JW, Sephens PH, Powell TPS, Bowen DM (1987): The dementia of Alzheimer's disease: An update. *J Neurol Neurosurg Psychiatr* 50:242–243

Freund TF, Antal M (1988): GABA-containing neurons in the septum control inhibitory interneurones in the hippocampus. *Nature* 366:170–173

Fuller TA, Russchen FT, Price JL (1987): Sources of presumptive glutamatergic/aspartergic afferents to the rat ventral striatopallidal region. *J Comp Neurol* 258:317–338

Gallagher M, Burwell RD, Kodsi MH, McKinney M, Southerland S, Vella-Rountree L, Lewis MH (1990): Markers for biogenic amines in the aged rat brain: Relationship to decline in spatial learning ability. *Neurobiol Aging* 11:507–514

Gaykema RPA, Gaál G, Traber J, Hersh LB, Luiten PGM (1991a): The basal forebrain cholinergic system: Efferent and afferent connectivity and long-term effects of lesions. *Acta Psychiatr Scand Suppl* 366:14–26

Gaykema RPA, van Weeghel R, Hersh LB, Luiten PGM (1991b): Prefrontal cortical projections to the cholinergic neurons in the basal forebrain. *J Comp Neurol* 303:563–583

Gentleman SM, Falkai P, Bogerts B, Herrero MT, Polak JM, Roberts GW (1989): Distribution of galanin-like immunoreactivity in the human brain. *Brain Res* 505:311–315

Gerfen CR (1991): Substance P (neurokinin-1) receptor mRNA is selectively expressed in cholinergic neurons in the striatum and basal forebrain. *Brain Res* 556:165–170

Gerfen CR, Baimbridge KG, Miller JJ (1985): The neostriatal mosaic: Compartmental distribution of calcium-binding protein and parvalbumin in the basal ganglia of the rat and monkey. *Proc Natl Acad Sci USA* 82:8780–8784

Gorry JD (1963): Studies on the comparative anatomy of the ganglion basale of Meynert. *Acta Anat* 55:51–104

Griffith WH, Sim JA, Matthews RT (1991): Electrophysiologic characteristics of basal forebrain neurons in vitro. In: *The Basal Forebrain: Anatomy to Function*, Napier TC, Kalivas PW, Hanin I, eds. New York, Plenum Press, pp 143–155

Haber SN, Nauta WJH (1983): Ramifications of the globus pallidus in the rat as indicated by patterns of immunohistochemistry. *Neuroscience* 9:245–260

Hallanger AE, Price SD, Steininger T, Wainer BH (1988): Mesopontine tegmental projections to the nucleus basalis of Meynert: An ultrastructural study. *Soc Neurosci Abst* 14:1184

Haring JH, Wang RY (1986): The identification of some sources of afferent input to the rat nucleus basalis magnocellularis by retrograde transport of horseradish peroxidase. *Brain Res* 366:152–158

Heimer L, de Olmos J, Alheid GF, Záborszky L (1991): "Perestroika" in the basal forebrain: Opening the border between neurology and psychiatry. *Prog Brain Res* 87:109–165

Henderson Z (1987): A small proportion of cholinergic neurons in the nucleus basalis magnocellularis of ferret appear to stain positively for tyrosine hydroxylase. *Brain Res* 412:363–369

Houser CR, Crawford GD, Salvaterra PM, Vaughn JE (1985): Immunocytochemical localization of choline acetyltransferase in rat cerebral cortex: A study of cholinergic neurons and synapses. *J Comp Neurol* 234:17–34

Hurley KM, Herbert H, Moga MM, Saper CB (1991): Efferent projections of the infralimbic cortex of the rat. *J Comp Neurol* 308:249–276

Ichimiya Y, Emson PC, Mountjoy CQ, Lawson DEM, Iizuka R (1989): Calbindin-immunoreactive cholinergic neurons in the nucleus basalis of Meynert in Alzheimer-type dementia. *Brain Res* 499:402–406

Illes P (1986): Mechanism of receptor-mediated modulation of transmitter release in noradrenergic, cholinergic and sensory neurones. *Neuroscience* 17:909–928

Ingham CA, Bolam JP, Wainer BH, Smith AD (1985): A correlated light and electron microscopic study of identified cholinergic basal forebrain neurons that project to the cortex in the rat. *J Comp Neurol* 239:176–192

Ingham CA, Bolam JP, Smith AD (1988): GABA-immunoreactive synaptic boutons in the rat basal forebrain: Comparison of neurons that project to the neocortex with pallidosubthalamic neurons. *J Comp Neurol* 273:263–282

Izquierdo I, Medina JH (1991): Memory: The role of endogenous benzodiazepines. *Trends in Pharm Sci* 12:260–265

Jasper HH, Tessier J (1971): Acetylcholine liberations from cerebral cortex during paradoxical (REM) sleep. *Science* 172:601–602

Jellinger K (1986): Overview of morphological changes in Parkinson's Disease. *Adv Neurol* 45:1–18

Joh TH, Baker H (1988): Molecular biology of nerve degeneration. In: *Central Nervous System Disorders of Aging: Clinical Intervention and Research*, Strong R, Wood WG, Burke WJ, eds. New York: Raven Press, pp 181–188

Jones BE, Cuello AC (1989): Afferents to the basal forebrain cholinergic cell area from the pontomesencephalic-catecholamine, serotonin, and acetylcholine-neurons. *Neuroscience* 31:37–61

Kanai T, Szerb TC (1965): Mesencephalic reticular activating system and cortical acetylcholine output. *Nature* 205:80–82

Khateb A, Serafin M, Jones BE, Alonso A, Mühlethaler M (1991): Pharmacological study of basal forebrain neurons in guinea pig brain slices. *Soc Neurosci Abst* 17:881

Koliatsos VE, Martin LJ, Walker LC, Richardson RT, DeLong MR, Price DL (1988): Topographic, non-collateralized basal forebrain projections to amygdala, hippocampus, and anterior cingulate cortex in the rhesus monkey. *Brain Res* 463:133–139

Koliatsos VE, Martin LJ, Price DL (1990): Efferent organization of the mammalian basal forebrain. In: *Brain Cholinergic Systems*, Steriade M, Biesold D, eds. New York: Oxford University Press, pp 120–152

Koliatsos VE, Price DL (1991): The basal forebrain cholinergic system: An evolving concept in the neurobiology of the forebrain. In: *Activation to Acquisition: Functional Aspects of the Basal Forebrain Cholinergic System*, Richardson RT, ed. Boston: Birkhäuser, pp 11–71

Kordower JH, Mufson EJ (1990): Galanin-like immunoreactivity within the primate basal forebrain: Differential staining patterns between humans and monkeys. *J Comp Neurol* 294:281–292

Kosaka T, Tauchi M, Dahl JL (1988): Cholinergic neurons containing GABA-like and/or glutamic acid decarboxylase-like immunoreactivities in various brain regions of the rat. *Exp Brain Res* 70:605–617

Kowall NW, Beal MF (1989): Galanin-like immunoreactivity is present in human substantia innominata and in senile plaques in Alzheimer's disease. *Neurosci Lett* 98:118–123

Lamour Y, Dutar P, Rascol O, Jobert A (1986): Basal forebrain neurons projecting to the rat frontoparietal cortex: Electrophysiological and pharmacological properties. *Brain Res* 362:122–131

Lamour Y, Dutar P, Jobert A, Dykes RW (1988): An iontophoretic study of single somatosensory neurons in rat granular cortex serving the limbs: A laminar analysis of glutamate and acetylcholine effects on receptive-field properties. *J Neurophysiol* 60:725–750

Lamour Y, Senut MC, Dutar P, Bassant MH (1989): Neuropeptides and septo-hippocampal neurons: Electrophysiological effects and distributions of immunoreactivity. *Peptides* 9:1351–1359

Lebrun C, Durkin TP, Marighetto A, Jaffard R (1990): A comparison of the working memory performances of young and aged mice combined with parallel measures of testing and drug-induced activations of septo-hippocampal and nBM-cortical cholinergic neurons. *Neurobiol Aging* 11:515–521

Leonard CS, Llinás RR (1990): Electrophysiology of mammalian pedunculopontine and laterodorsal tegmental neurons in vitro: Implications for the control of REM sleep. In: *Brain Cholinergic Systems*, Steriade M, Biesold D, eds. New York: Oxford University Press, pp 205–223

Leranth C, Frotscher M (1989): Organization of the septal region in the rat brain: Cholinergic-GABAergic interconnections and the termination of hippocampo-septal fibers. *J Comp Neurol* 289:304–314

Levey A, Kitt CA, Simonds WF, Price DL, Brann MR (1991): Identification and localization of muscarinic acetylcholine receptor proteins in brain with subtype-specific antibodies. *J Neurosci* 11:3218–3226

Levin ED, McGurk SR, Rose JE, Butcher LL (1990): Cholinergic-dopaminergic interactions in cognitive performance. *Behav Neural Biol* 54:271–299

Lindvall O, Björklund A (1979): Dopaminergic innervation of the globus pallidus by collaterals from nigrostriatal pathway. *Brain Res* 172:169–173

Loy R, Sheldon A (1987): Sexually dimorphic development of cholinergic enzymes in the rat septohippocampal system. *Dev Brain Res* 34:156–160

Luine VN (1985): Estradiol increases choline acetyltransferase activity in specific basal forebrain nuclei and projection areas of female rats. *Exp Neurol* 89:484–490

Lysakowski A, Wainer BH, Bruce G, Hersh LB (1989): An atlas of the regional and laminar distribution of choline acetyltransferase immunoreactivity in rat cerebral cortex. *Neuroscience* 28:291–336

Malthe-Sorenssen D, Cheney DL, Costa E (1978): Modulation of acetylcholine metabolism in the hippocampal cholinergic pathway by intraseptally injected substance P. *J Pharmacol Exp Therap* 206:21–28

Mann DMA (1988): Neuropathological and neurochemical aspects of Alzheimer's disease. In: *Handbook of Psychopharmacology*, Vol 22, Iversen LL, Iversen SD, Snyder SH, eds. New York: Plenum Press, pp 1–56

Marighetto A, Durkin T, Toumane A, Lebrun C, Jaffard R (1989): Septal α-noradrenergic antagonism in vivo blocks the testing-induced activation of septo-hippocampal cholinergic neurons and produces a concomitant deficit in working memory performance of mice. *Pharmacol Biochem Behav* 34:553–558

Markham JA, Vaughn JE (1990): Ultrastructural analysis of choline acetyltransferase-immunoreactive sympathetic preganglionic neurons and their dendritic bundles in rat thoracic spinal cord. *Synapse* 5:299–312

Martinez-Murillo R, Blasco I, Alavrez FJ, Villalba R, Solano ML, Montero-Caballero I, Rodrigo J (1988a): Distribution of enkephalin-immunoreactive nerve fibers and terminals in the region of the nucleus basalis magnocellularis of the rat: A light and electron microscopic study. *J Neurocytol* 17:361–376

Martinez-Murillo R, Semenenko F, Cuello AC (1988b): The origin of tyrosine hydroxylase immunoreactive fibers in the regions of the nucleus basalis magnocellularis of the rat. *Brain Res* 451:227–236

Martinez-Murillo R, Villalba RM, Rodrigo J (1990): Immunocytochemical localization of cholinergic terminals in the region of the nucleus basalis magnocellularis of the rat: A correlated light and electron microscopic study. *Neuroscience* 36:361–376

Maslowski RJ, Napier TC, Beck SG (1991): Rat ventral pallidal neurons recorded in vitro: Membrane properties and responses to dopamine. *Soc Neurosci Abst* 17:248

McEwen BS, Coirini H, Westlind-Danielsson A, Frankfurt M, Gould E, Schumacher M, Woolley C (1991): Steroid hormones as mediators of neural plasticity. *J Steroid Biochem Mol Biol* 39:223–232

McGeer PL, McGeer EG, Akiyama H, Itagaki S, Harrop R, Peppard R (1990): Neuronal degeneration and memory loss in Alzheimer's Disease and aging. In: *The Principles of Design and Operation of the Brain*, Eccles JC, Creutzfeldt O, eds. New York: Springer-Verlag, pp 411–431

Melander T, Staines WA (1986): A galanin-like peptide coexists in putative cholinergic somata of the septum-basal forebrain complex and in acetylcholinesterase-containing fibers and varicosities within the hippocampus in the owl monkey (*Aotus trivirgatus*). *Neurosci Lett* 68:17–22

Melander T, Hökfelt T, Rökaeus A (1986): Distribution of galanin-like immunoreactivity in rat central nervous system. *J Comp Neurol* 248:475–517

Mesulam MM, Geula C (1988): Nucleus basalis (Ch4) and cortical cholinergic innervation in the human brain: Observations based on the distribution of acetylcholinesterase and choline acetyltransferase. *J Comp Neurol* 275:216–240

Mesulam MM, Mufson EJ, Levey AI, Wainer BH (1983a): Cholinergic innervation of cortex by the basal forebrain: Cytochemistry and cortical connections of the septal area, diagonal band nuclei, nucleus basalis (substantia innominata), and hypothalamus in the rhesus monkey. *J Comp Neurol* 214:170–197

Mesulam MM, Mufson EJ, Wainer BH, Levey AI (1983b): Central cholinergic pathways in the rat: An overview based on an alternative nomenclature (Ch1-Ch6). *Neuroscience* 10:1185–1201

Mesulam MM, Volicer L, Marquis JK, Mufson EJ, Green RC (1986): Systematic regional differences in the cholinergic innervation of the primate cerebral cortex: distribution of enzyme activities and some behavioral implications. *Ann Neurol* 19:144–151

Miller JA, Chmielewski PA (1990): The regulation of high-affinity choline uptake in vitro in rat cortical and hippocampal synaptosomes by β-carbolines administered in vivo. *Neurosci Lett* 114:351–355

Milner TA (1991): Cholinergic neurons in the rat septal complex: Ultrastructural characterization and synaptic relations with catecholaminergic terminals. *J Comp Neurol* 314:37–54

Mitchell JF (1963): The spontaneous and evoked release of acetylcholine from the cerebral cortex. *J Physiol* 168:98–116

Moran PM, Kane JM, Moser PC (1992): Enhancement of working memory performance in the rat by MDL 26,479, a novel compound with activity at the $GABA_A$ receptor complex. *Brain Res* 569:156–158

Mugnaini E, Oertel WH (1985): An atlas of the distribution of GABAergic neurons and terminals in the rat CNS as revealed by GAD immunohistochemistry. In: *Handbook of Chemical Neuroanatomy: GABA and Neuropeptides in the CNS*, Vol 4, Björklund A, Hökfelt, eds. Amsterdam: Elsevier Science Publishers, pp 436–608

Mullin WJ, Phillis JW (1975): The effect of graded forelimb afferent volleys on acetylcholine release from cat sensorimotor cortex. *J Physiol* 244:741–756

Nakajima Y, Nakajima S, Obata K, Carlson CG, Yamaguchi K (1985): Dissociated cell culture of cholinergic neurons from nucleus basalis of Meynert and other basal forebrain nuclei. *Proc Natl Acad Sci USA* 82:6325–6329

Nakajima Y, Nakajima S, Inoue M (1988): Pertussis toxin-insensitive G protein mediates substance P-induced inhibition of potassium channels in brain neurons. *Proc Natl Acad Sci USA* 85:3643–3647

Nakajima Y, Stanfield PR, Yamaguchi K, Nakajima S (1991): Substance P excites cultured cholinergic neurons in the basal forebrain. In: *The Basal Forebrain: Anatomy to Function*, Napier TC, Kalivas PW, Hanin I, eds. New York: Plenum Press, pp 157–182

Napier TC, Potter PE (1989): Dopamine in the ventral pallidum/substantia innominata: Biochemical and electrophysiological studies. *Neuropharmacology* 28:757–760

Napier TC, Muench MB, Maslowski RJ, Battaglia G (1991): Is dopamine a neurotransmitter within the ventral pallidum/substantia innominata? In: *The Basal Forebrain: Anatomy to Function*, Napier TC, Kalivas PW, Hanin I, eds. New York: Plenum Press, pp 183–218

Nilsson OG, Kalén P, Rosengren E, Björklund A (1990): Acetylcholine release in the rat hippocampus as studied by microdialysis is dependent on axonal impulse flow and increases during behavioral activation. *Neuroscience* 36:325–338

Nitecka L, Frotscher M (1989): Organization and synaptic interconnections of GABAergic and cholinergic elements in the rat amygdaloid nuclei: Single- and double-immunolabeling studies. *J Comp Neurol* 279:470–488

Olton D, Markowska A, Voytko ML, Givens B, Gorman L, Wenk G (1991): Basal forebrain cholinergic system: A functional analysis In: *The Basal Forebrain: Anatomy to Function*, Napier TC, Kalivas PW, Hanin I, eds. New York: Plenum Press, pp 353–372

Palazzi E, Fisone G, Hökfelt T, Bartfai T, Consolo S (1988): Galanin inhibits the muscarinic stimulation of phosphoinositide turnover in rat ventral hippocampus. *Eur J Pharmacol* 148:479–480

Palkovits M (1984): Distribution of neuropeptide in the central nervous system: A review of biochemical mapping studies. *Prog Neurobiol* 23:151–189

Pasqualotto BA, Vincent SR (1991): Galanin and NADPH-diaphorase coexistence in cholinergic neurons of the rat basal forebrain. *Brain Res* 551:78–86

Patel AJ, Hayashi M, Hunt A (1988): Role of thyroid hormone and nerve growth factor in the development of choline acetyltransferase and other cell-specific marker enzymes in the basal forebrain of the rat. *J Neurochem* 50:803–811

Pearson RCA, Gatter KC, Brodal P, Powell TPS (1983): The projection of the basal nucleus of Meynert upon the neocortex in the monkey. *Brain Res* 259:132–136

Penny GR, Afsharpour S, Kitai ST (1986): The glutamate decarboxylase-leucine enkephalin-, methionine enkephalin- and substance P-immunoreactive neurons in the neostriatum of the rat and cat: Evidence for partial population overlap. *Neuroscience* 17:1011–1045

Pepeu G, Casamenti F, Giovannini MG, Vannucchi MG, Pedata F (1990): Principal aspects of the regulation of acetylcholine release in the brain. *Prog Brain Res* 84:273–278

Perry EK, Perry RH, Blessed G, Tomlinson BE (1977): Neurotransmitter enzyme abnormalities in senile dementia. *J Neurol Sci* 34:247–265

Pirch J, Rigdon G, Rucker H, Turco K (1991): Basal forebrain modulation of cortical cell activity during conditioning. In: *The Basal Forebrain: Anatomy to Function*, Napier TC, Kalivas PW, Hanin I, eds. New York: Plenum Press, pp 219–231

Price JL, Carnes KM (1991): Input/output relations of the magnocellular nuclei of the basal forebrain. In: *Activation to Acquisition: Functional Aspects of the Basal Forebrain Cholinergic System*, Richardson RT, ed. Boston: Birkhäuser, pp 87–113

Price JL, Stern R (1983): Individual cells in the nucleus basalis-diagonal band complex have restricted axonal projections to the cerebral cortex in the rat. *Brain Res* 269:352–356

Ramón-Moliner E, Nauta WJH (1966): The isodendritic core of the brain stem. *J Comp Neurol* 126:311–336

Reiner PB, Semba K, Fibiger HC, McGeer EG (1987): Physiological evidence for subpopulations of cortically projecting basal forebrain neurons in the anesthetized rat. *Neuroscience* 20:629–636

Richardson RT, DeLong MR (1991a): Functional implications of tonic and phasic activity changes in nucleus basalis neurons. In: *Activation to Acquisition: Functional Aspects of the Basal Forebrain Cholinergic System*, Richardson RT, ed. Boston: Birkhäuser, pp 135–166

Richardson RT, DeLong MR (1991b): Electrophysiological studies of the functions of the nucleus basalis in primates. In: *The Basal Forebrain: Anatomy to Function*, Napier TC, Kalivas PW, Hanin I, eds. New York: Plenum Press, pp 233–252

Robinson SE (1986a): Contribution of the dorsal noradrenergic bundle to the effect of amphetamine on acetylcholine turnover. *Adv Behav Biol* 30:43–50

Robinson SE (1986b): 6-Hydroxydopamine lesion of the ventral noradrenergic bundle blocks the effect of amphetamine on hippocampal acetylcholine. *Brain Res* 397:181–184

Robinson SE, Malthe-Sorenssen D, Wood PL, Commissiong J (1979): Dopaminergic control of the septal-hippocampal cholinergic pathway. *J Pharmacol Exp Therap* 208:476–479

Robinson SE, Hambrecht KL, Lyeth BG (1988): Basal forebrain carbachol injection reduces cortical acetylcholine turnover and disrupts memory. *Brain Res* 445:160–164

Rolls ET (1989): Information processing in the taste system of primates. *J Exp Biol* 146:141–164

Rye DB, Wainer BH, Mesulam M-M, Mufson EJ, Saper CB (1984): Cortical projections arising from the basal forebrain: A study of cholinergic and noncholinergic components employing combined retrograde tracing and immunohistochemical localization of choline acetyltransferase. *Neuroscience* 13:627–643.

Salvaterra PM, Vaughn JE (1989): Regulation of choline acetyltransferase. *Int Rev Neurobiol* 31:81–143

Saper CB, Wainer BH, German DC (1987): Axonal and transneuronal transport in the transmission of neurological disease: Potential role in system degenerations, including Alzheimer's disease. *Neuroscience* 23:389–398

Sarter M (1991): Taking stock of cognition enhancers. *Trends in Pharm Sci* 12:456–461

Sarter M, Bruno JP, Dudchenko P (1990): Activating the damaged basal forebrain cholinergic system: Tonic stimulation versus signal amplification. *Psychopharmacology* 101:1–17

Sato H, Hata Y, Hagihara K, Tsumoto T (1987): Effects of cholinergic depletion of neuron activities in the cat visual cortex. *J Neurophysiol* 58:781–794

Satoh K, Fibiger HC (1986): Cholinergic neurons of the laterodorsal tegmental nucleus: Efferent and afferent connections. *J Comp Neurol* 253:277–302

Semba K, Fibiger H (1989): Organization of central cholinergic systems. *Prog Brain Res* 79:37–63

Semba K, Reiner PB, McGeer EG, Fibiger H (1987): Morphology of cortically projecting basal forebrain neurons in the rat as revealed by intracellular iontophoresis of horseradish peroxidase. *Neuroscience* 20:637–651

Semba K, Reiner PB, McGeer EG, Fibiger HC (1988): Brainstem afferents to the magnocellular basal forebrain studied by axonal transport, immunohistochemistry, and electrophysiology in the rat. *J Comp Neurol* 267:433–453

Sesack SR, Deutch AY, Roth RH, Bunney BS (1989): Topographical organization of the efferent projections of the medial prefrontal cortex in the rat: An anterograde tract-tracing study with *Phaseolus vulgaris* leucoagglutinin. *J Comp Neurol* 290:213–242

Sillito AM, Kemp JA (1983): Cholinergic modulation of the functional organization of the cat visual cortex. *Brain Res* 89:143–155

Sim JA, Griffith WH (1991): Muscarinic agonists block a late-afterhyperpolarization in medial septum/diagonal band neurons in vitro. *Neurosci Lett* 129:63–68

Simasko SM, Horita A (1984): Localization of thyrotropin releasing hormone (TRH) receptors in the septal nucleus of the rat brain. *Brain Res* 296:393–395

Sofroniew MV, Pearson RCA, Powell TPS (1987): The cholinergic nuclei of the basal forebrain of the rat: Normal structure, development and experimentally induced degeneration. *Brain Res* 411:310–331

Spencer DG, Horváth E, Traber J (1986): Direct autoradiographic determination of M1 and M2 muscarinic acetylcholine receptor distribution in the rat brain: Relation to cholinergic nuclei and projections. *Brain Res* 380:59–68

Steriade M, Gloor P, Llinás RR, Silva FH, Mesulam M-M (1990): Basic mechanisms of cerebral rhythmic activities. *EEG Clin Neurophysiol* 76:481–508

Szerb JC (1967): Cortical acetylcholine release and electroencephalographic arousal. *J Physiol* 192:329–343

Szigethy E, Leonard K, Beaudet A (1990): Ultrastructural localization of [^{125}I]neurotensin binding sites to cholinergic neurons of the rat nucleus basalis magnocellularis. *Neuroscience* 36:377–391

Tamiya R, Hanada M, Inagaki S, Takagi H (1991): Synaptic relation between neuropeptide Y axons and cholinergic neurons in the rat diagonal band of Broca. *Neurosci Lett* 122:64–66

Toran-Allerand CD, Miranda RC, Sohrabji E (1991): Interactions of estrogen and nerve growth factor (NGF) on estrogen receptor mRNA expression in explants of the septum/diagonal band. *Soc Neurosci Abst* 17:221

Tucek S (1985): Regulation of acetylcholine synthesis in the brain. *J Neurochem* 44:11–24

Vizi S (1979): Presynaptic modulation of neurochemical transmission. *Progr Neurobiol* 12:181–190

Wainer BH, Mesulam M-M (1990): Ascending cholinergic pathways in the rat brain. In: *Brain Cholinergic Systems*, Steriade M, Biesold D, eds. New York: Oxford University Press, pp 65–119

Walaas I, Fonnum F (1979): The distribution and origin of glutamate decarboxylase and choline acetyltransferase in ventral pallidum and other basal forebrain regions. *Brain Res* 177:325–336

Walker LC, Koliatsos VE, Kitt CA, Richardson RT, Rökaeus Ä, Price DL (1989): Peptidergic neurons in the basal forebrain magnocellular complex of the rhesus monkey. *J Comp Neurol* 280:272–282

Wenk GL (1984): Pharmacological manipulations of the substantia innominata-cortical cholinergic pathway. *Neurosci Lett* 51:99–103

Westlind-Danielsson A, Gould E, McEwen BS (1991): Thyroid hormone causes sexually distinct neurochemical and morphological alterations in rat septal-diagonal band neurons. *J Neurochem* 56:119–128

Whitehouse PJ (1991): Pathology in the cholinergic basal forebrain: Implications for treatment. In: *The Basal Forebrain: Anatomy to Function*, Napier TC, Kalivas PW, Hanin I, eds. New York: Plenum Press, pp 447–452

Wilson FAW (1991): The relationship between learning, memory and neuronal responses in the primate basal forebrain. In: *The Basal Forebrain: Anatomy to Function*, Napier TC, Kalivas PW, Hanin I, eds. New York: Plenum Press, pp 253–266

Wilson FAW, Rolls ET (1990): Neuronal responses related to reinforcement in the primate basal forebrain. *Brain Res* 59:213–231

Wood PL, Cheney DL (1979): The effect of muscarinic receptor blockers on the turnover rate of acetylcholine in various regions of the rat brain. *Can J Physiol Pharmacol* 57:404–411

Wood PL, McQuade P (1986): Substantia innominata-cortical cholinergic pathway: Regulatory afferents. *Adv Behav Biol* 30:999–1006

Wood PL, Richard J (1982): GABAergic regulation of the substantia innominata-cortical cholinergic pathway. *Neuropharmacology* 21:969–972

Woolf NJ (1991): Cholinergic systems in mammalian brain and spinal cord. *Prog Neurobiol* 37:475–524

Woolf NJ, Butcher LL (1991): The cholinergic basal forebrain as a cognitive machine. In: *Activation to Acquisition: Functional Aspects of the Basal Forebrain Cholinergic System*, Richardson RT, ed. Boston: Birkhäuser, pp 347–380

Yau WM, Lingle PF, Youther ML (1983): Modulation of cholinergic neurotransmitter release from myenteric plexus by somatostatin. *Peptides* 4:49–53

Yau WM, Dorset JA, Youther ML (1986): Evidence for galanin as an inhibitory neuropeptide on myenteric cholinergic neurons in the guinea-pig small intestine. *Neurosci Lett* 3:305–308

Záborszky L (1989a): Afferent connections of the forebrain cholinergic projection neurons, with special reference to monoaminergic and peptidergic fibers. In: *Central Cholinergic Synaptic Transmission*, Frotscher M, Misgeld U, eds. Basel: Birkhäuser, pp 12–32

Záborszky L (1989b): Peptidergic-cholinergic interactions in the basal forebrain. In: *Alzheimer's Disease: Advances in Basic Research and Therapies*, Wurtman RJ, Corkin SH, Growdon JH, Ritter-Walker E, eds. Proc. Fifth Meeting Int. Study Group on the Pharmacology of Memory Disorders Associated with Aging, CBSMCT, Cambridge, Massachusetts, pp 521–528

Záborszky L, Braun A (1988): Peptidergic afferents to forebrain cholinergic neurons. *Soc Neurosci Abstr* 14:905

Záborszky L, Cullinan WE (1989): Hypothalamic axons terminate on forebrain cholinergic neurons: An ultrastructural double-labeling study using PHA-L tracing and ChAT immunocytochemistry. *Brain Res* 479:177–184

Záborszky L, Cullinan WE (1992): Projections from the nucleus accumbens to cholinergic neurons of the ventral pallidum: A correlated light and electron microscopic double-immunolabeling study in rat. *Brain Res* 570:92–101

Záborszky L, Luine VN (1987): Evidence for existence of monoaminergic-cholinergic interactions in the basal forebrain. *J Cell Biol Suppl* 11D:187

Záborszky L, Alheid GF, Alones V, Oertel WH, Schmechel DE, Heimer L (1982): Afferents of the ventral pallidum studied with a combined immunohistochemical-anterograde degeneration method. *Soc Neurosci Abstr* 8:218

Záborszky L, Eckenstein F, Leranth C, Oertel W, Schmechel D, Alones V, Heimer L (1984a): Cholinergic cells of the ventral pallidum: A combined electron microscopic immunocytochemical, degeneration and HRP study. *Soc Neurosci Abst* 10:8

Záborszky L, Leranth C, Heimer L (1984b): Ultrastructural evidence of amygdalofugal axons terminating on cholinergic cells of the rostral forebrain. *Neurosci Lett* 52:219–225

Záborszky L, Carlsen J, Brashear HR, Heimer L (1986a): Cholinergic and GABAergic afferents to the olfactory bulb in the rat with special emphasis on the projection neurons in the nucleus of the horizontal limb of the diagonal band. *J Comp Neurol* 243:488–509

Záborszky L, Heimer L, Eckenstein F, Leranth C (1986b): GABAergic input to cholinergic forebrain neurons: An ultrastructural study using retrograde tracing of HRP and double immunolabeling. *J Comp Neurol* 250:282–295

Záborszky L, Cullinan WE, Braun A (1991): Afferents to basal forebrain cholinergic projections neurons: An update. In: *The Basal Forebrain: Anatomy to Function*, Napier TC, Kalivas PW, Hanin I, eds. New York: Plenum Press, pp 43–100

Záborszky L, Luine VN, Cullinan WE, Heimer L (1992): Direct catecholaminergic-cholinergic interactions in the basal forebrain: Morphological and biochemical studies. *Neuroscience* (in press)

Záborszky L, Cullinan WE, Garzanna R (1993): Distribution of locus coeruleus axons in the rostral forebrain with special reference to innervation of the cholinergic projection neurons: A correlated light and electron microscopic double-labeling study (submitted)

4

Functional Pharmacology of Basal Forebrain Dopamine

T. CELESTE NAPIER

Several regions within the basal forebrain are known to be terminal sites for ascending fibers originating within the dopaminergic midbrain. Of concern in this chapter is a region only recently considered dopaminoceptive: the infracommissural extension of the external segment of the dorsal globus pallidus, i.e., the ventral pallidum (VP) and its caudal extension, the sublenticular substantia innominata (SI). The literature establishing dopamine (DA) as a neurotransmitter within the VP/SI has been overviewed elsewhere (Napier et al., 1991a) and only a brief highlight will be discussed here.

Ventral Pallidal Dopamine Anatomy and Biochemistry

Voorn and colleagues (1986), using specific antibodies against dopamine (DA), reported that the VP is "relatively sparsely innervated by DA fibers" that arise from the medial forebrain bundle and ramify into varicosities in the VP/SI. The presence of DA and its major metabolites within VP/SI tissue homogenates has been demonstrated biochemically using high-performance liquid chromagraphic separation with electrochemical detection techniques (Geula and Slevin, 1989; Napier and Potter, 1989). These biochemical assays concur with the anatomical conclusion that innervation of this region by DA is relatively sparse as compared to the striatum.

Anterograde and retrograde tracing techniques demonstrated that the source of the dopaminergic inputs is midbrain dopaminergic somata, including the ventral tegmental area, the substantia nigra (Fallon and Moore, 1978; Russchen et al., 1985; Haring and Wang, 1986; Grove, 1988; Martinez-Murillo et al., 1988; Semba et al., 1988; Zaborszky, 1989; Jones and Cuello, 1989), and the retrorubral field and zona incerta of the substantia nigra (Deutch et al., 1988; Jones and Cuello, 1989). When this system is disrupted by microinjections of the dopaminotoxin, 6-hydroxydopamine, DA and its major metabolites are greatly reduced in VP/SI tissue (Napier and Potter, 1989; Geula and Slevin, 1989). Thus, biochemical and anatomical approaches agree that a dopaminergic projection arising from the midbrain terminates within the VP/SI. This projection likely reflects collateralization of the massive ascending system that terminates within the striatum and nucleus accumbens.

Of the DA receptors subtypes known to exist, the two most characterized are D1 and D2, which demonstrate opposing effects on signal transduction. Using autoradiography, binding sites indicative of D_1 and D_2 DA receptors have been visualized for the VP/SI (Gehlert and Wamsley, 1985; Dawson et al., 1986; Contreras et al., 1987; Beckstead, 1988; Besson et al., 1988; Camps et al., 1989; Cortes et al., 1989; Richfield et al., 1989). Binding quantification using tissue homogenates indicated that receptors labeled with the D_1 ligand, SCH23390, and those labeled with the D_2 ligand, spiperone, occur in VP/SI tissue at about 30% of the striatum (Napier et al., 1991a). Thus, both receptor subtypes are present within the VP/SI and, like the concentration of the neurotransmitter itself, the receptor concentration is proportionally less than that observed within the striatum.

Functional Pharmacology

Electrophysiologic procedures allow for the functional evaluation of VP/SI DA at the level of the single cell. Initial studies characterized the pharmacologic profile of the direct agonist, apomorphine. When given in multiple, divided doses, apomorphine produced dose-related increases and decreases in firing (Napier et al., 1991b). The maximum responses and the dose that induced half-maximal responding are comparable to other dopaminoceptive brain regions (Skirboll et al., 1979; Rebec et al., 1979; Napier et al., 1991b). Apomorphine acts at both D_1 and D_2 DA receptor subtypes. To determine the influence of each on the apomorphine-mediated effect, response attenuation by DA receptor subtype-specific antagonists was characterized. Sulpiride, a D2-specific antagonist, only slightly attenuated apomorphine effects when given either after the agonist or as a pretreatment (Fig. 1A and 1C). SCH23390, a D_1-specific antagonist, was effective at attenuating the response (Fig. 1B). However, more complete antagonism of apomorphine occurred with the combination of D_1 and D_2 blockers (Fig. 1B and 1C). These data illustrate that both receptor subtypes are involved in the apomorphine-mediated rate changes in the VP/SI.

By combining single neuron recording procedures with microiontophoretic applications of a drug, one can ascertain the effects of bathing only the local milieu of a neuron with a drug while leaving the rest of the brain naive to the treatment. Thus, responses to iontophoretically applied drugs indicate that the receptor mediating the effect is in the immediate vicinity of the recorded neuron. Of the VP/SI neurons that responded to microiontophoretic applications of DA (43 of 102 neurons tested), 72% were suppressed and 28% were excited (Napier et al., 1991b; Fig. 2). It is noteworthy that only a portion (42%) of the encountered VP/SI neurons were sensitive to the neurotransmitter, an anticipated phenomenon given the relatively low concentration of DA and its receptors in VP/SI tissue.

Iontophoretically applied haloperidol antagonized rate changes observed with DA (Fig. 2), verifying that the responses observed during DA application reflects

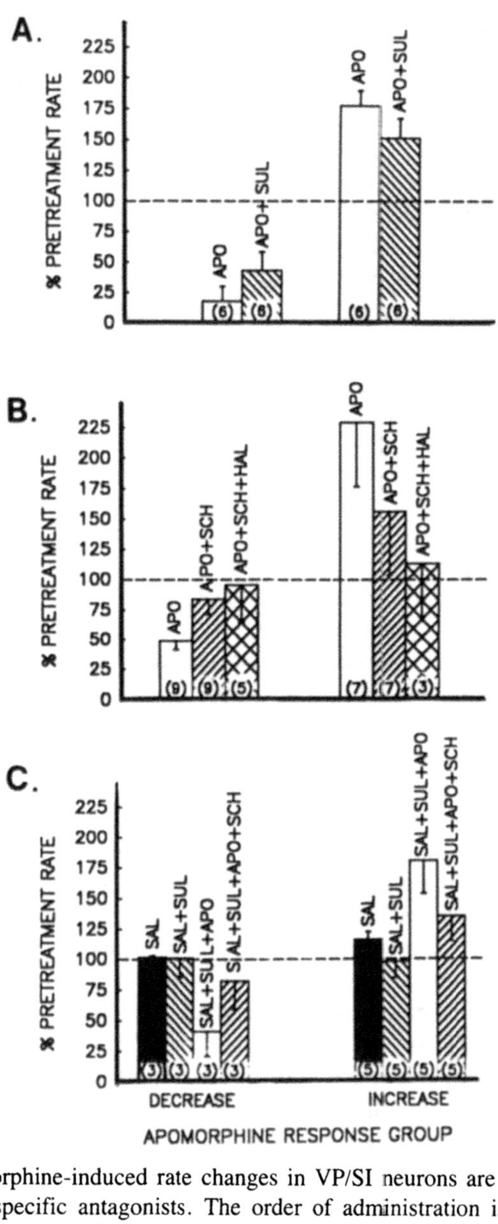

FIGURE 1. Apomorphine-induced rate changes in VP/SI neurons are attenuated by DA receptor subtype specific antagonists. The order of administration is indicated by the treatments listed above each bar graph. The numbers in parenthesis at the base of each bar indicate the number of neurons. Apomorphine (APO; 0.5 mg/kg i.v.); sulpiride (SUL; 12.5 mg/kg i.v.); SCH23390 (SCH; 0.1 mg/kg i.v.); haloperidol (HAL; 0.5 mg/kg i.v.). Data taken from Maslowski and Napier (1991a).

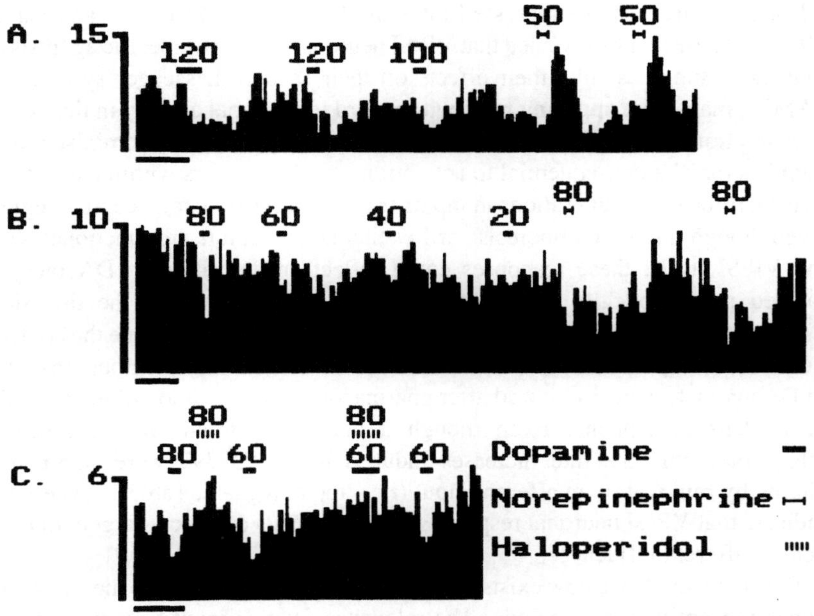

FIGURE 2. Effects of locally applied DA and verification of response specificity. Responses to DA were analyzed by varying the current level applied to the DA-containing pipette. DA-induced responses could be distinguished from those to norepinephrine by the response direction (A) or temporal characteristics (B). Reversal by a DA antagonist was indicated by coiontophoresing haloperidol (HAL) and DA, and demonstrating that the DA-induced effect was no longer present (C). Reprinted with permission from *Neuropharmacology*, Vol. 28 No. 7, by Napier TC and Potter PE, Dopamine in the rat ventral palladium/substantia innominata: Biochemical and electrophysiological studies, Copyright 1989, Pergamon Press plc.

an activation of DA receptors. Since the synaptic concentration of iontophoretic haloperidol is unknown, and at higher concentrations this antagonist blocks the D_1 receptor, as well as D_2, an additional study was conducted to determine the contribution of the receptor subtypes to the DA-mediated response. Like responses observed with apomorphine, DA was attenuated with systemically administered D_1 or D_2 antagonists; however, full blockade was most often reached when both were administered (Napier et al., 1991b).

These electrophysiologic studies demonstrate functionally that the VP/SI is a site of action for DA neurotransmission. It appears that both D_1 and D_2 receptor subtypes contribute to VP/SI responses to apomorphine and DA; however, from a pharmacologic (and potentially therapeutic) vantage point, it is of interest to ascertain if the individual subtypes independently mediate neuronal activity. Experiments with systemic administration of the D_1 agonist, SKF38393, demonstrated a dose-dependent increase in firing in 42 of 61 neurons tested (Maslowski and Napier, 1991b). In contrast, the D_2 agonist quinpirole increased firing in only

12 of 51 neurons, but suppressed rates in 30 neurons (Maslowski and Napier, 1991b). These results revealed that VP/SI neurons respond to specific agonists and that these subtypes, like their effects on their second messenger system (i.e., cAMP), may serve opposing roles with regard to neuronal activity in this region.

Alterations in VP/SI neuronal firing induced by systemically administered DA agonists may be consequential to activation of DA receptors within the VP/SI or may be reflective of alterations in inputs from sensitive dopaminoceptive regions. Even though microiontophoresis verifies that DA receptors are functional within the VP/SI itself, these responses could reflect an activation of DA receptors located presynaptically on afferent terminals that are within the diffusional distance of the iontophoretically applied agonist. Thus, to determine the contribution of two major afferent inputs from dopaminoceptive brain regions, responses to DA agonists were monitored after pharmacologic inactivation of the amygdala or nucleus accumbens. Even though a decreasing trend was observed for quinpirole, only the rate increases induced by SKF38393 were significantly altered by removal of an afferent input (i.e., the amygdala; Table 1). These data indicate that VP/SI neuronal responses to DA agonists can occur largely independent of afferent influences.

Sufficient evidence now exists for the inclusion of the VP/SI into the category of dopaminoceptive brain regions. The relevance for this input becomes apparent when one considers the gamut of behaviors attributed to the VP/SI.

Reward

Behavioral and electrophysiological studies suggest that the VP/SI plays an important role in positive reinforcement (DeLong, 1971; Linseman, 1974; Mora et al., 1976; Rolls et al., 1979, 1980; Richardson and DeLong 1986; Huston

TABLE 1. The Influence of Pharmacologic Inactivation of the Amygdala or Nucleus Accumbens on Ventral Pallidal Responses to DA Agonists

	Agonist—Predominant response to agonist[a]	
Procaine pretreatment	SKF38393[b]—Increase	Quinpirole[c]—Decrease
None	19/23 (83%)	8/13 (62%)
Amygdala	6/13[d] (31%)	6/16 (38%)
Accumbens	12/14 (86%)	8/23 (35%)

[a]Responses are defined as at least a 20% change in firing rate from preagonist baseline and are reported as number of neurons that responded to the agonist as compared to the number tested or that responded to the procaine pretreatment indicated.
[b]The SKF38393 dose was 3.2 mg/kg i.v.
[c]The quinpirole dose was 0.1 mg/kg i.v.
[d]Different from the portion of responding neurons obtain without procaine pretreatment. ($\chi^2 = 3.6$, $p = 0.029$ for a one-tailed test).
Data taken from Napier (1992).

et al., 1987; Hubner and Koob, 1990; Wilson and Rolls 1990). DA systems have long been known for their involvement in reward phenomenon (for review see Wise, 1980). Investigations are needed to determine the contribution of VP/SI DA neurotransmission to this process, for such studies may provide new insights into reward.

Locomotion

Studies using intracerebral microinjection techniques have demonstrated that intra-VP/SI treatments of GABAergic agents, and opioids can influence motoric behaviors (Jones and Mogenson, 1980; Mogenson and Nielsen, 1983; Mogenson et al., 1985; Napier and Marx, 1987; Swerdlow and Koob, 1987; Baud et al., 1988; Napier et al., 1988; Shreve and Uretsky, 1988, 1991; Will et al., 1988; Austin and Kalivas, 1990, 1991; Hoffman et al., 1991). To evaluate the possibility that VP/SI DA may also alter locomotion, rats were microinjected with DA into the VP/SI of both hemispheres and ambulations were quantified. DA produced a dose-related increase in locomotion (Fig. 3) that was blocked by pretreatment with DA antagonists (data not shown), substantiating the dopaminergic nature of this effect. The relevance of these observations is underscored by activating another neurotransmitter system within the VP/SI. VP/SI injections of the *mu*-specific opioid peptide, DAMGO, produced contralateral rotations that were attenuated by intra-VP/SI pretreatment with either a D_1 or a D_2-specific antagonist (Napier, unpublished results). Thus, VP/SI DA receptor also can moderate motor events, including those induced by VP/SI opioids, providing compelling evidence for a motoric role of VP/SI DA neurotransmission.

Cognition

The VP/SI contains a massive cortically directed cholinergic projection (for review see McGeer et al., 1986) that is thought to influence cognitive functioning (Murray and Fibiger, 1985; Hurlbut et al., 1987; Robinson et al., 1988; Ueki and Miyoshi, 1989). Zaborszky (1989) observed that terminals immunoreactive for tyrosine hydroxylase (a catecholamine synthetic enzyme) make monosynaptic contacts with VP/SI dendritic processes that are immunoreactive for choline acetyl transferase (a cholinergic synthetic enzyme). This provides for the possibility that DA may monosynaptically regulate VP/SI cholinergic efferents. Some of the DA-sensitive VP/SI neurons recorded in our laboratory may be cholinergic for the electrophysiologic characteristics of these neurons (Maslowski and Napier, 1991b; Napier et al., 1991b) are similar to those observed for VP/SI neurons antidromically activated by stimulation of cholinergic terminal regions in the cortex (Aston-Jones et al., 1985; Lamour et al., 1986; Reiner et al., 1987). Additional support for a DA mediation of VP/SI cholinergic neurons comes from studies of conditioned slow potential shifts recorded from the cortical surface that are generated by VP/SI cholinergic neurons (Pirch et al., 1986). These event-related slow potentials, reflective of associative processes, are enhanced by haloperidol (Pirch and Corbus, 1983) and are suppressed by amphetamine (Pirch

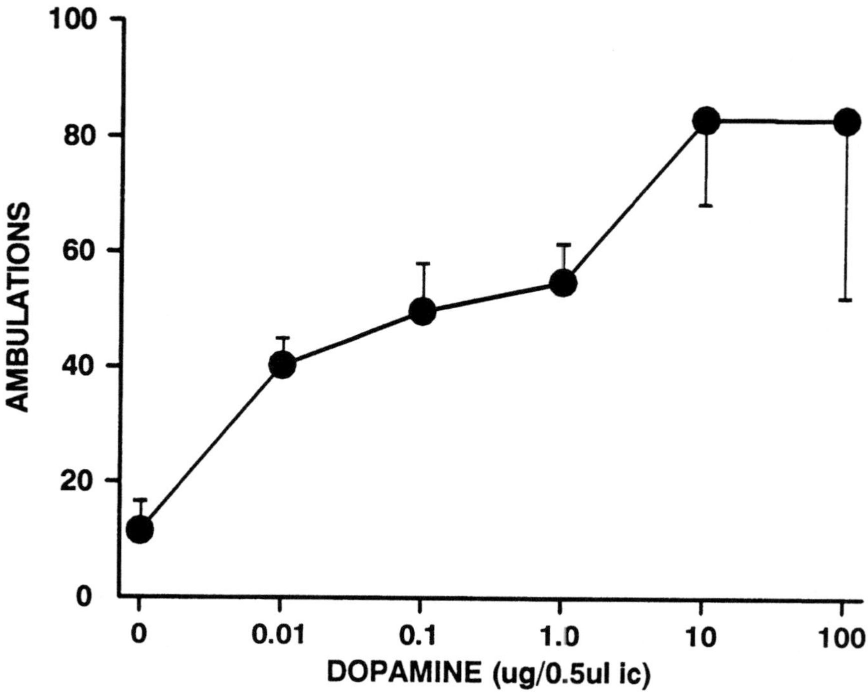

FIGURE 3. Effects of bilateral DA microinjections into the VP/SI on locomotor activity (ambulations) in rats. Data are presented as the mean ± SEM for the total number of ambulations obtained for each rat for each session (1 min observations taken every 5 min for 1 h) when averaged across the eight animals tested per treatment group (except for 100 μg dose; n = 7, see below). Treatment is presented as the infused intracerebral (i.c.) dose per hemisphere. As indicated by the increased variability for 100 μg DA, this dose is likely to be toxic. One rat demonstrated mild seizures early into the observation period, and its scores were not included in the data averages. Three additional rats demonstrated seizures immediately after being returned to their home cage, and these rats generally did not ambulate during the observation period. The remaining rats were hyperactive, often with scores above the average for the 10 μg DA dose. Results obtained with injections of the 0.3% ascorbic acid vehicle solution are indicated as 0 μg/0.5 μl (Napier and Chrobak, 1992).

1977a,b, 1980; Pirch et al., 1981a,b). To determine if VP/SI DA contributes to the cognitive functions attributed to the VP/SI, an experiment was conducted using local microinjections of DA, and the effects on the ability of rats to perform radial maze tasks were evaluated. Vehicle infusions, but not sham treatments, produced performance deficits that were dependent upon the mnemonic demands of the task (Chrobak et al., 1989; Chrobak et al., 1991). DA infusions differed from vehicle infusions only with regard to an increased latency of choices (Chrobak and Napier, unpublished results), the effective dose was 30 μg/0.5 μl, supporting previous observations of altered motoric function following intra-VP DA treatments (refer to Fig. 3). Thus, vehicle infusions into the VP/SI can cause a perturbation that becomes manifest in particular cognitive testing paradigms (a phenomenon not observed in the locomotor studies). These data concur with the biochemical and behavioral alterations observed following vehicle infusions in other basal forebrain regions (for review see Chrobak et al., 1991) and highlights the importance of uninjected controls in studies employing these approaches. This also suggests that an assessment of the effects of DA (and other treatments) on the cognitive processes engaged in the radial maze task cannot be performed using intracerebral microinfusions into VP/SI tissue. Therefore, the possibility of VP/SI dopaminergic neurotransmission influencing cognitive functions remains unresolved.

Conclusion

Functions attributed to the VP/SI are consequential to the summed output of neuronal communications within this region. Likewise, the function of DA neurotransmission within the VP/SI reflects the behaviors governed by this region. VP/SI neurons are sensitive to activation of VP/SI DA receptors, uniquely demonstrating an oppositional effect of stimulating D_1 vs. D_2 receptors. The behavioral consequences of this response may predominantly involve locomotion, so that DA may be most important in modifying responses to other VP/SI transmitters. Several reviews have discussed the role that the VP/SI has in integrating limbic functions with organized, coordinated motor responses (for review see Mogenson and Yang, 1991). DA may be critical in the motor component of this integration.

Acknowledgments. Work overviewed in this chapter was supported by USPHS grants MH45180 and DA05255; BASG and Potte Estate Loyola University Chicago Medical Center.

References

Aston-Jones G, Shaver R, Dinan TG (1985): Nucleus basalis neurons exhibit axonal branching with decreased impulse conduction velocity in rat cerebrocortex. *Brain Res* 325:271–285

Austin MC, Kalivas PW (1990): Enkephalinergic and GABAergic modulation of motor activity in the ventral pallidum. *J Pharmacol Exp Therap* 252:1370–1377

Austin MC, Kalivas PW (1991): Dopaminergic involvement in locomotion elicited from the ventral pallidum/substantia innominata. *Brain Res* 542:123–131

Baud P, Mayo W, LeMoal M, Simon H (1988): Locomotor hyperactivity in the rat after infusion of muscimol and [D-Ala^2]Met-enkephalin into the nucleus basalis magnocellularis. Possible interaction with cortical cholinergic projections. *Brain Res* 452:203–211

Beckstead RM (1988): Association of dopamine D1 and D2 receptors with specific cellular elements in the basal ganglia of the cat: The uneven topography of dopamine receptors in the striatum is determined by intrinsic striatal cells, not nigrostriatal axons. *Neuroscience* 27:851–863

Besson M-J, Graybiel AM, Nastuk MA (1988): [^{3}H]SCH 23390 binding to D1 dopamine receptors in the basal ganglia of the cat and primate: Delineation of striosomal compartments and pallidal and nigral subdivisions. *Neuroscience* 26:101–119

Camps M, Kelly PH, Palacios JM (1989): Autoradiographic localization of dopamine D1 and D2 receptors in the brain of several mammalian species. *J Neural Transm* 80:105–127

Chrobak JJ, An D, Napier TC (1989): Vehicle infusion into the basal forebrain produces task-specific cognitive deficits in the rat. *Soc Neurosci Abstr* 15:1173

Chrobak JJ, Napier TC, Hanin I, Walsh TJ (1991): The pharmacology of basal forebrain involvement in cognition. In: *The Basal Forebrain: Anatomy to Function*. Advances in Experimental Medicine and Biology, Vol 295, Napier TC, Kalivas PW, Hanin I, eds. New York: Plenum Press, pp 383–398

Contreras PC, Quirion R, Gehlert DR, Contreras ML, O'Donohue TL (1987): Autoradiographic distribution of non-dopaminergic binding sites labeled by [^{3}H]haloperidol in rat brain. *Neurosci Lett* 75:133–140

Cortes R, Gueye B, Pazos A, Probst A, Palacios JM (1989): Dopamine receptors in human brain: Autoradiographic distribution of D1 sites. *Neuroscience* 23:263–273

Dawson TM, Barone P, Sidhu A, Wamsley JK, Chase TN (1986): Quantitative autoradiographic localization of D-1 dopamine receptors in the rat brain: Use of the iodinated ligand [^{125}I]SCH23390. *Neurosci Lett* 68:261–266

DeLong MR (1971): Activity of pallidal neurons during movement. *J Neurophysiol* 34:414–427

Deutch AY, Goldstein M, Baldino F, Roth RH (1988): Telencephalic projections of the A8 dopamine cell group. In: *The Mesocorticolimbic Dopamine System, Annals of the New York Academy of Sciences*, Vol. 537, Kalivas PW, Nemeroff CB, eds. New York: The New York Academy of Sciences, pp 27–49

Fallon JH, Moore RY (1978): Catecholamine innervation of the basal forebrain IV. Topography of the dopamine projection to the basal forebrain and neostriatum. *J Comp Neurol* 180:545–580

Gehlert DR, Wamsley JK (1985): Dopamine receptors in the rat brain: Quantitative autoradiographic localization using [^{3}H]sulpiride. *Neurochem Int* 7:717–723

Geula C, Slevin JT (1989): Substantia nigra 6-hydroxydopamine lesions alter dopaminergic synaptic markers in the nucleus basalis magnocellularis and striatum of rats. *Synapse* 4:248–253

Grove EA (1988): Neural associations of the substantia innominata in the rat: Afferent connections. *J Comp Neurol* 277:315–346

Haring JH, Wang RY (1986): The identification of some sources of afferent input to the rat nucleus basalis magnocellularis by retrograde transport of horseradish peroxidase. *Brain Res* 366:152–158

Hoffman DC, West TEG, Wise RA (1991): Ventral pallidal microinjection of receptor-selective opioid agonists produce differential effects on circling and locomotor activity in rats. *Brain Res* 550:205–212

Hubner CB, Koob GF (1990): The ventral pallidum plays a role in mediating cocaine and heroin self-administration in the rat. *Brain Res* 508:20–29

Hurlbut BJ, Lubar JF, Switzer R, Dougherty J, Eisenstadt ML (1987): Basal forebrain infusion of HC-3 in rats: Maze learning deficits and neuropathology. *Physiol Behav* 39:381–393

Huston JP, Kiefer S, Buscher W, Monoz C (1987): Lateralized functional relationship between the preoptic area and lateral hypothalamic reinforcement. *Brain Res* 436:1–8

Jones BE, Cuello AC (1989): Afferents to the basal forebrain cholinergic cell area from pontomesencephalic-catecholamine, serotonin, and acetylcholine-neurons. *Neuroscience* 31:37–61

Jones DA, Mogenson GJ (1980): Nucleus accumbens to globus pallidus GABA projection subserving ambulatory activity. *Am J Physiol* 238:R65–R69

Lamour Y, Dutar P, Rascol O, Jobert A (1986): Basal forebrain neurons projecting to the rat frontoparietal cortex: Electrophysiological and pharmacological properties. *Brain Res* 362:122–131

Linseman MA'(1974): Inhibitory unit activity of the ventral forebrain during both appetitive and aversive Pavlovian conditioning. *Brain Res* 80:146–151

Martinez-Murillo R, Semenenko F, Cuello AC (1988): The origin of tyrosine hydroxylase-immunoreactive fibers in the regions of the nucleus basalis magnocellularis of the rat. *Brain Res* 451:227–236.

Maslowski RJ, Napier TC (1991a): Effects of D1 and D2 antagonists on apomorphine-induced responses of ventral pallidal neurons. *Neuroreport* 2:451–454

Maslowski RJ, Napier TC (1991b) D_1 and D_2 dopamine receptor agonists induce opposite changes in the firing rate of ventral pallidum neurons. *Eur J Pharm* 200:103–112

McGeer PL, McGeer EG, Kimura H, Peng J-F (1986): Cholinergic neurons and cholinergic projections in the mammalian CNS. In: *Dynamics of Cholinergic Function*. Advances in Behavioral Biology, Vol 30, Hanin I, ed. New York: Plenum Press, pp 11–21

Mogenson GR, Nielsen MA (1983): Evidence that an accumbens to subpallidal GABAergic projection contributes to locomotor activity. *Brain Res Bul* 11:309–314

Mogenson GJ, Yang CR (1991): The contribution of basal forebrain to limbic-motor integration and the mediation of motivation to action. In: *The Basal Forebrain: Anatomy to Function*. Advances in Experimental Medicine and Biology, Vol. 295, Napier TC, Kalivas PW, Hanin I, eds. New York: Plenum Publishing, pp 267–290

Mogenson GR, Swanson LW, Wu M (1985): Evidence that projections from substantia innominata to zona incerta and mesencephalic locomotor region contribute to locomotor activity. *Brain Res* 334:65–76

Mora F, Rolls ET, Burton MJ (1976): Modulation during learning of the responses of neurons in the lateral hypothalamus to the sight of food. *Exp Neurol* 53:508–519

Murray CL, Fibiger HC (1985): Learning and memory deficits after lesions of the nucleus basalis magnocellularis: Reversal by physostigmine. *Neuroscience* 14:1025–1032

Napier TC (1992): Contribution of the amygdala and nucleus accumbens to ventral pallidal responses to dopamine agonists. *Synapse* 10:110–119

Napier TC, Chrobak JJ (1992): Evaluations of ventral pallidal dopamine receptor activation in behaving rats. *Neuroreport* 3: in press

Napier TC, Marx K (1987): Enkephalin unilaterally microinjected into the ventral pallidum/nucleus basalis induces circling. *Soc Neurosci Abstr* 13:445

Napier TC, Potter PP (1989): Dopamine in the rat ventral pallidum/substantia innominata: Biochemical and electrophysiological studies. *Neuropharmacology* 28:757–760

Napier TC, An D, Austin MC, Kalivas PW (1988): Opiates microinjected into the ventral pallidum/substantia innominata (VP/SI) produce locomotor responses that involve dopaminergic systems. *Soc Neurosci Abstr* 14:293

Napier TC, Muench MB, Maslowski RJ, Battaglia, G (1991a): Is dopamine a neurotransmitter in the ventral pallidum/substantia innominata? In: *The Basal Forebrain: Anatomy to Function*. Advances in Experimental Medicine and Biology, Vol 295, Napier TC, Kalivas PW, Hanin I, eds. New York: Plenum Press, pp 183–196

Napier TC, Simson PE, Givens BS (1991b): Dopamine electrophysiology of ventral pallidum/substantia innominata neurons: Comparison with the dorsal globus pallidus. *J Pharmacol Exp Therap* 258:249–262

Pirch JH (1977a): Effects of amphetamine and chlorpromazine on brain slow potentials in the rat. *Pharmacol Res Commun* 9:669–674

Pirch JH (1977b): Amphetamine effects on brain slow potentials associated with discrimination in the rat. *Pharmac Biochem Behav* 6:697–700

Pirch JH (1980): Effects of dextroamphetamine on event-related potentials in rat cortex during a reaction time task. *Neuropharmacology* 19:365–370

Pirch JH, Corbus MJ (1983): Haloperidol antagonism of amphetamine-induced effects on event-related slow potentials from rat cortex. *Int J Neurosci* 18:137–142

Pirch JH, Corbus MJ, Napier TC (1981a): Auditory cue preceding intracranial stimulation induces event-related potential in rat frontal cortex: Alterations by amphetamine. *Brain Res Bull* 7:799–804

Pirch JH, Napier TC, Corbus MJ (1981b): Brain stimulation as a cue for event-related potentials in rat cortex: Amphetamine effects. *Int J Neurosci* 15:217–222

Pirch JH, Corbus MJ, Rigdon GC, Lyness WH (1986): Generation of cortical event-related slow potentials in the rat involves nucleus basalis cholinergic innervation. *Electroencephal Clin Neurophysiol* 63:464–475

Rebec GV, Bashore TR, Zimmerman KS, Alloway KS (1979): "Classical" and "atypical" antipsychotic drugs: Differential antagonism of amphetamine- and apomorphine-induced alterations of spontaneous neuronal activity in the neostriatum and nucleus accumbens. *Pharmacol Biochem Behav* 11:529–538

Reiner PB, Semba K, Fibiger HC, McGeer EG (1987): Physiological evidence for subpopulations of cortically projecting basal forebrain neurons in the anesthetized rat. *Neuroscience* 20:629–636

Richardson RT, DeLong MR (1986): Nucleus basalis of Meynert neuronal activity during a delayed response task in monkey. *Brain Res* 399:364–368

Richfield EK, Penney JB, Young AB (1989): Anatomical and affinity state comparisons between dopamine D1 and D2 receptors in the rat central nervous system. *Neuroscience* 30:767–777

Robinson SE, Hambrecht KL, Lyeth BG (1988): Basal forebrain carbachol injection reduces cortical acetylcholine turnover and disrupts memory. *Brain Res* 445:160–164

Rolls ET, Sanghera MK, Roper-Hall A (1979): The latency of activation of neurones in the lateral hypothalamus and substantia innominata during feeding in the monkey. *Brain Res* 164:12–135

Rolls ET, Burton MJ, Mora F (1980): Neurophysiological analysis of brain-stimulation reward in the monkey. *Brain Res* 194:339–357

Russchen FT, Amaral DG, Price JL (1985): The afferent connections of the substantia innominata in the monkey, *Macaca fascicularis*. J Comp Neurol 242:1–27

Semba K, Reiner PB, McGeer EG, Fibiger HC (1988): Brainstem afferents to the magnocellular basal forebrain studied by axonal transport, immunohistochemistry, and electrophysiology in the rat. *J Comp Neurol* 267:433–453

Shreve PE, Uretsky NJ (1988): Effect of GABAergic transmission in the subpallidal region on the hypermotility response to the administration of excitatory amino acids and picrotoxin into the nucleus accumbens. *Neuropharmacology* 27:1271–1277

Shreve PE, Uretsky NJ (1991): GABA and glutamate interact in the substantia innominata/lateral preoptic area to modulate locomotor activity. *Pharmacol Biochem Behav* 38:385–388

Skirboll LR, Grace AA, Bunney BS (1979): Dopamine auto- and postsynaptic receptors: Electrophysiological evidence for differential sensitivity to dopamine agonists. *Science* 206:80–82

Swerdlow NR, Koob GF (1987): Lesions of the dorsomedial nucleus of the thalamus, medial prefrontal cortex and pedunculopontine nucleus: Effects on locomotor activity mediated by nucleus accumbens-ventral pallidal circuitry. *Brain Res* 412:233–243

Voorn P, Jorritsma-Byham B, Van Dijk C, Buijs RM (1986): The dopaminergic innervation of the ventral striatum in the rat: A light- and electron-microscopical study with antibodies against dopamine. *J Comp Neurol* 251:84–99

Will BE, Toniolo G, Brailowski S (1988): Unilateral infusion of GABA and saline into the nucleus basalis of rats: 1. Effects on motor function and brain morphology. *Behav Brain Res* 27:123–129

Wilson FAW, Rolls ET (1990): Neuronal responses related to reinforcement in the primate basal forebrain. *Brain Res* 509:213–231

Wise RA (1980): The dopamine synapse and the notion of 'pleasure centers' in the brain. *Trends Neurosci* 3:91–94

Ueki A, Miyoshi K (1989): Effects of cholinergic drugs on learning impairment in ventral globus pallidus-lesioned rats. *J Neurol Sci* 90:1–21

Zaborszky L (1989): Afferent connections of the forebrain cholinergic projection neurons, with special reference to monoaminergic and peptidergic fibers. In: *Central Cholinergic Synaptic Transmission*. Frotscher M, Misgeld U, eds. Basel, Switzerland: Birkhäuser Verlag. pp 12–32

5

Cholinergic/Noradrenergic Interactions and Memory

MICHAEL W. DECKER

Studies into the role of acetylcholine (ACh) in learning in memory date back more than 20 years (Deutsch, 1971). However, in the last decade investigation of the role of cholinergic dysfunction in memory deficits has taken on new urgency with the discovery that cholinergic dysfunction is characteristic of aging (Decker, 1987) and that profound disruption of cholinergic function is found in Alzheimer's disease (AD) (Coyle et al., 1983). In AD, cholinergic dysfunction results from degeneration of cholinergic neurons in the nucleus basalis of Meynert, the source of cholinergic input to the neocortex and amygdala, and in the medial septal area/diagonal band of Broca (MSA/DBB), the source of cholinergic input to the hippocampus. Based on the hypothesis that this cholinergic dysfunction plays an important role in dementia (Bartus et al., 1982, 1985), experimentally induced disruption of cholinergic function has played a central role in the development of animal models of dementia (for reviews, see Olton and Wenk, 1987; Smith, 1988).

Cholinergic Involvement in Memory

Evidence for the involvement of cholinergic neurotransmission in learning and memory is substantial. Systemic administration of muscarinic cholinergic antagonists, such as scopolamine, disrupts the performance of experimental animals on a wide variety of learning and memory tasks, including inhibitory (passive) avoidance (Bammer, 1982), delayed response (Bartus and Johnson, 1976; Spencer et al., 1985; Pontecorvo et al., 1988), and spatial maze tasks, such as the Morris water maze (Whishaw et al., 1985; Whishaw, 1985; Buresova et al., 1986; Whishaw and Tomie, 1987) and the radial arm maze (Eckerman et al., 1980; Stevens, 1981; Buresova and Bures, 1982; Okaichi and Jarrard, 1982; Wirsching et al., 1984; Beatty and Bierley, 1986). In addition, blockade of nicotinic cholinergic neurotransmission impairs memory performance in many of these same tasks (Oliverio, 1966; Levin et al., 1987; Jackson et al., 1989; Riekkinen et al., 1990b; Decker and Majchrzak, in press).

Consistent with the results obtained with muscarinic and nicotinic antagonists, electrolytic lesions and axon-sparing ibotenic acid lesions of the nucleus basalis magnocellularis (NBM; the rodent homologue of the nucleus basalis of Meynert) or the MSA/DBB also produce cognitive impairments in experimental animals (for review, see Dekker et al., 1991). However, recent findings have revealed that quisqualic acid injected into the NBM reduces cholinergic input to the cortex more completely than does ibotenic acid, but produces only minimal memory deficits (Dunnett et al., 1987; Robbins et al., 1989; Wenk et al., 1989). Thus, independent manipulation of the cholinergic system in experimental animals may be an inadequate model of cognitive dysfunction. Consistent with this conclusion, cholinergic dysfunction found in aging and AD is accompanied by changes in other neurotransmitter systems, such as the noradrenergic system (Mann, 1983; Scarpace and Abrass, 1988), which may be important in memory modulation.

Adrenergic Involvement in Memory

Although phasic intracranial pharmacological manipulations of noradrenergic function affect memory performance (Gallagher et al., 1977; Liang et al., 1986; Marighetto et al., 1989), systemic noradrenergic manipulations and chronic norepinephrine (NE) depletion have only minimal consequences for memory function. For example, several tasks that are disrupted by cholinergic blockade are not affected by depletion of forebrain NE, including the Morris water maze (Hagan et al., 1983), the radial arm maze (Chrobak et al., 1985; Decker and Gallagher, 1987; Sara, 1989), and inhibitory (passive) avoidance (Archer et al., 1985; Decker and McGaugh, 1989). Furthermore, systemic administration of adrenergic antagonists does not disrupt inhibitory avoidance conditioning (McGaugh, 1989) or radial arm maze performance (Beatty and Rush, 1983; Hiraga and Iwasaki, 1984).

Given these relatively mild effects on learning and memory, it has been suggested that NE may not play an important role in dementia (Wenk et al., 1987; Pontecorvo et al., 1988). There is evidence, however, that noradrenergic dysfunction is involved in age-related cognitive impairments. Alpha-2 agonists such as clonidine have been reported to enhance memory in aged monkeys, an effect that appears to be mediated by postsynaptic receptors (Arnsten and Goldman-Rakic, 1985; Arnsten et al., 1988; Jackson and Buccafusco, 1991). Furthermore, impaired inhibitory avoidance performance and noradrenergic degeneration are highly correlated in aged mice (Leslie et al., 1985), and retention of inhibitory avoidance training in aged rodents can be improved by chronic intraventricular infusion of NE (Collier et al., 1988) or by direct stimulation of the locus coeruleus (LC), the source of forebrain NE (Zornetzer, 1985).

Cholinergic-Adrenergic Interactions

Although more extensive evidence is available regarding the independent roles of NE and acetylcholine (ACh) in learning and memory, there are now several studies suggesting that interactions between these two neurotransmitter systems may be of particular importance for memory modulation (Decker and McGaugh, 1991).

Muscarinic Cholinergic/Noradrenergic Interactions in Memory

Much of the evidence for the importance of cholinergic/noradrenergic interactions in learning and memory comes from studies in which muscarinic cholinergic and noradrenergic neurotransmission were concurrently disrupted. For example, we assessed the effect of NE depletion produced by injections of 6-OHDA into the dorsal noradrenergic bundle on the deficits in radial maze performance produced by muscarinic blockade (Decker and Gallagher, 1987). Although we found that even extensive depletion of NE input to forebrain did not alter performance on the radial arm maze, the disruptive effects of scopolamine were potentiated by NE depletion (Fig. 1). In contrast, we found that methylscopolamine, a muscarinic antagonist that does not readily cross the blood-brain barrier, did not alter the performance of NE-depleted animals. Thus, the NE depletion appeared to potentiate scopolamine's central effects. Our results are consistent with reports that NE depletion does not affect retention of active avoidance training or acquisition in the Morris water maze, but potentiates the disruptive effects of

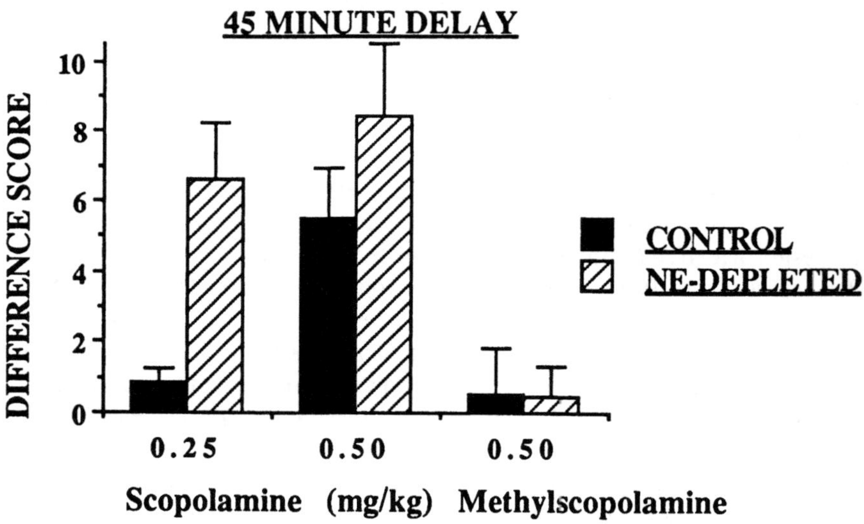

FIGURE 1. The effects of NE depletion on scopolamine-induced deficits in radial maze performance. Each animal (n = 6 for controls and n = 7 for NE depleted) received methylscopolamine and both doses of scopolamine. Each drug day was preceded by a saline day, and the number of errors under saline were subtracted from the errors under drug to derive the difference score. Scores represent the sum of two trials under each drug dose. Reprinted with permission of Elsevier Science Publishers from Decker MW, Gallagher M (1987): Scopolamine-disruption of radial arm maze performance: modification by noradrenergic depletion. *Brain Res* 417:59–69.

scopolamine on these tasks (Kruglikov, 1982; Riekkinen et al., 1990a) and that NE depletion potentiates scopolamine deficits in delayed-response performance in NBM-lesioned rats (Sahgal et al., 1990).

Recently we have found that potentiation of scopolamine amnesia can also be produced by β-adrenergic blockade (Decker et al., 1990). In these experiments, we assessed the effects of combined administration of scopolamine and propranolol, a β-adrenergic receptor antagonist, on two different learning and memory tasks. In one experiment, we found that a pretraining injection of 10 mg/kg of (±)propranolol did not alter the retention of an inhibitory avoidance response. However, when this dose of propranolol was combined with doses of scopolamine that did not impair retention under our experimental conditions, we observed markedly impaired retention (Fig. 2). A similar effect was observed when we administered propranolol and scopolamine concurrently during training on the

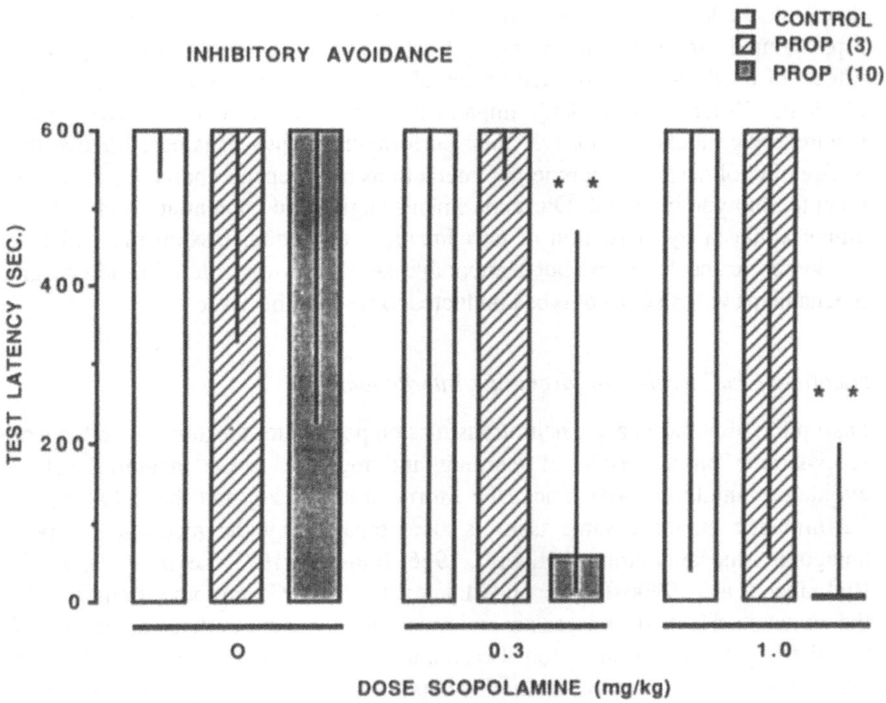

FIGURE 2. Median step-through latencies and interquartile ranges during a retention test conducted 24 hr after inhibitory avoidance training. Rats received either distilled water, scopolamine (0.3 or 1.0 mg/kg), ± propranolol (10.0 mg/kg), or a combination of scopolamine and propranolol prior training (n = 10–13 per group). **Significantly different from control group ($p < .005$, Mann-Whitney U test). Reprinted with permission of Elsevier Science Publishers from Decker MW et al. (1990): Concurrent muscarinic and β-adrenergic blockade in rats impairs place-learning in a water maze and retention of inhibitory avoidance. *Brain Res* 513:81–85.

Morris water maze. Here independent administration of either 10 mg/kg of (±)propranolol or 0.3 mg/kg of scopolamine before each of four training sessions did not alter the ability of rats to learn the location of a hidden platform in the water maze. Yet, when these two drugs were concurrently administered, the performance of the animals was significantly impaired. Thus, doses of propranolol and scopolamine that do not impair performance when administered independently can significantly disrupt performance when administered together.

Propranolol-induced spatial memory deficits have also been observed in rats with lesions of the medial septum. Harrell et al. (1990) found that propranolol impairs radial maze performance in septal-lesioned but not in control rats. Although septal lesions are by no means neurotransmitter specific, one of the principal consequences of medial septal damage is the disruption of cholinergic input to the hippocampus. Thus the finding that propranolol impairs memory in septal-lesioned rats parallels our finding that propranolol potentiates scopolamine-induced memory impairments. Taken together, these two sets of findings suggest that β-blockade-induced potentiation of scopolamine amnesia may involve septo-hippocampal circuitry. Interestingly, Harrell et al. (1990) also found that the effects of α-adrenergic blockade on septal rats are distinct from the effects of β blockade. Whereas β blockade impairs the performance of medial septal rats, α-adrenergic blockade *improves* their performance. Thus, it is possible that the nature of cholinergic/noradrenergic interactions may depend upon the adrenergic receptor subtype involved. Our experiments suggest that potentiation of scopolamine amnesia by disruption of noradrenergic neurotransmission may involve β-adrenergic mechanisms, but comparable studies on the role of α-adrenergic mechanisms will also need to be conducted to resolve this issue.

Nicotinic Cholinergic/Noradrenergic Interactions

Disruption of noradrenergic neurotransmission potentiates the amnestic effects of scopolamine on a variety of learning and memory tasks, including active avoidance, inhibitory avoidance, the Morris water maze, and the radial maze. Performance on these same tasks is also impaired by the nicotinic receptor antagonist, mecamylamine (Oliverio, 1966; Bammer, 1982; Levin et al., 1987; Riekkinen et al., 1990b; Decker and Majchrzak, 1992). Given some similarities in the amnestic effects of muscarinic and nicotinic blockade, perhaps disruption of noradrenergic neurotransmission also potentiates the memory-impairing effects of nicotinic blockade. We have tested this hypothesis in a preliminary experiment by assessing the effects of independent and concurrent administration of mecamylamine and propranolol on the radial maze performance of rats (Decker and Majchrzak, unpublished data).

Nine male, Long-Evans rats, previously used to test the effects of muscarinic cholinergic compounds (Anderson et al., 1991), were used in this experiment. The apparatus and training procedures used have been described previously (Anderson et al., 1991). In summary, four of the eight arms of the maze were blocked during the first half of the session, and the rat was allowed to retrieve food rewards from the ends of the four open arms. Forty-five minutes later, the rat was returned to the maze and allowed access to all eight arms. Only arms not visited during the first portion of the session were baited during this second part of the session. Entries into previously visited arms were scored as errors and were summed across both parts of the session. Rats not completing a session were assigned an error score by adding the number of unvisited arms to their repeat entries, as previously described (Decker and Gallagher, 1987). Mecamylamine HCl (0, 3, or 10 mg/kg) and (±)propranolol HCl (0 or 10 mg/kg) were administered in two separate i.p. injections (each 1.0 ml/kg) 30 min before training. A training session was conducted on Monday, Wednesday, and Friday of each week. Drugs were administered before the Monday and Friday sessions, but no injection was made before the Wednesday session. Each animal received each of the six possible drug combinations one time during the experiment, with the order of administration counterbalanced across animals.

As can be seen in Figure 3, mecamylamine dose dependently impaired

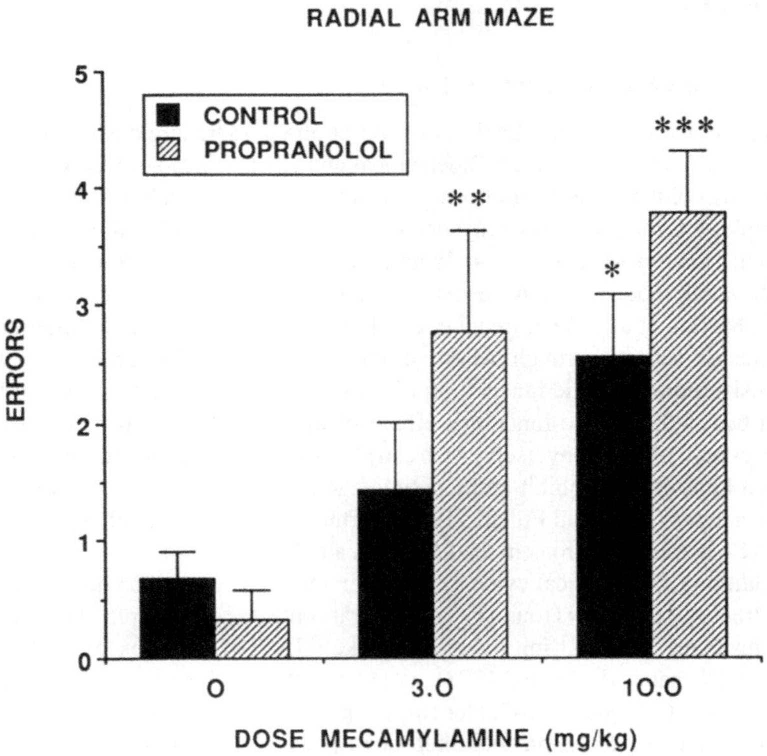

FIGURE 3. Radial maze errors by rats after separate or concurrent treatment with mecamylamine (0, 3, or 10 mg/kg) and propranolol (0 or 10 mg/kg). Each rat (n = 9) received one session under each treatment condition. Significantly different from water-injected controls: *p < .05, **p < .01, ***p < .001 From Decker and Majchrzak, unpublished data.

performance on the radial maze. A significant impairment was observed after a 10 mg/kg dose of mecamylamine, but not after a 3 mg/kg dose. A 10 mg/kg dose of propranolol, in contrast, did not affect the accuracy of animals on this task. However, when this dose or propranolol was administered concurrently with the subeffective dose of mecamylamine (3 mg/kg), a marked impairment was observed. A similar trend toward potentiation of mecamylamine effects on memory was also observed when propranolol was combined with the higher dose of mecamylamine. The propranolol-induced potentiation of mecamylamine's effects on radial maze performance in this experiment resembles the NE-depletion-induced potentiation of scopolamine's effects on this task shown in Figure 1. Since both mecamylamine and propranolol have profound autonomic effects, most notably a lowering of blood pressure, some caution must be exercised in interpreting these results. However, these preliminary findings suggest that it is possible that β-adrenergic blockade potentiates the amnestic effects of nicotinic blockade in much the same way as β-adrenergic blockade or NE depletion potentiates the amnestic effects of muscarinic blockade. Thus cholinergic/noradrenergic interactions may involve both muscarinic and nicotinic mechanisms.

Potential Mechanisms of Interaction

There are a variety of mechanisms by which the noradrenergic and cholinergic systems influence each other. Biochemical and electrophysiological interactions between cholinergic and noradrenergic systems have been well documented. For example, NE modulates the responsivity of hippocampal (Kruglikov, 1982) and cortical (Jones and Olpe, 1984; Waterhouse et al., 1981) neurons to ACh. In addition, NE reduces ACh release in the forebrain by presynaptic inhibition (Vizi, 1980; Moroni et al., 1983; Beani et al., 1986), but stimulates septohippocampal cholinergic activity through actions in the septal area (Robinson et al., 1978). Extensive noradrenergic innervation of NBM cholinergic neurons has also been described, although the functional effects of this input has not been established (Zaborszky, 1989). Conversely, ACh can affect noradrenergic neurotransmission. Muscarinic agonists inhibit and nicotinic agonists stimulate NE turnover in the hippocampus (Birch and Fillenz, 1986; Hörtnagl et al., 1987; Roth et al., 1982), and ACh activates neurons in the LC (Egan and North, 1985).

Additional biochemical evidence for interactions between the cholinergic and noradrenergic systems is found in a behavioral context. For example, the basal rate of hippocampal high-affinity choline uptake (HACU), an index of cholinergic activity (Kuhar and Murrin, 1978), is not affected by long-term NE depletion (Decker and Gallagher, 1987; Cornwell-Jones et al., 1989). However, we have observed that NE depletion does alter the response of the cholinergic system to a training experience (Cornwell-Jones et al., 1990). In this experiment, we trained rats in a one-way active avoidance task using mild footshock as the aversive stimulus. Consistent with previous reports, active avoidance training increased hippocampal HACU in normal rats. However, active avoidance training did not significantly stimulate hippocampal HACU in NE-depleted rats. Thus, the effects of NE depletion on cholinergic function were only evident after training. Similarly, Marighetto et al. (1989) found that intraseptal phenoxybenzamine, an α-adrenergic antagonist, blocks training-induced changes in septohippocampal cholinergic activity without affecting the basal level of septohippocampal cholinergic activity. Interestingly, there is now evidence that training-induced changes in hippocampal HACU are also attenuated in behaviorally impaired aged rodents (Decker et al., 1988; Lebrun et al., 1990), even though the basal rate of hippocampal HACU is unchanged in senescent rodents. The similarities between the effects of aging and experimentally induced NE dysfunction on septohippocampal cholinergic function suggest that an interaction between these two systems might also be of importance in age-related memory impairments.

Caveats Regarding Cholinergic/Noradrenergic Interactions

Although potentiation of the amnestic effects of cholinergic blockade in NE-depleted or propranolol-treated animals has now been demonstrated using a variety of tasks, the generality and relative importance of cholinergic/noradrenergic interactions in normal memory modulation remain to be determined. For example, evidence for NE-depletion-induced potentiation of scopolamine amnesia is not found in all learning and memory tasks. Spangler et al. (1990) did not find evidence of such an interaction using the Stone maze, and Pontecorvo and coworkers (Pontecorvo et al., 1988) have found negative results with a continuous nonmatching to sample operant task. It is possible that the importance of cholinergic/noradrenergic interactions in memory modulation varies depending on the nature of the task. In these experiments, however, it also notable that NE depletion was produced by systemic injection of the noradrenergic neurotoxin DSP-4. Similarly, we also found that DSP-4 does not potentiate scopolamine-induced deficits in the retention of inhibitory avoidance training, although we observed at least additive effects of DSP-4 and scopolamine in these same mice in a spatial water maze task (Decker and McGaugh, 1989). Since systemic administration of DSP-4 typically produces incomplete depletion of forebrain NE, the NE depletion may not have been extensive enough to produce substantial disruption of noradrenergic function in these experiments. Consistent with this hypothesis is the finding that in vivo NE release is not significantly affected unless NE depletion is nearly complete (Abercrombie et al., 1989).

The results from our radial maze study cited earlier (Decker and Gallagher, 1987) also suggest that NE-depletion-induced potentiation of scopolamine amnesia is highly dependent on the degree of NE depletion. In the experiment described earlier in which we found a profound and lasting potentiation of scopolamine effects on radial maze performance (Fig. 1), a 93% depletion of NE was obtained in the cortex and hippocampus. In that same study, however, when less complete NE depletion was obtained (83%), the potentiation of scopolamine effects was less impressive and largely disappeared when animals received additional training. As we have discussed earlier (Decker and Gallagher, 1987), recovery of noradrenergic function may explain the differences in these lesion effects.

Conclusions

One practical implication of the effect of β-adrenergic blockade on scopolamine and mecamylamine-treated animals is that β blockers, used in the treatment of hypertension, may have untoward effects on memory processes in elderly patients with compromised cholinergic function. A more general implication of the potentiation of the amnestic effects of cholinergic blockade induced by NE depletion or β blockade is the possibility that age-related memory impairments might result from concurrent degeneration of the cholinergic and noradrenergic systems. If the compensatory capacity of the noradrenergic system is such that near-complete destruction of the system is necessary to effect significant disruption of function, however, of what importance is the moderate cell loss in the LC found in aging and AD? The answer to this question is, of yet, unknown, but it should be noted that the compensatory capacity of this system found in young animals is not necessarily maintained in senescence (Greenberg et al., 1985). Thus, even moderate noradrenergic neuronal degeneration in aged animals might potentiate the effects of cholinergic dysfunction. Clearly, further work will be necessary to distinguish specific mechanisms and sites of action of cholinergic/noradrenergic interactions important in memory modulation and to determine if these interactions play a role in memory deficits associated with normal aging or AD. However, the results obtained to date suggest that exploration of neurotransmitter interactions in learning and memory will be critical in furthering our understanding of the biological basis of cognitive function.

References

Abercrombie ED, Zigmond MJ (1989): Partial injury to central noradrenergic neurons: Reduction of tissue norepinephrine content is greater than reduction of extracellular norepinephrine measured by microdialysis. *J Neurosci* 9:4062–4067

Anderson DJ, Decker MW, Americ SP, Cadman E, Buckley MJ, Vella-Rountree L, Williams M (1991): The aminopyridazine muscarinic agonist, SR 95639A, is a functional M2 receptor antagonist in rat brain. *Drug Dev Res* 24:107–118

Archer T, Jonsson G, Ross SB (1985): Active and passive avoidance following the administration of systemic DSP4, xylamine or p-chloro-amphetamine, *Behav Neural Biol* 43:238–249

Arnsten AFT, Goldman-Rakic PS (1985): α2-adrenergic mechanisms in prefrontal cortex associated with cognitive decline in aged nonhuman primates. *Science* 230:1273–1276

Arnsten AFT, Cai JX, Goldman-Rakic PS (1988): The alpha-2 adrenergic agonist guanfacine improves memory in aged monkeys without sedative or hypotensive side effects: Evidence for alpha-2 receptor subtypes. *J Neurosci* 8:4287–4298

Bammer G (1982): Pharmacological investigations of neurotransmitter involvement in passive avoidance responding: A review and some new results. *Neurosci Biobehav Rev* 6:247–296

Bartus RT, Johnson HR (1976): Short-term memory in the rhesus monkey: Disruption from the anti-cholinergic scopolamine. *Pharmacol Biochem Behav* 5:39–46

Bartus RT, Dean RL, Beer B, Lippa AS (1982): The cholinergic hypothesis of geriatric memory dysfunction. *Science* 217:408–417

Bartus RT, Dean RL, Pontecorvo MJ, Flicker C (1985): The cholinergic hypothesis: A historical overview, current perspective, and future directions. *Ann NY Acad Sci* 444:332–358

Beani L, Tanganelli S, Antonelli T, Bianchi C (1986): Noradrenergic modulation of cortical acetylcholine release is both direct and γ-aminobutyric acid-mediated. *J Pharmacol Exp Ther* 236:230–236

Beatty WW, Bierley RA (1986): Scopolamine impairs encoding and retrieval of spatial working memory in rats. *Physiol Psychol* 14:82–86

Beatty WW, Rush JR (1983): Spatial working memory in rats: Effects of monoaminergic antagonists. *Pharmacol Biochem Behav* 18:7–12

Birch PJ, Fillenz M (1986): Muscarinic receptor activation inhibits both release and synthesis of noradrenaline in rat hippocampal synaptosomes. *Neurochem Int* 8:171–177

Buresova O, Bures J (1982): Radial maze as a tool for assessing the effect of drugs on the working memory of rats. *Psychopharmacology* 77:268–271

Buresova O, Bolhuis JJ, Bures J (1986): Differential effects of cholinergic blockade on performance of rats in the water tank navigation task and in a radial water maze. *Behav Neurosci* 100:476–482

Chrobak JC, DeHaven DL, Walsh TJ (1985): Depletion of brain norepinephrine with DSP-4 fails to alter acquisition or performance of a radial arm maze task. *Behav Neural Biol* 44:144–150

Collier TJ, Gash DM, Sladek JR (1988): Transplantation of norepinephrine neurons into aged rats improves performance of a learned task. *Brain Res* 448:77–87

Cornwell-Jones CA, Decker MW, Chang JW, Cole B, Goltz KM, Tran T, McGaugh JL (1989): Neonatal 6-hydroxydopa, but not DSP4, elevates brainstem monoamines and impairs inhibitory avoidance learning in developing rats. *Brain Res* 493:258–268

Cornwell-Jones CA, Decker MW, Gianulli T, Wright EL, McGaugh JL (1990): Norepinephrine depletion reduces the effects of social and olfactory experience. *Brain Res Bull* 25:643–649

Coyle JT, Price DL, DeLong MR (1983): Alzheimer's disease: A disorder of cortical cholinergic innervation. *Science* 219:1184–1190

Decker MW (1987): The effects of aging on hippocampal and cortical projections of the forebrain cholinergic system. *Brain Res Rev* 12:423–438

Decker MW, Gallagher M (1987): Scopolamine-disruption of radial arm maze performance: Modification by noradrenergic depletion. *Brain Res* 417:59–69

Decker MW, Majchrzak MJ (1992): The effects of systemic and intracerebroventricular administration of mecamylamine, a nicotinic cholinergic antagonist, on spatial memory in rats. *Psychopharmacology* 107:530–534

Decker MW, McGaugh JL (1989): Effects of concurrent manipulations of cholinergic and noradrenergic function on learning and retention in mice. *Brain Res* 477:29–37

Decker MW, McGaugh JL (1991): The role of interactions between the cholinergic system and other neuromodulatory systems in learning and memory. *Synapse* 7:151–168

Decker MW, Pelleymounter MA, Gallagher M (1988): Effects of training on a spatial memory task on high affinity choline uptake in hippocampus and cortex in young adult and aged rats. *J Neurosci* 8:90–99

Decker MW, Gill TM, McGaugh JL (1990): Concurrent muscarinic and β-adrenergic blockade in rats impairs place-learning in a water maze and retention of inhibitory avoidance. *Brain Res* 513:81–85

Dekker AJAM, Connor DJ, Thal LJ (1991): The role of cholinergic projections from the nucleus basalis in memory. *Neurosci Biobehav Rev* 15:299–317

Deutsch JA (1971): The cholinergic synapse and the site of memory. *Science* 174:788–794

Dunnett SB, Whishaw IQ, Jones GH, Bunch ST (1987): Behavioural, biochemical and histochemical effects of different neurotoxic amino acids injected into nucleus basalis magnocellularis of rats. *Neuroscience* 20:653–669

Eckerman DA, Gordon WA, Edwards JD, MacPhail RC, Gage MI (1980): Effects of scopolamine, pentobarbital, and amphetamine on radial arm maze performance in the rat. *Pharmacol Biochem Behav* 12:595–602

Egan TM, North RA (1985): Acetylcholine acts on M2-muscarinic receptors to excite rat locus coeruleus neurones. *Br J Pharmac* 85:733–735

Gallagher M, Kapp BS, Musty RE, Driscoll PA (1977): Memory formation: Evidence for a specific neurochemical system in the amygdala. *Science* 198:423–425

Greenberg LH, Brunswick DJ, Weiss B (1985) Effect of age on the rate of recovery of beta-adrenergic receptors in the rat brain following desmethylimipramine-induced subsensitivity. *Brain Res* 328:81–88

Hagan JJ, Alpert JE, Morris RGM, Iversen SD (1983): The effects of central catecholamine depletions on spatial learning in rats. *Behav Brain Res* 9:83–104

Harrell LE, Peagler A, Parsons DS (1990): Adrenoreceptor antagonist treatment influences recovery of learning following medial septal lesions and hippocampal sympathetic ingrowth. *Pharmacol Biochem Behav* 35:21–28

Hiraga Y, Iwasaki T (1984): Effects of cholinergic and monoaminergic antagonists and tranquilizers upon spatial memory in rats. *Pharmacol Biochem Behav* 20:205–207

Hörtnagl H, Potter PE, Hanin I (1987): Effect of cholinergic deficit induced by ethylcholine aziridinium (AF64A) on noradrenergic and dopaminergic parameters in rat brain. *Brain Res* 421:75–84

Jackson WJ, Buccafusco JJ (1991): Clonidine enhances delayed matching-to-sample performance by young and aged monkeys. *Pharmacol Biochem Behav* 39:79–84

Jackson WJ, Elrod K, Buccafusco JJ (1989): Delayed matching-to-sample in monkeys as a model for learning and memory deficits: Role of brain nicotinic receptors. In: *Novel Approaches to the Treatment of Alzheimer's Disease*, Meyer EM, Simpkins JW, Yamamoto J, eds. New York: Plenum Publishing, pp 39–52

Jones RSG, Olpe HR (1984): Monoaminergic modulation of the sensitivity of neurones in the cingulate cortex to iontophoretically applied substance. P. *Brain Res* 311:297–305

Kruglikov RI (1982): On the interaction of neurotransmitter systems in processes of learning and memory. In: *Neuronal Plasticity and Memory Formation*, Marsan CA, Matthies H, eds. New York: Raven Press, pp 339–351

Kuhar MJ, Murrin LC (1978): Sodium-dependent high affinity choline uptake. *J Neurochem* 30:15–21

Lebrun C, Durkin TP, Marighetto A, Jaffard R (1990): A comparison of the working memory performances of young and aged mice combined with parallel measures of testing and drug-induced activations of septo-hippocampal and NBM-cortical cholinergic neurones. *Neurobiol Aging* 11:515–521

Leslie FM, Loughlin SE, Sternberg DB, McGaugh JL, Young LE, Zornetzer SF (1985): Noradrenergic changes and memory loss in aged mice. *Brain Res* 359:292–299

Levin ED, Castonguay M, Ellison GD (1987): Effects of the nicotinic receptor blocker, mecamylamine, on radial-arm maze performance in rats. *Behav Neural Biol* 48:206–212

Liang KC, Juler RG, McGaugh JL (1986): Modulating effects of posttraining epinephrine on memory: Involvement of the amygdala noradrenergic system. *Brain Res* 368:125–133

Mann DMA (1983): The locus coeruleus and its possible role in aging and degenerative disease of the human central nervous system. *Mech Aging Devel* 23:73–94

Marighetto A, Durkin T, Toumane A, Lebrun C, Jaffard R (1989): Septal-noradrenergic antagonism in vivo blocks the testing-induced activation of septo-hippocampal cholinergic neurones and produces a concomitant deficit in working memory performance of mice. *Pharmacol Biochem Behav* 34:553–558

McGaugh JL (1989): Involvement of hormonal and neuromodulatory systems in the regulation of memory stage. *Ann Rev Neurosci* 12:255–287

Moroni F, Tanganelli S, Antonelli T, Carla V, Bianchi C, Beani L (1983): Modulation of cortical acetylcholine and gamma-aminobutyric acid release in freely moving guinea pigs: Effects of clonidine and other adrenergic drugs. *J Pharmacol Exper Ther* 227:435–440

Okaichi H, Jarrard LE (1982): Scopolamine impairs performance of a place and a cue task in rats. *Behav Neural Biol* 35:319–325

Oliverio A (1966): Effects of mecamylamine on avoidance conditioning and maze learning of mice. *J Pharmacol Exper Ther* 154:350–356

Olton DS, Wenk GL (1987): Dementia: Animal models of the cognitive impairments produced by degeneration of the basal forebrain cholinergic system. In: *Psychopharmacology: The Third Generation of Progress*, Meltzer HY, ed. New York: Raven Press, pp 941–953

Pontecorvo M, Clissold DB, Conti LH (1988): Age-related cognitive impairments as assessed with an automated repeated measures memory task: Implications for the possible role of acetylcholine and norepinephrine in memory dysfunction. *Neurobiol Aging* 9:617–625

Riekkinen P Jr, Sirviö J, Valjakka A, Pitkänen A, Partanen J, Riekkinen P (1990a): The effects of concurrent manipulations of cholinergic and noradrenergic systems on neocortical EEG and spatial learning. *Behav Neural Biol* 54:204–210

Riekkinen P Jr, Sirviö J, Aaltonen M, Riekkinen P (1990b): The effects of concurrent manipulations of nicotinic and muscarinic receptors on spatial and passive avoidance learning. *Pharmacol Biochem Behav* 37:405–410

Robbins TW, Everitt BJ, Marston HM, Wilkinson J, Jones GH, Page KJ (1989): Comparative effects of ibotenic acid- and quisqualic acid-induced lesions of the substantia innominata on attentional function in the rat: Further implications for the role of the cholinergic neurons of the nucleus basalis in cognitive processes. *Behav Brain Res* 35:221–240

Robinson SE, Cheney DL, Costa E (1978): Effect of Nomifensine and other antidepressant drugs on acetylcholine turnover in various regions of rat brain. *N-S Arch Pharmacol* 304:263–269

Roth KA, McIntire SL, Barchas JD (1982): Nicotinic-catecholaminergic interactions in rat brain: Evidence for cholinergic nicotinic and muscarinic interactions with hypothalamic epinephrine. *J Pharmacol Exp Ther* 221:416–420

Sahgal A, Keith AB, Lloyd S, Kerwin JM, Perry EK, Edwardson JA (1990): Memory following cholinergic (NBM) and noradrenergic (DNAB) lesions made singly or in combination: Potentiation of disruption by scopolamine. *Pharmacol Biochem Behav* 37:597–605

Sara SJ (1989): Noradrenergic-cholinergic interaction: Its possible role in memory dysfunction associated with senile dementia. *Arch Gerontol Geriatr* Suppl. 1:99–108

Scarpace PJ, Abrass IB (1988): Alpha- and beta-adrenergic receptor function in the brain during senescence. *Neurobiol Aging* 9:53–58

Smith G (1988): Animal models of Alzheimer's disease: Experimental cholinergic denervation. *Brain Res Rev* 13:103–118

Spangler EL, Wenk GL, Chachich ME, Smith K, Ingram DK (1990): Complex maze performance in rats: Effects of noradrenergic depletion and cholinergic blockade. *Behav Neurosci* 104:410–417

Spencer DG Jr., Pontecorvo MJ, Heise GA (1985): Central cholinergic involvement in working memory: Effects of scopolamine on continuous nonmatching and discrimination performance in the rat. *Behav Neurosci* 99:1049–1065

Stevens R (1981): Scopolamine impairs spatial maze performance in rats. *Physiol Behav* 27:385–386

Vizi ES (1980): Modulation of cortical release of acetylcholine by noradrenaline released from nerves arising from the rat locus coeruleus. *Neuroscience* 5:2139–2144

Waterhouse BD, Moises HC, Woodward DJ (1981): Alpha receptor-mediated facilitation of somatosensory cortical neuronal responses to excitatory synaptic inputs and iontophoretically applied acetylcholine. *Neuropharmacology* 20:907–920

Wenk G, Hughey D, Boundy V, Kim A (1987): Neurotransmitters and memory: Role of cholinergic, serotonergic, and noradrenergic systems. *Behav Neurosci* 101:325–332

Wenk GL, Markowska AL, Olton DS (1989): Basal forebrain lesions and memory: Alterations in neurotensin, not acetylcholine, may cause amnesia. *Behav Neurosci* 103:765–769

Whishaw IQ (1985): Cholinergic receptor blockade in the rat impairs locale but not taxon strategies for place navigation in a swimming pool. *Behav Neurosci* 99:979–1005

Whishaw IQ, Tomie J (1987): Cholinergic receptor blockade produces impairments in sensorimotor subsystem for place navigation in the rat: Evidence from sensory, motor, and acquisition tests in a swimming pool. *Behav Neurosci* 101:603–613

Whishaw IQ, O'Connor WT, Dunnett SB (1985): Disruption of central cholinergic systems in the rat by basal forebrain lesions or atropine: Effects on feeding, sensorimotor behaviour, locomotor activity and spatial navigation. *Behav Brain Res* 17:103–115

Wirsching BA, Beninger RJ, Jhamandas K, Boegman RJ, El-Defrawy SR (1984): Differential effects of scopolamine on working and reference memory of rats in the radial maze. *Pharmacol Biochem Behav* 20:659–662

Zaborszky L (1989): Afferent connections of the forebrain cholinergic projection neurons, with special reference to monoaminergic and peptidergic fibers. In: *Central Cholinergic Synaptic Transmission*. Frotscher M, Misgeld U, eds. Basel: Birkhäuser, pp 12–32

Zornetzer SF (1985): Catecholamine system involvement in age-related memory dysfunction. *Ann NY Acad Sci* 444:242–254

6

The Role of the Noradrenergic System in Higher Cerebral Functions: Experimental Studies about the Effects of Noradrenergic Modulation on Electrophysiology and Behavior

PAAVO J. RIEKKINEN, JOUNI SIRVIO, PAAVO RIEKKINEN JR., ANTTI VALJAKKA, PEKKA JAKALA, ESA KOIVISTO, AND RISTO LAMMINTAUSTA

Introduction

The ascending systems (cholinergic, noradrenergic, dopaminergic, serotonergic, and histaminergic) have been proposed to regulate the function of information networks in the forebrain (Mesulam, 1990). The degeneration of these systems may lead to cognitive dysfunctions related to aging and to a greater extent in Alzheimer's disease (AD) (Reinikainen et al., 1990). Most of clinical drug trials of Alzheimer's disease have based on a "cholinergic hypothesis of geriatric memory dysfunction." Because of the limited therapeutic impact of this strategy (Kumar and Calache, 1991), other alternative treatment approaches are needed for the symptomatic treatment of patients with AD.

Extensive electrophysiological, anatomical, and behavioral data demonstrate that the noradrenergic system has an important role in the regulation of neocortical arousal and vigilance, attention, learning, and memory, as well as neural plasticity (Heal and Marsden, 1990). The locus coeruleus (LC) is the major source of the noradrenergic innervation (the dorsal noradrenergic bundle, DNAB) of the forebrain. Electrophysiological and neurochemical studies have shown that the firing rate of the neurons of LC and the release of noradrenaline is regulated by α_2 adrenergic autoreceptors. These receptors mediate the autoinhibition of noradrenergic neurons, and the activation of these receptors decreases the turnover of noradrenaline. On the other hand, the blockade of those receptors increases the release of noradrenaline. However, it is evident that the number of heterosynaptic α_2 adrenergic receptors exceeds the number of α_2 adrenergic receptors in the brain (Heal & Marsden, 1990).

As a part of our ongoing project about the role of ascending systems in higher cerebral functions, we are studying functional interactions between noradrenergic and cholinergic systems, since there is much biochemical and anatomical evidence for interactions between those systems (Jones and Cuello, 1989). We are also studying whether pharmacological stimulation of noradrenergic system by α_2 antagonist would have beneficial effects on cognitive functions in aging or AD. This paper is intended to briefly summarize our recent experimental results.

Methods

All the experiments have been carried out using laboratory rats. The methods have been described in our original publications cited in Results and Discussion.

Cortical electroencephalography (EEG) was used to measure cortical desynchronization/synchronization (Fig. 1). Increased cortical slow-wave activity, which reflects the summation of long-lasting hyperpolarization of pyramidal neurons in the fifth layer, is dependent on the dysfunction of basalo-cortical cholinergic neurons (Riekkinen et al., 1991). Furthermore, the thalamic oscillation, which is reflected as immobility-related high-voltage spindles in cortical EEG (Fig. 2), is dependent on the ascending systems (Steriade and Buzsaki, 1990). The rhythmic burst firing mode of thalamic neurons (thalamic oscillation) is proposed to interfere with the information transfer of thalamic relay neurons, in contrast to the tonic firing mode of thalamic neurons (McCormick, 1989).

The hippocampal formation is involved in the mechanisms of learning and memory (Buzsaki, 1989; Squire and Zola-Morgan, 1991). There is evidence to suggest relationships between hippocampal EEG (theta activity, sharp waves,

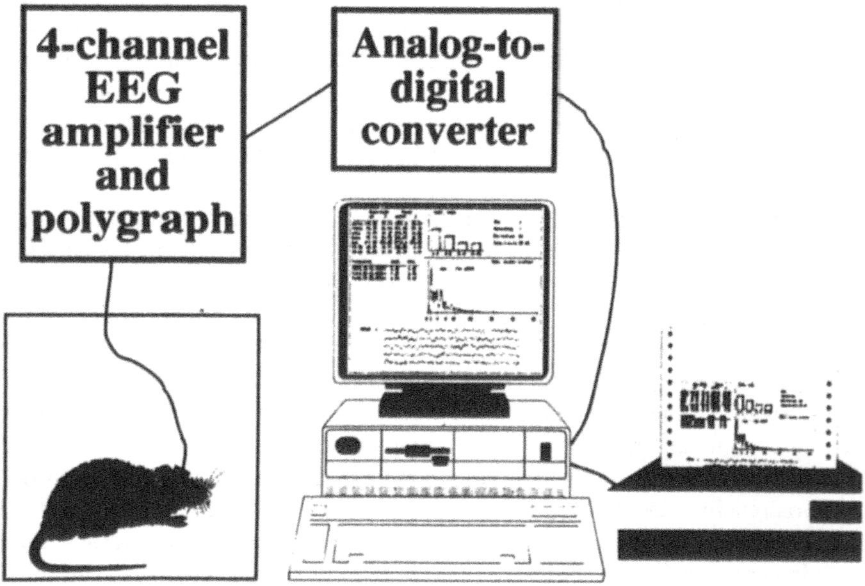

FIGURE 1. The analysis of cortical quantitative EEG. Recording electrodes are implanted bilaterally on the frontal and occipital cortex, and ground as well as indifferent electrodes are located in the midline above the cerebellum. For the fast Fourier transformation, five 4-sec artifact-free epochs of EEG are recorded. EEG samples are measured simultaneously from all four EEG sites. The spectrum is compressed into four bands (slow waves: delta = 1–4 Hz, theta = 4–8 Hz; fast waves: alpha = 8–12 Hz, beta = 12–20 Hz). The absolute amplitudes and relative amounts of different bands are calculated by the computer.

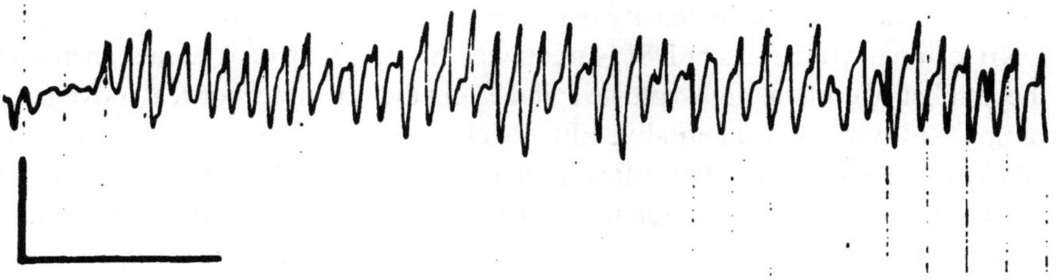

FIGURE 2. The analysis of cortical high-voltage spindle activity. A typical HVS epoch recorded from the neocortex of the rat is shown. HVS (frequency 6–10 Hz, bilaterally synchronous appearance, amplitude >2 times higher than desynchronized activity) number, and duration are measured from the polygraph charts using ruler. The time scale (horizontal bar = 1 sec) and amplitude scale (vertical bar = 300 μV) are shown on the right corner of the recording.

long-term potention), and learning/memory processes (Teyler, 1991). Spatial learning, which is critically dependent on intact hippocampal function, was assessed using the water-maze task.

Emotional coloring plays an important role in learning and memory. The retention of the passive avoidance task (aversively motivated task) was used to assess those aspects of cognitive functions. Previous data suggest that the modulation of amygdaloid function, e.g., by the noradrenergic system, is critical for responses to fear (McGaugh et al., 1990; Selden et al., 1990b).

Operant chamber tasks assessing selective attention (five-choice serial reaction time task) and spatial working memory (delayed nonmatching to position) in rats have also been employed. These tasks are somewhat analogous to those neuropsychological tests that are used for primates.

In many of the studies, dexmedetomidine (an α_2 agonist) or atipamezole (an α_2 antagonist) (Orion Corporation Farmos, R&D Pharmaceuticals, Turku, Finland) were used. Medetomidine (a recemic mixture of levo and dextro (active enantiomer) forms (Savola and Virtanen, 1991) has been shown dose dependently (1.0–300 μg/kg) to decrease the turnover of noradrenaline in brain after systemic administration (MacDonald et al., 1988). Atipamezole, which is a potent and selective α_2 adrenergic antagonist (Virtanen et al., 1989), has been shown to increase the turnover of noradrenaline in rat brain after systemic administration (Scheinin et al., 1988).

Results and Discussion

Cortical Electroencephalography

The partial noradrenaline depletion in brain (−70% in the cerebral cortex) induced by DSP-4 did not affect cortical EEG in young rats (Riekkinen Jr et al., 1990c,

1991c). However, more extensive depletions of noradrenaline of >90%, either by DNB lesion in young rats or DSP-4 treatment in a subgroup of aged rats, increased the number of cortical high-voltage spindles but did not affect cortical movement- or immobility-related quantitative EEG (Riekkinen Jr et al., 1992a). Furthermore, DNB lesion diminished the efficacy of tacrine, an anticholinesterase drug, in reversing nucleus-basalis lesion induced high-voltage spindle activity (Riekkinen Jr et al., 1992c).

Aged rats had a decreased amount of fast activity and an unchanged amount of slow-wave activity in the cortical quantitative EEG. The number of high-voltage spindles is increased during aging (Sirviö et al., 1989; Figure 3). Biochemical investigation revealed a slightly increased noradrenaline content in the cerebral cortex and hippocampus of aged rats, and the concentrations of biogenic amines or their metabolites did not correlate with the different parameters of cortical EEG in young and aged rats (Valjakka et al., 1990b).

The α_2 agonist, guanfacine, at the doses of 0.004, 0.02, and 0.1 mg/kg increased dose dependently the number of cortical high-voltage spindles (Riekkinen Jr et al., 1990b, 1991c). Clonidine, an α_2 agonist, at 1.0 mg/kg increased slow-wave activity (increased spectral amplitudes) during immobility and mobility. Clonidine also increased slow wave activity in the rats with ibotenic acid-induced lesion of the nucleus basalis (Riekkinen Jr et al., 1990a). When tested, the effects of an α_2 agonist could be reversed by atipamezole, a specific and potent α_2 antagonist. More importantly, atipamezole (1.0, 3.0, and 10.0 mg/kg) decreased dose dependently the number of high-voltage spindles in adult and aged rats, as well as in the nucleus-basalis lesioned rats (Riekkinen Jr et al., 1991a,b (Fig. 3). Furthermore, a high dose of atipamezole (10 mg/kg) decreased slow-wave activity but did not affect fast activity in the nucleus-basalis lesioned rats (Riekkinen Jr et al., 1990a). The combination of atipamazole with pilocarpine, a muscarinic agonist, or with tacrine, an anticholinesterase, produced a significantly greater decrease of high-voltage spindles in rats than either treatment alone (Riekkinen Jr et al., 1990b; 1991a,b,c). Our recent results suggest that the ability of atipamezole to suppress high-voltage spindle activity in adult rats persists during subchronic administration (Jäkälä et al., 1992b).

Altogether these results suggest that the noradrenergic system has important effects on cortical desynchronization that are at least in part mediated by α_2 adrenergic mechanisms. Further studies are needed to clarify the role of indirect activation of cholinergic system with regard to these effects. However, recent data by Buzsaki et al. (1991) suggest that α_2 adrenergic mechanisms in the thalamus are also involved; the activation of α_2 adrenergic receptors could hyperpolarize the thalamo-cortical neurons, leading to high-voltage spindles in the cerebral cortex. Interestingly, the stimulation of α_1 adrenergic receptors could have an opposite effect on thalamic oscillation. Thus, the effects of the α_2 adrenergic antagonist, atipamezole, on the suppression of high-voltage spindle activity could be mediated partly by the blockade of α_2 adrenoceptors in the thalamus and/or indirect stimulation of α_1 receptors in the thalamus subsequent to the increased release of noradrenaline.

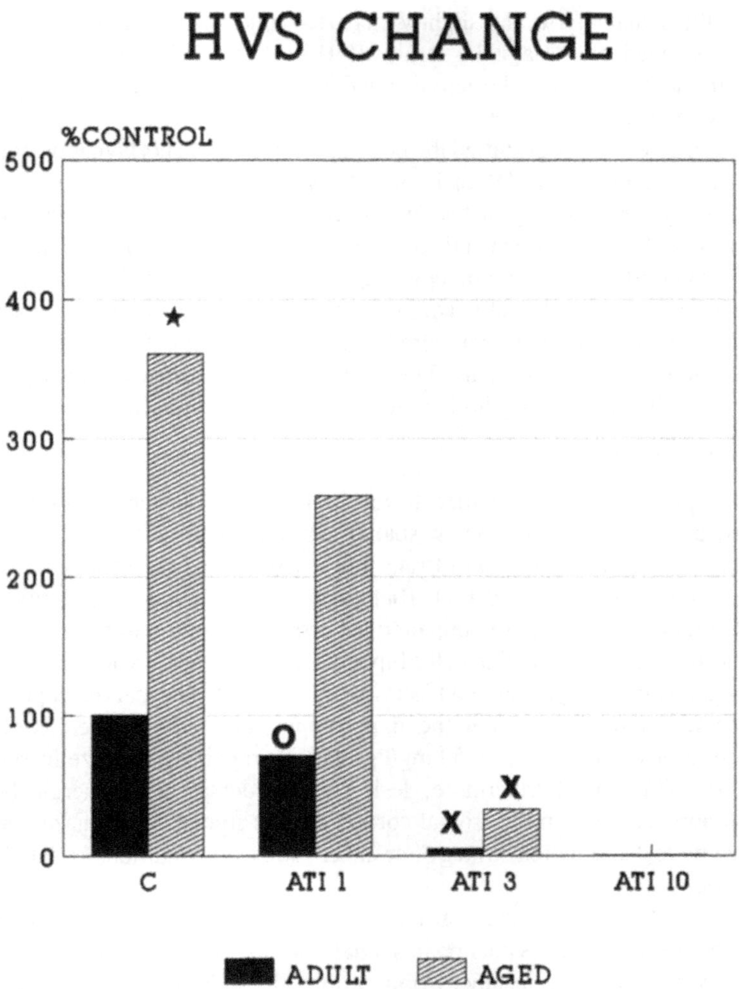

FIGURE 3. The effects of atipamezole (1, 3, or 10 mg/kg, s.c.) on the total duration of high voltage spindle activity in adult or aged rats. The mean value of adult control treatment is 100%. Note the increased amount of high-voltage spindle activity in aged rats (*$p < .001$). ○, $p < 0.05$ as compared to control treatment. X, $p < 0.01$ as compared to control treatment. These modified data have been taken from Riekkinen Jr et al. (1991a,c).

Hippocampal Electroencephalography

Partial noradrenaline depletion in brain (−40% in the hippocampus) induced by DSP-4 treatment did not affect the nonrhythmical hippocampal EEG (recorded during awake immobility) in young and aged rats (Valjakka et al., 1991c). Atipamezole (3 mg/kg) increased the baseline amplitude of the nonrhythmical

EEG in the dentate gyrus and slightly increased spike activity in the CA1 area in young and aged rats (Valjakka et al., 1991c). The baseline amplitude of the nonrhythmical EEG and spike activity the CA1 area in were decreased in aged rats (Valjakka et al., 1991c).

Atipamezole (1 mg/kg) shifted the population spike-field postsynaptic potential response curve towards the left in the dentate gyrus of adult rats, and this effect was abolished by the lesioning of the fimbria-fornix (Valjakka et al., 1991b). The population spike-field postsynaptic potential relationship in the hilus of the dentate gyrus did not differ between young and aged rats (Valjakka et al., 1991a).

The effects of systemically administered atipamezole on the hippocampal evoked potentials suggest that noradrenergic modulation facilitates neuronal transmission in the dentate gyrus. These effects may be mediated, at least in part, by the ascending aminergic/cholinergic afferents of the hippocampus.

Behavioral Studies

Partial depletion of noradrenaline in the brain (−70% in the cerebral cortex) induced by DSP-4 did not impair spatial learning/memory assessed using the water-maze task in young control rats, but it aggravated the scopolamine (0.8 mg/kg)-induced acquisition deficit (Riekkinen Jr et al., 1990c). In addition, the partial noradrenaline depletion impaired the acquisition of the water maze task in aged rats (−78% noradrenaline in the hippocampus), but not in young rats (−57% noradrenaline in the hippocampus) (Sirviö et al., 1991c). The marked depletion of brain noradrenaline (>−95% in the hippocampus) induced by three consecutive administrations of DSP-4 (3 × 50 mg/kg) did not impair water maze learning in young rats (Fig. 4). Furthermore, lesion of the dorsal noradrenergic bundle (−86% noradrenaline in the frontal cortex) did not impair water-maze learning assessed by an "alternation strategy" or under "stressful conditions" (Valjakka et al., 1990a).

These results suggests that partial, or even almost complete, depletion of noradrenaline in brain does not impair spatial learning/memory in young "normal" rats. However, in some instances (muscarinic blockade and aged brain), even partial depletion could induce a deficit in the acquisition of the water-maze task. The neuropsychological processes involved in those deficits and their neuronal substrates need further study. Interestingly, a recent study suggested that noradrenergic deficit in the forebrain impairs locally cued navigation while facilitating the spatially cued navigation of rats in the water-maze task (Selden et al., 1990a).

Our recent study investigated the effects of dexmedetomidine (0.3–9.0 μg/kg) on the acquisition and retention of passive avoidance and the water-maze task in young rats in order to elucidate the effects of pharmacologically decreased activity of noradrenergic neurones on learning and memory (Sirviö et al., 1992a). Pretraining administration of 0.3 or 0.9 μg/kg dexmedetomidine impaired the acquisition of the water-maze task, whereas higher doses (3.0 or 9.0 μg/kg) had no significant effect on spatial learning. Pretraining administration of 9.0 μg/kg dexmedetomidine impaired the retention of the passive avoidance task (assessed

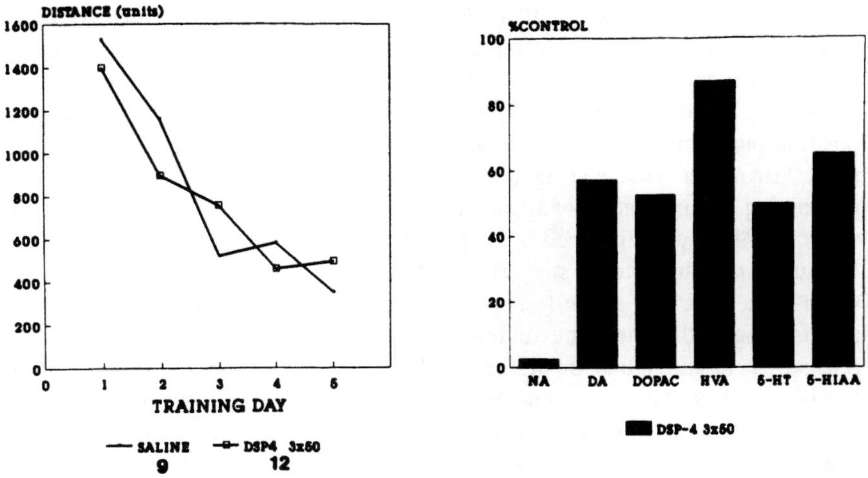

FIGURE 4. The effects of repeated administration (3 days) of noradrenergic neurotoxin, DSP-4 (50 mg/kg/day, i.p.) on the acquisition of the water-maze task (escape distance onto a hidden platform that was held in a constant position in the pool) and monoaminergic systems of the hippocampus in rats. This kind of treatment depleted noradrenaline in the hippocampus, and it slightly affected the dopaminergic (DA, DOPAC, HVA) and serotonergic (5-HT, 5-HIAA) systems. Water-maze learning did not differ between saline (n = 9) and DSP-4 treated (n = 12) rats [ANOVA: $F(1,354) = 0.6$, $p > 0.1$]. Unpublished results. See Riekkinen Jr et al. (1990d) for the description of the water maze apparatus.

24 hr after training), but it did not affect the training of this task. Smaller doses did not affect the training or retention of this aversively motivated task. These data agree with previous findings that pharmacological manipulation of the noradrenergic system affects the retention of aversively motivated tasks (McGaugh et al., 1990). The results of Sirviö et al. (1992a) suggest that the dose-response curve of the effects of dexmedetomidine (which decreases the release of noradrenaline) on the impairment of learning/memory differs between the passive avoidance and water-maze tasks. This possibly indicates the differences between the role of the noradrenergic system on conditioning to explicit or contextual cues of aversively motivated tasks and on the acquisition of the spatially or locally cued water maze task (Selden et al., 1990a,b,c).

Atipamezole (3.0 mg/kg) improved slightly the acquisition of the radial arm maze task, both in controls and in rats with the partial lesion of the fimbria-fornix lesion (unpublished data). However, the studies that investigated the effects of α_2 adrenergic drugs on the age-related deficit of water-maze learning have yielded negative results so far. A low dose (1 μg/kg) of guanfacine did not improve this deficit in aged rats (Sirviö et al., 1991b). Atipamezole (0.1 or 0.3 mg/kg) did not improve performance in old rats of the same colony (unpublished data). Furthermore, atipamezole (0.3 mg/kg) tended to impair the acquisition of the water-maze task in a subgroup of old rats that were screened to be markedly impaired in this task (Sirviö et al., 1992b). Because atipamezole (at doses >0.6 mg/kg) induces freezing behavior when rats are put into a water pool, high doses of atipamezole could not be tested using pretraining administration of the drug. However, posttraining administration of atipamezole (3.0 mg/kg) did not improve water-maze learning in young or aged rats (Riekkinen Jr et al., 1992b). In addition, pretraining administration of atipamezole (0.1 or 0.6 mg/kg) did not alleviate the water learning deficit induced by scopolamine (0.8 mg/kg) (Sirviö et al., 1992a). Because previous studies suggest that age-related learning impairment in the water-maze task may be, at least partly, due to a cholinergic deficit, the present results suggest that the acute administration of atipamezole, which increases the release of noradrenaline in the brain, does not alleviate this learning deficit. Further studies will determine whether atipamezole or other α_2 adrenergic antagonists are effective when administered chronically in aged intact rats or in aged rats with induced partial noradrenaline depletion (see Sirviö et al., 1991c for a further discussion about the involvement of the noradrenergic deficit in age-related spatial learning/memory impairment in rats).

In the delayed nonmatching to position task (assessing spatial working memory), a single dose of atipamezole (0.1–2.7 mg/kg) did not improve the performance of adult or aged rats (Sirviö et al., 1991a). In the "normal" version of a five-choice serial reaction time task (assessing selective attention), atipamezole (0.3 mg/kg) slightly improved the accuracy performance of the "impaired" subgroup of adult rats (Jäkälä et al., 1992a). Furthermore, atipamezole (1.0 mg/kg) slightly improved accuracy performance of adult rats when assessed using reduced stimulus intensity (Fig. 5). It remains to be studied whether an α_2 antagonist is effective in rats with a cholinergic and/or noradrenergic lesion in attentional tasks. The effects of atipamezole on selective attention could not be tested in aged rats, because aged rats (18 month old) did not learn the five-choice serial reaction time task when they were tested in the same manner as adult rats (unpublished data).

As discussed above, the noradrenergic system, possibly in the amygdala, is involved in the processing of emotion-related information, e.g., the retention of the passive avoidance task. Aged rats are impaired in this task. Interestingly, a high dose of atipamezole (3.0 mg/kg) administered before the retention test improved this retention deficit in aged rats (Riekkinen Jr et al., 1992b). However, it is important to know whether atipamezole or other α_2 adrenoceptor antagonists improve this deficit when the drugs are administered before training, especially as a chronic treatment.

Conclusions

Pharmacological studies suggest interactions between the cholinergic and noradrenergic system in the regulation of thalamo-cortical oscillation (cortical high-voltage spindle activity), and in the acquisition of the spatial navigation task, which is critically dependent on normal hippocampal function. Interestingly, the

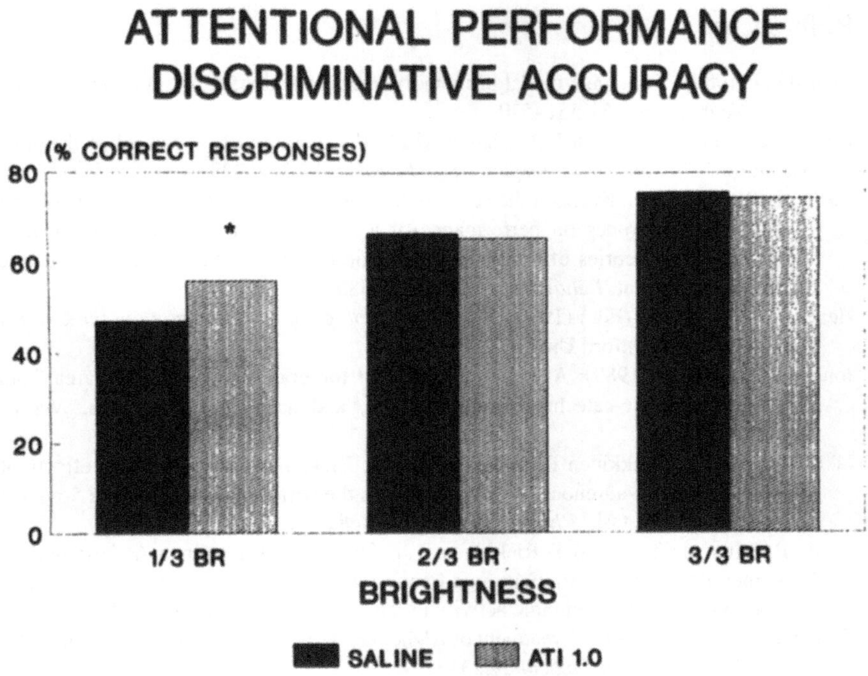

FIGURE 5. The effects of atipamezole on the choice accuracy of young rats tested in a five-choice serial reaction time task using different stimulus intensities. The reducing of stimulus intensity impairs the choice accuracy of rats. Atipamezole 1.0 mg/kg, s.c. (ATI 1.0) slightly improved choice accuracy (%correct responses) when tested using 1/3 stimulus intensity (⅓ BR) (MANOVA: $F(2,32) = 3.5$, $p < 0.05$). See Carli et al. (1983) for the original description of the task that is proposed to assess selective attention in rats.

effects of partial noradrenaline depletion on thalamo-cortical oscillation, and spatial navigation differs between young and aged rats.

An α_2-adrenergic antagonist when administered systemically stabilizes the age-related or the nucleus basalis-lesioned increase in the thalamo-cortical oscillation, which could facilitate information transfer in the thalamus. Furthermore, our electrophysiological data suggest that this treatment may facilitate information transfer in the hippocampus as well. At the behavioral level, a systemically administered α_2 antagonist has been found to improve the retention of aversively motivated task, accuracy performance in the attentional task, and the acquisition of the radial arm maze task.

Acknowledgments. This project has been financially supported by the Finnish Academy of Sciences, Medical Council, and the Ministry of Trade and Industry, Center for Technological Achievements.

References

Buzsáki G (1989): Two-stage model of memory trace formation: A role for "noisy" brain states. *Neuroscience* 31:551–570

Buzsáki G, Kennedy B, Solt VB, Ziegler M (1991): Noradrenergic control of thalamic oscillation: The role of α_2 receptors. *Eur J Neurosci* 3:222–229

Carli M, Robbins TW, Evenden JL, Everitt BJ (1983) Effects of lesions to ascending noradrenergic neurones on performance of a 5-choice serial reaction task in rats; implications for theories of dorsal noradrenergic bundle function based on selective attention and arousal. *Behav Brain Res* 9:361–380

Heal DJ, Marsden CA (Eds) (1990): *The Pharmacology of Noradrenaline in the Central Nervous System*, Oxford University Press

Jones BE, Cuello A (1989): Afferents to the basal forebrain cholinergic cell area from pontomesencephalic-catecholamine, serotonin, and acetylcholine-neurons. *Neuroscience* 31:37–61

Jäkälä P, Sirviö J, Riekkinen P Jr, Haapalinna A, Riekkinen P (1992a): The effects of atipamezole, an α_2-adrenoceptor antagonist, on the performance of rats in a 5-choice serial reaction time task. *Pharmacol Biochem Behav*, in press

Jäkälä P, Viitamaa T, Sirviö J, Riekkinen P Jr, Salonen J, Haapalinna A, Virtanen R, Riekkinen P (1992b): Continuous α_2-adrenoceptor blockade by atipamezole decreases neocortical high voltage spindle activity in rats. *Eur J Pharmacol*, revised

Kumar V, Calache M (1991): Treatment of Alzheimer's disease with cholinergic drugs. *Int J Clin Pharmacol Ther Toxicol* 29:23–37

MacDonald E, Scheinin H, Scheinin M (1988): Behavioural and neurochemical effects of medetomidine, a novel veterinary sedative. *Eur J Pharmacol* 158:119–127

McCormick DA (1989): Cholinergic and noradrenergic modulation of thalamocortical processing. *Trends Neurosci* 12:215–221

McGaugh JL, Introini-Collinson IB, Nagahara AH, Cahill L (1990): Involvement of the amygdaloid complex in neuromodulatory influences on memory storage. *Neurosci Biobehav Rev* 14:425–431

Mesulam MM (1990): Large-scale neurocognitive networks and distributed processing for attention, language, and memory. *Ann Neurol* 28:597–613

Reinikainen KJ, Soininen HS, Riekkinen PJ (1990): Neurotransmitter changes in Alzheimer's disease: Implications to diagnostics and therapy. *J Neurosci Res* 27:576–586

Riekkinen P, Buzsáki G, Riekkinen P Jr, Soininen H, Partanen J (1991): The cholinergic system and EEG slow waves. *EEG Clin Neurophysiol* 78:89–96

Riekkinen P Jr., Sirviö J, Jäkälä P, Lammintausta R, Riekkinen P (1990a): Effect of $alpha_2$ antagonists and an agonist on EEG slowing induced by scopolamine and lesion of the nucleus basalis. *Neuropharmacology* 29:993–999

Riekkinen P Jr., Sirviö J, Jäkälä P, Lammintausta R, Riekkinen P (1990b): Interaction between the $alpha_2$-noradrenergic and muscarinic systems in the regulation of neocortical high voltage spindles. *Brain Res Bull* 25:147–149

Riekkinen P Jr, Sirviö J, Valjakka A, Pitkänen A, Partanen J, Riekkinen P (1990c): The effects of concurrent manipulations of cholinergic and noradrenergic systems on neocortical EEG and spatial learning. *Behav Neural Biol* 54:204–210

Riekkinen P Jr, Sirviö J, Riekkinen P (1990d) Similar memory impairment found in medial septal-vertical diagonal band of Broca and nucleus basalis lesioned rats: Are memory defects induced by nucleus basalis lesions related to the degree of non-specific subcortical cell loss? *Behav Brain Res* 37:81–88

Riekkinen P Jr, Riekkinen M, Jäkälä P, Sirviö J, Lammintausta R, Riekkinen P (1991a): Combination of atipamezole and tetrahydroaminoacridine/pilocarpine treatment suppresses high voltage spindle activity in aged rats. *Brain Res Bull* 27:237–239

Riekkinen P Jr, Sirviö J, Valjakka A, Riekkinen M, Lammintausta R, Riekkinen P (1991b): Effects of atipamezole and tetrahydroaminoacridine on nucleus basalis lesion-induced EEG changes. *Brain Res Bull* 27:231–235

Riekkinen P Jr., Sirviö J, Jäkälä P, Riekkinen M, Lammintausta R, Riekkinen P (1991c): Effects of $alpha_2$-drugs and pilocarpine on the high-voltage spindle activity of young and aged control and DSP4-lesioned rats. *Physiol Behav* 50:955–959

Riekkinen P Jr., Riekkinen M, Valjakka A, Riekkinen P, Sirviö J (1992a): DSP-4, a noradrenergic neurotoxin, produce more severe biochemical and functional deficits in aged than young rats. *Brain Res* 570:293–299

Riekkinen P Jr., Sirviö J, Riekkinen M, Lammintausta R, Riekkinen P (1992b): Atipamezole, an alpha-2 antagonist, stabilizes age-related high voltage spindles and passive avoidance defects. *Pharmacol Biochem Behav*, 41:611–614

Riekkinen P Jr., Riekkinen M, Sirviö J, Riekkinen P (1992c): Neurophysiological consequences of combined cholinergic and noradrenergic lesions. *Exp Neurol*, 116:64–68

Savola J-M, Virtanen R (1991): Central α_2-adrenoceptors are highly stereoselective for dexmedetomidine, the dextro enantiomer of medetomidine. *Eur J Pharmacol* 195:193–199

Scheinin H, MacDonald E, Scheinin M (1988): Behavioural and neurochemical effects of atipamezole a novel α_2-adrenoceptor antagonist. *Eur J Pharmacol* 151:35–42

Selden NWR, Cole B, Everitt BJ, Robbins TW (1990a): Damage to coeruleo-cortical noradrenergic projections impairs locally cued, but enhances spatially cued water maze acquisition. *Behav Brain Res* 39:29–51

Selden NWR, Everitt BJ, Jarrard LE, Robbins TW (1990b): Complementary roles for the amygdala and hippocampus in aversive conditioning to explicit and contextual cues. *Neuroscience* 42:335–350

Selden NWR, Robbins RW, Everitt BJ (1990c): Dorsal noradrenergic bundle lesions impair conditioning to explicit but enhance conditioning to contextual aversive cues: Support for an attentional theory of central noradrenergic functions. *J Neurosci* 10:531–539

Sirviö J, Pitkänen A, Pääkkönen A, Partanen J, Riekkinen PJ (1989): Brain cholinergic enzymes and cortical EEG activity in young and old rats. *Comp Biochem Physiol* 94C:277–283

Sirviö J, Lukkarinen K, Riekkinen P Jr., Koivisto E, Virtanen R, Pennanen A, Valjakka A, Riekkinen PJ (1991a): The effects of atipamezole, an alpha-2 antagonist, on the performance of young and aged rats in the delayed nonmatching to position task. *Pharmacol Biochem Behav* 39:1015–1019

Sirviö J, Riekkinen P Jr, Vajanto I, Koivisto E, Riekkinen PJ (1991b): The effects of guanfacine, α-2 agonist, on the performance of young and aged rats in spatial navigation task. *Behav Neural Biol* 56:101–107

Sirviö J, Riekkinen P Jr, Valjakka A, Jolkkonen J, Riekkinen PJ (1991c): The effects of noradrenergic neurotoxin, DSP-4, on the performance of young and aged rats in spatial navigation task. *Brain Res* 563:297–302

Sirviö J, Riekkinen P Jr, Ekonsalo T, Lammintausta R, Riekkinen PJ (1992a): The effects of dexmedetomidine, an $alpha_2$ agonist, on learning and memory, assessed using passive avoidance and water maze tasks in rats. *Neuropharmacology* 31:163–168

Sirviö J, Riekkinen P Jr, MacDonald E, Airaksinen M, Lammintausta R, Riekkinen PJ (1992b): The effects of alpha-2 adrenoceptor antagonist, atipamezole, on spatial learning in scopolamine-treated and aged rats. *J Neural Transm* [P-P Sect] 4:99–106
Squire LR, Zola-Morgan S (1991): The medial temporal lobe memory system. *Science* 253:1380–1386
Steriade M, Buzsaki G (1990): Parallel activation of thalamic and cortical neurons by brainstem and basal forebrain cholinergic systems. In: *Brain Cholinergic Systems*, Steriade M, Biesold D, eds. New York: Oxford University Press
Teyler TJ (1991): Memory: Electrophysiological Analogs. In: Learning and Memory. *A Biological View*, 2nd ed, Martinez JL Jr, Kesner RP, eds. San Diego: Academic Press
Valjakka A, Riekkinen P Jr, Sirviö J, Nieminen S, Airaksinen M, Miettinen R, Riekkinen P (1990a): The effects of dorsal noradrenergic bundle lesions on spatial learning, locomotor activity, and reaction to novelty. *Behav Neural Biol* 54:323–329
Valjakka A, Sirviö J, Pitkänen A, Riekkinen PJ (1990b): Brain amines and neocortical EEG in young and aged rats. *Comp Biochem Physiol* 96C:299–304
Valjakka A, Koivisto E, Riekkinen P (1991a): Evoked field responses and awake immobility-related non-rhythmical EEG in hippocampus of young and aged rats. *Neurosci Res Commun* 8:29–35
Valjakka A, Lukkarinen K, Koivisto E, Lammintausta R, Airaksinen MM, Riekkinen P (1991b): Evoked field responses, recurrent inhibition, long-term potentiation and immobility-related nonrhythmical EEG in the dentate gyrus of fimbria-fornix-lesioned and control rats. *Brain Res Bull* 26:525–532
Valjakka A, Lukkarinen K, Koivisto E, Riekkinen P Jr., Miettinen P, Airaksinen MM, Lammintausta R, Riekkinen P (1991c): Modulation of EEG rhythmicity and spike activity in the rat hippocampus by systemically administered tetrahydroaminoacridine, scopolamine and atipamezole. *Brain Res Bull* 26:739–745
Virtanen R, Savola J-M, Saano V (1989): Highly selective and specific antagonism of central and peripheral α_2-adrenoceptors by atipamezole. *Arch Int Pharmacodyn Ther* 297:190–204

7

Septal Noradrenergic and Glutamatergic Influences on Hippocampal Cholinergic Activity in Relation to Spatial Learning and Memory in Mice

ROBERT JAFFARD, ALINE MARIGHETTO, AND JACQUES MICHEAU

The concept of central cholinergic intervention in learning and memory processes has been largely derived from pharmacological studies and has been reinforced by more recent research into dementing cognitive disorders (Bartus et al., 1982). Systemic pharmacological treatments, however, globally affect central cholinergic neurones so that little resolution can be gained by such approaches into the relative contribution of defined cholinergic pathways. Moreover, as pointed out by Decker and McGaugh (1991), recent research suggests that selective central cholinergic dysfunction cannot provide a complete account of the spectrum of age-related cognitive deficits and that age-related changes in cholinergic function typically occur within the context of changes taking place in several other neurotransmitter systems. Thus, together with the existence of different forms of memory involving different neural circuits and different neural mechanisms (Squire, 1987), the complex nature of interactions between central cholinergic pathways and other neuromodulatory systems has to be taken into consideration during different stages of learning and memory processes.

Nevertheless, if the postulate of a functional role of central cholinergic neurones in learning and memory is valid, then it should be possible to demonstrate that these neurones alter their activity during and/or after learning, and possibly also as a function of both the form of memory involved and the level of individual performance. Accordingly, our first aim was to provide a detailed performance-related description of changes in hippocampal high-affinity choline uptake (HACU) associated with training in an eight-arm radial maze. Our second aim was to attempt to identify the nature of the synaptic mechanisms that, at the septal level, might be involved in the control of these training-induced changes in septo-hippocampal cholinergic activity. Accordingly, the effects of intraseptal injections of noradrenergic and glutamatergic antagonists on the training-induced changes in hippocampal HACU were studied. Finally, previous behavioral pharmacological data from our group are examined and discussed with respect to both the behavioral specificity of the training-induced cholinergic alterations and the hypothesized mechanisms of control occurring at the septal level. Taken as a whole, the combination of such correlational and interventional approaches appears as a more optimal stategy to achieve a better understanding of the complex neurotransmitter interactions that underlie learning and memory processes. Furthermore, via an integration of data obtained from studies into the pathological states, these approaches may eventually provide pharmacological tools aimed at transsynaptically counteracting the adverse consequence of cholinergic hypofunction on learning and memory performance (Zornetzer, 1986; Sarter et al., 1988; Jaffard et al., 1989).

Description of the Effects of Training on Hippocampal High-Affinity Choline Uptake

In this study, a total of 164 adult male mice of the C57BL/6 strain were used. Animals were tested for their ability to acquire a spatial discrimination using three constantly reinforced arms of an automated eight-arm radial maze (mixed working-reference memory task) over nine successive days. On each daily session, mice were given six trials separated by a 1-min interval. Sodium-dependent high-affinity choline uptake (HACU) velocities in crude synaptosomal (P2) fractions of hippocampus (Atweh et al., 1975; Durkin et al., 1982) were systematically analyzed at two time points (30 sec and 15 min) following the first three and the last (ninth) session, and comparisons were made against quiet control mice.

The main results can be summarized as follows. As shown in Figure 1, each testing session induced an immediate (30 sec) increase in HACU, the magnitude of which did not significantly vary as learning progressed (first session: +31.1%; third: +29.1%; ninth: +25.0%). It is important to note, however, that animals trained for either two or eight sessions both displayed 24 hr later (i.e., before testing on days 3 and 9) lower hippocampal HACU than that in nontrained animals maintained in quiet conditions. Thus, in addition to the daily acute testing-induced increase in HACU in the hippocampus, it also appeared that even minimal training (i.e., 2 days) resulted in a persistent inhibition of this basal cholinergic activity. The time courses of the changes in HACU observed between 30 sec and 15 min following each session are shown in Figure 2. While the initial (30 sec) activation was maintained at an elevated level for at least 15 min after the second and third sessions, animals tested on the ninth session exhibited a rapid decrease of HACU reaching, at 15-min posttest, the same level as that observed before testing (i.e., lower HACU than for nontrained animals maintained in quiet conditions; see Figure 1). It is clear, however, that this shortening of the duration of posttraining cholinergic activation cannot be explained solely by the persistent inhibition previously mentioned, since disactivation was not observed at 15 min following the third session where a pretest reduction of HACU levels of the same magnitude as that observed on the ninth session was already present.

Since various behaviors have been reported to alter HACU (see Discussion), an experiment was designed to determine whether, and/or to what degree, the characteristics of the above-reported changes in cholinergic activity depended on the type of memory that was tapped. Accordingly, a group of mice was trained in

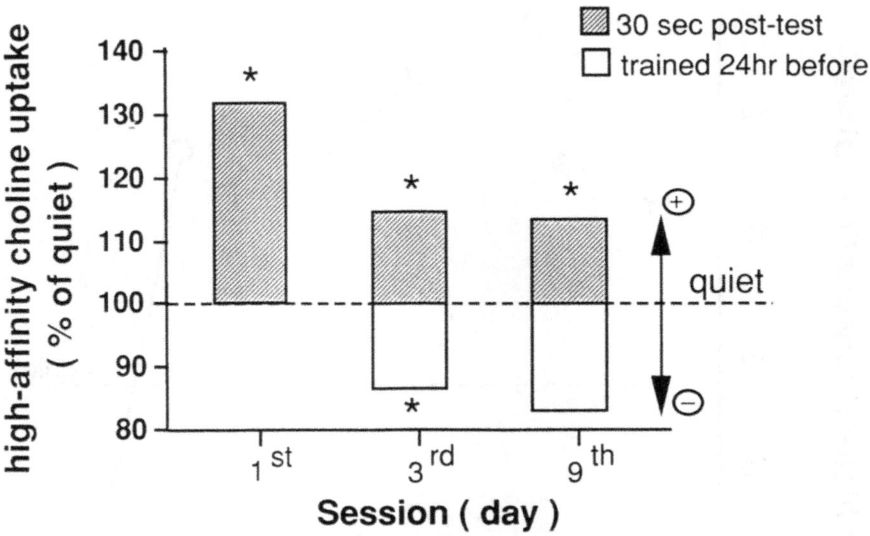

FIGURE 1. Bar diagram showing the amplitude of testing-induced increases in hippocampal high-affinity choline uptake (HACU) on the first, third, and ninth daily sessions of the mixed working-reference memory task. On the third and ninth sessions, the rate of HACU before testing (i.e., in mice trained for 2 and 8 days, respectively) was lower (−) than in quiet (untrained) controls but was increased above that basal level (+) as a result of testing. *Significantly different from quiet condition, $p < 0.05$.

the same radial maze as that previously used but on a specific working (delayed-nonmatching-to-place) memory task. The use of different list lengths (forced trials) followed by choice trials (recognition) allowed us to give animals a similar degree of training as in the previous conditions and to obtain identical levels of choice accuracy. As shown in Figure 3, one important difference between the effects of these two training procedures was that tested working memory did not induce the long-lasting (24 hr) inhibition of hippocampal HACU produced by the mixed working-reference memory task. It thus may be suggested that once the very first sessions of a set memory task have been given, a specific pattern of testing-induced hippocampal HACU changes emerges with respect to the type of memory that is tapped. Accordingly, the long-term decrease in HACU produced by spatial discrimination training could be selectively associated to the use of reference memory.

Correlations Between Training-Induced Changes in Hippocampal HACU and Performance

Having established the main characteristics of the changes in hippocampal activity associated with different phases of radial-maze discrimination training, we were

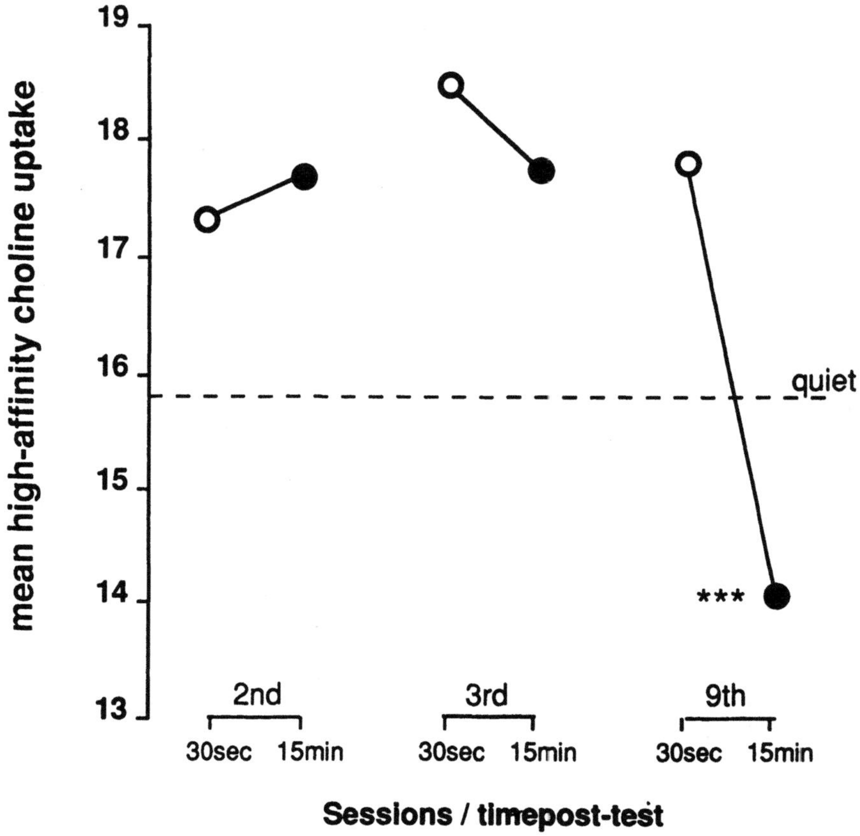

FIGURE 2. The time courses of testing-induced changes in hippocampal high-affinity choline uptake between 30 sec and 15 min posttest as a function of the degree of training (i.e., second, third, and ninth daily sessions) in the mixed working-reference memory task. ***Significantly different from the second and third sessions at 15 min, $p < 0.001$.

then able to examine whether any functional link exists between performance levels and the observed changes in HACU. A close examination of individual learning curves first showed that animals could be divided into two groups on the basis of their between vs. within-session progression of discrimination performance over the 9 days of training. In the first group, the individual mean decrease in reference memory errors was predominantly observed within each daily session (i.e., between the first three and last three trials of the same session), while in the second group this decrease was mainly observed between successive 24-hr-spaced daily sessions (i.e., in this group, 72% of the net improvement of acquisition performance was observed between the last three trials of each session and the first three trials of the following one). Linear regression analysis revealed a negative correlation between HACU velocities measured immediately (30 sec) after the

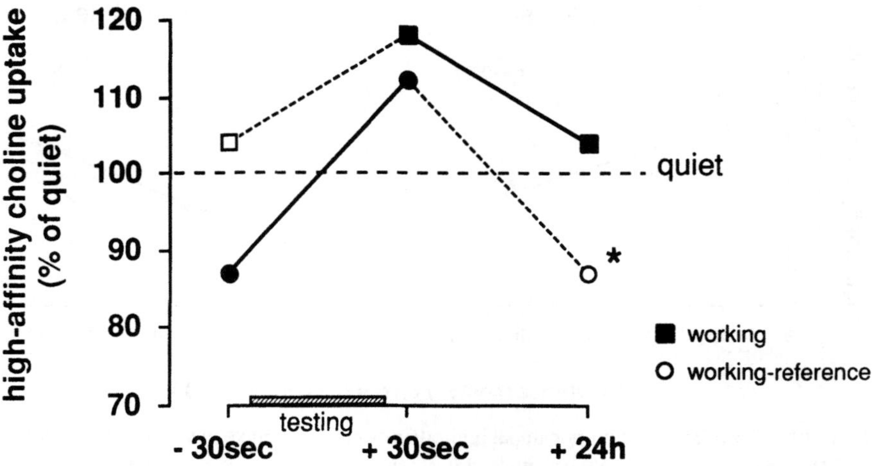

FIGURE 3. A comparison of the effects of specific working and mixed working-reference memory testing on changes in hippocampal high-affinity choline uptake (HACU). In both conditions animals were trained until they displayed comparable levels of performance. HACU was measured both before (working-reference) and either 30 sec (both cases) or 24 hr (working) following a test session for which, in each condition, the number of trials and duration were approximately the same. *Significantly different from working, $p < 0.05$.

ninth session and the within-session improvement of discrimination scores over the 9 days of training ($r = -0.71$; $p < 0.01$), while a positive correlation was found for between-session improvement ($r = +0.67$; $p < 0.02$). As the pattern of progression of each individual subject did not substantially change as training progressed, it may be suggested that reduced hippocampal HACU during the training session (as can be assumed from measures at 30-sec posttest), would facilitate acquisition over the six trials whilst, inversely, impairing subsequent 24-hr delayed retention test performance and vice versa. The hypothesis that response execution (within-session improvement) and memory storage (between-session improvement) functions are initiated by two reciprocal and inhibitory systems has already been formulated. Thus, in line with the present findings, it has been suggested that the primary indicator of which of these two functions predominates was the cholinergic-dependent hippocampal rhythmic slow activity (Routtenberg, 1971) and/or hippocampal choline acetyltransferase activity (Jaffard et al., 1985).

Animals that displayed between- rather than within-session improvement of performance learned the task faster, thereby exhibiting significantly less reference memory (R.M.) errors over the nine sessions compared to those that displayed within-session improvement (slow learners). Figure 4 shows the relationship between the number of these R.M. errors and hippocampal HACU measured either 30 sec or 15 min after the last (ninth) training session. As could be expected

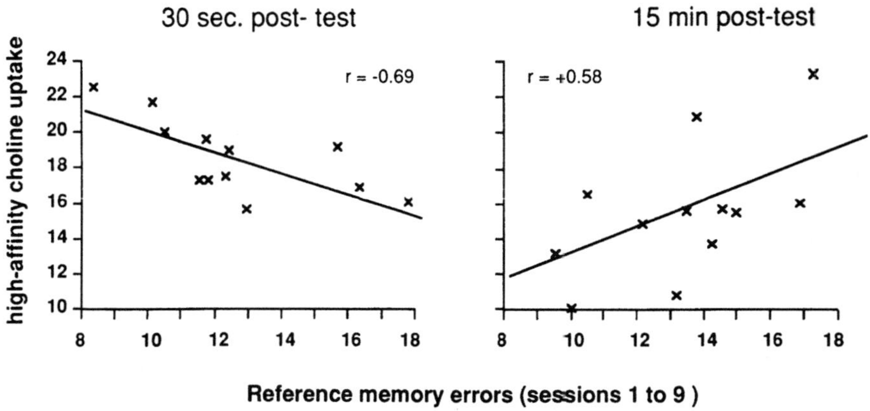

FIGURE 4. Correlations of hippocampal high-affinity choline uptake measured 30 sec (left) and 15 min (right) following the ninth testing session with pooled reference memory errors recorded in all (first to ninth) daily sessions. 30 sec: $r = -0.69$; $p < 0.02$; 15 min: $r = +0.58$; $p < 0.05$.

from the previous analysis, a significant *negative* relationship was found between R.M. errors and hippocampal HACU at 30-sec posttest ($r = .069$; $p < 0.02$), while at 15 min an inverse, that is, a significant *positive* relationship was found between these two measures ($r = +0.58$; $p < 0.05$). Taken as a whole, these results indicate that (a) animals that display between- rather than within-session improvement of discrimination performance learn faster than those with the inverse pattern; (b) the amplitudes of both the initial increase (30 sec) and subsequent decrease (15 min) of hippocampal HACU induced by training are larger in fast than in slow learners.

An Attempt to Identify Some Septal Transsynaptic Inputs Involved in the Training-Induced Changes in Hippocampal HACU

Previous pharmacological experiments using intraseptal injections of various agonists and antagonists of several neurotransmitters have provided evidence for the existence of a complex transsynaptic control of the cholinergic septo-hippocampal neurones. These studies were able to identify the net excitatory or inhibitory action of a number of identified neurotransmitters and also to determine whether their effect was phasic or tonic (see Costa et al., 1983 for a review). However, the mechanisms involved still remain to be clarified. Consequently, the following experiments were aimed at studying the effects of intraseptal injection of drugs that interfere with noradrenergic and glutamatergic neurotransmission on hippocampal HACU in both resting and trained animals. Our purpose was to determine whether, and if so how and when, these transsynaptic mechanisms may be involved in the observed training-induced changes in HACU.

Noradrenaline

Although both widespread norepinephrine (NE) depletion induced by neurotoxins and systemic administration of noradrenergic antagonists have been found to have little or no effect on several learning tasks, there is some evidence that intracerebral pharmacological manipulations of NE function can affect learning and memory processes (for a review, see Decker and McGaugh, 1991). Moreover, among the phasically active excitatory afferents to the septum (Costa et al., 1983; Cheney, 1984) that could mediate the observed daily acute testing-induced increase in hippocampal HACU (see Fig. 1), noradrenaline appeared as one major candidate in view of its implication in both learning and memory, and/or related functions (Harley, 1987; Quatermain, 1983; Robbins and Everitt, 1985; Zornetzer, 1986), and of previous data concerning its functional synergistic interaction with cholinergic neurones at the septal level (Decker and McGaugh, 1991). In a series of experiments, we have thus studied the effects of intraseptal injections (0.2 μl) of α-noradrenergic antagonists on hippocampal HACU measured in both quiet animals and in mice that were submitted to the working-reference memory test. On a whole, our results were much more complex than would be predicted from simple models; therefore, only the most significant ones will be presented here.

As shown in Figure 5, bilateral intraseptal injection of both the α-nonspecific receptor antagonist, phenoxybenzamine, and the α_2-specific antagonist, yohimbine, produced similar dose-related changes in hippocampal HACU measured in quiet animals. At the lower doses used (i.e., 100 ng and 0.5 ng, respectively), both drugs induced a slight nonsignificant decrease in HACU, while at higher doses (500 or 1000 and 50 ng, respectively), an opposite tendency towards an increase was observed. In contrast, intraseptal injections of the α_1-specific antagonist BE2254 at 100 and 500 ng produced no significant changes in HACU. This last result was of particular interest in view of the effects of this drug on the immediate (30 sec) training-induced increase in hippocampal HACU shown in Figure 6. Indeed, the intraseptal injection of BE2254 produced a dose-dependent specific reduction of the hippocampal HACU activation induced by testing, while having no effect on the basal level of this cholinergic activity. Taken together, these results suggest that a large part of the septal phasic excitatory input that is responsible for the observed testing-induced activation of hippocampal HACU would be mediated by α_1-noradrenergic receptors and that these receptors are not tonically involved in the control of basal cholinergic activity (i.e., in quiet conditions). Results with yohimbine (and to a lesser extent with phenoxybenzamine) might indicate that α_2-receptors could be weakly involved in this tonic inhibitory control if one assumes that the lower doses we have used mainly act presynaptically (see Fig. 5). At the higher doses used, however, both of these two drugs also significantly attenuated the testing-induced increase in HACU.

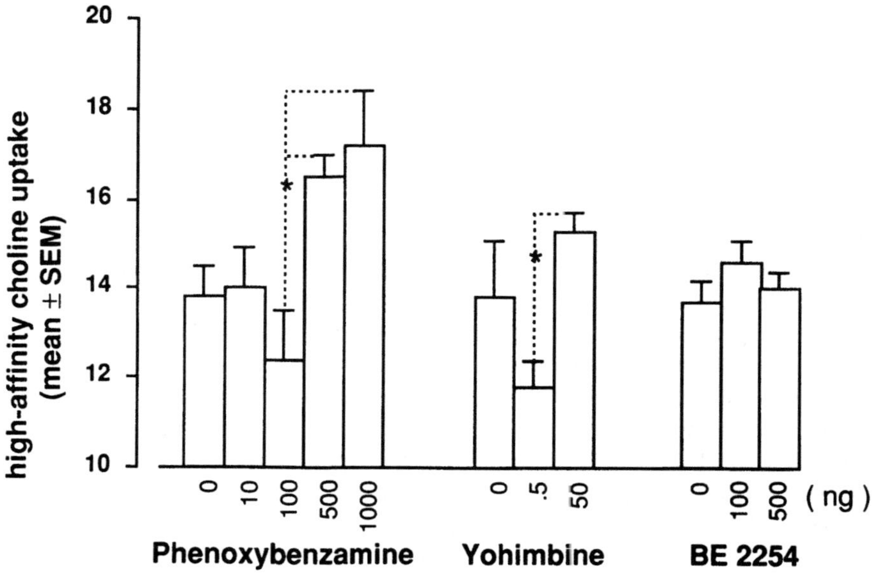

FIGURE 5. Effects of intraseptal injections (0.2 μl bilaterally) of different doses of α-noradrenergic antagonists on hippocampal high-affinity choline uptake measured 40 min after treatments in quiet control conditions. Both phenoxybenzamine (α nonspecific) and yohimbine (α_2 specific) produced a slight decrease followed by an increase in HACU as a function of increasing doses (*$p < 0.05$).

Glutamate

The lateral septum receives a dense innervation from the hippocampal formation via the fimbria fibers (Leranth and Frotscher, 1989). Recent experiments have confirmed that the projection is mediated by the excitatory amino acid glutamate (Joels and Urban, 1984; Stevens and Cotman, 1986) and that high-frequency fimbrial stimulation induced a long-term synaptic enhancement (LTE) in the lateral septum (Racine et al., 1983), which is thought to be N-Methyl-D-Aspartate (NMDA)-dependent (Van Den Hooff et al., 1989). In view of the posttraining maintenance of the daily testing-induced increase in hippocampal HACU observed at the beginning of acquisition (i.e., days 2 and 3; see Fig. 2), we speculated that once the putative initial NE-dependent cholinergic activation had occurred, it could subsequently be maintained by a septo-hippocampal-septal cholinergic-glutamatergic-cholinergic positive feedback loop involving an increase in the synaptic efficacity of septal glutamatergic synapses. Such a hypothesis implies that, as suggested by previous data (Costa et al., 1983; Cheney, 1984), septal glutamatergic terminals did mediate an excitatory effect on medial septal/diagonal band cholinergic neurons. Data from our experiments using intraseptal injections of glutamatergic agonists and antagonists suggest a different picture, as illustrated by the following results. When administered 20 min before either the first or the

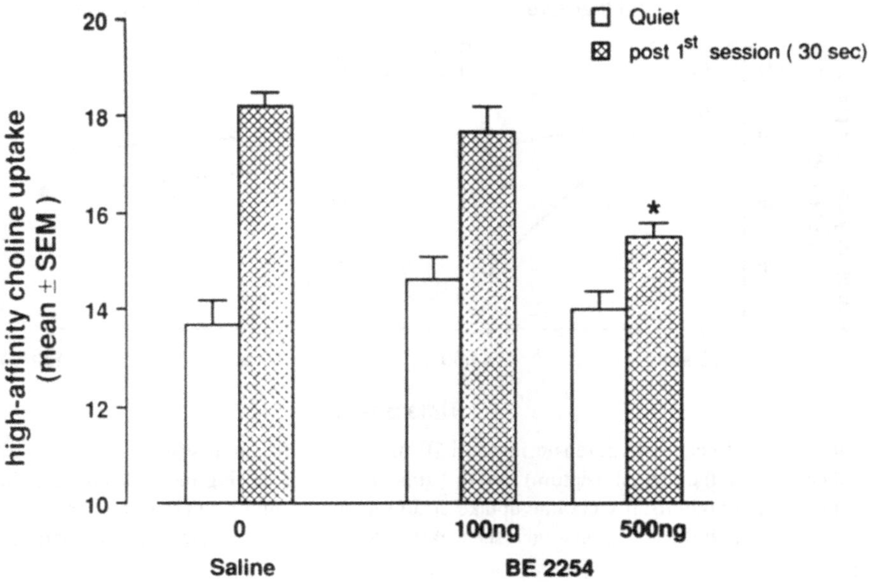

FIGURE 6. Hippocampal high-affinity choline uptake in saline and BE2254-injected groups (0.2 μl bilaterally in the lateral septum) in quiet conditions (open bars) and 30 sec following the first training session in the radial-maze (shaded bars). *Significantly different from saline, $p < 0.05$.

ninth training session, kynurenate (5 ng bilaterally), did not produce significant changes on the immediate (30 sec) testing-induced increase in hippocampal HACU. In contrast, experimental subjects sacrificed 15 min after the ninth session exhibited higher, although nonsignificant, rates of HACU as compared to controls (+8.3%). However, since fast learners but not slow learners had been shown to exhibit a substantial decrease of HACU between 30 sec and 15 min posttest (see previous section), we examined the effects of kynurenate on each subject of each of these subgroups. As shown in Figure 7, kynurenate prevents the posttraining 15-min delayed disactivation of hippocampal HACU in fast learners (treatment × delay interaction; $p < 0.05$) but has no significant effect in slow learners (treatment × delay × group; $p < 0.02$). These results would suggest that the posttraining (15 min) *disactivation* of HACU that characterizes fast learners is controlled by glutamatergic synapses in the lateral septum. This conclusion stands in apparent contrast to previous data suggesting that septal glutamatergic afferents mediate a net *excitatory* control of septo-hippocampal cholinergic neurons [according to Cheney (1984) this indirect excitatory control would require the existence of two distinct sets of GABA interneurons linked in series]. Moreover, in additional experiments performed in resting animals, we were unable to confirm the previous observation (Durkin et al., 1989) in favor of such an excitatory control. Indeed, using different but relatively low doses of glutamate, the only

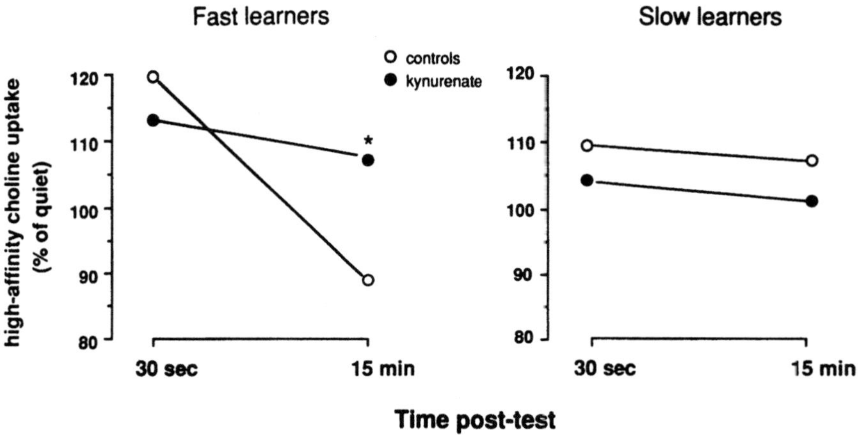

FIGURE 7. Effects of kynurenate injected 20 min before the ninth session (5 ng/0.2 μl bilaterally in the lateral septum) on the time course of testing-induced changes in hippocampal high-affinity choline uptake at 30 sec and 15 min posttest in fast and slow learners (for further explanations see text). *Significantly different from controls, $p < 0.05$.

significant effect obtained was actually a decrease in HACU (−14% at 2.5 ng); similarly, the NMDA receptor antagonist AP5 was found to produce only a slight but significant increase in HACU (+12% at both 100 ng and 1 μg) in resting animals. Consequently, and given the potentially important functional role of this septal glutamatergic/cholinergic interaction, it is evident that further pharmacological and neuroanatomical (Jakab and Leranth, 1990) experiments are required in order to acquire a better understanding of this transsynaptic interaction.

Are the Observed Training-Induced Changes in Hippocampal HACU Causally Related to Learning?

As pointed out in the introduction, our aim was to obtain a detailed description of hippocampal HACU alterations associated with spatial discrimination learning and, through a correlative approach, to attempt to provide some evidence for a functional link between these changes and performance. The correlations we have observed do not necessarily mean that the learning and memory processes engaged are causally related to these cholinergic alterations. Thus, having determined which septal neurotransmitter systems (among others) might be involved in the transsynaptic control of these training-induced HACU changes, it becomes feasible to assess their behavioral specificity, not only with respect to cholinergic changes, but also in relation to their hypothesized transsynaptic septal mechanisms of control. Together with pharmacological and biochemical investigations, behavioral pharmacological experiments are currently in progress. However, although not strictly related to the analysis and data presented in previous sections, some previous results from our group seem to weigh in favor of part of the above hypothesis. Concerning noradrenaline, it was shown that acute intraseptal injection of phenoxybenzamine (500 ng, bilaterally) interferes with spatial working memory in a delayed nonmatching-to-place task (Marighetto et al., 1989). Interestingly, this effect was difficulty-dependent, suggesting thereby that memory, rather than performance, would depend on this septal NE-cholinergic interaction. Given 30 min *before* the *fourth* daily session of the mixed working-reference learning task (see previous sections), phenoxybenzamine was also found to induce an increase in reference memory errors on the first three but not the last three trials of this (fourth) session (unpublished observations). This would support the idea that acute noradrenergic hypofunction interferes with retrieval processes (but see Sara, 1985) and is in accordance with the observation that slow learners, who would naturally exhibit a low level of phasic septal NE release, are characterized by within- rather than between-session improvement of discrimination performance. Concerning glutamate, we have shown that intraseptal injection of the NMDA receptor blocker MK801, given immediately *after* the second session of a spatial concurrent discrimination learning task, impaired performance on the subsequent 24-hr delayed (third) session (Durkin et al., 1989). Together with our previous data, this last result suggests that glutamatergic synapses of the lateral septum might be functionally involved in the process of memory consolidation. It seems, however, difficult in this context to reconcile the suggestion that the glutamatergic antagonist would act via a *disinhibition* of cholinergic activity (see previous section), with the observation that posttraining *muscarinic blockade* by the systemic injection of scopolamine produced, in the same task, the *same* retention impairment as MK801 (unpublished observations).

Discussion

As shown in the present experiment and previous experiments (Toumane et al., 1988) using the eight-arm radial maze, a *first exposure* to several behavioral situations, such as place training in a water tank (Decker et al., 1988), bar-press conditioning (Micheau et al., 1986), avoidance conditioning (Raaijmakers, 1982; Wenk et al., 1984), but also free exploration and/or locomotor activity (Beracochea et al., 1992), immobilization restraint (Lai et al., 1986) has been demonstrated to produce alterations in hippocampal HACU. Without considering that the directions, amplitudes, and kinetics of these changes are not related to some specific factors (especially stress), it seems quite clear that, at this stage, they are *a priori* unrelated to the specific response the animal has eventually to learn. In line with this idea, Decker et al. (1988) showed that, as compared to naive (quiet) animals, yoked and place-trained rats exhibited a similar decrease in hippocampal HACU 15 min following the first training session in a water tank (day 1), whereas following the fourth session (day 4), only place-trained subjects displayed lower HACU than the naive ones. Accordingly, these authors have suggested that nonspecific factors may largely account for HACU changes observed on day 1 but that these changes are specifically related to place-training acquisition on day 4. In some agreement with this interpretation we have observed that HACU measured 15 min following the first (day 1) training session was highly negatively correlated with the time needed to achieve the first trial ($r = -0.89$), probably reflecting anxiety; subsequently, as acquisition progressed changes in HACU evolved to depend more on both the form of memory that was tapped (i.e., working vs. reference) and on performance (i.e., fast vs. slow learners). This evidently does not mean that the immediate (30 sec) testing-induced activation of HACU observed following the first session (or that is observed following the first exposure to several situations) would not reflect the more or less preexisting capabilities of septo-hippocampal cholinergic neurones of a given subject to be activated by behavior. Indeed, both aging and chronic ethanol consumption have been reported to significantly reduce this immediate behavioral-induced HACU activation (Jaffard et al., 1989; Béracochea et al., 1992). In the context of the above-reported experiments, the question arises as to whether the HACU increase observed immediately after the first training session might be predictive both of subsequent training-related HACU changes and performance with respect to the specific task requirements. Thus one may speculate that the learning and memory impairments associated with either aging or alcohol intoxication are related to the diminished capabilities of cholinergic neurones to be activated by behavior. However, the problem still remains as to whether and how these preexisting capabilities, together with the task requirements, may (or not) give rise to a specific (task-related) pattern of hippocampal HACU alterations. The methodology we have used does not enable us to provide direct evidence for a link between the magnitude of the immediate testing-induced increase in hippocampal cholinergic presynaptic activity on day 1 and subsequent performance. However, given both the positive (30 sec) and negative (15 min) correlations between HACU measures performed following the ninth session and performances on acquisition, one would suspect the immediate (30 sec), and apparently constant, posttest HACU activation to be somehow linked to its subsequent disactivation (i.e., at 15 min post test). Data from pharmacological experiments reported earlier suggest that the immediate cholinergic activation could be mediated by a phasic septal noradrenergic input acting through α_1-NE receptors. Our present results also suggest that septal glutamatergic synapses could be actively involved in the disactivation observed at 15 min, even though it is not clear to what extent, or indeed whether, this phenomenon is also related to the long-term acquisition-dependent inhibition of HACU observed 24 hr following the last testing session. Whatever is the case, it seems possible to speculate that, as shown for hippocampal glutamatergic synapses (Stanton and Sarvey, 1986), the septal NE input might facilitate the induction of long-term synaptic enhancement of these septal glutamatergic afferents. At this point, it must be mentioned that the lack of effect of intraseptal kynurenate on the immediate training-induced increase in HACU (day 9) could possibly be explained by the inhibitory effect of NE release on glutamatergic neurotransmission. Indeed, it has been reported that iontophoretic application of NE depressed the amplitude of a fimbria-evoked field potential in the lateral septum (Marchand and Hagino, 1983).

The observation that daily working memory testing did not produce the long-term rebound reduction in HACU observed following the mixed working-reference memory test suggest that the pattern of these training-induced changes might be related to the form of memory that is involved. Such a possibility has been discussed by Decker et al. (1988) with regards to both the authors' findings showing that the acquisition of spatial reference memory resulted in a short-term delayed decrease in hippocampal HACU and those of Wenk et al. (1984), showing that intensive training on a working memory task actually produced an opposite pattern, that is, a long-term increase (i.e., for up to 20 days) in the velocities of this marker. Taken as a whole, these findings are congruent with our own data and also with the idea that, contrary to the information held in working memory, information stored in reference memory remains permanently valid and, as such, may (and must) be consolidated. The observation that posttraining intraseptal injection of MK801 produced retrograde amnesia suggests that septal glutamatergic synapses could play a role in consolidation, and possibly via the NMDA-dependent long-term synaptic enhancement described previously (Racine et al., 1983; Van Den Hooff et al., 1989).

Finally, since we observed that fast learner mice also exhibited less working (repetitive) memory errors in the mixed working-reference memory task, this provides additional (although still indirect) arguments for the hypothesis that septal NE synapses interacting with cholinergic neurones would play a pivotal role in both working (directly) and reference (as a priming mechanism) memories. It may be also noted that, since in both tasks spatial mapping (O'Keefe and Nadel, 1978; Nadel, 1991) and/or configural association processes (Sutherland and Rudy, 1989) are certainly involved, these processes could thus also be considered as the prime function linked to this septal noradrenergic/cholinergic interaction.

Acknowledgments. This research was part of the doctoral dissertation of A.M. and was supported by the "Centre National de la Recherche Scientifique" and the "Fondation pour la Recherche Médicale". A.M. holds a fellowship from the "Ministère de la Recherche et de la Technologie".

References

Atweh SF, Simon JR, Kuhar MJ (1975): Utilization of sodium-dependent high affinity choline uptake *in vitro* as measure of the activity of cholinergic neurons *in vivo*. *Life Sci* 17:1535–1544

Bartus RT, Dean RL, Beer B, Lippa AS (1982): The cholinergic hypothesis of geriatric memory dysfunction. *Science* 217:408–417

Beracochea D, Micheau J, Jaffard R (1992): Memory deficits following chronic alcohol consumption in mice: relationships with hippocampal and cortical cholinergic activities. *Pharmacol Biochem Behav* June 1992, 42:

Cheney DL (1984): Drug effects on transmitter dynamics: An overview. In: *Dynamics of Neurotransmitter Function*, I. Hanin, ed. New York: Raven Press

Costa E, Panula P, Thompson HK, Cheney DL (1983): The transsynaptic regulation of the septal hippocampal cholinergic neurons. *Life Sci* 32:165–179

Decker MW, McGaugh JL (1991): The role of interactions between the cholinergic system and other neuromodulatory systems in learning and memory. *Synapse* 7:151–168

Decker MW, Pelleymounter MA, Gallagher M (1988): Effects of training on a spatial memory task on high affinity choline uptake in hippocampus and cortex in young adult and aged rats. *J Neurosci* 8:90–99

Durkin TP, Hashem-Zadeh H, Mandel P, Ebel A (1982): A comparative study of the acute effects of ethanol on the cholinergic system in hippocampus and striatum of inbred mouse strains. *J Pharmacol Exp Ther* 220:203–208

Durkin T, Marighetto A, Lebrun C, Toumane A, Jaffard R (1989): Investigations into the role of septal-noradrenergic and NMDA receptors in the induction and maintenance of septo-hippocampal cholinergic activation induced by spatial memory testing in the mouse. *Soc Neurosci Abst* 15:80

Harley CW (1987): A role for norepinephrine in arousal, emotion and learning: Limbic modulation by norepinephrine and the Kety hypothesis. *Prog Neuro-Psychopharmacol Biol Psychiat* 11:419–458

Jaffard R, Galey D, Micheau J, Durkin T (1985): The cholinergic septo-hippocampal pathway, learning and memory. In: *Brain Plasticity Learning and Memory*, Will BE, Schmitt P, Dalrymple-Alford JC, eds. New York: Plenum Publishing

Jaffard R, Durkin T, Toumane A, Marighetto A, Lebrun C (1989): Experimental dissociation of memory systems in mice: Behavioral and neurochemical aspects. *Arch Gerontol Geriatr* S1:55–70

Jakab RL, Leranth C (1990): Catecholaminergic, GABAergic, and hippocamposeptal innervation of GABAergic "somatospiny" neurons in the rat septal area. *J Comp Neurol* 302:305–321

Joels M, Urban IJA (1984): Electrophysiological and pharmacological evidence in favor of aminoacid neurotransmission in fimbria-fornix fibers innervating the lateral septal complex of rats *Exp Brain Res* 54:455–462

Lai H, Zabawska J, Horita A (1986): Sodium-dependent high-affinity choline uptake in hippocampus and frontal cortex of the rat affected by acute restraint stress. *Brain Res* 372:366–369

Leranth C, Frotscher M (1989): Organization of the septal region in the rat brain: cholinergic-GABAergic interconnections and the termination of hippocampo-septal fibers. *J Comp Neurol* 289:304–314

Marchand JE, Hagino N (1983): Effects of iontophoretically applied norepinephrine and dopamine on fimbria-evoked activity in the lateral septum. *Exp Neurol* 82:683–697

Marighetto A, Durkin T, Toumane A, Lebrun C, Jaffard R (1989): Septal-noradrenergic antagonism in vivo blocks the testing-induced activation of septo-hippocampal cholinergic neurones and produces a concomitant deficit in working memory performance. *Pharmacol Biochem Behav* 34:553–558

Micheau J, Durkin T, Galey D, Destrade C, Rolland Y, Jaffard R (1986): Bar-press conditioning induced a specific increase in hippocampal high-affinity choline uptake: Effects of facilitative and disruptive pharmacological treatments. *Colloque Français des Neuroscience* (abstract in French): C58

Nadel L (1991): The hippocampus and space revisited. *Hippocampus* 1:221–229

O'Keefe J, Nadel L (1978): *The Hippocampus as a Cognitive Map*. Oxford: Clarendon

Quatermain D (1983): The role of catecholamines in memory processing. In: *Physiological Basis of Memory*, Deutsch JA, ed. New York: Academic Press

Raaijmakers WGM (1982): High affinity choline uptake in hippocampal synaptosomes and learning in the rat. In: *Neuronal Plasticity and Memory Formation*, Ajmone-Marsan C, Matthies H, eds New York: Raven Press

Racine RJ, Milgram NW, Hafner S (1983): Long-term potentiation in the rat limbic forebrain. *Brain Res* 260:217–231

Robbins TW, Everitt HB (1985): Noradrenaline and selective attention. In: *Brain Plasticity Learning and Memory*, Will BE, Schmitt P, Dalrymple-Alford JC, eds. New York: Plenum Publishing

Routtenberg A (1971): Stimulus processing and response execution: A neurobehavioral theory. *Physiol Behav* 6:589–596

Sara SJ (1985): Noradrenergic modulation of selective attention: Its role in memory retrieval. *Ann NY Acad Sci* 444:178–193

Sarter M, Schneider HH, Stephens DN (1988): Treatments strategies for senile dementia. *Trends Neurosci* 11:13–17

Squire LR (1987): *Memory and Brain* Oxford: Oxford University Press

Stanton PK, Sarvey JM (1987): Norepinephrine regulates long-term potentiation of both the population spike and dendritic EPSP in hippocampal dentate gyrus. *Brain Res Bull* 18:115–119

Stevens DR, Cotman C (1986): Excitatory amino-acid antagonists depress transmission in hippocampal projection to the lateral septum. *Brain Res* 382:437–440

Sutherland RJ, Rudy JW (1989): Configural association theory: The role of the hippocampal formation in learning, memory, and amnesia. *Psychobiology* 17:129–144

Toumane A, Durkin T, Marighetto A, Galey D, Jaffard R (1988): Differential hippocampal and cortical cholinergic activation during the acquisition, retention, reversal and extinction of a spatial discrimination in an 8-arm radial maze by mice. *Behav Brain Res* 30:225–234

Van Den Hooff P, Urban IJA, De Wied D (1989): Vasopressin maintains long-term potentiation in rat lateral septum slices. *Brain Res* 505:181–186

Wenk G, Hepler D, Olton D (1984): Behavior alters the uptake of (3H) choline into acetylcholinergic neurons of the nucleus basalis magnocellularis and medial septal area. *Behav Brain Res* 13:129–138

Zornetzer SF (1986): The noradrenergic locus coeruleus and senescent memory dysfunction. In: *Treatment Development Strategies for Alzheimer's Disease*. Madison, CT: Mark Powley Associates

8

Neurotransmitter Interactions and Responsivity to Cholinomimetic Agents

V. HAROUTUNIAN, A.C. SANTUCCI, AND K.L. DAVIS

Introduction

There is now overwhelming evidence for the involvement of forebrain cholinergic systems in Alzheimer's disease and the cognitive processes that subserve learning and memory (Perry et al., 1978; Bigl et al., 1987; Koshimura et al., 1987; Doucette et al., 1986; Ichimiya et al., 1986; Saper et al., 1985; Whitehouse, 1986; McGeer et al., 1984; Mann et al., 1986; Giacobini, 1990; Wenk et al., 1987). Support for the role of forebrain cholinergic systems in learning and memory can be gleaned from innumerable studies in which perturbation of forebrain cholinergic systems, whether through pharmacological means or lesion of the nucleus basalis of Meynert (nbM), has been demonstrated to profoundly impair learning and memory. This literature has been amply reviewed in recent articles and numerous chapters in this book (Wenk and Olton, 1987; Gold and Zornetzer, 1983; Dekker et al., 1991) and will not be reiterated here. In general, however, lesions of the nbM have been shown to impair performance on a very large variety of tasks and in a large and varied number of mammalian species. A smaller, but nevertheless voluminous, literature also attests to the ability of cholinomimetic agents such as physostigmine, oxotremorine, pilocarpine, etc. to at least partially reverse the learning and memory deficits induced by nbM lesions (Haroutunian et al., 1990c; Mandel et al., 1989; Fine et al., 1985; Bhat et al., 1990; Haroutunian et al., 1989b, 1990; Murray and Fibiger, 1985). The results of clinical studies of the effects of cholinomimetics and cholinesterase inhibitors on learning and memory in Alzheimer's disease (AD) patients have been more variable (Beller et al., 1985; Summers et al., 1981; Haroutunian et al., 1990c,1991). When significant improvements in learning and memory have been observed, these improvements have not been of the same order of magnitude as those observed in studies of animals with forebrain cholinergic lesions. A variety of practical, conceptual, and theoretical reasons can account for this discrepancy. For example, it has been especially difficult to achieve high levels of central cholinesterase inhibition in elderly patients without significant deleterious peripheral effects and side effects (Pomponi et al., 1990). In a similar vein, pharmacokinetics issues of absorption, central nervous system penetrance, and plasma half-life have significantly hampered treatment strategies based on cholinomimetic drugs. An additional variable has been the use of cholinesterase inhibitors within the context of a cholinergic degenerative process where the efficacy of the drug is dependent on the presence of substrate (i.e., acetylcholinesterase) and the presence of adequate cholinergic activity for the inhibition of catabolic enzymes to have an effect.

A second possible reason for the lack of consistent and pronounced beneficial effects of cholinomimetics in AD is the presence of other, noncholinergic lesions in this disease. The Alzheimer's disease literature is replete with studies demonstrating dramatic deficits in noradrenergic, serotonergic, somatostatinergic, and a host of other systems (Rossor et al., 1980; Davies and Terry, 1981; Davies et al., 1980; Beal et al., 1987; Perry et al., 1981; Cross et al., 1983; Adolfsson et al., 1979; Gottfries et al., 1989a,b; Mann et al., 1980, 1983; Palmer et al., 1987a,b; Yates et al., 1983; Shimohama et al., 1986; Bondareff et al., 1982; Bowen et al., 1983; Wilcock et al., 1988; Sparks, 1989; Sparks et al., 1988; Reinikainen et al., 1990). Thus, one of the reasons for the lack of correspondence between cholinomimetic treatment effects in lesioned animals and AD patients may be the failure of the animal model studies to adequately model some of these noncholinergic lesions present in AD. The studies outlined in this chapter constitute a series of experiments that have aimed to address this potential source of variance by studying the degree to which selected neurotransmitter systems interact with the forebrain cholinergic system to affect changes in responsivity to cholinomimetics. We will discuss the effects of combined noradrenergic and cholinergic lesions, somatostatinergic and cholinergic lesions, and serotonergic and cholinergic lesions.

Cholinergic/Noradrenergic Interactions

It is now apparent that a significant proportion of AD patients suffer considerable cortical noradrenergic depletion and decreased NE turnover (Cross et al., 1983; Forno et al., 1989; Adolfsson et al., 1979; Goedert et al., 1986; Perry et al., 1981; Bondareff et al., 1982; Yates et al., 1983; Marcyniuk et al., 1989; Mann et al., 1985, 1986; Arai et al., 1984; Ichimiya et al., 1986). These results, which have been replicated many times, provide a strong basis for the belief that noradrenergic deficits contribute to the pathophysiology of AD in at least a significant subpopulation of AD patients. The evidence that noradrenergic dysfunction influences the cognitive deficits of AD is less well established and is somewhat contradictory. Significant correlations between brain NE markers and performance on cognitive tests have been reported in AD patients. Animal studies involving lesions of the forebrain noradrenergic system, especially in rodents, however, fail to show a strong relationship between noradrenergic function and performance on learning and memory tasks; however, in monkeys lesions of forebrain catecholaminergic systems do result in cognitive impairments (Arnsten and Goldman-Rakic,

1985a,b; Arnsten et al., 1988). There is strong and almost unanimous evidence for the involvement of noradrenergic systems in the extinction of learned behaviors (Haroutunian et al., 1986; Mason and Fibiger, 1979; Mason, 1979; McNaughton and Mason, 1980), as well as in the ability of noradrenergic drugs to potentiate performance on memory tests in humans and lower animals (Arnsten and Goldman-Rakic, 1985a,b; Arnsten et al., 1988; McEntee and Mair, 1980; Mair and McEntee, 1986; Gold and Zornetzer, 1983). There also appears to be a relatively strong correlation between age-related forebrain NE depletion and performance on tests of learning and memory in monkeys. Furthermore, NE receptor stimulation by high doses of systemically administered clonidine can attenuate age-related cognitive deficits in monkeys and can prevent the amnesia that results from cycloheximide administration (Arnsten et al., 1988b; Quartermain and Botwinick, 1975; Quartermain et al., 1977, 1988; Botwinick and Quartermain, 1974). Thus, although it is clear that profound NE lesions occur in AD and that forebrain noradrenergic systems participate in the processes that subserve learning and memory, their precise function and contribution to cognitive deficits is not well understood.

Since both cholinergic and noradrenergic systems are involved in AD, the functional interrelationship between these systems must be considered. There are many potential sites, such as the cell bodies, the nbM, the terminal fields, and common target cells where interactions between ascending noradrenergic and cholinergic systems can occur (Beani et al., 1986; Andrade and Aghajanian, 1985; Engberg and Svensson, 1980; Madison and Nicoll, 1984; Egan et al., 1983; Egan and North, 1985; Vizi, 1980; Lewander et al., 1977; Aou et al., 1983b; Robinson, 1986; Robinson et al., 1978; Westfall 1974, and chapter by Zaborski et al., this volume). In addition to NE/ACh interactions at subcortical sites and the functional effects of NE and acetylcholine could ultimately depend upon the interaction of these transmitters with receptors on postsynaptic target cells. Electrophysiological, neurochemical, and behavioral data indicate that these transmitters may have synergistic effects. For example, norepinephrine and α-adrenergic agonists applied at doses that by themselves have no electrophysiological effects greatly potentiate the responsiveness of somatosensory cortex cells to acetylcholine (Waterhouse et al., 1980, 1981). In addition, the ability of cholinergic agonists to produce the behavioral syndrome of catalepsy is dependent upon the integrity of the ascending adrenergic projection to the cortex and hippocampus (Mason and Fibiger, 1979). Our studies of learning and memory employing physostigmine, the studies of Decker and colleagues (Decker and Gallager, 1987; Decker and McGaugh, 1991) employing scopolamine, the experiments of Huygens et al. (1980) with oxotremorine, and the work of Mason and Fibiger (Mason and Fibiger, 1979) using a variety of cholinomimetics show that noradrenergic lesions can block the behavioral effects of cholinergic stimulations. Aou et al. and others (Aou et al., 1983a,b; Lewander et al., 1977; Robinson, 1986; Robinson et al., 1978; Westfall, 1974) have documented that the cholinergic and noradrenergic systems are intimately involved in the electrophysiological and biochemical events occurring in the frontal cortex during learning and suggest that ACh may play a modulatory role in the learning process. These and other studies clearly demonstrate that cholinergic and noradrenergic systems interact at a variety of different sites and suggest that therapeutic approaches to AD can be enhanced by adopting a multitransmitter view of the AD process and recognizing the interdependence of forebrain cholinergic and noradrenergic systems.

We have investigated the interaction of forebrain cholinergic and noradrenergic systems from the perspective of (a) simple learning and memory using a one-trial passive-avoidance paradigm; (b) responsivity to the retention test performance enhancing properties of two cholinomimetics, the acetylcholinesterase inhibitor, physostigmine, and the muscarinic receptor agonist oxotremorine; and (c) the induction of catalepsy by oxotremorine. Portions of these studies have been reported previously (Haroutunian et al., 1986, 1990a). In a single surgical session different groups of rats received (Ns = 12) ibotenic acid-induced lesions of the nbM (Bregma −0.3, ML ± 2.8, DV = 7.5–8.0); 6-OHDA-induced lesions (8 μg/per side) of the ascending noradrenergic bundle (ANB) at stereotaxic coordinates Bregma −6.0, ML ± 0.8, DV = 6.5; combined lesions of the nbM and the ANB; or sham lesions. Two to three weeks after the lesioning procedure, the rats were tested in a standardized one-trial passive-avoidance paradigm with a 2-sec long 0.6-mA scrambled foot-shock serving as the reinforcer. The retention of passive avoidance was assessed 72 hr later using procedures identical to those used during training, except that shock was omitted. All rats were then sacrificed within a week of the passive-avoidance test and their cortices were assayed for ChAT, AChE, NE, and DA.

The results of these studies are presented in Figure 1 and in Table 1. As Figure 1 indicates ibotenic acid-induced lesions of the nbM led to significant deficits in the 72-hr retention of passive avoidance ($ps < 0.01$), while lesions of the ANB had no observable effect on passive-avoidance behavior. It is also evident that the noradrenergic lesion failed to significantly alter passive-avoidance retention-test performance beyond that expected from the nbM lesion alone. It is possible that the superimposition of the ANB lesion upon the nbM lesion does in fact lead to an additional impairment of passive avoidance, but that this effect is obscured by floor values. This possibility was partially rejected in an ancillary experiment in which the intensity of the shock administered during passive avoidance training was increased to 8.0 mA. This increase in shock intensity raised the passive avoidance crossthrough latencies of the nbM and nbM+ANB lesioned rats, but no differential effect of the ANB lesion emerged. The failure of forebrain noradrenergic lesions to significantly affect the learning and memory of this relatively simple task is not particularly surprising, since similar findings have been reported by a number of other investigators. The failure of forebrain noradrenergic depletion to further impair the performance of nbM lesioned rats has been confirmed by other investigators using other behavioral tasks and NE-depleting agents (Wenk and Olton, 1987; Decker and McGaugh, 1991; Decker and Gallagher, 1987).

Neurochemical analyses (Table 1) confirm and extend the behavioral findings. Lesions of the nbM led to significant cortical cholinergic marker deficits

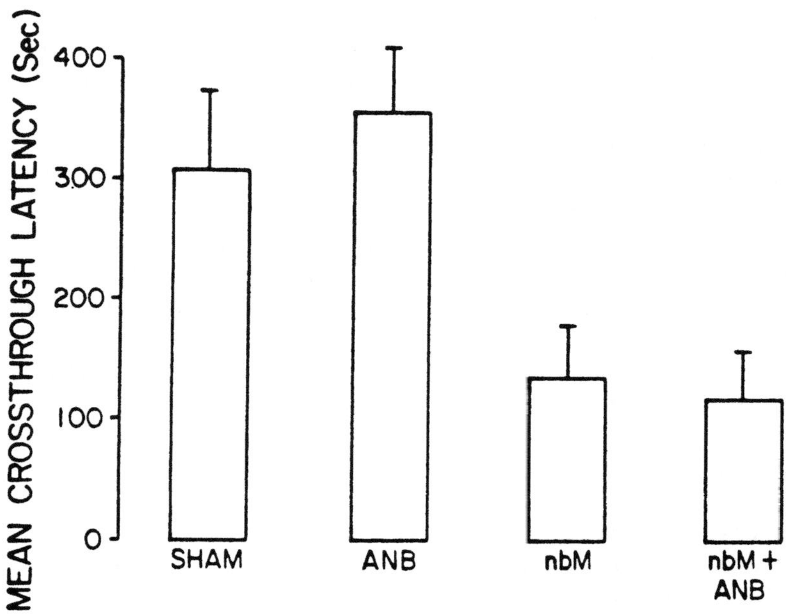

FIGURE 1. Effects of single or combined bilateral lesions of the nbM (NMDA-induced) or ANB (6-OHDA-induced) on the 72 hour retention of a single trial passive avoidance task. Mean crossthrough latencies for nbM and nbM+ANB lesioned groups were significantly (ps<0.01) reduced relative to sham operated rats.

($p < 0.01$), while 6-OHDA-induced lesions of the dorsal bundle profoundly affected cortical noradrenergic markers ($p < .01$). Levels of dopamine in the cortex were not affected by either lesion procedure, nor did the cholinergic and noradrenergic lesions interact to significantly alter the deficits produced by either lesion procedure alone.

As discussed in the introduction to this chapter, a significant number of studies have documented the efficacy of cholinomimetic agents to enhance the retention-test performance of nbM-lesioned rats. We will not reiterate and describe these studies in detail, but the results of one such experiment conducted in our laboratories is presented in Figure 2 (data relevant to the enhancement of passive-avoidance retention test performance by physostigmine in nbM-lesioned rats is also presented in subsequent sections of this chapter). This experiment, which employed lesion procedures and cognitive testing parameters identical to those described above, illustrates the enhancement of nbM lesion-induced passive-avoidance deficits by physostigmine and provides the basis for studying cholinomimetic effects on the retention-test performance of rats with combined cholinergic and noradrenergic lesions.

To study the effects of combined lesions of the forebrain cholinergic and noradrenergic lesions on the enhancement of retention-test performance by

TABLE 1. Effects of nbM (NMDA-induced) and ANB (6-OHDA-induced) lesions on cortical cholinergic and catecholaminergic markes. Results are expressed as mean/SEM(N).

	CAT (nmol ACH/min/mg prot.)	ACHE (nmol/min/mg prot.)	NE (ng/100 mg tissue)	DA (ng/100 mg tissue)
SHAM	1.66/0.06 (12)	2509/41 (12)	30.1/0.85 (12)	6.7/0.24 (12)
nbM	1.17/0.07 (14)	1804/83 (14)	29.2/0.92 (14)	6.3/0.28 (14)
% Depletion	29.5*	28.0*	3.1	5.2
ANB	1.70/0.05 (12)	2461/77 (12)	1.9/1.22 (12)	6.4/0.34 (12)
% Depletion	2.4	1.9	93.7*	3.8
nbM + ANB	1.19/0.06 (13)	1847/65 (13)	3.5/2.2 (13)	5.9/0.36 (13)
% Depletion	27.7*	26.4*	88.4*	11.2

*vs Sham operated controls, ps < 0.01.

physostigmine, we prepared two groups of rats. One group received combined nbM and ANB lesions as described above, while the second group received sham operations. Two weeks after the lesion procedure, each rat was trained in the one-trial passive-avoidance paradigm as described above. Immediately after the passive-avoidance training procedure, different rats in each of the two groups

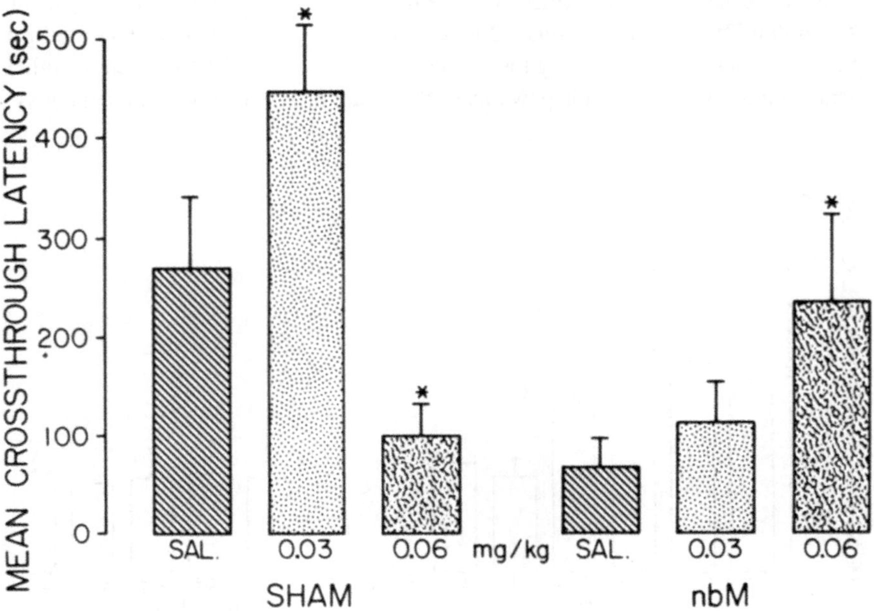

FIGURE 2. Effects of different post-acquisition doses of physostigmine on the 72 hour retention of a single trial passive avoidance task in nbM lesioned and sham operated rats. Retention test performance was significantly (ps<0.05) enhanced by 0.03 mg/kg and 0.06 mg/kg doses of physostigmine in sham operated and nbM lesioned rats respectively.

received one of six doses of physostigmine (0.0–0.24 mg/kg s.c., Ns = 11–12). Retention of passive avoidance was assessed 72 hr later. The results of this experiment are shown in Figure 3. Physostigmine at moderate doses enhanced the retention-test performance of sham-operated rats ($p < 0.05$), replicating earlier results (Haroutunian et al., 1990a,c, 1985; Murray and Fibiger, 1985), but failed to affect the retention-test performance of nbM + ANB lesioned rats at any of the doses investigated. It should be noted that the 0.24 mg/kg dose used in this study was the highest dose possible without inducing peripheral cholinesterase inhibition symptoms.

These results clearly suggest that physostigmine enhancement of retention test performance in nbM-lesioned rats is dependent upon the integrity of the forebrain noradrenergic system. A second study was conducted to test the generality of this finding and to assess whether noradrenergic deficits also affect responsivity to other cholinomimetic agents under different conditions. In this experiment we assessed the ability of 12.5–75 mg/kg doses of DSP-4 to block the cataleptic response induced by oxotremorine (0.2 mg/kg). One week following the administration of 0, 12.5, 25, 50, or 75 mg/kg DSP-4 (s.c., N = 6), each rat was injected (s.c.) with a 0.2 mg/kg dose of oxotremorine. Fifteen minutes later the forepaws of each rat were placed on a 3-mm diameter metal rod, which was elevated 7 cm above the surface of the test table, and the rats' latency to withdraw both paws from the rod was recorded. Three days later each animal was sacrificed and cortical NE levels were determined by HPLC. Oxotremorine caused signifiant catalepsy in the 0.0 mg/kg DSP-4 group of rats, as evidenced by the long latencies recorded for paw placement. Rats receiving the 50 and 75 mg/kg doses of DSP-4, on the other hand, readily removed their paws from the metal rod and showed no catalepsy.

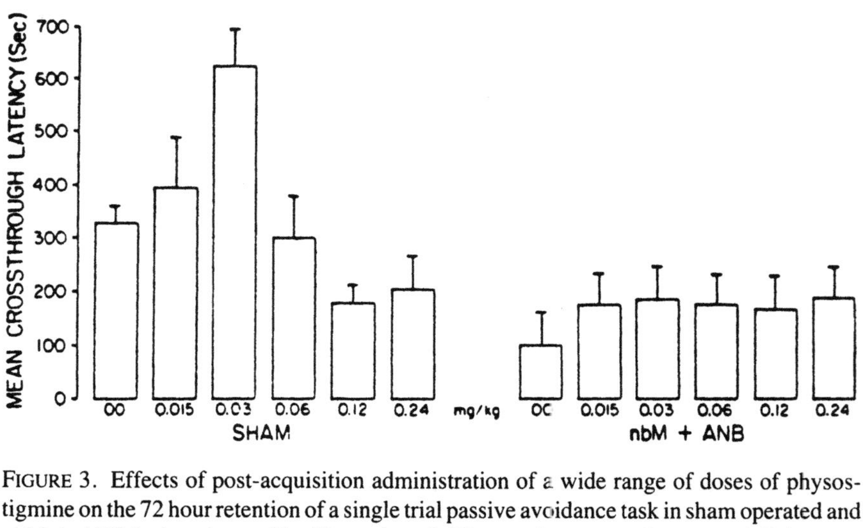

FIGURE 3. Effects of post-acquisition administration of a wide range of doses of physostigmine on the 72 hour retention of a single trial passive avoidance task in sham operated and nbM + ANB lesioned rats. Significant ($ps<0.01$) retention test performance deficits were evident in all nbM + ANB lesioned groups relative to sham operated controls.

Statistical analysis of these results yielded a highly significant effect of DSP-4 dose, with significant ($ps < 0.05$) blockade of catalepsy achieved at the 50 and 75 mg/kg doses. Examination of the neurochemical results revealed that the 50 mg/kg and 75 mg/kg doses of DSP-4 caused a 57.8% and 86.6% cortical noradrenergic deficit, respectively. These results suggest that the blockade of responsivity to cholinomimetics by central lesions of the NE system is a general phenomenon affecting responsivity to physostigmine and to oxotremorine in two completely independent and divergent behavioral paradigms.

The results of the studies described above show that the presence of noradrenergic lesions can significantly inhibit responsivity to cholinomimetic agents, raising the possibility that noradrenergic lesions in AD could substantially diminish the efficacy of cholinomimetic drugs used as therapeutic agents. In subsequent experiments we have examined whether noradrenergic agonists alone or in combination with cholinomimetics can reverse the passive-avoidance deficits induced by combined lesions of the nbM and the ascending noradrenergic bundle. Two groups of animals were surgically prepared with sham lesions or combined nbM+ANB lesions, as before. Two to three weeks later each animal received s.c. injections of saline, physostigmine (0.06 mg/kg), the α_2-receptor agonist clonidine (0.5 or 0.01 mg/kg), or a combination of physostigmine and clonidine immediately after passive-avoidance training. Retention of passive avoidance was again assessed 72 hr later. Clonidine was selected as the noradrenergic drug of choice, as some experiments (Arnsten and Goldman-Rakic, 1985b; Arnsten et al., 1988; Mair and McEntee, 1986; McEntee and Mair, 1980; Quartermain and Botwinick, 1975) had shown that clonidine effectively reversed the cognitive deficits evident in cycloheximidie-induced amnesic rats, aged rhesus monkeys suffering forebrain noradrenergic deficits, and Korsakoff's psychosis patients with compromised central noradrenergic systems.

Figure 4 demonstrates that neither physostigmine nor a low dose of clonidine alone were able to improve retention test performance in nbM+NE lesioned rats. The combination of a 0.06 mg/kg dose of physostigmine with a 0.01 mg/kg dose of clonidine, however, results in significant ($ps < 0.01$) improvements in 72-hr retention. Very similar results were observed when, in a separate replication of this study, a 0.005 mg/kg dose of clonidine was used in addition to the 0.01 mg/kg dose depicted in Figure 4. It is important to keep in mind that the nbM + ANB lesioned animals have such profound forebrain NE deficits (Table 1) that significant clonidine effects could only be exerted at postsynaptic sites. Figure 4 also shows that the very high dose of clonidine (0.5 mg/kg) when administered alone led to a significant ($p < 0.01$) improvement in retention-test performance. This dose was originally selected because of its potency in reversing cycloheximide-induced amnesia, but it is such a high dose that it could not be safely used in humans because of its hypotensive effects. It is also a rather nonspecific dose, affecting numerous other neurotransmitter systems.

A final study in this series tested the generality of this finding by assessing the ability of different doses (0, 0.01, 0.05, and 0.1 mg/kg) of the muscarinic agonist oxotremorine, administered alone or in combination with a 0.01 mg/kg dose of

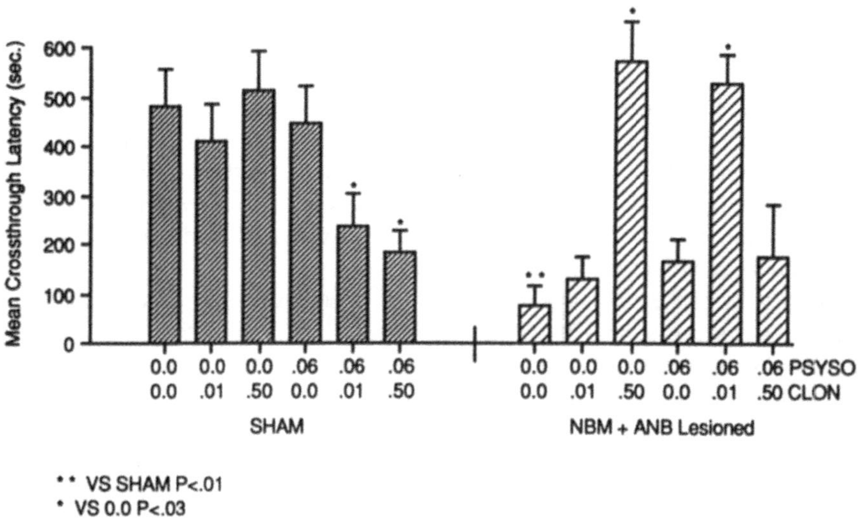

FIGURE 4. Effects of post-acquisition administration of single or combined doses of physostigmine and clonidine on the 72 hour retention of a single trial passive avoidance task in sham operated and nbM + ANB lesioned rats. Retention test performance was significantly (ps<0.05) impaired in sham operated rats receiving physostigmine and 0.01 mg/kg clonidine led to significant (ps<0.01) enhancement of retention test performance in the nbM + ANB lesioned rats. (* VS 0.0 P<.03, ** VS SHAM P<.01)

clonidine, to enhance passive-avoidance retention-test performance of nbM+ANB lesioned rats (Ns = 9–11). As was the case with physostigmine, oxotremorine alone (0.01 mg/kg) enhanced the retention-test performance of sham-operated rats, replicating earlier reports (Haroutunian et al., 1985a; Altman et al., 1987), but failed to affect the performance of ANB+nbM lesioned animals. Effective enhancement of 72-hr retention was attained in combined lesion rats receiving 0.01 mg/kg clonidine plus either 0.01 or 0.05 mg/kg doses of oxotremorine (Fs > 3.2, ps < 0.006).

One conclusion that can be reached from these studies is that while low doses of clonidine and cholinomimetic agents are not able to alleviate retention deficits in nbM+NE lesioned rats when administered alone, they are able to produce significant memory enhancement when administered together. These results, viewed together with the failure of even high doses of cholinomimetics to enhance the retention test performance of rats with combined forebrain cholinergic and noradrenergic lesions, support the conclusion that central cholinergic and noradrenergic systems interact dynamically, at least at the pharmacological level, and the abilities of cholinergic and noradrenergic drugs to affect memory are highly interdependent.

The results of the studies outline above raise the possibility that other central neurochemical systems may also influence and interact with central cholinergic systems to affect learning, memory, and responsivity to cholinergic agents. Two series of experiments have examined the roles of central somatostatinergic and central serotonergic systems in this context.

Cholinergic/Somatostatinergic Interactions

A significant number of postmortem studies of Alzheimer's disease (AD) have documented the susceptibility of central somatostatinergic systems in this disease (Rossor et al., 1980; Davies and Terry, 1981; Davies et al., 1980; Beal et al., 1987). Postmortem studies conducted in our own laboratories comparing 10 cortical Brodmann regions from 41 AD cases and 10 non-AD controls have revealed approximately 40–50% deficits in somatostatin-like immunoreactivity in the frontal, temporal, and parietal lobes of AD victims. The influence of central somatostatinergic systems on learning and memory have been studied in animals using the somatostatin-depleting agent cysteamine (Sagar et al., 1982; Bakhit et al., 1983; Brown et al., 1983). These and other experiments have documented the ability of cysteamine to deplete central somatostatin levels and to produce profound perturbation of learning and memory (Haroutunian et al., 1989b; Haroutunian et al., 1987; Vecsei et al., 1984).

The anatomic distribution of somatostatinergic systems is complex, as are the mechanisms through which the systemic or central administration of cysteamine deplete central somatostatin stores and affect learning and memory. Most studies, including those conducted in our laboratories (Haroutunian et al., 1987, 1989b) have found that the acute or subchronic administration of cysteamine results in an approximately 50% depletion of central somatostatin stores, and that the acute administration of cysteamine to rats can significantly impair performance on tests of learning and memory. The results of one such study are shown in Figure 5. The administration of a 150 mg/kg dose of cysteamine to rats immediately following the acquisition of a one-trial passive-avoidance task (described above) resulted in dramatic deficits in the 72-hr retention of passive-avoidance and depleted cortical somatostatin-like activity by 48%. Subsequent studies demonstrated that one of the mechanisms by which cysteamine produced deficits in somatostatin-like immunoreactivity and the retention of passive avoidance in this experiment was through the release of somatostatin from central stores (Haroutunian et al., 1987; Bakhit et al., 1983). That cysteamine-induced release of somatostatin was at least partly responsible for the depletion of somatostatin was supported by the finding that the cysteamine-induced depletion of somatostatin levels in the cortex was accompanied with parallel increases in CSF somatostatin-like activity. That this release of somatostatin contributed to cysteamine-induced amnesia was demonstrated by the finding that the subchronic pretreatment of rats with 150 mg/kg cysteamine for 3 consecutive days failed to affect their retention-test performance. Thus when rats were treated with cysteamine immediately after the acquisition of the passive-avoidance response, memory was impaired and cortical somatostatin levels were reduced. When rats were pretreated with cysteamine, cortical

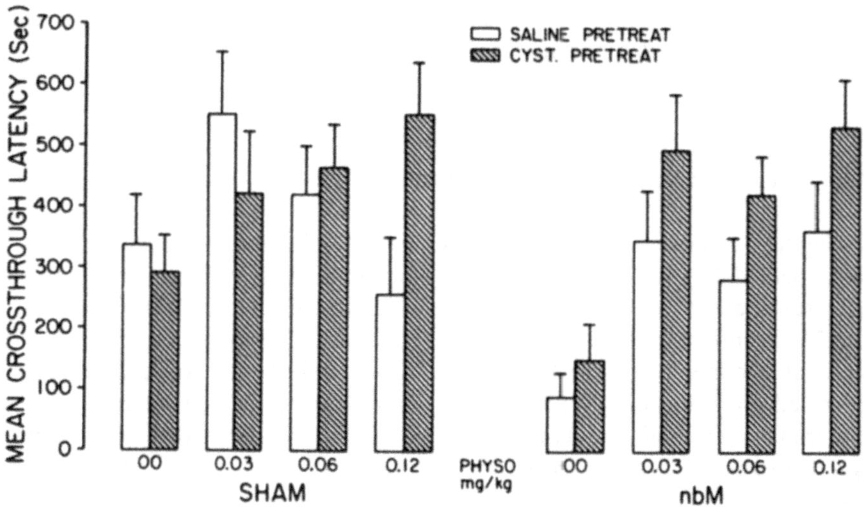

FIGURE 5. Effects of a wide range of doses of the somatostatin depleting agent cysteamine on the 72 hour retention of a single trial passive avoidance task. Retention test performance was significantly (ps<0.05) impaired in rats receiving 150-250 mg/kg doses.

somatostatin levels were reduced but passive-avoidance retention-test performance was unaffected.

Irrespective of the mechanism of action of cysteamine, a large body of evidence shows that (a) somatostatinergic lesions are present in AD and (b) under some circumstances somatostatinergic deficits can lead to significant learning and memory deficits. These considerations raise the possibility that, like noradrenergic deficits, somatostatinergic deficits may affect (hamper) responsivity to cholinomimetics in AD, normal, and nbM-lesioned rats. To test this possibility we prepared rats with ibotenic-acid-induced lesions of the nbM or with sham lesions. Two weeks after the nbM lesion, one half of the rats in each lesion group received s.c. injections of 150 mg/kg cysteamine for 3 consecutive days. On the following day each rat was trained in the previously described one-trial passive-avoidance paradigm and received one of four doses (0.0, 0.03, 0.06, and 0.12 mg/kg) of physostigmine (Ns = 10–12). Retention of passive avoidance was assessed 72 hr later.

As expected, ibotenic-acid-induced lesions of the nbM produced a significant ($p < 0.01$) passive-avoidance deficit (Fig. 6), and physostigmine significantly enhanced the retention test performance of sham-operated and nbM-lesioned rats ($ps < 0.05$), replicating earlier findings. As Figure 6 demonstrates, however, the subchronic pretreatment of neither sham nor nbM-lesioned rats with cysteamine had a significant effect on the retention-test performance of the rats. The determination of somatostatin-like immunoreactivity in the cortices of animals run in parallel and when sacrificed at the time of physostigmine administration

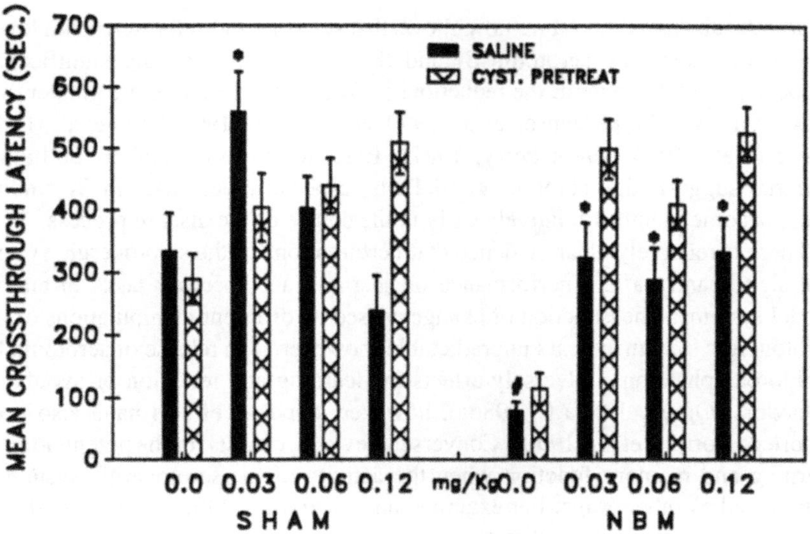

FIGURE 6. Effects of pre-acquisition treatment with cysteamine on physostigmine-induced potentiation of passive avoidance retention test performance in nbM and sham lesioned rats. Physostigmine significantly (ps<0.05) enhanced retention test performance in nbM lesioned rats irrespective of cysteamine treatment. (* vs 0.0 DOSE ps<0.01, # vs SHAM p<0.01)

demonstrated that the cysteamine treatments had produced a 48% depletion of cortical somatostatin.

These results support the conclusion that somatostatinergic deficits (a) unlike noradrenergic deficits, do not affect responsivity to cholinomimetics; and (b) somatostatinergic deficits do not augment the retention test deficits observed following forebrain cholinergic lesions. Unfortunately, because of the transient and unusual nature of the somatostatinergic deficits induced by cysteamine, it is even more difficult to draw conclusions regarding the interaction of different central systems in this case than in others. It is nevertheless clear that significant perturbation of the somatostatinergic system does not lead to clear and readily demonstrable changes in behavioral and pharmacological indices of cholinergic activity.

Cholinergic/Serotonergic Interactions

Serotonin-containing nerve terminals and serotonin receptors are distributed throughout the human cortex, hippocampus, and subcortical structures, including the substantia innominata basalis of Meynert area (Ichimiya et al., 1986; Yamamoto and Hirano, 1985; German et al., 1987; Cross et al., 1983; Gottfries, 1990; Gottfries et al., 1989a,b; Bowen et al., 1983; Palmer et al., 1987a, 1988; Reinikainen et al., 1988; Baker and Reynolds, 1989; Palacios et al., 1983; Cross,

1990). A substantial literature now shows that serotonergic cells, serotonin levels and metabolites, and serotonin S_1 and S_2 receptor numbers are significantly reduced in AD. In general, the reductions in S_2-type binding sites are greater than those of the S_1 subtype (Jansen et al., 1990; Dewar et al., 1990; Perry et al., 1984; Cross et al., 1984, 1986; Perry, 1987). Evidence from cortical tissue biopsy material suggests that serotonergic deficits, especially decreases in S_2 binding sites, become manifest relatively early in the course of the disease process.

There is relatively clear evidence that perturbations of the serotonergic system can significantly affect performance on learning and memory tasks in animal model systems. The direction of change caused by different manipulations of the serotonergic system appears unpredictable, however. The release of serotonin by p-chloroamphetamine adversely affects the learning and retention of avoidance behaviors (Ogren, 1985a,b, 1986a); however, opposite effects have also been reported (Normile et al., 1990). Conversely, evidence exists for the potentiation of learning and memory function when the activity of the serotonergic system is augmented by other drugs. For example, administration of the serotonin reuptake blockers 5-methoxy-N,N-dimethyl-tryptamine, fluoxetine, alaproclate, and zimeldine to rats can significantly enhance performance on a variety of measures of learning and memory (Quartermain et al., 1988; Altman et al., 1984; Flood and Cherkin, 1987). Paradoxically, the serotonergic receptor antagonists pirenperone, ketanserin, mianserin, methysergide, and metergoline also produce dose-dependent enhancement of retention-test performance in aversively motivated lick suppression and passive-avoidance tests (Altman and Normile, 1988; Ogren, 1985a, 1986c). Lesions of forebrain serotonergic systems lead to mixed results of either enhancement of performance, performance deficits, or no change, depending on the lesion method, postlesion interval, and the particular study in question (Ogren et al., 1985a, 1986a,b).

These inconsistent and often confusing results of studies examining the involvement of the serotonergic system in learning and memory may in part be explained by the neurophysiology of this system. The serotonergic system is generally inhibitory, diffuse, highly collateralized, tonically active, and has a long duration of action and axonal transmitter-releasing properties. It is quite possible that this system is involved in the modulation of attentional and behavioral (arousal) state (Fornal and Jacobs, 1988). This hypothesis becomes especially attractive when considered in light of electrophysiological data suggesting a level of sensitivity setting or gating role for the serotonergic innervation of the neocortex (Waterhouse et al., 1986; Lakoski and Aghajanian, 1985; Ragawski and Aghajanian, 1980; Fornal and Jacobs, 1988; Aghajanian et al., 1987). Additional suggestive evidence for a primarily "state" modulating role for serotonin can be gleaned from the existing, though meager, results of clinical trials with serotonergic drugs in AD. For example, the administration of the relatively selective serotonin reuptake blockers, alaproclate (Bergman et al., 1983) and zimeldine (Cutler et al., 1985), leads to modest positive effects in AD patients, but these improvements are attributable to enhanced emotional function, rather than alleviation of memory deficits.

There is little question that serotonergic and cholinergic systems interact in forebrain structures such as the cerebral cortex, hippocampus, striatum, and central structures (Kohler, 1984; Waterhouse et al., 1986; Altman et al., 1987; Ogren et al., 1985a; Cross and Deakin, 1985; Quirion et al., 1985; Wenk and English, 1986; Mitchell et al., 1984; Waterhouse et al., 1986; Bergman et al., 1983; Robinson et al., 1986). One general conclusion that can be reached from these *in vitro* studies is that serotonin inhibits cholinergic activity presynaptically and dampens the postsynaptic effects of acetylcholine. Electrophysiological studies of the intact rat somatosensory cortex are in agreement with this general conclusion (Waterhouse et al., 1986). There is also strong evidence for the presence of serotonergic projections to regions of origin of forebrain cholinergic cells and their terminal areas (Mitchell et al., 1984; Pazos and Palacios, 1985; Pazos et al., 1985). The results of studies of the effects of forebrain cholinergic lesions on the serotonergic system have been mixed. Serotonin binding sites have been reported to show an increase (Wenk and English, 1986), a decrease (Quirion et al., 1985), or a selective decrease in S_1 but not S_2 receptors (Cross and Deakin, 1985).

Although the *in vitro* studies of cholinergic/serotonergic interaction have yielded generally consistent results, these results do not lend themselves to simple explanations from a behavioral or Alzheimer's disease perspective. At face value, the *in vitro* results suggest that serotonergic hypofunction should result in the release of cholinergic cells from inhibition, leading to the enhancement of cholinergic function. If this hypothesis was correct, a negative correlation between serotonergic and cognitive deficits would be predicted in AD. This is clearly not the case. Furthermore, pharmacological data suggest a synergistic relationship between cholinergic and serotonergic functions. Serotonergic drugs appear to potentiate the effects of muscarinic agonists. Overt behavioral responses attributed to muscarinic receptor stimulation (e.g., purposeless chewing, salivation, tremor) are dependent upon and are enhanced by serotonergic activity (Altman et al., 1987; Ogren et al., 1985b). A similar potentiation of cholinomimetic activity by serotonin reuptake blockade has recently been reported in a learning and memory paradigm (Altman et al., 1987). In view of the conflict between the *in vitro* and *in vivo* studies, it must be assumed that the cholinergic and serotonergic systems interact in complex ways that are as yet unpredictable.

We have begun our analysis of forebrain cholinergic/serotonergic interactions at a simple behavioral and pharmacological level by studying the effects of serotonergic depletions produced by (a) the systemic administration of p-chloroamphetamine (PCA); (b) the infusion of the serotonergic toxin 5-7-DHT into the nbM; and (c) the infusion of 5-7-DHT into the dorsal raphe nucleus.

Our initial studies of the effects PCA administration on passive-avoidance behavior (Santucci et al., 1990) generally confirmed previous results showing that the pretreatment of rats with the PCA 60–30 min prior to the acquisition phase of a passive-avoidance response led to significant retention tests deficits 72 hr later. These results were obtained, however, only when PCA administration preceded passive-avoidance training and not when PCA was administered immediately following acquisition. These findings suggested that the passive-avoidance deficits observed with PCA could be a result of the well-known serotonin-releasing property of PCA (Ogren, 1986b; Adell et al., 1989), rather than of the depletion of serotonin per se.

Time course studies of cortical 5-HT depletion with PCA (Table 2) showed that serotonin levels decrease significantly within 60 min of PCA (2.5 mg/kg) administration and remain depressed for at least 2 weeks. To determine whether PCA-induced passive-avoidance deficits were due to the depletion of central 5-HT or to the release of 5-HT by PCA, we studied seven groups of rats. Each rat received a s.c. injection of either saline or 2.5 mg/kg PCA. Different groups of rats (Ns = 8) were then trained on the one-trial passive-avoidance task 15, 30, 45, and 120 min and 1 and 7 days after PCA administration. Only those rats that received PCA within 45 min of training showed passive-avoidance retention deficits, whereas Table 2 indicates that serotonin levels were significantly depleted in all groups. These results show that it is not the depletion of serotonin that is important in producing passive-avoidance deficits but rather the timing of this depletion relative to the acquisition of passive avoidance. This experiment provides at least partial support for the hypothesis that PCA-induced amnesia is mediated through the rapid release of serotonin by PCA within a time frame to interfere with the acquisition of the passive-avoidance response.

Autoradiographic studies of the serotonergic system in the rat have revealed a high concentration of 5-HT receptors in the area of the nucleus basalis of Meynert (nbM) (Pazos and Palacios, 1985; Pazos et al., 1985; Mitchell et al., 1984). Cells in the nucleus basalis of Meynert have been known to degenerate in AD, and 5-HT_2 receptor binding has been shown to be significantly decreased in AD brains. These findings suggest that serotonergic cholinergic interactions in the nbM may be of importance to the cognitive deficits observed in AD and to the retention-test deficits produced by PCA. We hypothesized that, irrespective of the exact mechanisms responsible for PCA-induced passive-avoidance deficits, perturbation of serotonergic activity at the level of the nbM could be of particular importance to learning and memory. To test this hypothesis we conducted a study to determine whether lesions of the 5-HT system at this site would affect passive-avoidance retention-test performance. To understand further the site of action and the nature of PCA-induced disruption of memory, the effects of PCA were evaluated in rats with serotonergic lesions of the nbM. Serotonergic lesioning

TABLE 2. Effects of PCA (2.5 mg/kg) on cortical serotonergic markers at various times post-administration. Values are expressed as percent of saline injected controls.

	Post PCA Injection Time in Minutes						Days	
	0.0	15	30	45	60	120	3	14
5-HT	−9.2	+10.1	+8.5	−3.7	−31.6*	−52.2*	−45.8*	−44.1*
5-HIAA	−18.2	−22.1*	−12.4	−10.8	−15.5*	−28.8*	−31.6*	−39.9*

*vs saline, $ps < 0.05$.

of the nbM was achieved by the stereotaxic infusion of the serotonergic neurotoxin 5-7-DHT into the nbM. Approximately 2 weeks following 5-7-DHT lesions, lesioned and sham-operated rats received saline or 2.5 mg/kg PCA 30 min prior to passive-avoidance training. All rats were tested for the retention of passive avoidance 72 hr later. The rats were then sacrificed, cortices and the region of the nbM were dissected, and they were assayed for 5-HT.

The results of this study demonstrated that the systemic administration of PCA and the infusion of 5-7-DHT into the region of the nbM led to profound and statistically equivalent depletions of cortical and nbM levels of 5-HT ($ps < 0.001$). Analysis of the passive-avoidance results, on the other hand, revealed that passive-avoidance deficits were evident only in those animals that had received PCA ($ps < 0.01$). Thus, despite equivalent depletions of cortical and nbM levels of serotonin in PCA and 5-7-DHT treated rats, only those rats receiving PCA were impaired on the passive-avoidance task. These results suggest that the effects of PCA on the retention of passive avoidance are mediated through central nervous system structures other than the nbM or the cortex. If this reasoning is correct, then PCA-induced passive-avoidance retention-test deficits would be expected to be blocked by lesions of the serotonergic system that affect the central nervous system more broadly.

To test this hypothesis directly we lesioned the serotonergic system centrally by infusing 2 μl of 5-7-DHT (15 μg/μl) into the dorsal raphe nucleus. Two weeks later different groups of lesioned and sham-operated rats (Ns = 10–12) received s.c. injections of either saline or 2.5 mg/kg PCA and were trained on the passive-avoidance response 30 min later. Upon retention testing 72 hr later, only those rats that had received sham lesions showed significant PCA-induced retention-test deficits. Raphe-lesioned rats were not impaired on this passive-avoidance task. These results (a) confirm our earlier hypothesis that PCA-induced passive-avoidance deficits are caused by the release caused by PCA rather than the resultant depletion of serotonin and (b) demonstrated that the amnesic effects of PCA are achieved through its direct effects on serotonergic systems rather than through an unknown postsynaptic effect. At the same time this study confirmed our prediction that PCA-induced passive avoidance deficits could be blocked by more global lesions of central 5-HT systems.

Irrespective of the central nervous system site where serotonergic deficits lead to learning and memory deficits, there is strong evidence in favor of cholinergic serotonergic interactions (see above and chapters by Nomile and by Vanderwolf in this book and Normile et al., 1990), and it is possible that the serotonergic system influences the properties and actions of cholinomimetic drugs. In a study design similar to that used earlier to assess cholinergic noradrenergic interactions, we (Santucci et al., 1990) tested the effects of combined nbM and PCA-induced serotonergic lesions on physostigmine enhancement of passive-avoidance retention test performance. Different groups of rats received nbM or sham lesions of the nbM using 50 nM NMDA as the neurotoxin. Two weeks later nbM-lesioned rats received either saline or 2.5 mg/kg PCA 30 min prior to passive-avoidance training. Immediately after passive-avoidance training, one half of the rats in each

PCA dose condition received 0.06 mg/kg physostigmine s.c. The remaining rats were injected with an equivalent volume of saline. Retention of passive avoidance was assessed 72 hr later. The results of this study are presented in Figure 7. Analysis of variance and subsequent *post hoc* tests revealed that PCA+nbM lesioned rats performed poorly on the passive-avoidance retention test relative to sham-lesioned rats ($ps < 0.05$). The administration of 0.06 mg/kg physostigmine enhanced retention-test performance in nbM-lesioned rats ($p < 0.05$; as expected from previous studies), but not in rats receiving PCA. Since no dose-response studies have been conducted as yet, it is premature to draw any firm conclusions from these data. These results do suggest, however, that PCA, at the very least, alters the dose(s) of physostigmine that can enhance passive-avoidance retention test performance.

Two additional findings were of note in this study. Firstly, the administration of PCA to nbM-lesioned rats led to passive-avoidance retention test deficits that were significantly more pronounced than the retention-test deficits observed in nbM-lesioned but saline-pretreated rats. Secondly, there was a significant effect of PCA

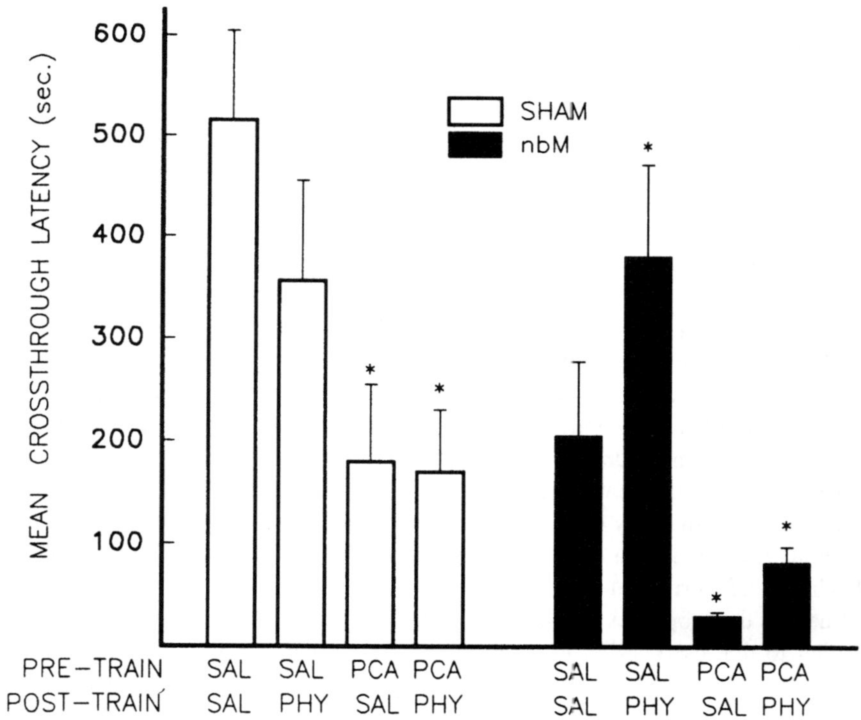

FIGURE 7. Effects of physostigmine (0.06 mg/kg) and PCA (2.5 mg/kg) administration on the 72 hour retention of a single trial passive avoidance task in rats with sham or nbM lesions. The retention test performance enhancing effects of physostigmine were blocked by pretreatment with PCA (* vs SAL/SAL groups, ps<0.05).

administration on cortical cholinergic markers. The administration of PCA to nbM-lesioned rats led to a 21% and 15% exaggeration of the ChAT and AChE deficits, respectively, compared to the deficits produced by nbM lesions alone ($ps < 0.05$). Although a similar pattern of PCA effects was observed in sham-lesioned rats, these differences did not reach statistical significance. Thus, the administration of PCA to nbM-lesioned rats leads to the exaggeration of nbM lesion-induced passive-avoidance deficits and to the blockade of responsivity to a dose of physostigmine, which enhanced the retention-test performance of nbM-lesioned rats not receiving PCA. In addition, the administration of PCA led to more pronounced cortical cholinergic deficits than those produced by nbM lesions alone. It is clear, therefore, that, at least with respect to the actions of PCA, forebrain serotonergic and cholinergic systems interact such that perturbations of the serotonergic system increase the deficits produced by lesions of the nbM.

In a final study on the effects of combined forebrain cholinergic/serotonergic lesions on the efficacy of cholinomimetics to enhance passive-avoidance retention-test performance, we have conducted physostigmine dose-response studies in rats with combined ibotenic-acid-induced lesions of the nbM and 5-7-DHT-induced lesions of the dorsal raphe. Two groups of rats were prepared. One group consisted of sham-operated rats, while the second received ibotenic-acid-induced lesions of the nbM and 5-7-DHT-induced lesions of the dorsal raphe nucleus in a single surgical session. Two to three weeks later each rat was trained on the one-trial passive-avoidance task described earlier and received an s.c. injection of one of five doses of physostigmine (0.0, 0.03, 0.06, 0.12, and 0.24) immediately following training. Retention of passive avoidance was tested 72 hr later and the rats were sacrificed and their cortices were assayed for ChAT and serotonin.

Analysis of the passive-avoidance results (Fig. 8) revealed a significant lesion effect [$F\ (1/82) + 11.9$, $p = 0.001$], a significant drug effect [$F\ (4/82) = 4.9$, $p = 0.002$), and a significant lesion by drug interaction term [$F\ (4/82) = 3.5$, $p = 0.01$]. As expected, the retention-test performance of lesioned rats receiving posttraining saline injections was significantly impaired relative to sham-operated rats. Retention-test performance was enhanced by the 0.03 and 0.06 mg/kg doses of physostigmine in sham-operated rats and by the 0.06 and 0.12 mg/kg doses in nbM+DR lesioned rats. Analysis of cortical cholinergic and serotonergic marker values showed significant ChAT and 5-HT deficits in the cortices of lesioned rats. In the anterior cortex, ChAT was decreased by 38.4% ($p < 0.001$), whereas 5-HT and 5-HIAA levels were reduced by 83.8% and 70.6%, respectively ($ps < 0.001$). Indices of noradrenergic and dopaminergic activity were not significantly altered by these lesions.

The results of this study clearly indicate that, in contrast to the effects of combined lesions of the nbM and forebrain noradrenergic lesions, combined lesions of the nbM and the dorsal raphe serotonergic system fail to significantly alter responsivity to the cholinomimetic physostigmine. In general, our studies of central cholinergic/serotonergic systems have shown a complex pattern of interaction. Both systems have been found to affect learning and memory in the passive-avoidance paradigm, although the release rather than the depletion of serotonin stores appears to be of greater consequence. Perturbation of the serotonergic system with PCA has been found to inhibit the responsivity of forebrain cholinergic lesioned rats to at least one dose of physostigmine.

General Conclusions

We opened this chapter with the hypothesis that the efficacy of cholinomimetics in alleviating memory deficits caused by lesions of the forebrain cholinergic systems is dependent upon the integrity of other transmitter systems. We tested this general hypothesis by studying forebrain cholinergic/noradrenergic, cholinergic/somatostatinergic, and cholinergic/serotonergic interactions at a behavioral and pharmacological level. It is clear that each of the four systems investigated in these studies independently affect behavior, learning, and memory profoundly, and as such, interact to the extent that they impinge upon and affect the same behavioral and cognitive processes. Within the constraints of the methodologies used, it is possible to conclude that a strong pharmacological interaction exists between cholinergic and noradrenergic systems such that noradrenergic lesions can block the memory-enhancing actions of at least two cholinomimetic agents acting through diverse mechanisms. We can also conclude that the partial depletion of central somatostatinergic systems fails to affect the actions of these same cholinomimetics in identical tests of retention. Finally, it is possible to conclude that serotonergic and cholinergic systems interact sufficiently to inhibit the efficacy of at least some doses of physostigmine in enhancing retention-test performance.

References

Adell A, Sarna GS, Hutson PH, Cruzon G (1989): An in vivo dialysis and behavioural study of the release of 5-HT by p-chloroamphetamine in resperine-treated rats. *Br J Pharmacol* 97:206–212

Adolfsson R, Gottfries CG, Roos BE, Winblad B (1979): Changes in the brain catecholamines in patients with dementia of Alzheimer type. *Br J Psychiat* 135:216–223

Aghajanian GK, Sprouse JS, Rasmussen K (1987): Physiology of the midbrain serotonin system: In: *Psychopharmacology: The Third Generation of Progress*, Meltzer HY, ed. New York: Raven Press, pp 141–149

Altman HJ, Normile H (1988): What is the nature of the role of the serotonergic nervous system in learning and memory: Prospects for development of an effective treatment strategy for senile dementia. *Neurobiol Aging* 9:627–638

Altman HJ, Nordy DA, Ogren SO (1984): Role of serotonin in memory: Facilitation by alaproclate and zimeldine. *Psychopharmacology* 84:496–502

Altman HJ, Stone WS, Ogren SO (1987): Evidence for a possible functional interaction between serotonergic and cholinergic mechanisms in memory retreival. *Behav Neural Biol* 48:49–62

Andrade R, Aghajanian GK (1985): Opiate- and alpha$_2$-adrenoceptor-induced hyperpolarization of locus ceruleus neurons in brain slices: Reversal by cyclic adenosine 3′:5′-monophosphate analogues. *J Neurosci* 5:2359–2364

Aou S, Oomura Y, Nishino H (1983a): Influence of acetylcholine on neuronal activity in monkey orbital cortex during bar press feeding task. *Brain Res* 275:178–182

Aou S, Oomura Y, Nishino H, Inokuchi A, Mizuno Y (1983b): Influence of catecholamines on reward-related neuronal activity in monkey orbitofrontal cortex. *Brain Res* 267:165–170

Arai H, Kosaka K, Iizuka R (1984): Changes of biogenic amines and their metabolites in postmortem brains from patients with Alzheimer's-type dementia. *J Neurochem* 43:388–393

Arnsten AFT, Goldman-Rakic PS (1985a): Catecholamines and cognitive decline in aged nonhuman primates. *Ann NY Acad Sci* 444:218–234

Arnsten AFT, Goldman-Rakic PS (1985b): α_2-adrenergic mechanisms in prefrontal cortex associated with cognitive decline in aged nonhuman primates. *Science* 230:1273–1279

Arnsten AFT, Cai JX, Goldman-Rakic PS (1988): The alpha-2 adrenergic agonist guanfacine improves memory in aged monkeys without sedative or hypotensive side effects: Evidence for alpha-2 receptor subtypes. *J Neurosci* 8:4287–4298

Baker GB, Reynolds GP (1989): Biogenic amines and their metabolites in Alzheimer's disease: Noradrenaline, 5-hydroxytryptamine and 5-hydroxyindole-3-acetic acid depleted in hippocampus but not in substantia innominata. *Neurosci Lett* 100:335–339

Bakhit C, Benoit R, Bloom FE (1983): Effects of cysteamine on pro-somatostatin related peptides. *Regul Pept* 6:169–177

Beal MF, Svendsen CN, Bird ED, Martin JB (1987): Somatostatin and neuropeptide Y are unaltered in the amygdla in schizophrenia. *Neurochem Pathol* 6:169–176

Beani L, Tanganelli T, Bianchi C (1986): Noradrenergic modulation of cortical acetylcholine release in both direct and gamma-aminobutyric acid-mediated. *J PET* 236:230–236

Beller SA, Overall JE, Swann AC (1985): Efficacy of oral physostigmine in primary degenerative dementia. *Psychopharmacology* 87:147–151

Bergman I, Brane G, Gottfries CG, Jostell KG, Karlsson I, Svennerholm L (1983): Alaproclate: A pharmacokinetic and biochemical study in patients with dementia of the Alzheimer type. *Psychopharmacology* 80:279–283

Bhat RV, Turner SL, Marks MJ, Collins AC (1990): Selective changes in sensitivity to cholinergic agonists and receptor changes elicited by continuous physostigmine infusion. *J Pharmacol Exp Ther* 255:187–196

Bigl V, Arendt T, Fischer S, Werner M, Arendt A (1987): The cholinergic system in aging. *Gerontology* 33:172–180

Bondareff W, Mountjoy CQ, Roth M (1982): Loss of neurons of origin of the adrenergic projection to cerebral cortex (nucleus locus ceruleus in senile dementia). *Neurology* 32:164–168.

Botwinick CY, Quartermain D (1974): Recovery from amnesia induced by pre-test injection of monoamine oxidase inhibitors. *Pharm Biochem Behav* 2:375–379

Bowen DM, Allen SJ, Benton JS, Goodhardt MJ, Haan EA, Palmer AM, Sims NR, Smith CC, Spillane JE, Esiri MM, Neary D, Snowden JS, Wilcock GK, Davison AN (1983): Biochemical assessment of serotonergic and cholinergic dysfunction and cerebral atrophy in Alzheimer's disease. *J Neurochem* 41:266–272

Brown MR, Fisher LA, Sawchenko PE, Swanson LW, Vale W (1983): Biological effects of cysteamine: Relationship to somatostatin depletion. *Regul Pept* 5:163–174

Cross AJ (1990): Serotonin in Alzheimer-type dementia and other dementing illnesses. *Ann NY Acad Sci* 600:405–451

Cross AJ, Deakin JFW (1985): Cortical serotonin receptor subtypes after lesions of ascending cholinergic neurons in rat. *Neurosci Lett* 60:261–265

Cross AJ, Crow TJ, Johnson JA, Joseph MH, Perry EK, Perry RH, Blessed G, Tomlinson BE (1983): Monoamine metabolism in senile dementia of Alzheimer's type. *J Neurol Sci* 60:383–392

Cross AJ, Crow TJ, Ferrier IN, Johnson JA, Bloom SR, Crosellis JA (1984): Serotonin receptor changes in dementia of the Alzheimer type. *J Neurochem* 43:1574–1581

Cross AJ, Crow TJ, Ferrier IN, Johnson JA (1986): The selectivity of the reduction of serotonin S2 receptors in Alzheimer-type dementia. *Neurobiol Aging* 7:3–7

Cutler NR, Haxby J, Kay AD, Narang PK, Lesko LJ, Costa JL, Ninos M, Linnoila M, Potter WZ, Renfrew JW, et al. (1985): Evaluation of zimeldine in Alzheimer's disease. Cognitive and biochemical measures. *Arch Neurol* 42:744–748

Davies P, Terry RD (1981): Cortical somatostatin-like immunoreactivity in cases of Alzheimer's disease and senile dementia of Alzheimer's type. *Neurobiol Aging* 2:9–14

Davies P, Katzman R, Terry RD (1980): Reduced somatostatin-like-immunoreactivity in cerebral cortex from cases of Alzheimer's disease Alzheimer's senile dementia. *Nature (London)* 288:279–280

Decker MW, Gallager M (1987): Scopolamine-disruption of radial arm maze performance: Modification by noradrenergic depletion. *Brain Res* 417:59–69

Decker MW, McGaugh JL (1991): The role of interactions between the cholinergic system and other neuromodulatory systems in learning and memory. *Synapse* 7:151–168

Dekker AJAM, Connor DJ, Thal LJ (1991): The role of cholinergic projections from the nucleus basalis in memory. *Neurosci Biobehav Rev* 15:299–317

Dewar D, Graham DI, McCulloch J (1990): 5 HT2 receptors in dementia of Alzheimer type: A quantitative autoradiographic study of frontal cortex and hippocampus. *J Neural Transm Park Dis Dement Sect* 2:129–137

Doucette R, Fisman M, Hachinski VC, Mersky H (1986): Cell loss from the nucleus basalis of Meynert in Alzheimer's disease. *Can J Neurol Sci* 13:435–440

Egan TM, North RA (1985): Acetylcholine acts on m2-muscarinic receptors to excite rat locus coeruleus neurons. *Br J Pharmac* 85:733–735

Egan TM, Henderson G, North RA, Williams JT (1983): Noradrenaline-mediated synaptic inhibition in rat locus coeruleus neurons. *J Physiol* 345:477–488

Engberg G, Svensson TH (1980): Pharmacological analysis of a cholinergic receptor mediated regulation of brain norepinephrine neurons. *J Neural Trans* 49:137–150

Fine A, Dunnett SB, Bjorklund A, Iversen SD (1985): Cholinergic ventral forebrain grafts into the neocortex improve passive avoidance memory in a rat model of Alzheimer disease. *Proc Natl Acad Sci USA* 82:5227–5230

Flood JF, Cherkin A (1987): Fluoxetine enhances memory processing in mice. *Psychopharmacology* 93:36–43

Fornal CA, Jacobs BL (1988): Physiological and behavioral correlates of serotonergic single-unit activity. In: *Neuronal Serotonin*, Osborn NN, Hamon M, eds. New York: John Wiley, pp 305–345

Forno LS, Eng LF, Selkoe DJ (1989): Pick bodies in the locus ceruleus. *Acta Neuropathol. (Berlin)* 79:10–17

German DC, White CL, Sparkman DR (1987): Alzheimer's disease: Neurofibrillary tangles in nuclei that project to the cerebral cortex. *Neuroscience* 21:305–312

Giacobini E (1990): The cholinergic system in Alzheimer disease. *Prog Brain Res* 84:321–332

Goedert M, Fine A, Hunt SP, Ullrich A (1986): Nerve growth factor mRNA in peripheral and central rat tissue and in the human central nervous system. Lesion effects in the rat brain and levels in Alzheimer's disease. *Soc Neurosci Abst* 12.1:299

Gold, PE, Zornetzer SF (1983): The mnemon and its juices: Neuromodulation of memory processes. *Behav Neural Biol* 38:151–189

Gottfries CG (1990a): Disturbance of the 5-hydroxytryptamine metabolism in brains from patients with Alzheimer's dementia. *J Neural Trans [Suppl]* 30:33–43

Gottfries CG (1990b): Brain monoamines and their metabolites in dementia. *Acta Neurol Scand Suppl* 129:8–11

Gottfries CG, Adolfsson R, Aquilonius SM, Carlsson A, Eckernas SA, Nordberg A, Oreland L, Svennerholm L, Wiberg A, Winblad B (1989): Biochemical changes in dementia disorders of the Alzheimer type (AD/SDAT). *Neurobiol Aging* 4:261–271

Haroutunian V, Barnes E, Davis KL (1985a): Cholinergic modulation of memory in rats. *Psychopharmacology* 87:266–271

Haroutunian V, Kanof P, Davis KL (1985b): Pharmacological alleviation of cholinergic lesions induced memory deficits in rats. *Life Sci* 37:945–952

Haroutunian V, Kanof PD, Davis KL (1989a): Attenuation of nucleus basalis of Meynert lesion-induced cholinergic deficits by nerve growth factor. *Brain Res* 487:200–203

Haroutunian V, Kanof PD, Davis KL (1989b): Interactions of forebrain cholinergic and somatostatinergic systems in the rat. *Brain Res* 496:98–104

Haroutunian V, Kanof PD, Tsuboyama GK, Campbell GA, Davis KL (1986): Animal models of Alzheimer's disease: Behavior, pharmacology, transplants. *Can J Neurol Sci* 13:385–393

Haroutunian V, Mantin R, Campbell GA, Tsuboyama GK, Davis KL (1987): Cysteamine-induced depletion of central somatostatin-like immunoactivity: Effects on behavior, learning, memory and brain neurochemistry. *Brain Res* 403:234–242

Haroutunian V, Kanof PD, Tsuboyama G, Davis KL (1990a): Restoration of cholinomimetic activity by clonidine in cholinergic plus noradrenergic lesioned rats. *Brain Res* 507:261–266

Haroutunian V, Mantin R, Kanof PD (1990b): Frontal cortex as the site of action of physostigmine in nbM-lesioned rats. *Physiol Behav* 47:203–206

Haroutunian V, Santucci AC, Davis KL (1990c): Implications of multiple transmitter system lesions for cholinomimetic therapy in Alzheimer's disese. *Prog Brain Res* 84:333–346

Haroutunian V, Wallace WC, Davis KL (1991): Nucleus basalis lesions and recovery. In: *Cholinergic Basis for Alzheimer Therapy*, Giacobini E, Becker R, eds. Boston: Birkhäuser, pp 120–125

Huygens P, Baratti CM, Gardella JL, Filinger E (1980): Brain catecholamine modifications. The effects on memory facilitation induced by oxotremorine in mice. *Psychopharmacology* 69:291–294

Ichimiya Y, Arai H, Kosaka K, Iizuka R (1986): Morphological and biochemical changes in the cholinergic and monoaminergic systems in Alzheimer-type dementia. *Acta Neuropathol (Berl)* 70:112–116

Jansen KL, Faull RL, Dragunow M, Synek BL (1990): Alzheimer's disease: changes in hippocampal N-methyl-D-aspartate, quisqualate, neurotensin, adenosine, benzodiazepine, serotonin and opioid receptors—an autoradiographic study. *Neuroscience* 39:613–627

Kohler C (1984): The distribution of serotonin binding sites in the hippocampal region of the rat brain. An autoradiographic study. *Neuroscience* 13:667–680

Koshimura K, Kato T, Yohyama I, Nakamura S, Kameyama M (1987): Correlation of choline acetyltransferase activity between the nucleus basalis of Meynert and the cerebral cortex. *Neurosci Res* 4:330–336

Lakoski JM, Aghajanian GK (1985): Effects of ketanserin on neuronal responses to serotonin in the prefrontal cortex, lateral geniculate and dorsal raphe nucleus. *Neuropharmacology* 24:265–273

Lewander T, Joh TH, Reis DJ (1977): Tyrosine hydroxylase: Delayed activation in central noradrenergic neurons and induction in adrenal medulla elicited by stimulation of central cholinergic receptors. *J PET* 200:523–534

Madison DV, Nicoll RA (1984): Control of the repetitive discharge of CA1 pyramidal neurons in vitro. *J Physiol* 354:319–331

Mair RG, McEntee WJ (1986): Cognitive enhancement in Korsakoff's psychosis by clonidine: A comparison with l-dopa and ephedrine. *Psychopharmacology* 88:374–380

Mandel RJ, Chen AD, Connor DJ, Thal LJ (1989): Continuous physostigmine infusion in rats with excitotoxic lesions of the nucleus basalis magnocellularis: Effects on performance in the water maze task and cortical cholinergic markers. *J Pharmacol Exp Ther* 251:612–619

Mann DM, Lincoln J, Yates PO, Stamp JE, Toper S, Maruyama Y, Oshima T, Nakajima EI (1980): Changes in the monoamine containing neurons of the human CNS in senile dementia. Simultaneous determination of catecholamines in rat brain by reversed-phase liquid chromatography with electrochemical detection. *Br J Psychiat Life Sci* 26:1115–1120

Mann DMA, Lincoln J, Yates PO, Stamp JE, Toper S (1983): Changes in the monoamine containing neurones of the human CNS in senile dementia. *J Neurol Neurosurg Psychiat* 46:96–102

Mann DM, Yates PO, Marcyniuk B (1985): Correlation between senile plaque and neurofibrillary tangle counts in cerebral cortex and neuronal counts in cortex and subcortical structures in Alzheimer's disease. *Neurosci Lett* 56:51–55

Mann DM, Yates PO, Marcyniuk B (1986a): A comparison of nerve cell loss in cortical and subcortical structures in Alzheimer's disease. *J Neurol Neurosurg Psychiat* 49:310–312

Marcyniuk B, Mann DM, Yates PO (1989): The topography of nerve cell loss from the locus caeruleus in elderly persons. *Neurobiol Aging* 10:5–9

Mason ST (1979): Noradrenaline: Reward or extinction? *Neurosci Biobehav Rev* 3:1–10

Mason ST, Fibiger HC (1979): Possible behavioral function for noradrenaline-acetylcholine interaction in brain. *Nature (London)* 277:396–397

McEntee WJ, Mair RG (1980): Memory enhancement in Korsakoff's psychosis by clonidine: Further evidence for a noradrenergic deficit. *Ann Neurol* 7:466–470

McGeer PL, McGeer EG, Suzuki J, Dolman CE, Nagai T (1984): Aging, Alzheimer's disease, and the cholinergic system of the basal forebrain. *Neurology* 34:741–745

NcNaughton N, Mason ST (1980): The neuropsychology and neuropharmacology of the dorsal ascending noradrenergic bundle—A review. *Prog Neurobiol* 14:157–219

Mitchell IJ, Stuart AM, Slater P, Unwin HP, Crossman AR (1984): Autoradiographic demonstration of 5HT1 binding sites in the primate basal nucleus of Meynert. *Eur J Pharmacol* 104:189–190

Murray CL, Fibiger HC (1985): Learning and memory deficits after lesions of the nucleus basalis magnocellularis: Reversal by physostigmine. *Neuroscience* 19:1025–1032

Normile HJ, Jenden DJ, Kuhn DM, Wolf WA, Altman HJ (1990): Effects of combined serotonin depletion and lesions of the nucleus basalis magnocellularis on acquisition of a complex spatial discrimination task in the rat. *Brain Res* 536:245–250

Ogren SO (1985a): Evidence for a role of brain serotonergic neurotransmission in avoidance learning. *Acta Physiol Scand Suppl* 544:1–71

Ogren SO (1986a): Analysis of the avoidance learning deficit induced by serotonin releasing compound p-chloroamphetamine. *Brain Res Bull* 16:645–660

Ogren SO (1986b): Serotonin receptor involvement in the avoidance learning deficit caused by p-chloroamphetamine-induced serotonin release. *Acta Physiol Scand* 126:449–462

Ogren SO, Johansson C, Magnusson O (1985a): Forebrain serotonergic involvement in avoidance learning. *Neurosci Lett* 58:305–309

Ogren SO, Nordstrom O, Danielsson E, Peterso LL, Bartfai T (1985b): In vivo and in vitro studies on the potentiation of muscarinic receptor stimulation by alaprocite, a selective 5-HT uptake blocker. *J Neural Trans* 61:1–20

Palacios JM, Probst A, Cortes R (1983): The distribution of serotonin receptors in the human brain: High density of [3H]LSD binding sites in the raphe nuclei of the brainstem. *Brain Res* 274:150–155

Palmer AM, Francis PT, Bowen DM, Neary JS, Mann DMA, Snowden JS (1987a): Catecholaminergic neurons assessed ante-mortem in Alzheimer's disease. *Brain Res* 414:365–375

Palmer AM, Wilcock GK, Esiri MM, Francis PT, Bowen DM (1987b): Monoaminergic innervation of the frontal and temporal lobes in Alzheimer's disease. *Brain Res* 401:231–238

Palmer AM, Stratmann GC, Procter AW, Bowen DM (1988): Possible neurotransmitter basis of behavioral changes in Alzheimer's disease. *Ann Neurol* 23:616–620

Pazos A, Palacios JM (1985): Quantitative autoradiographic mapping of serotonin receptors in the rat brain. I. Serotonin-1 receptors. *Brain Res* 346:205–230

Pazos A, Cortes R, Palacios JM (1985): Quantitative autoradiographic mapping of serotonin receptors in the rat brain. II. Serotonin-2 receptors. *Brain Res.* 346:231–249

Perry EK (1987): Cortical neurotransmitter chemistry in Alzheimer's disease. In: *Psychopharmacology: The Third Generation of Progress*, Meltzer HY, ed. New York: Raven Press, pp 887–896

Perry EK, Tomlinson BE, Blessed G, Bergmann K, Gibson PH, Perry RH (1978): Correlation of cholinergic abnormalities with senile plaques and mental test scores in senile dementia. *Br Med J* 2:1457–1459

Perry EK, Tomlinson BE, Blessed G, Perry RH, Cross AJ, Crow TJ (1981): Neuropathological and biochemical observations on the noradrenergic system in Alzheimer's disease. *J Neurol Sci* 51:279–337

Perry EK, Perry RH, Candy JM, Fairbairn AF, Blessed G, Dick DJ, Tomlinson BE (1984): Cortical serotonin-S2 receptor binding abnormalities in patients with Alzheimer's disease: Comparisons with Parkinson's disease. *Neurosci Lett* 51:353–357

Pomponi M, Giacobini E, Brufani M (1990): Present state and future development of the therapy of Alzheimer disease. *Aging (Milano)* 2:125–153

Quartermain D, Botwinick CY (1975): Role of biogenic amines in the reversal of cycloheximide-induced amnesia. *J Comp Physiol Psychol* 88:386–401

Quartermain D, Freedmen LS, Botwinick CY, Gutwein BM (1977): Reversal of cycloheximide-induced amnesia by adrenergic receptor stimulation. *Physiol Biochem Behav* 7:259–267

Quartermain D, Judge ME, Leo P (1988): Attenuation of forgetting by pharmacological stimulation of aminergic neurotransmitter systems. *Pharm Biochem Behav* 30: 77–81

Quirion R, Richard J, Dam TV (1985): Evidence for the existance of serotonin type 2 receptors on cholinergic terminals in rat cortex. *Brain Res* 33:345–349

Ragawski MA, Aghajanian GK (1980): Norepinephrine and serotonin: Opposit effects on the activity of lateral geniculate neurons evoked by optic pathway stimulation. *Exp Neurol* 69:678–694

Reinikainen KJ, Paljarvi L, Huuskonen M, Soininen H, Laakso M, Riekkinen PJ (1988): A post-mortem study of noradrenergic, serotonergic and GABAergic neurons in Alzheimer's disease. *J Neurol Sci* 84:101–116

Reinikainen KJ, Soininen H, Riekkinen PJ (1990): Neurotransmitter changes in Alzheimer's disease: Implications to diagnostics and therapy. *J Neurosci Res* 27:576–586

Robinson S (1986): 6-hydroxydopamine lesions of the ventral noradrenergic bundle blocks the effects of amphetamine on hippocampal acetylcholine. *Brain Res* 397:181–184

Robinson SE, Cheney DL, Costa E (1978): Effects of normifensine and other antidepressant drugs on acetylcholine turnover in various regions of rat brain. *Naunyn Schmiedebergs Arch Pharmacol* 304:263–269

Robinson SE, Rice MA, Hambrecht KL (1986): Effect of intrastriatal injection of diisopropylfluorophosphate on acetylcholine, dopamine and serotonin metabolism. *J Neurochem* 46:1632–1638

Rossor MN, Emson PC, Montjoy CQ, Roth M, Iversen LL (1980): Reduced amounts of immunoreactive somatostatin in the temporal cortex in senile dementia of Alzheimer's type. *Neurosci Lett* 20:373–377

Sagar SM, Landry D, Millard WJ, Badger TM, Arnold MA, Martin JB (1982): Depletion of somatostatin-like immunoreactivity in the rat central nervous system by cysteamine. *J Neurosci* 2:225–231

Santucci AC, Kanof PD, Haroutunian V (1990): Serotonergic modulation of cholinergic systems involved in learning and memory in rats. *Dementia* 1:151–155

Saper CB, German DC, White CL (1985): Neuronal pathology in the nucleus basalis and associated cell groups in senile dementia of the Alzheimer's type: Possible role in cell loss. *Neurology* 35:1089–1095

Shimohama S, Taniguchi T, Fujiwara M, Kameyama M (1986): Biochemical characterization of a-adrenergic receptors in human brain and changes in Alzheimer-type dementia. *J Neurochem* 47:1294–1301

Sparks DL (1989): Aging and Alzheimer's disease. Altered cortical serotonergic binding. *Arch Neurol* 46:138–140

Sparks DL, DeKosky ST, Markesbery WR (1988): Alzheimer's disease. Aminergic-cholinergic alterations in hypothalamus. *Arch Neurol* 45:994–999

Summers WK, Viesselman JO, Marsh GM, Candelora K (1981): Treatment of Alzheimer-like dementia: Pilot study in twelve patients. *Biol. Psychiatr* 16:145–153

Vecsei L, Kiraly C, Bollok I, Nagy A, Verga J, Penke B, Telegdy G (1984): Comparative studies with somatostatin and cysteamine in different behavioral tests with rats. *Pharmacol Biochem Behav* 21:833–837

Vizi ES (1980): Modulation of cortical release of acetylcholine by noradrenaline released from nerve terminals arising from the rat locus coeruleus. *Neuroscience* 5:2139–2144

Waterhouse BD, Moises HC, Woodward D (1980): Noradrenergic modulation of somatosensory cortical neuronal responses to iontophoretically applied putative neurotransmitters. *Exper Neurol* 69:30–49

Waterhouse BD, Moises HC, Woodward D (1981): Alpha-receptor-mediated facilitation of somatosensory cortical neuronal responses to excitatory synaptic inputs and iontophoretically applied acetylcholine. *Neuropharmacology* 20:907–920

Waterhouse BD, Moises HC, Woodward DJ (1986): Interaction of serotonin with somatosensory cortical neuronal responses to afferent synaptic inputs and putative neurotransmitters. *Brain Res Bull* 17:507–518

Wenk GL, English KL (1986): [3H]Ketanserin (serotonin type 2) binding increases in rat cortex following basal forebrain lesions with ibotenic acid. *J Neurochem* 47:845–850

Wenk GL, Olton DS (1987): Basal forebrain cholinergic neurons and Alzheimer's disease. In: *Animal Models of Dementia: A Synaptic Neurochemical Perspective*, Coyle JT, ed. New York: Alan R. Liss, pp 81–102

Wenk G, Hughey D, Boundy V, Kim A, Walker L, OltonD (1987): Neurotransmitters and memory: Role of cholinergic, serotonergic, and noradrenergic systems. *Behav Neurosci* 3:325–332

Westfall T (1974): Effect of muscarinic agonists on the release of 3H-norepinephrine and 3H-dopamine by potassium and electrical stimulation from rat brain slices. *Life Sci* 14:1641–1652

Whitehouse PJ (1986): Clinical and neurochemical consequences of neuronal loss in the nucleus basalis of Meynert in Parkinson's disease and Alzheimer's disease. *Adv Neurol* 45:393–397

Wilcock GK, Esiri MM, Bowen DM, Hughes AO (1988): The differential involvement of subcortical nuclei in senile dementia of Alzheimer's type. *J Neurol Neurosurg Psychiatry* 51:842–849

Yamamoto T, Hirano A (1985): Nucleus raphe dorsalis in Alzheimer's disease: Neurofibrillary tangles and loss of large neurons. *Ann Neurol* 17:573–577

Yates CM, Simpson J, Gordon A, Maloney AFJ, Allison Y, Ritchie IM, Urquhart A (1983): Catecholamines and cholinergic enzymes in pre-senile and senile Alzheimer-type dementia and Down's syndrome. *Brain Res* 280:119–126

9

The Importance of D_1 and D_2 Interactions with Nicotinic and Muscarinic Systems for Working Memory Function

EDWARD D. LEVIN AND JED E. ROSE

Introduction

It is clear that disruption of cholinergic systems can cause impairments of cognitive function (Bartus et al., 1987). For example, the muscarinic cholinergic antagonist scopolamine has been found by numerous studies to impair working memory performance (Levin, 1988). A wide variety of experiments have also shown that nicotinic cholinergic antagonists impair learning (Chiappeta and Jarvik, 1969; Diltz and Berry, 1967; Oliverio, 1966; Sansone et al., 1990) and memory function (Decker and Majchrzak, 1992; Dougherty et al., 1981; Elrod and Buccafusco, 1988, 1991; Flood et al., 1981; Hodges et al., 1991; Levin et al., 1987a, 1989a,b; Levin and Rose, 1990; McGurk et al., 1989a; Osborne et al., 1991; Riekkinen et al., 1990) and that muscarinic and nicotinic receptors have important interactions (Levin et al., 1989b, 1990b; Riekkinen et al., 1990) (see also Levin and Russell, this volume). However, it has also become increasingly clear that cholinergic systems do not function in isolation; they have important interactions with other transmitter systems in the neural bases of cognitive function (Decker and McGaugh, 1991; Levin et al., 1990a).

Dopamine systems have also been found to be important for memory function (Baratti et al., 1983; Beatty and Rush, 1983; Beninger, 1989; Levin et al., 1987b; McGurk et al., 1988, 1989a). Cholinergic systems have direct interactions with dopamine systems in several areas of the brain that are known to be important for memory function, such as the septum, the ventral tegmental area, the basal forebrain, the striatum, and the frontal cortex (for review see Levin et al., 1990a).

Cholinergic-dopaminergic interactions have long been known to be important for motor function (Barbeau, 1962; McGeer et al., 1961), and we wondered whether these two systems were functionally related in the neural substrate of cognitive function. Some preliminary information indicated that this may be true. Baratti et al. (1983) demonstrated that the ACh agonist oxotremorine can facilitate memory performance but that this effect is dependent upon intact DA mechanisms. Conversely, DA inhibition with haloperidol has been found to impair conditioned-avoidance performance (Arnt and Christensen, 1981) and to decrease operant responding (Ljungberg, 1988), effects that are attenuated by the muscarinic blocker scopolamine. In a series of studies, we have examined the interactions of D_1 and D_2 systems with nicotinic and muscarinic systems. Specifically, we investigated the actions of D_1 and D_2 agonists and antagonists on the working memory impairments caused by nicotinic or muscarinic receptor blockade. These investigations not only help elucidate how neural systems perform memory functions, they also point to novel avenues for therapeutic treatment of cognitive disorders.

Pharmacological Interactions

Nicotinic Blockade

In an initial study, we found that mecamylamine, a nicotinic antagonist, impaired spatial working memory performance of rats in the radial-arm maze (Levin et al., 1987a). This seemed to be a central effect of mecamylamine because no choice accuracy deficits were seen after peripheral nicotinic blockade with hexamethonium. In several succeeding experiments, we have seen the same impairment in radial-arm maze choice working memory performance caused by acute doses of mecamylamine (Levin et al., 1989a,b; Levin and Rose, 1990; McGurk et al., 1989a). A rather high dose of 10 mg/kg of mecamylamine was needed to produce a deficit in these studies. Clarke and Fibiger did not find significant memory deficits on a spatial alternation task from mecamylamine doses up to 5 mg/kg (Clarke and Fibiger, 1990). They pointed out that the high doses found to be effective in other studies may have their actions mediated via NMDA receptors. It is true that high doses of mecamylamine appear to interact with NMDA receptors (MacLeod et al., 1984; O'Dell and Christensen, 1988). However, Decker and Majchrzak (1992) found that the behavioral nature of performance disruption caused by ICV mecamylamine differed from the effects of the classic NMDA antagonist MK-801. Whereas MK-801 disrupts performance on both spatial memory and cued versions of a water-maze task (Robinson et al., 1989), mecamylamine was found to have a more specific effect of impairing spatial memory but not cued water-maze performance (Decker and Majchrzak, 1992). In addition, it has been found that high doses of mecamylamine are not necessarily required for producing working memory impairments in the radial-arm maze. Hodges et al. (Hodges et al., 1991) have found deficits at a dose of 2 mg/kg. We have found that a low dose of 2.5 mg/kg of mecamylamine causes radial-arm maze working memory deficits when given in combination with subthreshold doses of the muscarinic antagonist scopolamine (Levin et al., 1990b) or the DA antagonists haloperidol (McGurk et al., 1989a) or raclopride (McGurk et al., 1989b) (see below). Finally, Elrod and coworkers examined the amnestic effects of a high dose of 25 mg/kg of mecamylamine and found that it correlated quite well with the inhibition of ACh synthesis (Elrod and Buccafusco, 1991). Thus, antagonistic actions of mecamylamine at NMDA receptors seems less likely than its antagonistic actions at nicotinic receptors to be responsible for its effect of impairing working memory performance in the radial-arm maze.

There are significant interactions of DA ligands with the mecamylamine-induced working memory impairment. Low doses of 0.04 mg/kg of the DA antagonist haloperidol or 2.5 mg/kg of mecamylamine were not found to cause significant working memory impairments in the radial-arm maze when given separately. However, when given together they did cause a significant deficit (McGurk et al., 1989a).

The interaction of DA antagonists with the mecamylamine-induced working memory deficit seems to dependent upon specific actions at a particular DA receptor subtype. The selective D_2 antagonist raclopride (0.5 mg/kg) by itself did not cause a significant working memory deficit in the radial-arm maze. However, a combination of this dose of raclopride with a subthreshold dose of 2.5 mg/kg of mecamylamine did cause a significant working memory impairment (McGurk et al., 1989b). In contrast, coadministration of the D_1 antagonist SCH 23390 (0.05 mg/kg) with 2.5 mg/kg of mecamylamine did not produce a significant deficit.

This finding of a significant relationship between a D_2 antagonist and mecamylamine led to a study of the interactions of D_1 and D_2 agonists with the mecamylamine-induced deficit. The prediction was that since the D_2 antagonist potentiated the mecamylamine-induced deficit, a D_2 agonist should attenuate it. As we had seen previously, mecamylamine (10 mg/kg) in this study was found to cause a significant working memory deficit in the radial-arm maze. Coadministration of the D_2 agonist quinpirole (0.05 mg/kg) with mecamylamine reversed this deficit (Levin et al., 1989a). This facilitation occurred even though quinpirole by itself did not facilitate working memory performance and has previously been found to impair it (Levin and Bowman, 1986). The D_1 agonist SKF 38393 (3 mg/kg) had no interactive effect with mecamylamine. Recently Osborne and coworkers have also seen this effect of quinpirole reversing a mecamylamine-induced working memory deficit in the radial-arm maze (Osborne et al., 1991).

This interaction of nicotinic systems with D_2 receptors is complicated by the recent discovery of DA receptor subtypes in addition to D_1 and D_2 (Andersen et al., 1990), and the finding that quinpirole shows high-affinity binding to D_3 as well as D_2 receptors (Gehlert et al., 1991). It remains to be seen whether D_3 agonist actions of quinpirole contribute to its efficacy in reversing the mecamylamine-induced memory deficit.

Muscarinic Blockade

Scopolamine, a muscarinic antagonist, has been found by a great many studies to impair working memory performance in the radial-arm maze (Levin, 1988). Haloperidol, a DA antagonist, also impairs working memory performance in the radial-arm maze (Beatty and Rush, 1983; Levin et al., 1987b; McGurk et al., 1988, 1989a). Nevertheless, in two studies we have found that haloperidol (0.04–0.0625 mg/kg) is effective in attenuating the impairment of radial-arm maze working memory performance caused by 0.15 mg/kg of scopolamine (McGurk et al., 1988, 1989a).

Interestingly, haloperidol has been found to facilitate performance in a latent inhibition model of attentiveness (Feldon and Weiner, 1991; Weiner and Feldon, 1987). A variety of evidence points to a critical effect of scopolamine being an impairment of attentional processes (Eckerman and Bushnell, 1992). Perhaps in our studies haloperidol helped reverse the scopolamine-induced radial-arm maze choice accuracy deficit because it attenuated the attentional deficit caused by muscarinic blockade.

In contrast to nicotinic systems, we have found that muscrinic systems seem to more closely interact with D_1 than with D_2 systems. In a study following up the finding of the haloperidol-induced reversal of the scopolamine effect, we showed that the D_1 antagonist SCH 23390 (0.10 mg/kg) reversed the radial-arm maze working memory deficit caused by scopolamine. SCH 23390 by itself did not affect choice accuracy. The D_2 antagonist raclopride (5 mg/kg) was not found to significantly interact with the scopolamine effect.

Consistent with our previous study with D_1 and D_2 antagonists, in a study of D_1 and D_2 agonists we found that a D_1 (SKF 38393) but not a D_2 agonist (quinpirole) had significant interactions with the scopolamine-induced working memory deficit in the radial-arm maze (Levin and Rose, 1991). However, surprisingly, the D_1 agonist SKF 38393 had a similar effect as the D_1 antagonist SCH 23390 in attenuating the scopolamine-induced choice accuracy deficit. The low dose of 4 mg/kg of SKF 38393 cut by about half the working memory deficit caused by scopolamine (0.05 mg/kg), while the high dose of 8 mg/kg nearly eliminated it. This higher dose of SKF 38393 also reduced the pronounced working memory deficit caused by a higher dose of scopolamine (0.15 mg/kg). By itself SKF 38393 did not improve choice accuracy performance. The D_2 agonist quinpirole (0.04 and 0.08 mg/kg) did not significantly affect the scopolamine-induced choice accuracy deficit in the radial-arm maze.

Interestingly, Osborne and coworkers (Osborne et al., 1991) have found that a lower dose of quinpirole (0.01 mg/kg) is effective in counteracting the choice accuracy deficit caused by a low dose of 0.04 mg/kg of scopolamine. Perhaps lower doses of quinpirole are effective in this regard. Inspection of our data set shows that the lower dose of 0.04 mg/kg of quinpirole showed a trend toward attenuating the deficit caused by the lower dose of scopolamine (0.05 mg/kg). It may be the case that lower doses of quinpirole, like other DA agonists such as apomorphine, act more preferentially at presynaptic sites and reduce DA efflux. This would have a similar effect as an antagonist in reducing net DA actions. Thus, a low dose of quinpirole could have a similar action as the DA antagonists haloperidol and SCH 23390 in counteracting the scopolamine-induced radial-arm maze working memory deficit.

Drug–Lesion Interactions

Another way to investigate DA-ACh interactions is to examine the effects of DA ligands on impairments caused by lesions of cholinergic projections. Rats given knife-cut lesions of the medial ACh projection from the basal forebrain to the cortex show a pronounced choice accuracy deficit over the next 5 days of testing and then recovery near-baseline performance (McGurk et al., 1991, 1992). A recent study by McGurk, Levin, and Butcher (McGurk et al., 1992) examined the interaction of D_1 and D_2 ligands with the effects of this lesion on working memory performance. During the first 5 days after the lesion, the controls had 6.56 ± 0.27 (mean $\pm$ SEM) entries to repeat, while the untreated lesioned group had 4.33 ± 0.31 entries to repeat, a significant deficit ($p < 0.005$). The same D_1 and D_2 ligands that had been found to counteract the working memory deficits caused by nicotinic and muscarinic antagonists were tested for their effects in counteracting the working memory deficits caused by this lesion. Daily injection of the D_2 agonist quinpirole, which had been found to reverse the mecamylamine-induced deficit, attenuated the lesion-induced deficit. The lesioned rats given quinpirole had 6.07 ± 0.53 entries to repeat, a significant elevation in working memory performance over untreated lesioned rats ($p < 0.005$). As had been seen with mecamylamine administration, the D_2 antagonist raclopride had no beneficial effect in reversing the lesion-induced working memory deficit (mean = 4.88 ± 0.22). However, unlike the case with mecamylamine raclopride did not potentiate the deficit. Administration of the D_1 antagonist SCH 23390, which had been found to reverse the scopolamine-induced deficit, also attenuated the lesion-induced deficit. The lesion + SCH 23390 group had 5.64 ± 0.36 entries to repeat, a significant improvement over the untreated lesioned group ($p < 0.01$). As with scopolamine, the D_1 agonist SKF 38393 showed surprising efficacy. It was the most effective treatment, completely reversing the lesion-induced deficit. Lesioned rats given SKF 38393 had 6.62 ± 0.24 entries to repeat, significantly higher than the untreated lesioned rats ($p < 0.005$). In summary, the very same DA ligands that counteracted nicotinic and muscarinic blockade-induced working memory deficits were effective in reversing the deficit due to a lesion of the medial cholinergic projection from the basal forebrain to the cortex.

Anatomic Considerations

Likely sites for the DA-ACh interactions that we have seen may involve septohippocampal, basalocortical, striatal, and midbrain cholinergic systems. Each of these areas has important interactions with DA systems. Information regarding their relative importance for memory function varies from system to system. Best established is the importance of the septohippocampal ACh system. The basalocortical ACh system has received considerable attention, but there is debate concerning the necessity of this projection for adequate memory performance. Striatal ACh neurons were once thought to be mainly engaged in motor function, but recent studies have provided information about its possible importance for memory function. Little is known regarding the importance of the midbrain pedunculopontine ACh system, but its critical influence over the midbrain DA neurons in the ventral tegmental area and the substantia nigra suggets that it may be important for DA-ACh interactins with regard to memory function.

The DA innervation of the septum may be a critical site for the observed interactions between DA and muscarinic antagonists. Dopaminergic neurons from the ventral tegmental area send projections to the medial margin of the lateral septum. From there, probably via GABA interneurons, they exert an inhibitory influence on septohippocampal cholinergic cells located in the medial septal nucleus (Costa et al., 1983; Robinson et al., 1979). There is good evidence that septohippocampal ACh neurons are involved in the scopolamine-induced cognitive disruption in the radial-arm maze and other memory tasks (Dunnett et al., 1981). Local application of the dopaminergic antagonist haloperidol into the lateral septum stimulates efflux of ACh in the hippocampus (Durkin et al., 1986). This was paralleled by increased hippocampal pyramidal cell activity and accelerated extinction of a conditioned behavioral response. This group also found that selective lesions of the DA projections to the septum increased hippocampal ACh release, facilitated spontaneous alternation performance, and caused improvements in acquisition and reversal of T-maze spatial discrimination (Galey et al., 1985). A similar facilitation of reversal learning has been seen by Harrell et al. (Harrell et al., 1984) after 6-OHDA lesions in the medial septum. However, the effects of these types of lesions are complex in that another group has found that 6-OHDA lesions of the lateral septum caused impaired memory performance (Simon et al., 1986; Taghzouti et al., 1986).

DA inhibition of septohippocampal ACh cells may explain our finding that DA blockade is effective in reversing the scopolamine-induced choice accuracy deficit, DA blockade would release the septohippocampal ACh cells from inhibition, and the resultant increased ACh activity could help to overcome the effects of postsynaptic muscarinic blockade. The dopaminergic blockers used in our studies, haloperidol and SCH 23390, may have been acting in this manner to antagonize the dopamine inhibition of septohippocampal cholinergic cells. The increased firing of this cholinergic projection would then serve to overcome the understimulation of cholinoceptive cells in the hippocampus that was caused by the muscarinic antagonist scopolamine.

It is possible that the observed DA-muscarinic interactions take place directly in the hippocampus. There is DA innervation of the hippocampus (Bischoff, 1986; Bischoff et al., 1979; Ishikawa et al., 1982) from the VTA as well as the substantia nigra (Scatton et al., 1980). However, DA concentrations in the hippocampus are much lower than those in the septum (approximately 1:60) and the striatum (approximately 1:900) (Scatton et al., 1980). There seems to be a biphasic effect of DA in the hippocampus. DA applied directly to hippocampal slices causes an initial decrease in the population response of CA 1 pyramidal cells (Benardo and Prince, 1982; Gribkoff and Ashe, 1984), followed by a prolonged potentiation, which can be blocked by spiroperidol (Gribkoff and Ashe, 1984). There is binding of both D_1 and D_2 receptors in the hippocampus (Bischoff, 1986; Boyson et al., 1986; Bruinink and Bischoff, 1986; Dawson et al., 1985; Dubois et al., 1986). These receptor subtypes have opposite effects, with D_1 receptor activation inhibiting the firing of CA 1 pyramidal cells and D_2 receptor activation promoting the firing of CA 1 pyramidal cells (Smialowski and Bijak, 1987).

However, behavioral evidence indicates that the critical effect of DA antagonists in reversing the scopolamine-induced deficit is probably not mediated by their effect directly in the hippocampus. Brito and Brito (1991) (see also Brito in this volume) have found that the D_1 antagonist SCH 23390 administered into the hippocampus did not counteract the adverse effects of scopolamine on working memory performance in a T maze. In fact, coadministration of scopolamine and SCH 23390 into the hippocampus caused an impairment in reference memory performance in the T maze that was not apparent with either drug alone. Therefore, it appears the beneficial effects of DA antagonists in reversing scopolamine-induced working memory deficits was much more likely mediated via actions on ACh cells in the septum rather than ACh terminals in the hippocampus. A critical effect in the hippocampus may more likely explain the D_1 agonist reversal of the scopolamine-induced working memory deficit.

Another possible site for DA-ACh interactions is the striatum. This area has classically been thought to be mainly involved in motor function, but recently compelling evidence has been gathered to demonstrate its involvement in cognitive function as well (Packard et al., 1989; Pisa et al., 1980; Whishaw et al., 1987). Lesions in the striatum impair passive-avoidance learning and retention (Sanberg et al., 1978), T-maze spatial alternation and reversal learning (Pisa et al., 1980; Thompson and Yang, 1982), and Morris water-maze performance (Whishaw et al., 1987). 6-OHDA lesions of the DA innervation of the striatum increased errors on a conditioned discrimination task (Robbins and Everitt, 1987), Striatal lesions do not impair working memory performance when the radial-arm maze is utilized in the typical win-shift paradigm; however, when the test is run in a win-stay version, rats with striatal lesions show substantial deficits in working memory performance (Packard et al., 1989). Critical to understanding the deficits seen after striatal lesions may be the distinction between egocentric (position relative to the body) and allocentric (position invariant in space) memory. Cook and Kesner (Cook and Kesner, 1988) found that rats with striatal lesions show deficits on radial-arm maze paradigms requiring egocentric memory but not on tasks requiring allocentric memory. The substantia nigra also has been found to be important for cognitive function. Rats with substantia nigra lesions show deficits in learning and retention of spatial reversal problem in an aversively motivated T maze (Thompson and Yang, 1982).

In the striatum, DA projections from the substantia nigra inhibit cholinergic interneurons (Sethy and Wenger, 1985). This inhibition seems to preferentially involve the actions of D_2 receptors (Arbilla et al., 1987). There are also connections from the ACh interneurons on the terminals of the DA projections (De Belleroche et al., 1979). This ACh interaction with DA release is different for the subtypes of cholinergic receptors. DA release can be stimulated by either a muscarinic antagonist (Fernando et al., 1986) or a nicotinic agonist (Giorguieff-Chesselet et al., 1979). The D_2-mediated inhibition of striatal ACh cells may underlie the observed interactions between mecamylamine and DA ligands. Mecamylamine may act to reduce DA release by a presynaptic action. The result of this effect (decreased D_2 receptor stimulation) would be augmented by a D_2 antagonist and attenuated by a D_2 agonist.

The midbrain dopaminergic nuclei, the substantia nigra, and the ventral area (VTA) may be important for DA-ACh interactions in several ways. The substantia nigra provides DA input that affects ACh cells in the striatum. The VTA, as described above, provides DA innervation of the septum, influencing ACh projections to the hippocampus. In addition, both the nigra and the VTA have high densities of nicotinic receptors (Clarke and Pert, 1985). The source of this innervation has been identified as the pedunculopontine nucleus (Woolf and Butcher, 1986). Nicotinic stimulation has excitatory influences on the activity of these DA cells and increases DA release (Clarke et al., 1985; Meru et al., 1987), while nicotinic antagonist administration has been found to inhibit DA release (Ahtee and Kaakkola, 1978; Haikala and Ahtee, 1988). The choice accuracy deficits seen after mecamylamine administration may involve the underactivity of dopamine neurons originating in the substantia nigra and/or VTA. The efficacy of the D_2 agonist quinpirole in reversing the mecamylamine-induced deficit may result from overcoming the dopaminergic underactivity with postsynaptic agonist actions. Conversely, application of the D_2 antagonist raclopride would act to augment this effect.

The basal forebrain is another likely site for DA-ACh interactions. Nonspecific lesions of neurons in the nucleus basalis have been found to impair cognitive performance of rats in the radial-arm maze, the T-maze, and the Morris water maze (Dekker et al., 1991). However, several recent studies have found that the basalocortical ACh projection may not by itself be critical for accurate performance of working memory tasks (Robbins et al., 1989; Wenk et al., 1989; Whishaw et al., 1987). DA fibers from the substantia nigra and VTA innervate this area (Jones and Cuello, 1989; Vertes, 1988) and have an excitatory influence on the basalocortical ACh projection (Casamenti et al., 1986). This mechanism for DA-ACh interactions involving basal forebrain systems may account for the increase in muscarinic receptor number in the cortex following chronic administration of haloperidol (Pazo et al., 1987). The density of ACh cells in the nucleus basalis in the rat is lower than in primates, including humans (Saper, 1984); however, the impairment of memory for discrete events seen after basalis lesions in rats is similar to deficits seen in monkeys with these lesions and in humans with Alzheimer's disease, who have substantial damage in this area (Bartus et al., 1986; Kesner et al., 1986).

The effectiveness of the D_1 and D_2 ligands in reversing the adverse effects of the medial basalocortical cholinergic lesions shows that their action is not mediated via actions influencing the medial cholinergic pathway. A more likely pathway mediating the effect of the D_1 agonist SKF 38393 is DA projection to the frontal cortex. This innervation has been shown in primates to be important for working memory function (Brozoski et al., 1979). Recently it has been shown that D_1 receptor stimulation is especially important for working memory function in monkeys (Sawaguchi and Goldman-Rakic, 1991). Frontal DA innervation has been found to be important for working memory in rats as well; however, its relevance to performance in uninterrupted tasks in the radial-arm maze, such as we have used, is less clear (Bubser and Schmidt, 1990).

Conclusions

There are potent interactions between cholinergic and dopaminergic systems with regard to working memory performance. These interactions and their nature are specifically related to the different receptor subtypes of these systems: nicotinic and muscarinic cholinergic receptors and D_1 and D_2 dopaminergic receptors. Sites of DA-ACh interactions have been identified that are likely to be substrates for the observed interactions. Lesion studies and site-specific drug infusion studies will help determine which of these systems are critical for the observed systemic drug interactions.

An interesting outcome of these studies was that both the D_1 agonist SKF 38393 and the D_1 antagonist SCH 23390 were effective in counteracting the working memory deficits caused by the muscarinic antagonist scopolamine or a knife-cut lesion of the medial basalocortical cholinergic projection. There are several possible explanations for these results. Drug specificity may be an issue. SCH 23390 may be effective because of its activity as a 5-HT_2 antagonist (Bischoff et al., 1986). Serotonin inhibits ACh release in the hippocampus, an effect that is reversed by 5-HT_2 blockers (Muramatsu et al., 1988). Another possibility is that there is some idiosyncrasy of the radial-arm maze by which either D_1 hyperfunction or hypofunction is beneficial. It also may be the case that SCH 23390 and SKF 38393 do not act on exactly the same population of D_1 receptors. The observed results could be explained by one of the drugs preferentially acting at a certain subclass of D_1 receptors and the other preferentially acting at others. Inasmuch as these potentially different D_1 receptor subtypes may have different anatomic localization, the proposed studies may discover different critical loci for the therapeutic actions of SCH 23390 and SKF 38393 in the local infusion studies. As reviewed above, DA input has an inhibitory influence over septohippocampal ACh cells. Application of a D_1 antagonist in this area should help overcome anticholinergic effects. In contrast, DA stimulation in the nucleus accumbens (Taghzouti et al., 1986) and frontal cortex (Brozoski et al., 1979; Sawaguchi and Goldman-Rakic, 1991) is important for working memory function. D_1 receptors in the frontal cortex are particularly important for working memory performance (Sawaguchi and Goldman-Rakic, 1991). These may be likely sites for the beneficial effects of the D_1 agonist SKF 38393.

The efficacy of the D_1 and D_2 ligands in reversing the working memory deficits caused by muscarinic or nicotinic ACh blockade suggests that an approach to therapy using DA drugs may be useful in treating the memory impairment in disorders characterized by cholinergic underactivity such as Alzheimer's disease.

Acknowledgments. We would like to thank the support of the Alzheimer's Association/ Neil Bluhm Pilot Research Program and The Council for Tobacco Research USA for grants awarded to E.D.L. and the Medical Research Service of the Dept. of Medical Affairs and NIDA (DA 02665) for grant awarded to J.E.R.

References

Ahtee L, Kaakkola S (1978): Effect of mecamylamine on the fate of dopamine in striatal and mesolimbic areas of rat brain: Interaction with morphine and haloperidol. *Br J Pharmacol* 62:213–218

Andersen PH, Gingrich JA, Bates MD, Dearry A, Falardeau P, Senogles SE, Caron MG (1990): Dopamine receptors: Beyond the D1/D2 classification. *Trend Pharmacol Sci* 11:231–236

Arbilla S, Nowak JZ, Langer SZ (1987): Dopamine receptor mediated inhibition of ^{3}H-acetylcholine release. In: *Neurotransmitter Interactions in the Basal Ganglia*, Sandler M, Feuerstein C, Scatton B, eds. New York: Raven Press, pp 111–120

Arnt J, Christensen AV (1981): Differential reversal by scopolamine and THIP of the antistereotypic effects of neuroleptics. *Eur J Pharmacol* 69:107–111

Baratti CM, Introini IB, Huygens P, Gusovsky F (1983): Possible cholinergic-dopaminergic link in memory facilitation induced by oxotremorine in mice. *Psychopharmacology* 80:161–165

Barbeau A (1962): The pathogenesis of Parkinson's disease: A new hypothesis. *Can Med Assoc J* 87:802–807

Bartus RT, Flicker C, Dean RL, Fisher S, Pontecorvo M, Figueiredo J (1986): Behavioral and biochemical effects of nucleus basalis magnocellularis lesions: Implications and possible relevance to understanding or treating Alzheimer's disease. *Prog Brain Res* 70:345–361

Bartus RT, Deal RL, Flicker C (1987): Cholinergic psychopharmacology: An integration of human and animal research on memory. In: *Psychopharmacology: The Third Generation of Progress*, Meltzer HY, ed. New York: Raven Press, pp 219–232

Beatty WW, Rush RA (1983): Spatial working memory in rats: Effects of monoaminergic antagonists. *Pharmacol Biochem Behav* 18:7–12

Benardo LS, Prince DA (1982): Dopamine action on hippocampal pyramidal cells. *J Neurosci* 2:415–423

Beninger RJ (1989): Dissociating the effects of altered dopaminergic function on performance and learning. *Brain Res Bull* 23:365–371

Bischoff S (1986): Mesohippocampal dopamine system. In: *The Hippocampus*, Isaacson RL, Pribram KH, eds. New York: Plenum Press

Bischoff S, Scatton B, Korf J (1979): Biochemical evidence for a transmitter role of dopamine in the rat hippocampus. *Brain Res* 165:161–165

Bischoff S, Heinrich M, Sontag J, Kraus J (1986): The D-1 dopamine receptor antagonist SCH 23390 also interacts potently with brain serotonin (5-HT2) receptors. *Eur J Pharmacol* 129:367–370

Boyson SJ, McGonigle P, Molinoff PB (1986): Quantitative autoradiographic localization of the D_1 and D_2 subtypes of dopamine receptors in rat brain. *J Neurosci* 6: 3177–3188

Brito GNO, Brito LSO (1991): DA-ACh interactions in the hippocampus and memory in the rat. *Soc Neurosci Abstr* 17:138

Brozoski TJ, Brown RM, Rosvold HE, Goldman PS (1979): Cognitive deficit caused by regional depletion of dopamine in prefrontal cortex of rhesus monkey. *Science* 205:929–931

Bruinink A, Bischoff S (1986): Detection of dopamine receptors in homogenates of rat hippocampus and other brain areas. *Brain Res* 386:78–83

Bubser M, Schmidt WJ (1990): 6-Hydroxydopamine lesion of the rat prefrontal cortex increases locomotor activity, impairs acquisition of delayed alternation tasks but does not affect uninterrupted tasks in the radial maze. *Behav Brain Res* 37:157–168

Casamenti F, Deffenu G, Abbamondi AL, Pepeu G (1986): Changes in cortical acetylcholine output induced by modulation of the nucleus basalis. *Brain Res Bull* 16:689–695

Chiappeta L, Jarvik ME (1969): Comparison of learning impairment and activity depression produced by two classes of cholinergic blocking agents. *Arch Int Pharmacodyn* 179:161–166

Clarke PBS, Fibiger HC (1990): Reinforced alternation performance is impaired by muscarinic but not by nicotinic receptor blockade in rats. *Behav Brain Res* 36:203–207

Clarke PBS, Pert A (1985): Autoradiographic evidence for nicotine receptors on nigrostriatal and mesolimbic dopaminergic neurons. *Brain Res* 348:355–358

Clarke PBS, Hommer DW, Pert A, Skirboll LR (1985): Electrophysiological actions of nicotine on substantia nigra single units. *Br J Pharmacol* 85:827–835

Cook D, Kesner RP (1988): Caudate nucleus and memory for egocentric localization. *Behav Neural Biol* 49:332–343

Costa E, Panula P, Thompson HK, Cheney DL (1983): The transsynaptic regulation of the septo-hippocampal cholinergic neurons. *Life Sci* 32:165–179

Dawson T, Gehlert D, Yamamura H, Barnett A, Wamsley JK (1985): D1 dopamine receptors in the rat brain: Autoradiographic localization using [3H]SCH 23390. *Eur J Pharmacol* 108:323–325

De Belleroche J, Lugmani Y, Bradford HF (1979): Evidence for presynaptic cholinergic receptors on dopaminergic terminals: Degeneration studies with 6-hydroxydopamine. *Neurosci Lett* 11:209–213

Decker MW, Majchrzak MJ (1992): The effects of systemic and intracerebroventricular administration of mecamylamine, a nicotinic cholinergic antagonist, on spatial memory in rats. *Psychopharmacology*, 107:530–534

Decker MW, McGaugh JL (1991): The role of interactions between the cholinergic system and other neuromodulatory systems in learning and memory. *Synapse* 7:151–168

Dekker AJAM, Connor DJ, Thal LJ (1991): The role of cholinergic projections from the nucleus basalis in memory. *Neurosci Biobehav Rev* 15:299–317

Diltz SL, Berry CA (1967): Effect of cholinergic drugs on passive avoidance in the mouse. *J Pharmacol Exp Ther* 158:279–285

Dougherty J, Miller D, Todd G, Kostenbauder HB (1981): Reinforcing and other behavioral effects of nicotine. *Neurosci Biobehav Rev* 5:487–495

Dubois A, Avasta M, Curet O, Scatton B (1986): Autoradiographic distribution of the D1 antagonist [3H]SKF 38393 in the rat brain and spinal cord. Comparison with the distribution of D2 dopamine receptors. *Neuroscience* 19:125–138

Dunnett SB, Low W, Bunch ST, Thomas SR, Iversen SD, Lewis SD, Björklund PR, Stenevi U (1981): Septal transplant reinnervation of the hippocampus: Cholinergic enhancement of radial maze performance. *Behav Brain Res* 2:258–259

Durkin T, Galey D, Micheau J, Beslon H, Jaffard R (1986): The effects of intraseptal injection of haloperidol in vivo in hippocampal cholinergic function in the mouse. *Brain Res* 376:420–424

Eckerman DA, Bushnell PJ (1992): The neurotoxicology of Cognition: Attention, learning and memory. In: *Neurotoxicology*, Tilson HA, Mitchell CL, eds. New York: Raven Press, pp 213–270

Elrod K, Buccafusco JJ (1988): Selective central nicotinic receptor blockade inhibits passive but not active blockade in rats. *Soc Neurosci Abstr* 14:57

Elrod K, Buccafusco JJ (1991): Correlation of the amnestic effects of nicotinic antagonists with inhibition of regional brain acetylcholine synthesis in rats. *J Pharmacol Exp Ther* 258:403–409

Feldon J, Weiner I (1991): An animal model of attention deficit. In: *Neuromethods, Vol. 18: Animal Models in Psychiatry I*, Boulton A, Baker G, Martin-Iverson M, eds. Clifton, NJ: Humana Press, pp 313–361

Fernando JC, Hoskins B, Ho IK (1986): The role of dopamine in behavioral supersensitivity to muscarinic antagonists following cholinesterase inhibition. *Life Sci* 39:2169–2176

Flood JF, Landry DW, Jarvik ME (1981): Cholinergic receptor interactions and their effects on long-term memory processing. *Brain Res* 215:177–185

Galey D, Durkin T, Sifakis G, Kempf E, Jaffard R (1985): Facilitation of spontaneous and learned spatial behaviours following 6-hydroxydopamine lesions of the lateral septum: A cholinergic hypothesis. *Brain Res* 340:171–174

Gehlert DR, Gackenheimer SL, Seeman P, Schaus JM (1991): Localization of ^{3}H-quinpirole binding to D-2 and D-3 receptors in rat brain. *Soc Neurosci Abstr* 17:820

Giorguieff-Chesselet MF, Kemel ML, Wandscheer D, Glowinski J (1979): Regulation of dopamine release by presynaptic nicotinic receptors in rat striatal slices: Effect of nicotine in a low concentration. *Life Sci* 25:1257–1262

Gribkoff VK, Ashe JH (1984): Modulation by dopamine of population responses and cell membrane properties of hippocampal CA1 neurons in vitro. *Brain Res* 292:327–338

Haikala H, Ahtee L (1988): Antagonism of the nicotine-induced changes of the striatal dopamine metabolism in mice by mecamylamine and pempidine. *Naunyn-Schmiedberg's Arch Pharmacol* 338:169–173

Harrell LE, Barlow TS, Miller M, Haring JH, Davis JN (1984): Facilitated reversal learning of a spatial-memory task by medial septal injections of 6-hydroxydopamine. *Exp Neurol* 85:69–77

Hodges H, Gray JA, Allen Y, Sinden J (1991): The role of the forebrain cholinergic projection system in performance in the radial-arm maze in memory-impaired rats. In: *Effects of Nicotine on Biological Systems*, Adlkofer F, Thurau K, eds. Boston: Berkhäuser Verlag, pp 389–399

Ishikawa K, Ott T, McGaugh JL (1982): Evidence for dopamine as a transmitter in dorsal hippocampus. *Brain Res* 232:222–226

Jones BE, Cuello AC (1989): Afferents to the basal forebrain cholinergic cell area from pontomesencephalic-catecholamine, serotonin, and acetylcholine-neurons. *Neuroscience* 31:37–61

Kesner RP, Crutcher KA, Measom MO (1986): Medial septal and nucleus basalis magnocellularis lesions produce order memory deficits in rats which mimic symptomatology of Alzheimer's disease. *Neurobiol Aging* 7:287–295

Levin ED (1988): Psychopharmacological effects in the radial-arm maze. *Neurosci Biobehav Rev* 12:169–175

Levin ED, Bowman RE (1986): Effects of the dopamine D_2 receptor agonist, LY 171555, on radial arm maze performance in rats. *Pharmacol Biochem Behav* 25:83–88

Levin ED, Rose JE (1990): Anticholinergic sensitivity following chronic nicotine administration as measured by radial-arm maze performance in rats. *Behav Pharmacol* 1:511–520

Levin ED, Rose JE (1991): Interactive effects of D_1 and D_2 agonists with scopolamine on radial-arm maze performance. *Pharmacol Biochem Behav* 38:243–246

Levin ED, Castonguay M, Ellison GD (1987a): Effects of the nicotinic receptor blocker, mecamylamine, on radial arm maze performance in rats. *Behav Neural Biol* 48:206–212

Levin ED, Galen D, Ellison GD (1987b): Chronic haloperidol effects on radial arm maze performance and oral movements in rats. *Pharmacol Biochem Behav* 26:1–6

Levin ED, McGurk SR, Rose JE, Butcher LL (1989a) Reversal of a mecamylamine-induced cognitive deficit with the D_2 agonist, LY 171555. *Pharmacol Biochem Behav* 33:919–922

Levin ED, McGurk SR, South D, Butcher LL (1989b): Effects of combined muscarinic and nicotinic blockade on choice accuracy in the radial-arm maze. *Behav Neural Biol* 51:270–277

Levin ED, McGurk SR, Rose JE, Butcher LL (1990a): Cholinergic-dopaminergic interactions in cognitive performance. *Behav Neural Biol* 54:271–299

Levin ED, Rose JE, McGurk SR, Butcher LL (1990b): Characterization of the cognitive effects of combined muscarinic and nicotinic blockade. *Behav Neural Biol* 53:103–112

Ljungberg T (1988): Scopolamine reverses haloperidol-attenuated lever-pressing for water but not haloperidol-attenuated water intake in the rat. *Pharmacol Biochem Behav* 29:205–208

MacLeod NK, James TA, Starr MS (1984): Muscarinic action of acetylcholine in the rat ventromedial thalamic nucleus. *Exp Brain Res* 55:553–561

McGeer PL, Boulding JE, Gibson WC, Foulkes RG (1961): Drug-induced extra-pyramidal reactions. *JAMA* 177:665–670

McGurk S, Levin ED, Butcher LL (1988): Cholinergic-dopaminergic interactions in radial-arm maze performance. *Behav Neural Biol* 49:234–239

McGurk SR, Levin ED, Butcher LL (1989a): Nicotinic-dopaminergic relationships and radial-arm maze performance in rats. *Behav Neural Biol* 52:78–86

McGurk SR, Levin ED, Butcher LL (1989b): Radial-arm maze performance in rats is impaired by a combination of nicotinic-cholinergic and D_2 dopaminergic drugs. *Psychopharmacology* 99:371–373

McGurk SR, Levin ED, Butcher LL (1991): Impairment of radial-arm maze performance in rats following lesions of the cholinergic medial pathway: Reversal by arecoline and differential effects of muscarinic and nicotinic antagonists. *Neuroscience*, 44:137–147

McGurk SR, Levin ED, Butcher LL (1992): Dopaminergic drugs reverse the impairment of radial-arm maze performance caused by lesions involving the cholinergic medial pathway. *Neuroscience*, Vol 50

Meru G, Yoon KP, Boi V, Gessa GL, Naes L, Westfall TC (1987): Preferential stimulation of ventral tegmental area dopaminergic neurons by nicotine. *Eur J Pharmacol* 141:395–399

Muramatsu M, Tamaki-Ohashi J, Usuki C, Araki H, Aihara H (1988): Serotonin-2 receptor-mediated regulation of release of acetylcholine by minaprine in cholinergic nerve terminal of hippocampus of rat. *Neuropharmacology* 27:603–609

O'Dell TJ, Christensen BN (1988): Mecamylamine is a selective non-competitive antagonist of N-methyl-D-aspartate-and aspartate-induced currents in horizontal cells dissociated from the catfish retina. *Neurosci Lett* 94:93–98

Oliverio A (1966): Effects of mecamylamine on avoidance conditioning and maze learning of mice. *J Pharmacol Exp Ther* 154:350–356

Osborne B, Liddell R, Feeley J (1991): The effects of scopolamine, mecamylamine and quinpirole on radial-arm maze performance of rats with fornix transections. *Soc Neurosci Abstr* 17:136

Packard MG, Hirsh R, White NM (1989): Differential effects of fornix and caudate nucleus lesions on two radial maze tasks: Evidence for multiple memory systems. *J Neurosci* 9:1465–1472

Pazo JH, Levi de Stein M, Jerusalinsky D, Novas ML, Raskovsky S, Tumilasci OR, Medina JH, De Robertis E (1987): Selective increase of α_1-adrenoceptors and muscarinic cholinergic receptors in rat cerebral cortex after chronic haloperidol. *Brain Res* 414:405–408

Pisa M, Sanberg PR, Fibiger HC (1980): Locomotor activity, exploration and spatial alternation learning in rats with striatal injections of kainic acid. *Physiol Behav* 24:11–19

Riekkinen PJ, Sirviö J, Aaltonen M, Riekkinen P (1990): Effects of concurrent manipulations of nicotinic and muscarinic receptors on spatial and passive avoidance learning. *Pharmacol Biochem Behav* 37:405–410

Robbins TW, Everitt BJ (1987): Comparative functions of the central noradrenergic, dopaminergic and cholinergic systems. *Neuropharmacology* 26:893–901

Robbins TW, Everitt BJ, Ryan CN, Marston HM, Jones GH, Page KJ (1989): Comparative effects of quisqualic acid and ibotenic acid-induced lesions of the substantia innominata and globus pallidus on the acquisition of a conditional visual discrimination: Differential effects on cholinergic mechanisms. *Neuroscience* 28:337–352

Robinson GS, Jr., Crooks GB, Jr., Shinkman PG, Gallagher M (1989): Behavioral effects of MK-801 mimic deficits associated with hippocampal damage. *Psychobiology* 17:156–164

Robinson SE, Malthe-Sørenssen D, Wood PL, Commissiong J (1979): Dopaminergic control of the septo-hippocampal cholinergic pathway. *J Pharmacol Exp Ther* 208:476–479

Sanberg PR, Lehmann J, Fibiger HC (1978): Impaired learning and memory after kainic acid lesions of the striatum: A behavioral model of Huntington's disease. *Brain Res* 149:546–551

Sansone M, Castellano C, Battaglia M, Ammassari-Teule M (1990): Oxiracetam prevents mecamylamine-induced impairment of active, but not passive, avoidance learning in mice. *Pharmacol Biochem Behav* 36:389–392

Saper CB (1984): Organization of cerebral cortical afferent systems in the rat. II. Magnocellular basal nucleus. *J Comp Neurol* 222:313–342

Sawaguchi T, Goldman-Rakic PS (1991): D1 dopamine receptors in prefrontal cortex: Involvement in working memory. *Science* 251:947–950

Scatton B, Simon H, LeMoal M, Bischoff S (1980): Origin of dopaminergic innvervation of the rat hippocampal formation. *Neurosc Lett* 18:125–131

Sethy VH, Wenger D (1985): Effect of dopaminergic drugs on striatal acetylcholine concentration. *J Pharm Pharmacol* 37:73–74

Simon H, Taghzouti K, Le Moal M (1986): Deficits in spatial-memory tasks following lesions of septal dopaminergic terminals in the rat. *Behav Brain Res* 19:7–16

Smialowski A, Bijak M (1987): Excitatory and inhibitory action of dopamine on hippocampal neurons in vitro. Involvement of D_2 and D_1 receptors. *Neuroscience* 23:95–101

Taghzouti K, Simon H, Le Moal M (1986): Disturbances in exploratory behavior and functional recovery in the Y and radial mazes following dopamine depletion of the lateral septum. *Behav Neural Biol* 45:48–56

Thompson R, Yang S (1982): Retention of individual spatial reversal problems in rats with nigral, caudoputaminal, and reticular formation lesions. *Behav Neural Biol* 34:98–103

Vertes RP (1988): Brainstem afferents to the basal forebrain in the rat. *Neuroscience* 24:907–935

Weiner I, Feldon J (1987): Facilitation of latent inhibition by haloperidol. *Psychopharmacology* 91:248–253

Wenk GL, Markowska AL, Olton DS (1989): Basal forebrain lesions and memory: Alterations in neurotensin, not acetylcholine, may cause amnesia. *Behav Neurosci* 103:765–769

Whishaw IQ, Mittleman G, Bunch ST, Dunnett SB (1987): Impairments in the acquisition, retention and selection of spatial strategies after medial caudate-putamen lesions in rats. *Behav Brain Res* 24:125–138

Woolf NJ, Butcher LL (1986): Cholinergic systems in the rat brain: III. Projections from the pontomesencephalic tegmentum to the thalamus, tectum, basal ganglia, and basal forebrain. *Brain Res Bull* 16:603–637

10

Neurotransmitter Systems in Hippocampus and Prelimbic Cortex, Dopamine-Acetylcholine Interactions in Hippocampus, and Memory in the Rat

GILBERTO N.O. BRITO

Introduction

The neurobiology of memory continues to attract the attention of researchers in several disciplines. Memory is being studied at the molecular, cellular, systems, and organismic levels in experimental animals (Matthies, 1989). Additionally, advances in neuroimaging techniques have provided significant insights into the brain systems organization of human memory (Squire et al., 1990). Yet our understanding of the neurology of memory is far from complete.

Studies on the neuropsychology of memory suggest that memory itself is not unitary but is comprised of distinct types. Recent taxonomies of memory include episodic vs. semantic (Tulving, 1972), taxon vs. locale (O'Keefe and Nadel, 1978), working vs. reference (Olton et al., 1979), declarative vs. procedural (Cohen and Squire, 1980), and representational vs. dispositional (Thomas, 1984). Importantly, these different types of memory appear to have a distinct neurological organization (Squire, 1987).

In the first section of this chapter, I discuss evidence for the critical role of the septohippocampal system and the prelimbic sector of the frontal cortex in mnemonic function in the rat. In the second part, I examine the effects of direct manipulations of several neurotransmitter systems in the hippocampus and prelimbic cortex on memory function. The third section addresses the issue of dopamine (DA)-acetylcholine (Ach) interactions at the level of the hippocampus in relation to memory processes. In the final section, the implications of research on the behavioral correlates of neurotransmitter interactions for the clinical neurosciences are discussed.

The Neuroanatomy of Memory: The Role of the Septohippocampal and Prelimbic Cortex Systems

The prelimbic sector of the medial frontal cortex, as defined by Krettek and Price (1977), has extensive efferents to the hippocampal formation both in the rat (Beckstead, 1979; Sesack et al., 1989) and in the cat (Room et al., 1985). In primates, reciprocal connections between dorsolateral frontal cortex and the hippocampal formation have been described (Goldman-Rakic et al., 1984). Additionally, the prelimbic cortex projects to medial aspects of the lateral septal nucleus in the rat (Beckstead, 1979; Sesack et al., 1989), and a major projection of the lateral septal nucleus is to the medial septum (Swanson and Cowan, 1979). In addition to this indirect route to the medial septal nucleus, it is possible that terminals from neurons in the prelimbic cortex impinge upon dendritic extensions into the lateral septal region from neurons in the medial septum (Swanson and Cowan, 1976), thus providing a more direct route from the prelimbic cortex to the medial septal nucleus. The medial septal nucleus, in turn, sends a dense efferent pathway to the hippocampus (Meibach and Siegel, 1977; Swanson and Cowan, 1979). There is anatomical and electrophysiological evidence for a direct projection from the hippocampus to the medial frontal area (Ferino et al., 1987). Recent data indicate that the hippocampal projection to the medial frontal region is restricted to the prelimbic cortex (Jay et al., 1989). Furthermore, the hippocampo-prelimbic projection is excitatory and can undergo long-term potentiation after repeated stimulation of the hippocampus (Laroche et al., 1990).

The anatomical and physiological evidence discussed above suggests that the septohippocampal system and the prelimbic sector of medial frontal cortex comprise a functional system for the mediation of behavior, as proposed by Brito and collaborators (1982). Extensive bidirectional projections between the hippocampal formation and the principal sulcus, critical areas for the mediation of representational memory in the monkey, indicate that a similar functional system may also exist in primates (Goldman-Rakic, 1987).

On the basis of the close anatomical and physiological relationship between the septohippocampal system and the prelimbic cortex, Brito and Brito (1990), as well as others (e.g., Poucet, 1990), argued that lesions in these structures would have similar functional consequences. Both lesions in the septohippocampal system and large lesions of the medial frontal cortex impair spatial behavior (Sutherland et al., 1982; Kesner, 1990) and the performance of spatial delayed alternation in a T-maze (Thomas and Brito, 1980; Markowska et al., 1989). The effects of large medial frontal cortical lesions on spatial delayed T-maze alternation were reproduced by lesions restricted to the prelimbic sector of the medial frontal cortex (Brito et al., 1982; Thomas and Spafford, 1984) but not by lesions of the dorsal anterior cingulate cortex (Thomas and Brito, 1980). Therefore, I would suggest that large lesions of the medial frontal wall interfere with critical behavioral processes because of the involvement of the prelimbic cortex and its reciprocal connections to the hippocampal formation.

The effects of lesions in the septohippocampal system and medial frontal cortex have been interpreted in terms of deficits in the coding of temporal order of information (Kesner, 1990), spatial working memory (Markowska et al., 1989), and representational memory (Thomas and Spafford, 1984).

Brito and Brito (1990) suggested that the septohippocampal system and the prelimbic sector of medial frontal cortex were involved in mechanisms of working/representational memory (WRM), but not reference/dispositional memory (RDM). They went on to demonstrate that lesions in the septohippocampal system and lesions restricted to the prelimbic cortex impaired the performance of contingently reinforced and schedule-specific alternation (delayed nonmatching to sample-DNMS) in a T-maze, tasks thought to involve WRM mechanisms (Brito and Thomas, 1981; Thomas, 1984), as shown in panels A and C of Fig. 1. However, these lesions did not alter the performance of visual (brightness) and olfactory discrimination, tasks considered to involve RDM (Olton et al., 1979; Thomas, 1984), in the same maze, as indicated in panels B and D of Fig. 1. Additionally, Brito and Brito (1990) showed that rats with lesions in the prelimbic cortex, but not rats with septohippocampal lesions, were able to master the alternation tasks at brief, but not long, intervals with continued practice. Furthermore, rats with lesions in the septohippocampal system were still impaired on the DNMS task 4 months after the first postoperative test. Brito and Brito (1990) interpreted these data to be consistent with the hypothesis that the septohippocampal system and the prelimbic sector of the medial frontal cortex are important for WRM mechanisms and suggested a hierarchy such that the septohippocampal system plays a larger role than the prelimbic cortex.

The evidence discussed above should not be construed as indicating that lesions in the septohippocampal system and the prelimbic cortex impair WRM mechanisms by interfering with the same behavioral processes. As emphasized by Weiskrantz (1968), behavioral performance depends on several factors, i.e., behavior is multideterminant. Brito and Brito (1990) demonstrated that the

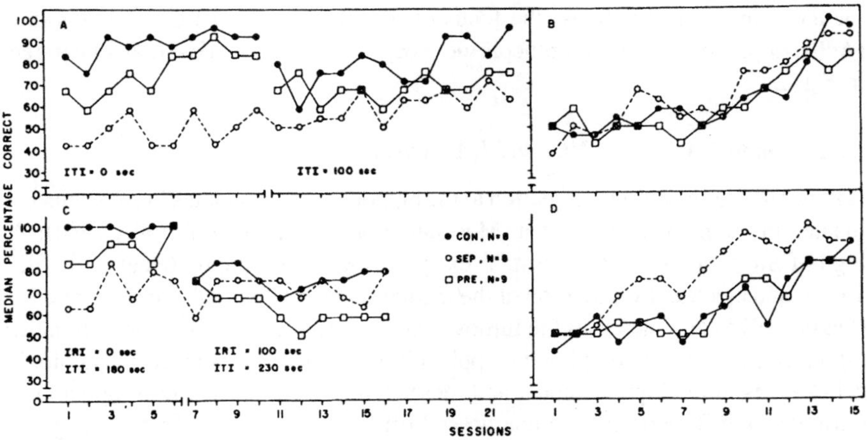

FIGURE 1. Median percentage correct as a function of postoperative acquisition sessions in four behavioral tasks in the T-maze. A: contingently reinforced alternation; B: visual discrimination; C: delayed nonmatching to sample; D: olfactory discrimination. IRI = interrun interval; ITI = intertrial interval. Reprinted with permission of Elsevier Science Publishers from Brito GNO and Brito LSO (1990): Septohippocampal system and the prelimbic sector of frontal cortex: A neuropsychological battery analysis in the rat. *Behav Brain Res* 36:127–146.

neuropsychological profiles of rats with lesions in the septohippocampal system and rats with lesions in the prelimbic cortex are similar, but not equivalent, as has been suggested by other investigators (e.g., Kolb, 1984; Poucet, 1990). Rats with lesions in the prelimbic cortex had difficulty in shifting cognitive strategies depending on task requirements (Brito and Brito, 1990). This conclusion is consistent with recent formulations about the neuropsychological effects of frontal lobe lesions in humans (Owen et al., 1990). However, it is entirely feasible that the septohippocampal system and the prelimbic cortex constitute a functional system for the mediation of WRM in the sense expressed by Flourens (1842), i.e., functional units are not tantamount to anatomical units. The cognitive requirements of behavior may interact with other components (e.g., nonassociative factors) to determine the degree of involvement of a brain structure in the performance of that behavior.

Having defined the septohippocampal system and the prelimbic sector of frontal cortex as the critical structures for WRM, but not RDM, mechanisms, we can now attempt to dissect the mnemonic role of different neurotransmitter systems in the hippocampus and prelimbic cortex.

Neurotransmitter Systems of the Hippocampus and Prelimbic Cortex: Evidence for a Specific Role of the Cholinergic System in WRM

Although several neurotransmitter systems are present in the hippocampus and medial frontal cortical areas, the focus of this chapter will be on acetylcholine (Ach), dopamine (DA), norepinephrine (NE), and histamine (HA), as illustrated in Fig. 2.

Hippocampus: Ach, DA, NE, and HA Systems

A major neurotransmitter projection to the hippocampus originates in Ach neurons located in the medial septal area (Mesulam et al., 1983; Woolf et al., 1984). A high-affinity choline uptake system exists in the hippocampus (Bagnoli et al., 1981), and electrical stimulation of the medial septal nucleus enhances the release (Dudar, 1975) and increases the turnover rate (Moroni et al., 1978) of Ach in the hippocampus. Microiontophoretic application of Ach induces electrophysiological changes in the hippocampus, and these changes are blocked by scopolamine (Krnjević and Ropert, 1981; Benardo and Prince, 1982). These findings suggest that the septohippocampal Ach system is muscarinic. Neurochemical studies support this suggestion (Yamamura and Snyder, 1974; Dooley and Bittiger, 1982) and demonstrate the presence of M_1, M_2, non-M_1, and non-M_2 muscarinic receptors in the hippocampus with a predominance of the M_1 type (Ehlert and Tran, 1990).

The hippocampus also receives a DA projection from the ventral tegmental area/substantia nigra complex (VTA/SN) via the fimbria-fornix (Bischoff, 1986;

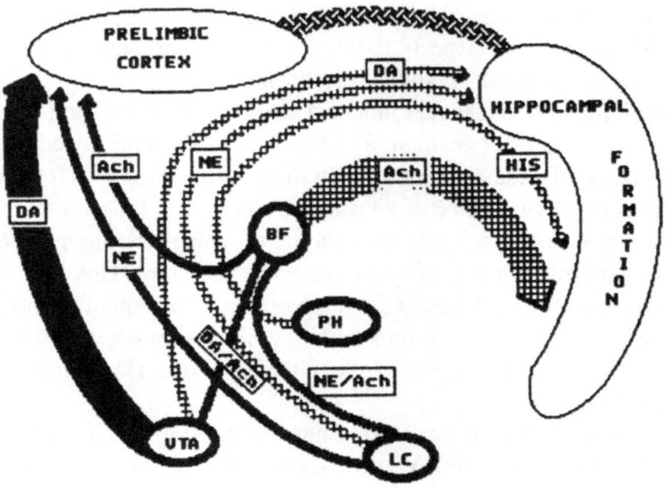

FIGURE 2. Schematic drawing to illustrate some of the neurotransmitter systems and connections of the prelimbic cortex and hippocampal formation. Ach = acetylcholine; BF = basal forebrain; DA = dopamine; HIS = histamine; LC = locus coeruleus; NE = norepinephrine; PH = posterior hypothalamus; VTA = ventral tegmental area.

Ishikawa et al., 1982; Oades and Halliday, 1987). Electrical stimulation of the VTA/SN produces inhibition of population spikes in the CA1 pyramidal layer (Spencer and Wheal, 1990). Additionally, both D_1 and D_2 dopaminergic receptors have been found in the hippocampus (Bruinink and Bischoff, 1986; Scatton and Dubois, 1985), and DA has been shown to induce excitation and inhibition of hippocampal neurons, mediated by D_2 and D_1 receptors, respectively (Śmialowski and Bijak, 1987).

Another neurotransmitter system that reaches the hippocampus originates in the locus coeruleus, LC (see review in Moore and Bloom, 1979) and is noradrenergic (NE). Stimulation of the locus coeruleus and application of NE produce inhibition of hippocampal pyramidal cells in vivo and in hippocampal slices (Moore and Bloom, 1979; Mynlieff and Dunwiddie, 1988). Both alpha and beta receptors have been found in the hippocampus (Dooley and Bittiger, 1982). Additionally, histaminergic (HA) fibers from the posterior hypothalamus project to the hippocampus (see review in Prell and Green, 1986). HA potentiates the effects of excitatory inputs to the hippocampus (Haas and Greene, 1986), and H_1, H_2, and possibly, H_3 histaminergic receptors have been described in the hippocampus (Prell and Green, 1986; Haaksma et al., 1990; Ruat et al., 1990).

Hippocampus: Specific Role of the Ach System in Memory

The hypothesis that central Ach mechanisms are involved in memory is not new (e.g., Deutsch, 1971). However, most of these studies used peripheral or intracerebroventricular injections of anticholinergic drugs. Therefore, these reports did not address the issue of where in the brain the drugs act to impair memory. Moreover, new aspects in the taxonomy of animal memory were not taken into account in these previous experiments.

In 1983, Brito and colleagues injected the muscarinic blocker scopolamine directly into the dorsal hippocampus and demonstrated that the injections impaired the performance of a delayed spatial alternation task in a T-maze but did not interfere with the performance of a visual discrimination task in the same maze. Therefore, it was concluded that the septohippocampal cholinergic system was involved in working, but not reference, memory processes. However, this study did not provide definitive evidence for a specific involvement of the hippocampal Ach system because other neurotransmitter systems were not directly evaluated. Here, I report evidence for a specific role of the Ach system in the hippocampus in WRM processes.

Young adult male Wistar rats were trained to perform either a DNMS task, a task considered to involve WRM, or a visual (brightness) discrimination (VD) task, a task thought to involve RDM (see discussion above), in the same maze. Behavioral procedures are described in Brito and Brito (1990). After the rats had mastered the task, they were subjected to surgery and implanted stereotaxically with 22-gauge guide cannulae aimed at a site 1.0 mm dorsal to the dorsal hippocampus so that the tip of the injection cannula (28 gauge) would be located inside the hippocampus proper. Following a postoperative recovery period, the rats were tested again for a few sessions to "refresh" their memory of the task before injection procedures began. Three experiments with a different order of drug injection were conducted, as described below. All injections were bilateral in the volume of 1.0 μl per site. Vehicle (VEH) was a Krebs-Ringer solution. Injection procedures can be found elsewhere (Brito et al., 1983).

Experiment 1: Ach − DA
Drugs: Scopolamine (SCO), muscarinic Ach blocker, 9 and 18 μg/μl
Sulpiride (SUL), D_2 DA antagonist, 5 and 10 μg/μl
Order: VEH − SCO − VEH − SUL
VEH − SUL − VEH − SCO

Experiment 2: NE
Drugs: Propranolol (PRO), β antagonist, 5 and 10 μg/μl
Prazosin (PRA), α_1 antagonist, 7 and 14 μg/μl
Yohimbine (YOH), α_2 antagonist, 6.5 and 13 μg/μl
Order: VEH − PRO − VEH − PRA − VEH − YOH
VEH − YOH − VEH − PRA − VEH − PRO

Experiment 3: HA
Drugs: Antazoline (ANT), H_1 antagonist, 0.45 and 0.90 μg/μl
Cimetidine (CIM), H_2 antagonist, 0.75 and 1.5 μg/μl
Beta-histine (HIS), H_3 antagonist, 0.25 and 0.5 μg/μl
Order: VEH − CIM − VEH − ANT − VEH − HIS
VEH − HIS − VEH − ANT − VEH − CIM

After completion of testing and immediately before perfusion, the rats were injected with India ink to verify cannula placement, and their brains were processed for histology. Only the behavioral data for rats with cannulae located within the boundaries of the hippocampus were analyzed. The order of injection had no effect on behavioral performance. Therefore, the behavioral data were collapsed across the order of injection.

As can be seen from Figure 3, the drugs used in Experiments 2 (NE) and 3 (HA) had no effect on the performance of either the WRM task delayed nonmatching to sample (DNMS) or the RDM task visual discrimination (VD). In Experiment 1, SUL had no effect on the performance of either type of task. SCO at the 18 μg/μl dose impaired the performance of both tasks due to prolongation of the response (entering the goal box of the maze) latencies, which led to abortion of several trials. However, SCO at the 9 μg/μl dose significantly interfered only with the DNMS task.

The results from these experiments replicate previous findings from our laboratory (Brito et al., 1983) that the performance of WRM tasks is sensitive to the blockade of cholinergic mechanisms in the hippocampus, whereas the performance of VD, a RDM task, is not. Additionally, our results demonstrate that intrahippocampal injections of D_2, NE, and HA antagonists at the doses used do not interfere with the performance of either task. The present data strengthen the theory that the cholinergic septohippocampal system is critically involved in WRM, but not RDM, mechanisms (Brito et al., 1983).

The deleterious effects of intrahippocampal injections of scopolamine on memory processes observed in the present study are consistent with reports that peripheral administration of scopolamine impair memory in human subjects (Drachman and Leavitt, 1974) and in experimental animals (Bartus and Johnson,

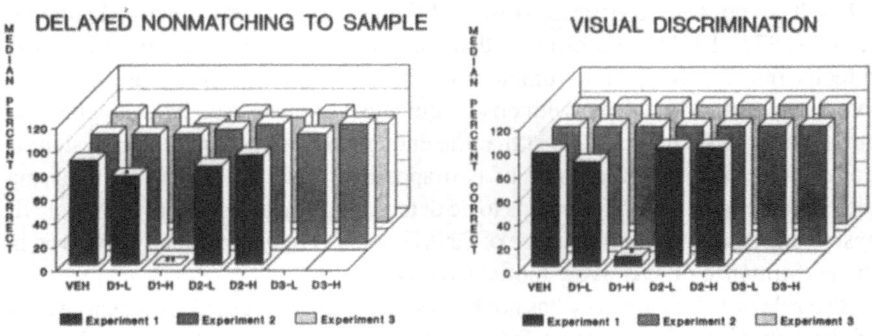

FIGURE 3. Median percentage correct on delayed nonmatching to sample (left panel) and visual discrimination (right panel) after bilateral hippocampal injections of Krebs-Ringer (VEH) and neurotransmitter antagonists. Experiment 1: D1-L and D1-H, SCO (9 and 18 μg/μl, respectively); D2-L and D2-H, SUL (5 and 10 μg/μl). Experiment 2: D1-L and D1-H, PRO (5 and 10 μg/μl); D2-L and D2-H, PRA (7 and 14 μg/μl); D3-L and D3-H, YOH (6.5 and 13 μg/μl). Experiment 3: D1-L and D1-H, ANT (0.45 and 0.90 μg/μl); D2-L and D2-H, CIM (0.75 and 1.5 μg/μl); D3-L and D3-H, HIS (0.25 and 0.5 μg/μl). SCO = scopolamine; SUL = sulpiride; PRO = propranolol; PRA = prazosin; YOH = yohimbine; ANT = antazoline; CIM = cimetidine; HIS = B-histine. Statistical significance is indicated by an asterisk. See text for further explanation.

1976; Beatty and Bierley, 1985, 1986). Our data indicate that a probable site of action of scopolamine in the brain to impair memory is the hippocampus. Furthermore, our results demonstrate that intrahippocampal injections of scopolamine specifically impair one type of memory, i.e., WRM. This hypothesis is consistent with findings that high-affinity choline uptake (HACU) in the hippocampus is raised by performance of a WRM task (Olton et al., 1991). However, the observation that HACU in the hippocampus is also raised by performance of a RDM task (Olton et al., 1991) would not be expected. I do not have an explanation for this discrepancy. However, task differences or behavioral and physiological correlates of HACU may reconcile the two findings.

Intracerebroventricular injections of SUL have been reported to induce impairment in the performance of a two-way active-avoidance task (Nishibe et al., 1982). It is possible that central injections of SUL specifically interfere with the performance of aversively motivated behavior, but not with the performance of appetitively motivated behavior, as used in experiments reported here. Peripheral injections of DA antagonists in the rat have been shown to impair spatial RDM (Whishaw and Dunnett, 1985) and WRM (McGurk et al., 1988). The data presented here would suggest that peripheral administration of DA antagonists interferes with the performance of these tasks by actions at brain sites different from the hippocampus.

The role of NE in memory is well known (Quatermain, 1983). Although some studies suggest a peripheral site of action of NE in mnemonic modulation (Gold and Zornetzer, 1983), other investigators consider that a peripheral site of action cannot account for the role of NE in memory processes (Quatermain, 1983). Reports that neurotoxic lesions of the dorsal NE bundle, which substantially decrease hippocampal NE levels, impair T-maze alternation (Mason and Fibiger, 1978) have not been confirmed (Pisa and Fibiger, 1983). Gallagher and collaborators (1977, 1981) demonstrated that microinjections of α and β antagonists directly into the amygdalar complex of rats modulate memory in an inhibitory avoidance task. My results, with doses equivalent to those used by Gallagher and colleagues (1977, 1981) suggest that the endogenous hippocampal NE system is not critical for the performance of two appetitive tasks in a T-maze involving different memory types. It remains to be determined whether the hippocampal NE system modulates the performance of DNMS and VD or is specifically involved in the performance of aversively motivated tasks.

The role of HA in memory has not been studied in much detail. A recent review of the behavioral functions of HA concluded that HA and its antagonists are involved in drinking, spontaneous locomotion, sexual behavior, and avoidance responding (White and Rumbold, 1988). Alvarez and collaborators reported that intrahippocampal injections of histamine inhibited locomotor and rearing behaviors in the open field (Alvarez et al., 1982) but did not change exploratory behavior in a hole-board task (Alvarez and Banzan, 1985). Peripheral injections of H1 blockers were found to depress performance in an active avoidance task (Kamei et al., 1990). Immediate posttraining intracerebroventricular injection of HA facilitated retention of a step-down inhibitory avoidance task (De Almeida and

Izquierdo, 1986) and habituation to an open field (De Almeida and Izquierdo, 1988). Combined, but not singly, injections of H_1 and H_2 antagonists blocked the effects of HA in both tasks (De Almeida and Izquierdo, 1986, 1988). My results indicate that endogenous hippocampal HA is not specifically involved in the mechanisms of WRM or RDM, although a modulatory role cannot be excluded and needs scrutiny.

In conclusion, the results discussed above indicate that only the hippocampal Ach system, in contrast to the DA, NE, and HA systems, is critically involved in a specific type of memory, i.e., WRM. Since it can be presumed that intrahippocampal injections of DA, NE, and HA antagonists also induce neurotransmitter imbalance in the hippocampus, it is probable that the imbalance induced by SCO injections is functionally more significant.

Prelimbic Cortex: Ach, DA, NE, and HA Systems

The specific neurotransmitter innervation pattern of the prelimbic cortex is not as well known as that of the hippocampus, probably because the prelimbic cortex is usually considered together with other medial frontal areas in most anatomical studies. Despite this caveat, it seems that the prelimbic cortex receives cholinergic projections from the basal forebrain (McKinney et al., 1983; Woolf et al., 1983) and the laterodorsal tegmental nucleus (Satoh and Fibiger, 1986). Electrophysiological information indicates that Ach has an excitatory effect in frontal cortical neurons (Swartz and Woody, 1979), which can be altered by muscarinic antagonists (Rigdon and Pirch, 1986).

A major projection system to the prelimbic cortex is dopaminergic and originates in the VTA/SN (Moore and Bloom, 1978; Oades and Halliday, 1987), and stimulation of the VTA/SN area has been reported to inhibit the spontaneous activity of prefrontal cortical neurons (Mantz et al., 1988). Additionally, the prelimbic cortex receives NE projections from LC (Moore and Bloom, 1979). Projections from the LC and VTA/SN converge with projections from the mediodorsal thalamus to the medial prefrontal cortex (Oades and Halliday, 1987; Mantz et al., 1988). The prelimbic cortex probably receives HA projections from the posterior hypothalamus (Prell and Green, 1986) as part of the diffuse cortical innervation pattern of the HA system.

Prelimbic Cortex: Role of the Ach and DA Systems on Memory

On the basis of the connectivity pattern of the prelimbic cortex and hippocampus, it could be speculated, from a neurobiological standpoint, that the Ach input to the hippocampus is more significant than the DA input. Conversely, it would appear that the DA input to prelimbic cortex from the VTA/SN complex is more significant than the Ach input from the basal forebrain and tegmentum. In agreement with this hypothesis, I presented data in the previous section that demonstrate that intrahippocampal injections of SCO, but not SUL, impair WRM. It would be important to demonstrate double dissociation via injections localized in the prelimbic cortex, i.e., SUL, but not SCO, impairing the performance of the same memory task affected by SCO, but not SUL, injections in the hippocampus.

The subjects of this study were young adult male Wistar rats. They were trained to perform the same VD and DNMS tasks described previously in this chapter. Subsequently, the rats underwent surgery to implant 22-gauge double guide cannulae in the prelimbic sector of frontal cortex. Following a postoperative recovery period, the rats were initially tested in the DNMS task and then in the VD task, i.e., in this experiment each rat was tested in both tasks. The order of drug injection was VEH (Krebs-Ringer), SUL (10 μg/μl), VEH , and SCO (18 μg/μl) for all rats within each task. Only behavioral data for rats with cannulae located in the prelimbic cortex were subjected to statistical analysis. The results from this experiment have appeared elsewhere (Brito et al., 1989) and are shown in Table 1. The data indicate that SCO injection into the prelimbic cortex significantly impaired the performance of the DNMS task, but had no effect on the performance of the VD task. SUL injection had no effect on either task.

The data presented in Table 1 suggest that D_2 mechanisms in the prelimbic sector of frontal cortex are not involved in either WRM or RDM. It has been reported that neurotoxic lesions of the DA system in the prefrontal cortex induce deficits in a delayed alternation task similar to the one used here (Simon, 1981). However, it is not clear whether these injections were placed in the prelimbic sector, and other neurotransmitter systems (e.g., NE) may have been affected because the use of an NE uptake blocker was not reported. Furthermore, neurotoxic lesions, given its chronic nature, may induce neurochemical changes not found after acute blockade by injection of antagonist drugs.

The involvement of the Ach system of the prelimbic cortex in WRM is consistent with findings that the nucleus basalis Ach system is involved in the generation of conditioned neural responses in the frontal cortex of the rat (Rigdon and Pirch, 1986). Additionally, the frontal cortex has been reported to be the site of the ameliorative action of physostigmine (an indirect Ach agonist) on memory of an inhibitory avoidance task in rats with lesions in the nucleus basalis (Haroutunian et al., 1990).

TABLE 1. Effects of Microinjections of Sulpiride and Scopolamine into the Prelimbic Cortex on Performance of a DNMS and a VD Task

	DNMS			VD		
Rat	VEH	SUL	SCO	VEH	SUL	SCO
1	83	92	75	67	83	83
2	83	83	50	92	83	75
3	71	58	50	75	75	100
4	92	92	58	92	100	83
5	83	67	92	75	83	100
6	79	83	75	100	100	92
Median	83	83	67	84	83	88

Adapted from Brito et al. (1989).

Although the data presented here suggest that the cholinergic system of the prelimbic cortex is specifically involved in WRM, other experiments involving other neurotransmitter systems are certainly required before a definitive conclusion is reached.

DA-Ach Interactions in the Hippocampus: Effects on Memory

The possibility that neurotransmitter systems interact in the mediation of memory has been receiving considerable attention from investigators in the field, as reviewed by Levin and collaborators (1990), and Decker and McGaugh (1990). Although the mechanism of such interactions is still speculative, research on regulation of neurotransmitter release by presynaptic autoreceptors and heteroreceptors (e.g., Kalsner, 1990) certainly provides a possible rationale for studies on neurotransmitter interactions in cognitive function.

In a series of studies, Levin and colleagues (1990) demonstrated that the cholinergic and dopaminergic systems interact to mediate the performance of a radial-arm maze task, a task considered to involve WRM (see discussion above). Performance of this task is impaired by peripheral injections of nicotinic and muscarinic Ach antagonists. Deficits induced by muscarinic blockade are reversed by injections of D_1 (but not D_2) DA blockers, whereas the impairment induced by nicotinic antagonists is potentiated by D_2 (but not D_1) blockers and is obliterated by D_2 agonists.

The evidence reviewed (Levin et al., 1990; Levin and Rose, this volume) clearly demonstrates that the Ach and DA systems interact to mediate WRM. However, the issue of the location in the brain where the interaction occurs has not been addressed. As suggested by Levin and colleagues (1990), there are several possible sites of Ach-DA interaction in the brain. One such possible site is the septohippocampal system, both at the level of the septal area and the hippocampus proper. Peripheral administration of DA agonists decreases the turnover rate of Ach in the hippocampus, whereas intraseptal injection of DA blockers or neurotoxic lesions of the VTA/SN complex enhance the turnover rate of Ach in the hippocampus (Robinson et al., 1979). There is evidence to suggest that DA exerts an inhibitory effect on Ach septal neurons through a GABA-ergic mechanism (Costa et al., 1983). It is interesting to note that injection of GABA agonists into the medial septal area impair performance of a WRM task (Chrobak et al., 1989; Givens and Olton, 1990).

DA and Ach may also interact directly in the hippocampus. As discussed previously in this chapter, the hippocampus receives both Ach and DA projections, and it has been reported that DA receptors modulate Ach release in slices of rabbit hippocampus (Strittmatter et al., 1982).

Brito and colleagues (1983) demonstrated the importance of the cholinergic septohippocampal system for WRM processes. However, they raised the possibility that cholinergic and catecholaminergic systems interact to mediate memory. Additionally, it was argued that this interaction occurs possibly in the hippocampus as well as in other areas of the brain, such as the medial frontal cortex. I will now present evidence for Ach-DA interactions in the hippocampus in the mediation of memory.

As in previous experiments, young adult male Wistar rats were used as subjects. Preoperative and postoperative behavioral procedures, as well as surgery, histology, and injection methodology, were as described in the section on the specific role of the hippocampal cholinergic system on memory (see above). Two experiments, each with a different group of rats, were conducted. Rats were trained and tested in one of two tasks, DNMS or VD in a T-maze. The first experiment addressed the interaction between intrahippocampal injections of SCO (a muscarinic Ach antagonist) and SCH 23390 (a D1 antagonist), whereas the second experiment evaluated the interactions between SCO and SUL (a D_2 antagonist), as summarized below. Doses of SCO low enough to interfere with performance of the DNMS task without unduly increasing the response latency were selected so possible interaction effects could be ascertained.

Experiment 1: Ach − D_1 interaction
Drugs: SCO, 7.5 μg/μl
SCH 23390, 1.5 μg/μl
Order: VEH − SCO − VEH − SCH 23390
VEH − SCH 23390 − VEH − SCO
VEH − SCO + SCH 23390 − VEH − SCO + SCH 23390
Experiment 2: Ach − D_2 interaction
Drugs: SCO, 9.0 μg/μl
SUL, 5.0 μg/μl
Order: VEH − SCO
VEH − SCO + SUL

Only behavioral data for rats with cannulae located within the boundaries of the hippocampus were analyzed. Since the order of injection in the first two order groups had no effect on performance, the data for SCO and SCH were collapsed across order groups.

The results are illustrated on the left (Experiment 1) and right (Experiment 2) panels of Figure 4. Intrahippocampal injections of a low dose of SCO had a statistical, but minor, effect on choice accuracy in the DNMS task as compared to VEH injections. SCH 23390 had no effect on behavioral performance. Combined injections of SCO and SCH 23390 interfered with performance of the DNMS task as much as injections of SCO alone. Single injections of SCO and SCH 23390 had no effect on the performance of the VD task as compared to VEH injections. However, combined injections significantly impaired task performance.

Results for Experiment 2 are shown on the right panel of Figure 4. It can be seen from the figure that a dose of SCO just a little higher than the dose used in Experiment 1 (7.5 vs. 9.0 μg/μl) produced a major impairment in the performance of the DNMS task but had no effect on performance of the VD task. As described before in this chapter, intrahippocampal injections of SUL were without effect on

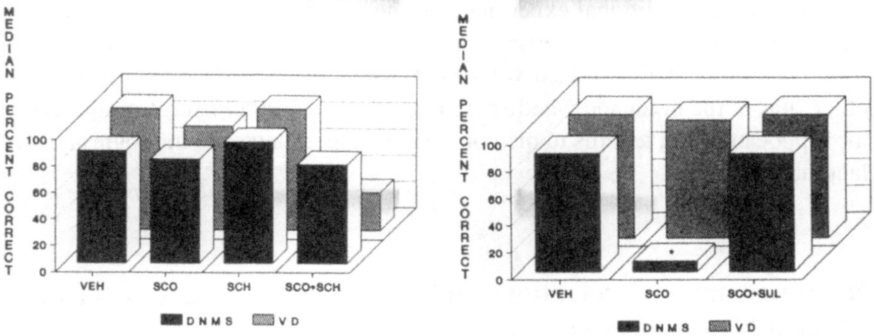

FIGURE 4. Median percentage correct on delayed nonmatching to sample (DNMS) and visual discrimination (VD) after bilateral hippocampal injections. The left panel shows the results for Experiment 1 (Krebs-Ringer, VEH; Scopolamine, SCO; SCH 23390, SCH; SCO combined with SCH). The right panel illustrates the results for Experiment 2 (VEH; SCO; SCO combined with sulpiride, SUL). See text for additional information.

either task. However, combined injections of SCO and SUL reversed the effect of SCO on the DNMS task.

The finding that intrahippocampal injections of low-dose SCO impair performance of DNMS, a WRM task, but do not alter the performance of VD in the same maze, a RDM task, is again replicated. Additionally, intrahippocampal injections of SCH 23390 were found to have no effect on the performance of either task. However, SCH 23390 interacts with SCO to impair RDM, but not WRM. Furthermore, intrahippocampal injections of SUL obliterate the deleterious effect of SCO on WRM.

Levin and colleagues (1990) reported that peripheral administration of D_1 blockers reverses the impairment in WRM induced by peripheral injections of SCO. Additionally, they pointed out that D_2 antagonists potentiate the impairment in WRM produced by peripheral injections of nicotinic antagonists. In the hippocampal model system used in the experiments described in this chapter, however, I found that injection of D_2, but not D_1, antagonists obliterates the impairment induced by SCO in WRM and that injection of a D_1, but not a D_2, blocker, interacted with SCO to impair RDM. Therefore, the findings described here are not consistent with data reported by Levin and colleagues (1990). Differences in drug levels (in the hippocampus) and behavioral tasks (radial arm vs. T-maze) could partly explain the discrepancy.

It is possible for a drug to have different behavioral effects when injected systemically as compared to local infusions into the brain. Systemic injections can be assumed to have direct and indirect effects on several neurotransmitter systems in different areas of the brain. Local infusions probably alter the balance between several neurotransmitter systems, but in a restricted area of the brain, thus minimizing "distance" effects of the drug. The present data suggest that Ach-DA interactions at the level of the hippocampus are distinct, in regards to the mediation of memory, from interactions observed after systemic manipulations. However, additional experiments using a wider dose range and involving other DA antagonists are necessary before this conclusion is accepted. Furthermore, it should be determined whether Ach-DA interactions may also exist in other sites in the brain and whether Ach interacts with other neurotransmitters in the hippocampus in regards to memory. These experiments are in progress in my laboratory.

Neurotransmitter Interactions and Behavior: Implications for the Clinical Neurosciences

The organization of the brain, especially the extensive degree of connectivity between neural structures, would make us presume that there is no such a thing as an isolated lesion or even dysfunction in the brain. Flourens (1842) argued that many neural "centers" are involved in the mediation of any behavior and that any neural "center" participates in many behaviors. Therefore, it is unlikely that degeneration or dysfunction in a certain area of the brain does not change the function of other areas with which it is connected directly or indirectly. The notion that a lesion in the brain interferes with functioning of a site distant from the location of the lesion was advanced by von Monakow in 1914 and has received support from PET studies in head-injured patients (Uzzell, 1986).

Acceptance of the idea that several brain systems interact in the mediation of behavior should not be taken to imply that the contribution of each brain system for that behavior is equivalent. On the contrary, it is clear that there is a hierarchy such that some structures are crucial for some behaviors, even though other systems may also play a role. It would appear to me that the same holds true in respect to neurotransmitter systems of the brain, as the data presented earlier in this chapter suggest. Therefore, degeneration or dysfunction of a neurotransmitter system in the human brain may be presumed to induce functional (and compensatory) changes in other systems. In addition, the coexistence of different neurotransmitters in the same neuron (Hokfelt et al., 1978) and autoregulation of neurotransmitter release by heteroreceptors and autoreceptors (Kalsner, 1990) add to the possible mechanisms of neurotransmitter (and neuromodulator) interactions in the central nervous system.

In this section I will discuss the relevance of neurotransmitter interactions for the clinical neurosciences, using Alzheimer's disease as a prototypical nosological entity. It is, of course, true that the analysis to be presented would have applications for other neuropsychiatric disorders with complex neurochemical pathologies, such as Parkinson's disease (Agid and Javoy-Agid, 1985), Schizophrenia (Roberts et al., 1983), Huntington's disease (Spokes, 1981), epilepsy (Morselli and Lloyd, 1985), Korsakoff's syndrome (Joyce, 1987), Tourette's syndrome (Leckman et al., 1988), autism (Anderson and Hoshino, 1987), anxiety disorders (Kidman, 1989), and depression (Gold et al., 1988).

Alzheimer's Disease (AD)

A considerable amount of research supports the cholinergic hypothesis of cognitive decline in AD (Bartus et al., 1982). Cholinergic neurons in the basal forebrain have been shown to degenerate (Whitehouse et al., 1981); presynaptic cholinergic markers are decreased, especially in hippocampus and temporal cortex (Perry et al., 1977); and losses of nicotinic (Whitehouse et al., 1986) and muscarinic (Flynn et al., 1990; Reinikainen et al., 1987) receptors have been reported.

Although the cholinergic system is clearly involved in the neurochemical pathology of AD, changes in other neurotransmitter systems have also been described. One such system is the NE system. A substantial loss of LC neurons has been found in AD (Chan-Palay and Asan, 1989; Mann et al., 1984). In addition, NE levels are decreased in several regions, including the hippocampus and frontal cortex (Reinikainen et al., 1988a); concentrations of the NE metabolite MHPG are reduced in the hippocampus (Cross et al., 1983); and the number of β-adrenergic receptors altered (Kalaria et al., 1989; Shimohama et al., 1988).

Degenerative changes typical of AD have been reported in the SN/VTA area (Gibb et al., 1989), which would indicate the involvement of the DA system in AD pathology. In addition, DA levels are reduced in the temporal cortex and hippocampus (Reinikainen et al., 1988b) and in cerebrospinal fluid (Bareggi et al., 1982), and the number of DA receptors is decreased in the nucleus basalis (Sparks et al., 1986) and frontal cortex (De Keyser et al., 1990).

The serotoninergic (5HT) system is also affected in AD, as reflected by a decrease in neuronal numbers in the raphe (see Hardy et al., 1985), and reduced concentrations of 5HT and its metabolite in the hippocampus (Reinikainen et al., 1990) and in cerebrospinal fluid (Volicer et al., 1985). $5HT_2$ receptor binding in frontal cortex and hippocampus, however, were reported to be unaltered (Dewar et al., 1990).

Other systems reported to be altered in AD include the amino acid (Hyman et al., 1987; Penney et al., 1990), benzodiazepine (Ferrarese et al., 1990; Shimohama et al., 1988), and adenosine (Kalaria et al., 1990) systems. Furthermore, several peptidergic systems have been shown to be altered in AD, including somatostatin (Joynt and McNeill, 1984), corticotropin releasing factor (Whitehouse et al., 1987), vasopressin (Mazurek et al., 1986), β-endorphin (Kaiya et al., 1983), neuropeptide Y (Chan-Palay et al., 1986), and galanin (Chan-Palay et al., 1989).

It is important to note that the complex neurochemical pathology of AD would suggest several possible sites of interaction between neurotransmitter systems. A decrease in the number of DA receptors in the nucleus basalis (Sparks et al., 1986) would be expected to further change Ach metabolism in cortical areas. The reported degeneration of the Ach innervation of the LC in AD (Strong et al., 1991) most certainly would interfere with NE mechanisms in widespread areas of the brain. Hyperinnervation of basal forebrain areas by galanin (Chan-Palay et al., 1989) would be expected to suppress the remaining activity in Ach neurons (Chan-Palay, 1990). In addition, degeneration of DA neurons in SN/VTA (Gibb et al., 1989) may induce changes in the regulation of β-adrenergic receptors in the frontal cortex (Hervé et al., 1990), degeneration of 5HT neurons could interfere with hippocampal and prefrontal cortical function (Mantz et al., 1990), a reduction in vasopressin levels can be expected to change the turnover rate of catecholamines in the brain (Kovács et al., 1977), and decreased levels of Ach in the hippocampus could alter the binding of glutamate (Deutsch et al., 1990).

The complexity of the neurochemical pathology and the multiple possible interactions between the neurotransmitter systems involved would suggest that pharmacological therapies that focus on a single neurotransmitter are bound to have limited success in the treatment of AD. This may explain the disappointment in the response of AD patients to Ach precursors, acetylcholinesterase (AchE) inhibition (Fitten et al., 1990), and Ach agonists (Mouradian et al., 1988), although there are a few reports of a mild effect of intracranial infusion of Ach agonists (Harbaugh et al., 1984; Read et al., 1990) and long-term administration of an AchE inhibitor (Harrell et al., 1990) in groups of AD patients. The heterogeneity of AD further adds to the complexity of its treatment.

The research reviewed here would suggest that approaches involving several neurotransmitter systems may offer a better perspective in the treatment of AD. It has been reported that the monoamine oxidase inhibitor (MAOI) deprenyl induces significant behavioral and, to a lesser extent cognitive, changes in AD patients (Tariot et al., 1987). It remains to be determined whether an association of a MAOI with Ach drugs would have a synergistic effect on AD patients. In any event, pharmacological therapies that emphasize the possibility of interaction between neurotransmitter systems may offer hope for the treatment of AD patients. It is, of course, probable that which transmitter systems should be manipulated will depend on the stage in the natural history of AD.

Conclusions

The use of microinjections in specific areas of the brain should continue to provide critical information as to how and where neurotransmitter systems interact to mediate cognitive function. The implications of this information for the clinical neurosciences are of much relevance. It is hoped that the development of recent experimental models that bear a closer pathobiological resemblance to human disease, such as the transgenic mouse model for AD (Wirak et al., 1991), together with current research methodology in the behavioral neurosciences, will lead to a better understanding of the mechanism of cognitive dysfunction in neurological disease.

Acknowledgments. The author wishes to thank Valeria F. Silva for expert technical assistance, Dr. Vilma A. Silva for comments on the manuscript, and Dr. J.C. Seggia and R.A. Siggia for assistance in computer-aided graphics, and Lepetit Laboratorios do Brazil Ltda for providing sulpiride. The research was supported by CNPq Research Career Award (30.0197/82) and grants from CNPq (40.6021/90), the Sandoz Foundation for Gerontological Research and FUNPENE.

References

Agid Y, Javoy-Agid F (1985): Peptides and Parkinson's disease. *Trends in Neurosciences* 8:30–35

Alvarez EO, Banzan AM (1985): Further evidence that histamine in hippocampus affects the exploratory behavior in the rat. *Physiol Behav* 34:661–664

Alvarez EO, Guerra FA (1982): Effects of histamine microinjections into the hippocampus on open-field behavior in rats. *Physiol Behav* 28:1035–1040

Anderson GM, Hoshino Y (1987): Neurochemical studies of autism. In: *Handbook of Autism and Pervasive Developmental Disorders*, Cohen DJ, Donnellan AM, eds. New York: John Wiley & Sons

Bagnoli P, Beudet A, Stella M, Cuénod M (1981): Selective retrograde labeling of cholinergic neurons with ^{3}H choline. *J Neurosci* 1:691–695

Bareggi SR, Franceschi M, Bonini L, Zecca L, Smirne S (1982): Decreased CSF concentrations of homovanillic acid and gamma-aminobutyric acid in Alzheimer's disease. *Arch Neurol* 39:709–712

Bartus RT, Johnson HR (1976): Short-term memory in the rhesus monkey: Disruption from the anticholinergic scopolamine. *Pharmacol Biochem Behav* 5:39–46

Bartus RT, Dean RL III, Beer B, Lippa AS (1982): The cholinergic hypothesis of geriatric memory dysfunction. *Science* 217:408–417

Beatty WW, Bierley RA (1985): Scopolamine degrades spatial working memory but spares spatial reference memory: Dissimilarity of anticholinergic effect and restriction of distal visual cues. *Pharmacol Biochem Behav* 23:1–6

Beatty WW, Bierley RA (1986): Scopolamine impairs encoding and retrieval of spatial working memory in rats. *Physiol Psychol* 14:82–86

Beckstead RM (1979): An autoradiographic examination of corticocortical and subcortical projections of the medio-dorsal projection (prefrontal) cortex in the rat. *J Comp Neurol* 184:43–62

Benardo LS, Prince DA (1982): Cholinergic pharmacology of mammalian hippocampal pyramidal cells. *Neuroscience* 7:1703–1712

Bischoff S (1986): Mesohippocampal dopamine system: Characterization, functional and clinical implications. In: The Hippocampus, Vol 3, Isaacson RL, Pribram KH, eds. New York: Plenum Press

Brito GNO, Brito LSO (1990): Septohippocampal system and the prelimbic sector of frontal cortex: A neuropsychological battery analysis in the rat. *Behav Brain Res* 36:127–146

Brito GNO, Thomas GJ (1981): T-maze alternation, response patterning, and septohippocampal circuitry in rats. *Behav Brain Res* 3:319–340

Brito GNO, Thomas GJ, Davis BJ, Gingold SI (1982): Prelimbic cortex, mediodorsal thalamus, septum and delayed alternation. *Exp Brain Res* 46:52–58

Brito GNO, Davis BJ, Stopp LC, Stanton ME (1983): Memory and the septo-hippocampal cholinergic system in the rat. *Psychopharmacology* 81:315–320

Brito GNO, Silva SP, Brito LSO (1989): The cholinergic and dopaminergic systems of the prelimbic sector of frontal cortex and memory in the rat. *Brazil J Med Biol Res* 22:1155–1158

Bruinink A, Bischoff S (1986): Detection of dopamine receptors in homogenates of rat hippocampus and other brain areas. *Brain Res* 386:78–83

Chan-Palay V, Asan E (1989): Alterations in catecholamine neurons of the locus coeruleus in senile dementia of the Alzheimer type and in Parkinson's disease with and without dementia and depression. *J Comp Neurol* 287:373–392

Chan-Palay V, Lang W, Haesler U, Kohler C, Yasargil G (1986): Distribution of altered hippocampal neurons and axons immunoreactive with antisera against neuropeptide Y in Alzheimer-type dementia. *J Comp Neurol* 248:376–394

Chrobak JJ, Stackman RW, Walsh TJ (1989): Intraseptal administration of muscimol produces dose-dependent memory impairments in the rat. *Behav Neural Biol* 52:357–369

Cohen NJ, Squire LR (1980): Preserved learning and retention of pattern-analyzing skill in amnesia: Dissociation of knowing how and knowing that. *Science* 210:207–209

Costa E, Panula P, Thompson HK, Cheney DL (1983): The transsynaptic regulation of the septal-hippocampal cholinergic neurons. *Life Sci* 32:165–179

Cross AJ, Crow TJ, Johnson JA, Joseph MH, Perry EK, Perry RH, Blessed G, Tomlinson BE (1983): Monoamine metabolism in senile dementia of Alzheimer type. *J Neurol Sci* 60:383–392

De Almeida MAMR, Izquierdo I (1986): Memory facilitation by histamine. *Arch Int Pharmacodyn* 283:193–198

De Almeida MAMR, Izquierdo I (1988): Intracerebroventricular histamine, but not 48/80, causes posttraining memory facilitation in the rat. *Arch Int Pharmacodyn* 291:202–207

Decker MW, McGaugh JL (1990): The role of interactions between the cholinergic system and other neuromodulatory systems in learning and memory. *Synapse* 7:151–168

De Keyser J, Ebinger G, Vauquelin G (1990): D1-dopamine receptor abnormality in frontal cortex points to a functional alteration of cortical cell membranes in Alzheimer's disease. *Arch Neurol* 47:761–763

Deutsch JA (1971): The cholinergic synapse and the site of memory. *Science* 174:788–794

Deutsch SI, Panchision DM, Rosse RB, Novitzki MR, Miller LP, Mastropaolo J (1990): Interaction of cholinergic and glutamatergic transmission in the hippocampus: An in vivo autoradiographic receptor analysis. *Neurosci Lett* 118:124–127

Dewar D, Graham DI, McCulloch J (1990): 5HT-2 receptors in dementia of the Alzheimer type: A quantitative autoradiographic study of frontal cortex and hippocampus. *J Neural Transm (P-D Sect)* 2:129–137

Dooley DJ, Bittiger H (1982): Characterization of neurotransmitter receptors in the rat hippocampal formation. *J Neurochem* 38:1621–1626

Drachman DA, Leavitt J (1974): Human memory and the cholinergic system. *Arch Neurol* 30:113–121

Dudar JD (1975): The effect of septal nuclei stimulation on the release of acetylcholine from the rabbit hippocampus. *Brain Res* 83:123–133

Ehlert FJ, Tran LLP (1990): Regional distribution of M1, M2 and non-M1, non-M2 subtypes of muscarinic binding sites in rat brain. *J Phcrmacol Exp Therap* 255:1148–1157

Ferino F, Thierry AM, Glowinski J (1987): Anatomical and electrophysiological evidence for a direct projection from Ammon's horn to the medial prefrontal cortex in the rat. *Exp Brain Res* 65:421–426

Ferrarese C, Appollonio I, Frigo M, Meregalli S, Piolti R, Tamma F, Frattola L (1990): Cerebrospinal fluid levels of diazepam-binding inhibitor in neurodegenerative disorders with dementia. *Neurology* 40:632–635

Fitten LJ, Perryman KM, Gross PL, Fine H, Cummins J, Marshall C (1990): Treatment of Alzheimer's disease with short- and long-term oral THA and lecithin: A double-blind study. *Am J Psychiatry* 147:239–242

Flourens P (1842): *Recherches Expérimentales sur les Propriétés et les Fonctions du Système Nerveux dans les Animaux Vertébrés*. Paris: Baillière

Flynn DD, Weinstein DA, Mash DC (1990): Loss of high-affinity binding to M1 muscarinic receptors in Alzheimer's disease: Implications for the failure of cholinergic replacement therapies. *Ann Neurol* 29:256–262
Gallagher M, Kapp BS (1981): Effect of phentolamine administration into the amygdala complex of rats on time-dependent memory processes. *Behav Neural Biol* 31:90–95
Gallagher M, Kapp BS, Musty RE, Driscoll PA (1977): Memory formation: Evidence for a specific neurochemical system in the amygdala. *Science* 198:423–425
Gibb WRG, Montjoy CQ, Mann DMA, Lees AJ (1989): The substantia nigra and ventral tegmental area in Alzheimer's disease and Down's syndrome. *J Neurol Neurosurg Psychiat* 52:193–200
Givens BS, Olton DS (1990): Cholinergic and GABAergic modulation of medial septal area: Effect on working memory. *Behav Neurosci* 104:849–855
Gold PE, Zornetzer SF (1983): The mnemon and its juices: Neuromodulation of memory processes. *Behav Neural Biol* 38:151–189
Gold PW, Goodwin FK, Chrousos GP (1988): Clinical and biochemical manifestations of depression. *N Engl J Med* 319:348–353
Goldman-Rakic PS, Selemon LD, Schwartz ML (1984): Dual pathways connecting the dorsolateral prefrontal cortex with the hippocampal formation and parahippocampal cortex in the rhesus monkey. *Neuroscience* 12:719–743
Goldman-Rakic, PS (1987): Circuitry of primate prefrontal cortex and regulation of behavior by representational memory. In: *Handbook of Physiology*, Mountcastle, VB, ed. Bethesda, MD: American Physiological Society
Haaksma EE, Leurs R, Timmerman H (1990): Histamine receptors: Subclasses and specific ligands. *Pharmac Ther* 47:73–104
Haas HL, Greene RW (1986): Effects of histamine on hippocampal pyramidal cells of the rat in vitro. *Exp Brain Res* 62:123–130
Harbaugh RE, Roberts DW, Coombs DW, Saunders RL, Reeder TM (1984): Preliminary report: Intracranial cholinergic drug infusion in patients with Alzheimer's disease. *Neurosurgery* 15:514–518
Hardy J, Adolfsson R, Alafuzoff I, Bucht G, Marcusson J, Nyberg P, Perdahl E, Wester P, Winblad B (1985): Transmitter deficits in Alzheimer's disease. *Neurochem Int* 7:545–563
Haroutunian V, Mantin R, Kanof PD (1990): Frontal cortex as the site of action of physostigmine in nbM-lesioned rats. *Physiol Behav* 47:203–206
Harrell LE, Callaway R, Morere D, Falgout J (1990): The effect of long-term physostigmine administration in Alzheimer's disease. *Neurology* 40:1350–1354
Hervé D, Trovero F, Blanc G, Vezina P, Glowinski J, Tassin JP (1990): Involvement of dopamine neurons in the regulation of B-adrenergic receptor sensitivity in rat prefrontal cortex. *J Neurochem* 54:1864–1869
Hokfelt T, Ljungdahl A, Steinbusch H, Verhofstad A, Nilsson G, Broden E, Pernow B, Goldstein M (1978): Immunohistochemical evidence of substance P-like immunoreactivity in some 5-hydroxytryptamine-containing neurons in the rat central nervous system. *Neuroscience* 3:517–538
Hyman BT, Van Hoesen GW, Damasio AR (1987): Alzheimer's disease: Glutamate depletion in the hippocampal perforant pathway zone. *Ann Neurol* 22:37–40
Ishikawa K, Ott T, McGaugh JL (1982): Evidence for dopamine as a transmitter in dorsal hippocampus. *Brain Res* 232:222–226
Jay TM, Glowinski J, Thierry AM (1989): Selectivity of the hippocampal projection to the prelimbic area of the prefrontal cortex in the rat. *Brain Res* 505:337–340

Joyce EM (1987): The neurochemistry of Korsakoff's syndrome. In: *Cognitive Neurochemistry*, Stahl SM, Iversen SD, Goodman EC, eds. Oxford: Oxford University Press
Joynt RJ, McNeill TH (1984): Neuropeptides in aging and dementia. *Peptides* 5 (Suppl 1):269–274
Kaiya H, Tanaka T, Takeuchi K, Morita K, Adachi S, Shirakawa H, Ueki H, Namba M (1983): Decreased level of β-endorphin-like immunoreactivity in cerebrospinal fluid of patients with senile dementia of Alzheimer type. *Life Sci* 33:1039–1043
Kalaria RN, Andorn AC, Tabaton M, Whitehouse PJ, Harik SI, Unnerstall JR (1989): Adrenergic receptors in aging and Alzheimer's disease: Increased B2-receptors in prefrontal cortex and hippocampus. *J Neurochem* 53:1772–1781
Kalaria RN, Sromek S, Wilcox BJ, Unnerstall JR (1990): Hippocampal adenosine A1 receptors are decreased in Alzheimer's disease. *Neurosci Lett* 118:257–260
Kalsner S (1990): Heteroreceptors, autoreceptors, and other terminal sites. *Ann NY Acad Sci* 604:1–6
Kamei C, Chung YH, Tasaka K (1990): Influence of certain H1-blockers on the step-through active avoidance response in rats. *Psychopharmacology* 102:312–318
Kesner RP (1990): Memory for frequency in rats: Role of the hippocampus and medial prefrontal cortex. *Behav Neural Biol* 53:402–410
Kidman A (1989): Neurochemical and cognitive aspects of anxiety disorders. *Prog Neurobiol* 32:391–402
Kolb B (1984): Functions of the frontal cortex of the rat: A comparative review. *Brain Res Rev* 8:65–98
Kovács GL, Vécsel L, Szabó G, Telegdy G (1977): The involvement of catecholaminergic mechanisms in the behavioural action of vasopressin. *Neurosci Lett* 5:337–344
Krettek JE, Price JL (1977): The cortical projections of the mediodorsal nucleus and adjacent thalamic nuclei in the rat. *J Comp Neurol* 171:157–192
Krnjević K, Ropert N (1981): Septo-hippocampal pathway modulates hippocampal activity by a cholinergic mechanism. *Can J Physiol Pharmacol* 59:911–914
Laroche S, Jay TM, Thierry AM (1990): Long-term potentiation in the prefrontal cortex following stimulation of the hippocampal CA1/subicular region. *Neurosci Lett* 114:184–190
Leckman JF, Riddle MA, Cohen DJ (1988): Pathobiology of Tourette's syndrome. In: *Tourette's Syndrome & Tic Disorders*, Cohen DJ, Bruun RD, Leckman JF, eds. New York: John Wiley & Sons
Levin ED, McGurk SR, Rose JE, Butcher LL (1990): Cholinergic-dopaminergic interactions in cognitive performance. *Behav Neural Biol* 54:271–299
Mann DMA, Yates PO, Marcyniuk B (1984): A comparison of changes in the nucleus basalis and locus coeruleus in Alzheimer's disease. *J Neurol Neurosurg Psychiat* 47:201–203
Mantz J, Milla C, Glowinski J, Thierry AM (1988): Differential effects of ascending neurons containing dopamine and noradrenaline in the control of spontaneous activity and of evoked responses in the rat prefrontal cortex. *Neuroscience* 27:517–526
Mantz J, Godbout R, Tassin JP, Glowinski J, Thierry AM (1990): Inhibition of spontaneous and evoked unit activity in the rat medial prefrontal cortex by mesencephalic raphe nuclei. *Brain Res* 524:22–30
Markowska AL, Olton DS, Murray EA, Gaffan D (1989): A comparative analysis of the role of fornix and cingulate cortex in memory: Rats. *Exp Brain Res* 74:187–201
Mason ST, Fibiger HC (1978): Noradrenaline and spatial memory. *Brain Res* 156:382–386

Matthies H (1989): Neurobiological aspects of learning and memory. *Ann Rev Psychol* 40:381–404

Mazurek MF, Growdon JH, Beal MF, Martin JB (1986): CSF vasopressin concentration is reduced in Alzheimer's disease. *Neurology* 36:1133–1137

McGurk SR, Levin ED, Butcher LL (1988): Cholinergic-dopaminergic interactions in radial-arm maze performance. *Behav Neural Biol* 49:234–239

McKinney M, Coyle JT, Hedreen JC (1983): Topographic analysis of the innervation of the rat neocortex and hippocampus by the basal forebrain cholinergic system. *J Comp Neurol* 217:103–121

Meibach RC, Siegel A (1977): Efferent connections of the septal area in the rat: An analysis utilizing retrograde and anterograde transport methods. *Brain Res* 119:1–20

Mesulam MM, Mufson EJ, Wainer BH, Levey AI (1983): Central cholinergic pathways in the rat: An overview based on an alternative nomenclature (Ch1–Ch6). *Neuroscience* 10:1185–1201

Moore RY, Bloom FE (1978): Central catecholamine neuron systems: Anatomy and physiology of the dopamine systems. *Ann Rev Neurosci* 1:129–169

Moore RY, Bloom FE (1979): Central catecholamine neuron systems: Anatomy and physiology of the norepinephrine and epinephrine systems. *Ann Rev Neurosci* 2:113–168

Moroni F, Malthe-Sorenssen D, Cheney DL, Costa E (1978): Modulation of Ach turnover in the septal hippocampal pathway by electrical stimulation and lesioning. *Brain Res* 150:333–341

Morselli PL, Lloyd KG (1985): Mechanism of action of antiepileptic drugs. In: *The Epilepsies*, Porter RJ, Morselli PL, eds. London: Butterworths

Mouradian MM, Mohr E, Williams JA, Chase TN (1988): No response to high-dose muscarinic agonist therapy in Alzheimer's disease. *Neurology* 38:606–608

Mynlieff M, Dunwiddie TV (1988): Noradrenergic depression of synaptic responses in hippocampus of rat: Evidence for mediation by alpha 1 receptors. *Neuropharmacology* 27:391–398

Nishibe Y, Matsuo Y, Yoshizaki T, Eigyo M, Shiomi T, Hirose K (1982): Differential effects of sulpiride and metoclopramide on brain: Homovanillic acid levels and shuttle box avoidance after systemic and intracerebral administration. *Naunyn-Schmiedeberg's Arch Pharmacol* 321:190–194

Oades RD, Halliday GM (1987): Ventral tegmental (A10) system: Neurobiology. 1. Anatomy and connectivity. *Brain Res Rev* 12:117–165

O'Keefe J, Nadel L (1978): *The Hippocampus as a Cognitive Map*. Oxford: Clarendon Press

Olton DS, Becker JT, Handelmann GE (1979): Hippocampus, space, and memory. *Behav Brain Sci* 2:313–365

Olton DS, Givens BS, Markowska AL, Shapiro M, Golski S (1991): Mnemonic functions of the cholinergic septohippocampal system. In: *Memory: Organization and Locus of Change*, Squire LR, Weinberger JL, McGaugh JL, eds. New York: Oxford University Press

Owen A, Downes JJ, Sahakian BJ, Polkey CE, Robbins TW (1990): Planning and spatial working memory following frontal lobe lesions in man. *Neuropsychologia* 28:1021–1034

Penney JB, Maragos WF, Greenamyre JT, Debowey DL, Hollingsworth Z, Young AB (1990): Excitatory amino acid binding sites in the hippocampal region of Alzheimer's disease and other dementias. *J Neurol Neurosurg Psychiat* 53:314–320

Perry EK, Gibson PH, Blessed G, Perry RH, Tomlinson BE (1977): Neurotransmitter enzyme abnormalities in senile dementia. *J Neurol Sci* 34:247–265

Pisa M, Fibiger HC (1983): Evidence against a role of the rat's dorsal noradrenergic bundle in selective attention and place memory. *Brain Res* 272:319–329

Poucet B (1990): Septum and medial frontal cortex contribution to spatial problem-solving. *Behav Brain Res* 37:269–280

Prell GD, Green JP (1986): Histamine as a neuroregulator. *Ann Rev Neurosci* 9:209–254

Quatermain D (1983): The role of catecholamines in memory processing. In: *The Physiological Basis of Memory*, 2nd ed., Deutsch JA, ed. New York: Academic Press

Read SL, Frazee J, Shapira J, Smith C, Cummings JL, Tomiyasu U (1990): Intracerebroventricular bethanechol for Alzheimer's disease. *Arch Neurol* 47:1025–1030

Reinikainen KJ, Riekkinen PJ, Halonen T, Laakso M (1987): Decreased muscarinic receptor binding in cerebral cortex and hippocampus in Alzheimer's disease. *Life Sci* 41:453–461

Reinikainen KJ, Paljarvi L, Huuskonen M, Soininen H, Laakso M, Riekkinen OJ (1988a): A post-mortem study of noradrenergic, serotonergic and GABAergic neurons in Alzheimer's disease. *J Neurol Sci* 84:101–116

Reinikainen KJ, Paljarvi L, Halonen T, Malminen O, Kosma VM, Laakso M, Riekkinen PJ (1988b): Dopaminergic system and monoamine oxidase-B activity in Alzheimer's disease. *Neurobiol Aging* 9:245–252

Reinikainen KJ, Soininen H, Riekkinen PJ (1990): Neurotransmitter changes in Alzheimer's disease: Implications to diagnostics and therapy. *J Neurosci Res* 27:576–586

Rigdon GC, Pirch JH (1986): Nucleus basalis involvement in conditioned neural responses in the rat frontal cortex. *J Neurosci* 6:2535–2542

Roberts GW, Ferrier IN, Lee Y, Crow TJ, Johnstone EC, Owens DGC, Bacarese-Hamilton AJ, McGregor G, O'Shaughnessey D, Polak JM, Bloom SR (1983): Peptides, the limbic lobe and schizophrenia. *Brain Res* 288:199–211

Robinson SE, Malthe-Sorenssen D, Wood PL, Commissiong J (1979): Dopaminergic control of the septal-hippocampal cholinergic pathway. *J Pharmacol Exp Therap* 208:476–479

Room P, Russchen FT, Groenewegen HJ, Lohman AHM (1985): Efferent connections of the prelimbic (area 32) and the infralimbic (area 25) cortices: An anterograde tracing study in the cat. *J Comp Neurol* 242:40–55

Ruat M, Traiffort E, Bouthenet ML, Schwartz JC, Hirschfeld J, Buschauer A, Schunack W (1990): Reversible and irreversible labeling and autoradiographic localization of the cerebral histamine H2 receptor using ^{125}I iodinated probes. *Proc Natl Acad Sci USA* 87:1658–1662

Satoh K, Fibiger HC (1986): Cholinergic neurons of the laterodorsal tegmental nucleus: Efferent and afferent connections. *J Comp Neurol* 253 277–302

Scatton B, Dubois A (1985): Autoradiographic localization of D1 dopamine receptors in the rat brain with ^{3}H SKF 38393. *Eur J Pharmacol* 111:145–146

Sesack SR, Deutch AY, Roth RH, Bunney BS (1989): Topographical organization of the efferent projections of the medial prefrontal cortex in the rat: An anterograde tract-tracing study with *Phaseolus vulgaris* leucoagglutinin. *J Comp Neurol* 290:213–242

Shimohama S, Taniguchi T, Fujiwara M, Kameyama M (1988): Changes in benzodiazepine receptors in Alzheimer-type dementia. *Ann Neurol* 23:404–406

Simon H (1981): Neurones dopaminergiques A10 et systéme frontal. *J Physiol (Paris)* 77:81–95

Śmialowski A, Bijak M (1987): Excitatory and inhibitory action of dopamine on hippocampal neurons in vitro. Involvement of D2 and D1 receptors. *Neuroscience* 23:95–101

Sparks DL, Markesbery WR, Slevin JT (1986): Alzheimer's disease: Monoamines and spiperone binding reduced in nucleus basalis. *Ann Neurol* 19:602–604

Spencer PM, Wheal HV (1990): Synaptic inhibition in the rat hippocampus in vivo following stimulation of the substantia nigra and ventral tegmentum. *J Physiol* 423:77–90

Spokes EGS (1981): The neurochemistry of Huntington's chorea. *Topics Neurosci* 4:115–118

Squire LR (1987): *Memory and Brain*. New York: Oxford University Press

Squire LR, Amaral DG, Press GA (1990): Magnetic resonance imaging of the hippocampal formation and mammillary nuclei distinguish medial temporal lobe and diencephalic amnesia. *J Neurosci* 10:3106–3117

Strittmatter H, Kackish R, Hertting G (1982): Role of dopamine receptors in the modulation of acetylcholine release in the rabbit hippocampus. *Naunyn-Schmiedeberg's Arch Pharmacol* 321:195–200

Strong R, Huang JS, Huang SS, Chung HD, Hale C, Burke WJ (1991): Degeneration of the cholinergic innervation of the locus coeruleus in Alzheimer's disease. *Brain Res* 542:23–28

Sutherland RJ, Kolb B, Whishaw IQ (1982): Spatial mapping: Definitive contribution by hippocampal or medial frontal cortical damage in the rat. *Neurosci Lett* 31:271–276

Swanson LW, Cowan WM (1976): Autoradiographic studies of the development and connections of the septal area in the rat. In: *The Septal Nuclei*, De France JF, ed. New York: Plenum Press

Swanson LW, Cowan WM (1979): The connections of the septal region in the rat. *J Comp Neurol* 186:621–656

Swartz BE, Woody CD (1979): Correlated effects of acetylcholine and cyclic guanosine monophosphate on membrane properties of mammalian neocortical neurons. *J Neurobiol* 10:465–488

Tariot PN, Cohen RM, Sunderland T, Newhouse PA, Yount D, Mellow AM, Weingartner H, Mueller EA, Murphy DL (1987): L-deprenyl in Alzheimer's disease. *Arch Gen Psychiatry* 44:427–433

Thomas GJ (1984): Memory: Time-binding in organisms. In: *Neuropsychology of Memory*, Butters N, Squire LR, eds. New York: Guilford Press

Thomas GJ, Brito GNO (1980): Recovery of delayed alternation in rats after lesions in medial frontal cortex and septum. *J Comp Physiol Psychol* 94:808–818

Thomas GJ, Spafford PS (1984): Deficits for representational memory induced by septal and cortical lesions (singly and combined) in rats. *Behav Neurosci* 98:394–404

Tulving E (1972): Episodic and semantic memory. In: *Organization of Memory*, Tulving E, Donaldson W, eds. New York: Academic Press

Uzzell BP (1986): Pathophysiology and behavioral recovery. In: *Clinical Neuropsychology of Intervention*, Uzzell BP, Gross Y, eds. Boston: Martinus Nijhoff

Volicer L, Langlais PJ, Matson WR, Mark KA, Gamache PH (1985): Serotoninergic system in dementia of the Alzheimer type. *Arch Neurol* 42:1158–1161

Weiskrantz L (1968): Some traps and pontifications. In: *Analysis of Behavioral Change*, Weiskrantz L, ed. New York: Harper & Row

Whishaw IQ, Dunnett SB (1985): Dopamine depletion, stimulation or blockade in the rat disrupts spatial navigation and locomotion dependent upon beacon or distal cues. *Behav Brain Res* 18:11–29

White JM, Rumbold GR (1988): Behavioural effects of histamine and its antagonists: A review. *Psychopharmacology* 95:1–14

Whitehouse PJ, Price DL, Clark AW, Coyle JT, DeLong MR (1981): Alzheimer disease: Evidence for selective loss of cholinergic neurons in the nucleus basalis. *Ann Neurol* 10:122–126

Whitehouse PJ, Martino AM, Antuono PG, Lowenstein PR, Coyle JT, Price DL, Kellar KJ (1986): Nicotinic acetylcholine binding sites in Alzheimer's disease. *Brain Res* 371:146–151

Whitehouse PJ, Vale WW, Zweig RM, Singer HS, Mayeux R, Kuhar MJ, Price DL, De Souza EB (1987): Reductions in corticotropin releasing factor-like immunoreactivity in cerebral cortex in Alzheimer's disease, Parkinson's disease, and progressive supranuclear palsy. *Neurology* 37:905–909

Wirak DO, Bayney R, Ramabhadran TV, Fracasso RP, Hart JT, Hauer PE, Hsiau P, Pekar SK, Scangos GA, Trapp BD, Unterbeck AJ (1991): Deposits of amyloid B protein in the central nervous system of transgenic mice. *Science* 253:323–325

Woolf NJ, Eckenstein F, Butcher LL (1983): Cholinergic projections from the basal forebrain to the frontal cortex: A combined fluorescent tracer and immunohistochemical analysis in the rat. *Neurosci Lett* 40:93–98

Woolf NJ, Eckenstein F, Butcher LL (1984): Cholinergic systems in the rat brain: I. Projections to the limbic telencephalon. *Brain Res Bull* 13:751–784

Yamamura HI, Snyder SH (1974): Postsynaptic localization of muscarinic cholinergic receptor binding in rat hippocampus. *Brain Res* 78:320–326

11

Nicotinic-Muscarinic Interactions in Cognitive Function

EDWARD D. LEVIN AND ROGER W. RUSSELL

Introduction

Nicotinic and muscarinic receptor subtypes of the cholinergic system each have a long history of study. However, investigation has mainly concentrated on their separate actions. Their potential interactions have received very little attention, despite the fact that they both are stimulated by the same endogenous ligand. There are a variety of mechanisms by which these two types of cholinergic receptors might act in concert with regard to normal function, disease processes, and the effects of drugs and toxicants. In this chapter we review neurochemical, anatomical, and behavioral mechanisms for the interaction of nicotinic and muscarinic effects and what consequences these interactions may have on cognitive function. This oldest and best established case of receptor types for a single transmitter provides a good example of how the study of the function of such structures can easily get tracked into how they function separately, ignoring the critical investigation of how they interact.

Much of the recent development in the field of neuropsychopharmacology has revolved around the discovery and characterization of subtypes of neurotransmitter receptors. Receptor subtypes have been described for norepinephrine, dopamine, serotonin, acetylcholine, GABA, and a variety of other transmitters. As investigation progresses, it has become evident that a variety of receptors probably exist for most transmitters found in the nervous system. In the process of characterizing receptor multiplicity, it is most common to focus on the differential functions of the subtypes. This is important, especially in the early phases of investigation when the separate identity of the putative subtypes has yet to be proven.

Once the subtypes have been clearly delineated and selective agonists and antagonists become available, this process of differentiating separate functions of the receptor subtypes typically continues. Thus, the focus of investigation can become fragmented, with attention restricted to the function of a single transmitter system and, indeed, to a single receptor subtype within that system. However, it is important to complement the knowledge gained from such a reductionistic approach with examination of integrative processes. This is particularly true in the case of receptor subtypes. Because the different receptor subtypes use the same transmitter, it is clear that the processes of synthesis, release, and inactivation of the transmitter are the same for all of them. Drugs, disease processes, or normal functional mechanisms that act on these processes would express actions through all receptor subtypes. However, also of potential importance may be interactions of subtypes that occur because of their anatomical juxtaposition or their involvement in the same or complementary behavioral processes. By studying the interactions of receptor subtypes, it becomes possible to understand more fully the integrative mechanisms of the nervous system in behavior.

It is the goal of this chapter to consider the mechanisms for interaction between the two major subtypes of cholinergic receptors, nicotinic and muscarinic, and the consequences of the interactions. There exists a long and rich literature concerning the neurobehavioral roles of each of these receptors. However, relatively little attention has been given to interactions between them. This is unfortunate because there are a variety of mechanisms by which nicotinic and muscarinic systems can readily interact, and the consequences of interactions may be involved in many important neurobehavioral functions. Initial evidence indicates that these systems may act together in terms of cognitive functioning, the behavioral output emphasized in this text. Such information may be critically important in the understanding of such disorders as dementia resulting from cholinergic underactivation and of the development of effective therapeutic treatments. The information may also be essential to the control of toxicants that affect the metabolic cycle of acetylcholine, e.g., acetylcholinesterase inhibitors used in pesticides, which have effects mediated via the two systems.

Nicotinic and Muscarinic Receptor Systems

The oldest established case of more than one receptor type for a single endogenous transmitter, acetylcholine (ACh), is the division of its receptors into two types: nicotinic and muscarinic. Subdivision of nicotinic and muscarinic receptors into two or more subtypes has more recently been proposed, and selective agonists and antagonists have been sought. Further investigation is needed to determine the significance of these and possibly further subdivisions (Watson et al., 1987). During recent years much has been learned about the molecular structure of nicotinic and muscarinic cholinergic receptors, and about how the receptors translate the binding of ACh into changes in ion conductance across the postsynaptic membrane, thereby initiating communication between cells (McCarthy, 1986). In general, the nicotinic receptor action is fast and fleeting, while the muscarinic response is slow and persistent.

Relatively selective agonists and antagonists for nicotinic and muscarinic receptors have provided the pharmacological tools necessary for investigating their involvement in neurobehavioral functions. As the names imply, the prototypic agonists for the two subtypes are nicotine and muscarine. Nicotine has been popular in pharmacological studies; for muscarinic stimulation, pilocarpine, carbachol, and oxotremorine are probably the most widely used ligands. Common among antagonists are mecamylamine for nicotinic receptors, and scopolamine and atropine for muscarinic receptors. Scopolamine and atropine are competitive blockers, while mecamylamine has both competitive and noncompetitive aspects.

Mechanisms for Interaction

Two principal mechanisms by which nicotinic and muscarinic systems may interact deserve special attention: neurochemical processes and neuroanatomic processes. Neurochemical mechanisms for interaction between nicotinic and muscarinic receptors consist mainly of the metabolic pathway of their common ligand, acetylcholine. The neuroanatomical sites for interaction consist of the regions in the nervous system where nicotinic and muscarinic receptors are colocalized, as well as neural circuits in which both receptor types are found.

Neurochemical Interactions

RECEPTOR REGULATION. Any manipulation of ACh metabolism, either in terms of synthesis or degradation, would be expected to have effects on both muscarinic and nicotinic systems, ACh being the common ligand of both. The behavioral outcomes of either hypo- or hyperfunctioning of the cholinergic system, e.g., decreasing the availability of choline for ACh synthesis or decreasing the rate of ACh breakdown by acetylcholinesterase (AChE) inhibitors, is most likely due to the combination of effects on both systems. For example, it was reported early in the study of these systems that chronic inhibition of AChE activity by the inhibitor diisopropyl fluorophosphate (DFP) caused functional changes indicative of downregulation of both nicotinic and muscarinic receptors (Overstreet et al., 1974). Subsequently, investigators confirmed that chronic DFP administration caused downregulation of nicotinic (Schwartz and Kellar, 1983, 1985) and muscarinic receptors (McDonald, et al., 1988; Russell, et al., 1989; Sivam, et al., 1983; Smolen, et al., 1986; Yamada, et al., 1983a, 1983b). These studies highlight the importance of considering both nicotinic and muscarinic systems in the investigation of exposure to such exogenous compounds as organophosphate pesticides, as well as the therapeutic effects of drugs like physostigmine.

Chronic nicotine administration produces an upregulation of nicotinic ACh receptors in several brain areas, including the striatum and cortex (Ksir et al., 1985, 1987; Marks and Collins, 1983; Marks and Collins, 1985; Schwartz and Kellar, 1983, 1985). This stands in contrast to the typical effect of chronic administration of an agonist. It is known that the nicotine receptor can become desensitized after stimulation (Ochoa et al., 1989). It has been speculated that the effects of nicotine results in an upregulation of receptors because it can in effect act as an antagonist by causing desensitization or depolarization blockade of nicotinic receptors (Marks and Collins, 1983; Schwartz and Kellar, 1985; Wonnacott et al.,

1989). Thus, the systems involved become understimulated and manufacture additional receptors to provide additional stimulation. The appearance and extent of nicotinic receptor upregulation has been found in some cases to correspond to the sensitization of the locomotor stimulant effect of acute nicotine doses after chronic nicotine administration (Ksir et al., 1985, 1987). The delayed onset of nicotinic receptor upregulation and persistence after nicotine withdrawal has a time course similar to the enhancement of cognitive function following chronic nicotine.

In studies of acute antagonist effects, nicotinic and muscarinic drugs act in an additive fashion in suppressing accurate memory performance (Levin, et al., 1989, 1990). It is possible that chronic nicotine administration has some of its effects on cognitive function via indirect actions on muscarinic systems. After withdrawal from chronic nicotine administration, rats show hypersensitivity to the working memory impairments caused by either nicotinic or muscarinic receptor antagonists (Levin and Rose, 1990). However, there is some controversy as to whether changes in muscarinic receptor binding are involved. Some studies have reported that chronic nicotine administration does not alter binding of acetylcholine to muscarinic receptors (Marks and Collins, 1985; Schwartz and Kellar, 1985), while others have found that it reduces the binding affinity of muscarinic agonists and antagonists but does not alter receptor numbers (Yamanaka et al., 1985, 1987).

There may be regional selectivity of the receptor downregulation during the development of tolerance to chronic AChE inhibition. Chronic exposure to sublethal doses of organophosphates has been shown to result in downregulation of muscarinic receptors in only certain brain areas (Churchill et al., 1984). Decreases were seen in the cortex, striatum, hippocampus, septum, superior colliculus, and central grey, whereas little or no change was seen in the thalamus, hypothalamus, reticular formation, and cerebellum. Chronic soman exposure was needed for downregulation of muscarinic receptors in the hippocampus. Exposure for only 40 hr did not cause this effect (Aas et al., 1987).

PRESYNAPTIC RELEASE. Regulation of ACh release is one of the key processes in the chain of events from the synthesis of the transmitter to effects on behavior (Russell, 1988). Essential to the regulation is the presence of autoreceptors on central presynaptic nerve endings. These are sensitive to ACh and have the capability of inhibiting release of the transmitter into the synapse. Detailed experimentation has now provided support for this "negative feedback" model. The research has involved pharmacological challenges with both muscarinic and nicotinic (Wonnacott et al., 1989) agonists and antagonists. Results of the former have led one author to comment: "Presynaptic receptors are an amazingly diversified mechanism for the physiological regulation of neurons. They are even more diversified mechanisms for pharmacological manipulation . . ." (Starke, 1981). Although nicotinic and muscarinic autoreceptors have been identified, it is particularly pertinent to the effects of both on behavior that the majority of brain nicotinic receptors are located presynaptically (Wonnacott et al., 1989).

Consequently the influence of nicotine or ACh release may be of special significance for behavioral neuropharmacology. Furthermore, it has been shown that release is differentially affected by acute and chronic nicotine treatments: Acute administration facilitates release (Beani et al., 1989; Rowell and Winkler, 1984; Wonnacott et al., 1989), while chronic treatment decreases release (Lapchak et al., 1989).

Knowledge about the dynamics of cholinergic function and of its sites of action opened opportunities to study cholinergic substrates of behavior experimentally by the use of cholinergic agonists and antagonists, and by punctate antomical manipulation of the system in specific brain regions. Of particular interest have been relations between concentration of an endogenous or exogenous compound and occupation of its receptor sites, and between the occupation of these sites and events that followed, including behavior.

Anatomical Interactions

Neurons containing the markers for acetylcholine (ChAT and AChE) are found in a variety of locations in the brain, major concentrations of cholinergic cells being located in the medial septal nucleus, basal forebrain, striatum, and brainstem (Butcher and Woolf, 1986). Principle projection areas for cholinergic neurons include the cortex, hippocampus, striatum, substantia nigra, and medial habenula (Woolf and Butcher, 1985, 1986; Woolf et al., 1984). Nicotinic and muscarinic receptors show distinct distributions, but there are several areas of overlap and areas where they are closely involved in the same neural circuits.

OVERLAPPING DISTRIBUTIONS. Binding studies have mapped distinct but sometimes overlapping distributions for the two types of receptors (Schwartz, 1986). Nicotinic receptors have their greatest concentration in the interpeduncular nucleus, thalamic nuclei, medial habenula, presubiculum, and superior colliculus. Muscarinic receptors have been found to be concentrated in brainstem nuclei, the pretectal area, anteroventral thalamic nucleus, the nucleus accumbens, striatum, and the olfactory cortex. The areas of overlap may be important sites for the interaction of nicotinic and muscarinic mechanisms. Significant overlap is seen in several thalamic nuclei, the interpeduncular nucleus, the superior colliculus, and the cerebral cortex (Schwartz, 1986).

PROXIMITY. There are several areas in the nervous system where nicotinic and muscarinic mechanisms come into close proximity. In the peripheral nervous system, the parasympathetic branch of the autonomic nervous system is a good example of how different receptor types can interact by being in the same neural circuit. The preganglionic synapses of both the sympathetic and parasympathetic branches use nicotinic receptors. The postganglionic synapse in the parasympathetic system uses muscarinic receptors. There is some evidence that both nicotinic and muscarinic receptors are located at the preganglionic site (Watson, 1984).

There also appear to be specific areas of nicotinic-muscarinic interaction in the brain. For example, the combination of nicotinic and muscarinic antagonists significantly decrease the development of kindled seizures in the amygdala, whereas either drug alone is ineffective (Meyerhoff, 1985).

Nicotinic and muscarinic systems interact in opposite ways with DA systems in the striatum. DA efflux is stimulated by either nicotinic activation (Giorguieff-Chesselet et al., 1979) or muscarinic blockade (Fernando et al., 1986), and DA neurons subsequently act to inhibit interstitial ACh cells (Sethy and Wenger, 1985). Interactive effects of ACh systems with DA mechanisms are reviewed in the chapter by Levin and Rose in this volume.

Behavioral Consequences of Interactions

Neuropharmacology

DRUGS AFFECTING CHOLINE. Precursor loading of choline should provide increased stimulation of both nicotinic and muscarinic receptors. However, the impact of this type of treatment is limited by the fact that the availability of choline for ACh synthesis in the brain is determined to the greatest degree by the rate of efflux of Ch out of the brain.

Another method of manipulating the availability of choline is to substitute a false precursor for choline in the diet. This results in the manufacture of an ineffective false transmitter instead of ACh. When rats are chronically given this diet, they show a progressive decrease in brain ACh levels and a progressive cognitive impairment (Russell, 1990). The involvement of muscarinic receptor systems in this deterioration have been extensively examined. Muscarinic receptors show an initial increased number in compensation for the decline in ACh stimulation (Russell, 1990). The involvement of nicotinic receptor systems in the effects of aminodenol have not yet been characterized. Given the fact that ACh would be lower in nicotinic as well as muscarinic synapses, there is every reason to believe that nicotinic receptor systems would be involved as well.

ACHE INHIBITORS. Drugs that inhibit AChE increase the longevity of ACh and consequently result in increased stimulation of both nicotinic and muscarinic receptors. Curiously these compounds have been investigated both as instruments of death for insects and people and possible therapeutic treatments for syndromes of cholinergic underactivity such as Alzheimer's disease. As discussed earlier, the prototypic AChE inhibitor DFP causes functional changes indicative of downregulation of both nicotinic and muscarinic receptors (Overstreet et al., 1974).

DRUGS SPECIFIC TO MUSCARINIC OR NICOTINIC RECEPTORS. The involvement of the muscarinic branch of the cholinergic system in cognitive function has been extensively documented. Roles for the nicotinic branch, though less widely studied, have also been established. Surprisingly, interactions of the two in affecting cognitive functions have scarcely been examined. Recent reports (Levin et al., 1989, 1990) have indicated that there are at least additive, and probably synergistic, effects of muscarinic and nicotinic antagonists in causing working memory deficits. Coadministration of nicotinic and muscarinic receptor blockers has been shown to cause severe disruption of memory function, performance accuracy being reduced to chance levels. For example, doses of scopolamine (a muscarinic blocker) and mecamylamine (a nicotinic blocker), which do not by themselves cause impairments, when given together cause cognitive deficits in treated rats. On the other hand, nicotinic and muscarinic agonists do not appear to summate in the improvement of working memory, which they each, but themselves, cause. Acute doses of nicotine and pilocarpine each improve choice accuracy performance in the radial-arm maze, but when given together they do not have an additive effect (Levin and Rose, 1991). Interestingly, both the nicotinic and muscarinic agonist effects were reversed by the appropriate nicotinic or muscarinic blocker, they were also reversed by an antagonist of the opposite receptor as well.

Nicotinic-muscarinic interactions have also been observed in other laboratories. Muscarinic receptor blockade inhibits the initial depressant effect that nicotine has on operant VI bar-pressing behavior, while nicotinic blockade blocks both the initial depression and subsequent increase in bar-pressing behavior (Morrison et al., 1969). Crosstolerance has been reported between nicotinic and muscarinic agonists in terms of effects on heart rate and body temperature (Marks and Collins, 1985).

Nicotinic-muscarinic receptor changes are seen with manipulations that specifically affect the individual receptor subtypes. Upregulation of cortical nicotinic receptors has been seen after chronic scopolamine treatment (Vigé and Briley, 1988). Conversely, cortical muscarinic agonist binding is downregulated after chronic nicotine administration (Yamanaka et al., 1987). After withdrawal from chronic nicotine administration, rats are hypersensitive to the amnestic effects of the muscarinic antagonist, scopolamine, as well as the nicotinic antagonist, mecamylamine (Levin and Rose, 1990).

NEUROTOXICOLOGY. Most study concerning the relationship of nicotinic and muscarinic systems in the field of neurotoxicology had been in the area of acetylcholinesterase (AChE) inhibitors. These have been used as therapeutic agents, as insecticides, and as chemical weapons. Since acetylcholine is catabolized by AChE in both nicotinic and muscarinic synapses, one would expect that it would lead to an increased stimulation of both receptors. Because of differences in anatomy and physiology, sometimes differential effects occur.

Chronic administration of the organophosphate insecticide disulfoton decreased the number of both muscarinic and nicotinic receptors, and it markedly reduced the antinociceptive effect of nicotine (Costa and Murphy, 1983; McDonald et al., 1988). Acute treatment with soman enhances nicotinic depolarizations and depresses hyperpolarizations in parasympathetic neurons (Kumamoto and Shinnick-Gallagher, 1988).

Involvement in Behavioral Disorders

Muscarinic-nicotinic interactions are also important in disease states such as Alzheimer's disease. A wide variety of studies have found that Alzheimer patients have a substantial reduction in nicotinic receptors in the cortex and hippocampus (Araujo et al., 1988; DeSarno et al., 1988; Flynn and Mash, 1986; Giacobini et al., 1988, 1989; Kellar and Wonnacott, 1990; Ksir et al., 1987; London et al., 1989; Nordberg et al., 1988; Nordberg and Winblad, 1986; Perry et al., 1985, 1987; Quiron et al., 1986; Rinne et al., 1991; Schröder et al., 1991; Shimohama et al., 1986; Sugaya et al., 1990; Whitehouse et al., 1986, 1988a, 1988b). The status of muscarinic receptors in Alzheimer's disease is more controversial, but it appears that there is a decline in cortical M_2 muscarinic receptors as well (Perry et al., 1987).

The amnesia resulting from scopolamine, a muscarinic blocker, has been proposed as a model of Alzheimer's disease. However, if both muscarinic and nicotinic receptor levels are decreased, combined administration of scopolamine and mecamylamine might be a more accurate model. This interaction may also be important in the development of therapeutic treatments for Alzheimer's disease. The use of muscarinic agonists alone does not seem to be effective. Perhaps combined administration of nicotinic and muscarinic agonists might be more effective.

Interactions with Other Organ Systems

As effects of direct interactions between the muscarinic and nicotinic neurotransmitter systems move "downstream," changes important to neurobehavioral functions may extend into other organ systems. Overt bodily responses to sensory-perceptual processes in the CNS may begin with muscarinic activity and end with changes in the skeletal muscles initiated by nicotinic events at the myoneural junction. The nicotinic receptor plays an important role in regulating endocrine function (Fuxe et al., 1990). For example, acute exposure to nicotine increases the secretion of corticosterone, with tolerance developing when exposure is chronic. Our present information suggests that hormonal systems can influence behavior, including such cognitive functions as memory storage, through their effects on neuromodulatory systems localized in specific brain areas: ". . . recent findings provide increasing evidence of interactions and convergences among the influences of the various memory-modulating treatments (McGaugh, 1990). Interactions between such organ systems and cholinergic functions may be reciprocal in the sense that the latter may be affected by prior actions of the former. For example, ACh release is inhibited by opioid agonists and is activated by opioid antagonists (Jhamandas and Sutak, 1983). Simultaneous posttraining administration of oxotremorine and naloxone at dose levels too low to produce effects independently may enhance memory (Baratti et al., 1984).

As we commented earlier, many events are involved between the time that an endogenous or exogenous chemical binds to its muscarinic and/or nicotinic receptor sites and the consequent effects at the behavioral level. The preceding examples illustrate only a few of the interactions with organ systems that may be involved, and are yet to be discovered.

Conclusions

In the elucidation of the functional roles of cholinergic systems, it is of critical importance to not only investigate both muscarinic and nicotinic mechanisms but also to examine their interaction. Both of these receptors use the same endogeneous ligand ACh. It is immediately apparent that any drug-induced alteration of the metabolic cycle of ACh choline supplementation or depletion or AChE inhibition would have effects mediated via both nicotinic and muscarinic receptor systems. Nicotinic-muscarinic interactions are also probably important in disease states such as Alzheimer's disease, in which both muscarinic and nicotinic receptors are known to be affected. But it is most important to keep in mind the interactions of nicotinic and muscarinic systems in normal brain function.

Despite the long tradition of separate investigation of nicotinic and muscarinic receptors, it is evident that in many ways these receptor subtypes act together in the synthesis of neurobehavioral function. Addressing this interaction cannot only help elucidate the actions of drugs and toxicants, and the mechanisms of disease processes, it can also provide insights into the basic functions of cholinergic systems in the brain.

References

Aas P, Veiteberg TA, Fonnum F (1987): Acute and sub-acute inhalation of an organophosphate induce alteration of cholinergic muscarinic receptors. *Biochem Pharmacol* 36:1261–1266

Araujo DM, Lapchak PA, Robitaille Y, Gauthier S, Quirion R (1988): Differential alteration of various cholinergic markers in cortical and subcortical regions of human brain in Alzheimer's disease. *J Neurochem* 50:1914–1923

Baratti CM, Introini IB, Huygens P (1984): Possible interaction between central cholinergic muscarinic and opioid peptidergic systems during memory consolidation in mice. *Behav Neural Biol* 40:155–169

Beani L, Bianchi C, Ferraro L, Nilsson L, Nordberg A, Romanelli L, Spalluto P, Sundwall A, Tanganelli S (1989): Effect of nicotine on the release of acetylcholine and amino acids in the brain. *Prog Brain Res* 79:149–155

Butcher LL, Woolf NJ (1986): Central cholinergic systems: Synopsis of anatomy and overview of physiology and pathology. In: *The Biological Substrates of Alzheimer's Disease*, Scheibel AB, Weschler AF, eds. New York: Academic Press, pp. 73–86

Churchill L, Pazdernik TL, Jackson JL, Nelson SR, Samson FE, McDonough JH, Jr. (1984): Topographical distribution of decrements and recovery in muscarinic receptors from rat brains repeatedly exposed to sublethal doses of soman. *J Neurosci* 4:2069–2079

Costa LG, Murphy SD (1983): [^{3}H]Nicotine binding in rat brain: Alteration after chronic acetylcholinesterase inhibition. *J PharmacolExp Ther* 226:392–397

DeSarno P, Giacobini E, Clark B (1988): Changes in nicotinic receptors in human and rat CNS. *Fed Proc* 2:364

Fernando JC, Hoskins B, Ho IK (1986): The role of dopamine in behavioral supersensitivity to muscarinic antagonists following cholinesterase inhibition. *Life Sci* 39:2169–2176

Flynn DD, Mash DC (1986): Characterization of L-[^{3}H]nicotine binding in human cerebral cortex: Comparison between Alzheimer's disease and the normal. *J Neurochem* 47:1948–1954

Fuxe K, Andersson K, Härfstrand A, Eneroth P, de la Mora MP, Agnati LF (1990): Effects of nicotine on synaptic transmission in the brain. In: *Nicotine Psychopharmacology: Molecular, Cellular and Behavioural Aspects*, Wonnacott S, Russell MAH, Stolerman IP, eds. New York: Oxford University Press, pp. 194–225

Giacobini E, DeSarno P, McIlhany M, Clark B (1988): The cholinergic receptor system in the frontal lobe of Alzheimer's patients. In: *Nicotinic Acetylcholine Receptors in the Nervous System*, Clementi F, Gotti C, Sher E, eds. Berlin: Springer-Verlag, pp. 367–378

Giacobini E, DeSarno P, Clark B, McIlhany M (1989): The cholinergic receptor system of the human brain. Neurochemical and pharmacological aspects of aging and Alzheimer. In: *Progress in Brain Research*, Nordberg A, ed. Amsterdam: Elsevier, pp. 335–343

Giorguieff-Chesselet MF, Kemel ML, Wandscheer D, Glowinski J (1979): Regulation of dopamine release by presynaptic nicotinic receptors in rat striatal slices: Effect of nicotine in a low concentration. *Life Sci* 25:1257–1262

Jhamandas J, Sutak M (1983): Stereo-specific enhancement of evoked release of brain acetylcholine by narcotic antagonists. *Br J Pharmacol* 78:433–440

Kellar KJ, Wonnacott S (1990): Nicotinic cholinergic receptors in Alzheimer's disease. In: *Nicotine Psychopharmacology: Molecular, Cellular, and Behavioral Aspects*, Wonnacott S, Russell MAH, Stolerman IP, eds. Oxford Oxford University Press, pp. 341–373

Ksir C, Hakan R, Hall DP, Kellar KJ (1985): Exposure to nicotine enhances the behavioral stimulant effect of nicotine and increases binding of ^{3}H-acetylcholine to nicotinic receptors. *Neuropharmacology* 24:527–531

Ksir C, Hakan R, Hall DP, Kellar KJ (1987): Chronic nicotine and locomotor activity: Influences of exposure dose and test dose. *Psychopharmacology* 92:25–29

Kumamoto E, Shinnick-Gallagher P (1988): Soman enhances nicotinic depolarizations, and depresses muscarinic hyperpolarization in parasympathetic neurons. *Brain Res* 458:151–156

Lapchak PA, Araujo DM, Quirion R, Collier B (1989): Effect of chronic nicotine treatment on nicotinic autoreceptor function and N-[^{3}H]methylcarbamylcholine binding sites in the rat brain. *J Neurochem* 52:483–491

Levin ED, Rose JE (1990): Anticholinergic sensitivity following chronic nicotine administration as measured by radial-arm maze performance in rats. *Behav Pharmacol* 1:511–520

Levin ED, Rose JE (1991): Nicotinic and muscarinic interactions and choice accuracy in the radial-arm maze. *Brain Res Bull*, 27:125–128

Levin ED, McGurk SR, South D, Butcher LL (1989): Effects of combined muscarinic and nicotinic blockade on choice accuracy in the radial-arm maze. *Behav Neural Biol* 51:270–277

Levin ED, Rose JE, McGurk SR, Butcher LL (1990): Characterization of the cognitive effects of combined muscarinic and nicotinic blockade. *Behav Neural Biol* 53:103–112

London ED, Ball MJ, Waller SB (1989): Nicotinic binding sites in cerebral cortex and hippocampus in Alzheimer's disease. *Neurochem Res* 14:745–750

Marks MJ, Collins AC (1985): Tolerance, cross-tolerance, and receptors after chronic nicotine or oxotremorine. *Pharmacol Biochem Behav* 22:283–291

Marks MJ, Burch JB, Collins AC (1983): Effects of chronic nicotine infusion on tolerance development and nicotine receptors. *J Pharmacol Exp Ther* 226:817–825

McCarthy MP, Earnest JP, Young EF, Choe S, Stroud RM (1986): The molecular neurobiology of the acetylcholine receptor. *Ann Rev Neurosci* 9:383–413

McDonald BE, Costa LG, Murphy SD (1988): Spatial memory impairment and central muscarinic receptor loss following prolonged treatment with organophosphates. *Toxicol Lett* 40:47–56

McGaugh J (1990): Significance and remembrance: The role of neuromodulatory systems. *Psychobiol Sci* 1:15–25

Meyerhoff JL, Bates VE (1985): Combined treatment with muscarinic and nicotinic cholinergic anatagonists slows development of kindled seizures. *Brain Res* 339:386–389

Morrison CF, Goodyear JM, Sellers CM (1969): Antagonism by antimuscarinic and ganglion-blocking drugs of some of the behavioural effects of nicotine. *Psychopharmacologia* 15:341–350

Nordberg A, Winblad B (1986): Reduced number of [^{3}H]nicotine and [^{3}H]acetylcholine binding sites in the frontal cortex of Alzheimer brains. *Neurosci Lett* 72:115–119

Nordberg A, Adem A, Hardy J, Winblad B (1988): Change in nicotinic receptor subtypes in temporal cortex of Alzheimer brains. *Neurosci Lett* 88:317–321

Ochoa ELM, Chattopadhyay A, McNamee MG (1989): Desensitization of the nicotinic acetylcholine receptor: Molecular mechanisms and effect of modulators. *Cell Mol Neurobiol* 9:141–178

Overstreet DH, Russell RW, Vasquez BJ, Dalglish FW (1974): Involvement of muscarinic and nicotinic receptors in behavioral tolerance to DFP. *Pharmacol Biochem Behav* 2:45–54

Perry EK, Curtis M, Dick DJ, Candy JM, Atack JR, Bloxham CA, Blessed G, Fairbairn A, Tomlinson BE, Perry RH (1985): Cholinergic correlates of cognitive impairment in Parkinson's disease: Comparisons with Alzheimer's disease. *J Neurol Neurosurg Psychiat* 48:413–421

Perry EK, Perry RH, Smith CJ, Dick DJ, Candy JM, Edwardson JA, Fairbairn A, Blessed G (1987): Nicotinic receptor abnormalities in Alzheimer's and Parkinson's diseases. *J Neurol Neurosurg Psychiat* 50:806–809

Quiron R, Martel JC, Robitaille Y, Etienne P, Wood P, Nair NPV, Gauthier S (1986): Neurotransmitter and receptor deficits in senile dementia of the Alzheimer's type. *Can J Neurol Sci* 13:503–510

Rinne JO, Myllykylä T, Lönnberg P, Marjamäki P (1991): A postmortem study of brain nicotinic receptors in Parkinson's and Alzheimer's disease. *Brain Res* 547:167–170

Rowell PP, Winkler DL (1984): Nicotinic stimulation of [^{3}H]acetylcholine release from mouse cerebral cortical synaptosomes. *J Neurochem* 43:1593–1598

Russell R (1988): Behavioral correlates of presynaptic events in the cholinergic neurotransmitter system. *Prog Brain Res* 32:43–130

Russell RW, Jenden DJ, Booth RA, Lauretz SD, Roch M, Rice KM (1990): Global in vivo replacement of choline by N-aminodeanal. Testing a hypothesis about progressive degenerative dementia. II. Physiological and behavioral effects. *Pharmacol Biochem Behav* 37:811–820

Russell RW, Booth RA, Smith CA, Jenden DJ, Roch M, Rice KM, Lauretz SD (1989): Roles of neurotransmitter receptors in behavior: Recovery of function following decreases in muscarinic receptor density induced by cholinesterase inhibition. *Behav Neurosci* 103:881–892

Schröder H, Giacobini E, Struble RG, Zilles K, Maelicke A (1991): Nicotinic cholinoceptive neurons of the frontal cortex are reduced in Alzheimer's Disease. *Neurobiol Aging* 12:259–262

Schwartz RD (1986): Autoradiographic distribution of high affinity muscarinic and nicotinic cholinergic receptors labeled with [^{3}H]acetylcholine in rat brain. *Life Sci* 38:2111–2119

Schwartz RD, Kellar KJ (1983): Nicotinic cholinergic binding sites in brain: Regulation in vivo. *Science* 220:214–216

Schwartz RD, Kellar KJ (1985): In vivo regulation of [^{3}H]acetylcholine recognition sites in brain by nicotinic cholinergic drugs. *J Neurochem* 45 427–433

Sethy VH, Wenger D (1985): Effect of dopaminergic drugs on striatal acetylcholine concentration. *J Pharm Pharmacol* 37:73–74

Shimohama S, Taniguchi T, Fujiwara M, Kameyama M (1986): Changes in nicotinic and muscarinic receptors in Alzheimer-type dementia. *J Neurochem* 46:288–293

Sivam SP, Norris JC, Lim DK, Hoskins B, Ho IK (1983): Effect of acute and chronic cholinesterase inhibition with diisopropyl fluorophosphate on muscarinic, dopamine and GABA receptors in the rat striatum. *J Neurochem* 40:1414–1422

Smolen TN, Smolen A, Collins AC (1986): Dissociation of decreased numbers of muscarinic receptors from tolerance to DFP. *Pharmacol Biochem Behav* 25:1293–1301

Starke K (1981): Presynaptic receptors. *Ann Rev Pharmacol Toxicol* 21:7–30

Sugaya K, Giacobini E, Chiappinelli VA (1990): Nicotinic acetylcholine receptor subtypes in human frontal cortex: Changes in Alzheimer's disease. *J Neurosci Res* 27:349–359

Vigé X, Briley M (1988): *Upregulation of Nicotinic Receptors by Chronic Scopolamine and Chronic Nicotine Treatments: Dependence on Presynaptic Integrity*. International Symposium on Nicotinic Receptors in the CNS: Their Role in Synaptic Transmission. Uppsala, Sweden.

Watson M, Roeske WR, Johnson PC, Yamamura HI (1984): [^{3}H] Pirenzapine identifies putative M_1 muscarinic receptors in human stellate ganglia. *Brain Res* 290:179–182

Watson M, Roeske WR, Yamamura HI (1987): Cholinergic receptor heterogeneity. In: *Psychopharmacology: The Third Generation of Progress*, Meltzer HY, ed. New York: Raven Press, pp. 241–248

Whitehouse PJ, Martino AM, Antuono PG, Lowenstein PR, Coyle JT, Price DL, Kellar KJ (1986): Nicotinic acetylcholine binding sites in Alzheimer's disease. *Brain Res* 371:146–151

Whitehouse PJ, Martino AM, Marcus KA, Zweig RM, Singer HS, Price DL, Kellar KJ (1988a): Reductions in acetylcholine and nicotine binding in several degenerative diseases. *Arch Neurol* 45:722–724

Whitehouse PJ, Martino AM, Wagster MV, Price DL, Mayeux R, Atack JR, Kellar KJ (1988b): Reductions in [^{3}H]nicotinic acetylcholine binding in Alzheimer's disease and Parkinson's disease: An autoradiographic study. *Neurology* 38:720–723

Wonnacott S, Irons J, Rapier C, Thorne B, Lunt GG (1989): Presynaptic modulation of transmitter release by nicotinic receptors. *Prog Brain Res* 79:157–163

Woolf NJ, Butcher LL (1985): Cholinergic systems in the rat brain: II. Projections to the interpeduncular nucleus. *Brain Res Bull* 14:63–83

Woolf NJ, Butcher LL (1986): Cholinergic systems in the rat brain: III. Projections from the pontomesencephalic tegmentum to the thalamus, tectum, basal ganglia and basal forebrain. *Brain Res Bull* 16:603–637

Woolf NJ, Eckenstein F, Butcher LL (1984): Cholinergic systems in the rat brain: I. Projections to the limbic telencephalon. *Brain Res Bull* 13:751–784

Yamada S, Isogai M, Okudaira H, Hayashi E (1983a): Correlation between cholinesterase inhibition and reduction in muscarinic receptors and choline uptake by repeated diisopropylfluorophosphate administration: Antagonism by physostigmine and atropine. *J Pharmacol Exp Ther* 226:519–525

Yamada S, Isogai M, Okudaira H, Hayashi E (1983b): Regional adaptation of muscarinic receptors and choline uptake in brain following repeated administration of diisopropylfluorophosphate and atropin. *Brain Res* 268:315–320

Yamanaka K, Oshita M, Muramatsu I (1985): Alterations of alpha and muscarinic receptors in rat brain and heart following chronic nicotine treatment. *Brain Res* 348:241–248

Yamanaka K, Muramatsu I, Kigoshi S (1987): Muscarinic agonist binding in rat brain following chronic nicotine treatment. *Brain Res* 409:395–397

12

The Influence of Neurotensin Upon Cholinergic Function

GARY L. WENK

Anatomical Considerations

Magnocellular cholinergic neurons are widely distributed throughout the ventral pallidum and substantia innominata within the basal forebrain of rodents and humans (Bigl et al., 1982; Butcher, this volume). These neurons provide the primary source of cholinergic innervation to the entire cortical mantle (Bigl et al., 1982; Luiten et al., 1987; Wenk et al., 1980; Woolf et al., 1984). Although there are many cholinergic systems within the brain (Butcher, this volume), this chapter will focus only upon those cholinergic cells within the nucleus basalis.

Neurotensin (NT) is a tridecapeptide that has been localized to various parts of the brain and gut. NT has numerous peripheral effects that appear to either directly or indirectly influence glucose metabolism (Bissette et al., 1978). Although NT has "kinin-like" properties (i.e., contraction of guinea pig ileum and rat uterus), its many unique biological properties (cutaneous vasodilation and cyanosis after i.v. administration), as well as its amino acid sequence, distinguish it from other known hypothalamic peptides. Immunoreactivity for NT has been localized to the central nucleus of the amygdala (Tay et al., 1989), where NT immunoreactive terminals form symmetric axoaxonal contracts with each other, suggesting the presence of local NT-to-NT circuits, as well as asymmetric axodendritic contacts. In addition to the amygdala, the concentration of NT is particularly high throughout the entire mammalian limbic system, including the medial hypothalamus, septal area, stria terminalis, and piriform cortex (Mai et al., 1987). In addition, a dense network of NT-containing fibers has been identified in the medial prefrontal and anterior cingulate cortex of monkeys (Satoh and Matsumura, 1990). No NT-containing cell bodies have been observed in the frontal cortex.

The correspondence between the distribution of NT immunoreactive cells and fibers within the brains of rodents and humans is very high. A majority of the large neurons in the neonatal human and rodent subiculum are immunoreactive for NT and project axons via the fimbria and fornix into the mammillary bodies. However, adult human and rodent brains contain relatively few NT immunoreactive cells and fibers within these regions. The presence of high concentrations of

NT neurons during the neonatal period within the limbic system suggests that this neurotransmitter system may have a prominent role during a period of rapid learning and emotional information processing (Kiyama et al., 1986; Sakamoto et al., 1986). An *in situ* hybridization study using a nucleotide probe to identify the mRNA that encodes both NT and neuromedin N, a related peptide with similar structure, identified many cells throughout the forebrain, including the septum, preoptic area, bed nucleus of the stria terminalis, hypothalamus, amygdala, nucleus accumbens, caudate, putamen, and piriform cortex in adult male rats (Alexander et al., 1989). The regional distribution of the mRNA corresponded to regions previously identified as being immunoreactive for NT cell bodies and fibers. One interesting exception to this correspondence involved area CA-1 of the hippocampus; this region contained large pyramidal cells that labeled intensely for the NT mRNA but not for NT immunoreactivity (i.e., the NT neuropeptide). A second significant discrepancy between the presence of high levels of NT mRNA and no evidence of NT immunoreactivity was discovered in the subiculum (Alexander et al., 1989). Taken together, these studies suggest that the NT mRNA precursor is processed to yield NT in the neonate human and rodent, and later is processed to yield another neuropeptide product in the adult brain. The identity of this new peptide product is unknown. Two recent studies, using *in situ* hybridization and immunocytochemistry, have also identified a continuum of NT-containing neurons that extend medially from the rostral amygdala, across the substantia innominata, into the lateral hypothalamus and bed nucleus of the stria terminalis (Alexander et al., 1989; Martin et al., 1991). The location of this continuum of cells provides ample opportunity for potential interaction between NT- and acetylcholine-containing neurons within the basal forebrain.

Functional Interactions

Nucleus Basalis

Recent studies using combined autoradiography and histochemistry in the rodent nucleus basalis have identified a high number of [^{125}I]NT binding sites in areas that exhibit intense acetylcholinesterase staining, suggesting an interaction of NT and acetylcholine neurons within this brain region (Moyse et al., 1987; Szigethy et al., 1990). To confirm this interaction, unilateral lesions of the nucleus basalis were made by injection of the neurotoxin ibotenic acid. This neurotoxin destroyed cell bodies near the injection site and significantly reduced the level of acetylcholinesterase activity and the number of [^{125}I]NT binding sites within the nucleus basalis (Szigethy et al., 1989). Initial autoradiographic investigations suggested that [^{125}I]NT receptors were associated with the plasma membrane of large cholinergic neurons. However, more recent investigations have suggested that these [^{125}I]NT binding sites are actually within the cell bodies and are being transported to axonal terminals within the cortex.

In a series of unpublished studies, I have further investigated the potential interactions between NT and acetylcholine neurons within the nucleus basalis. NT

(14.9 nM/μl, pH 7.4) was microinjected unilaterally into the nucleus basalis of rats via a chronic indwelling cannula. The injection solution also contained bacitracin (0.22 U/μl) to prolong the half-life of NT in the brain. The influence of the NT on the activity of cholinergic neurons in the nucleus basalis was determined by monitoring the release of acetylcholine via a microdialysis probe inserted into the frontal cortex.

Microdialysis has become a very useful technique for determining the concentration of selected neurotransmitters and their metabolites in the extracellular compartment of discrete brain regions in freely moving animals. Regional concentrations of acetylcholine and choline can then be determined from brain dialysates using a high-performance liquid chromatography system equipped with a postcolumn enzyme reactor that converts acetylcholine and choline into betaine and hydrogen peroxide for electrochemical detection. The microdialysis probe was perfused with a Ringer's solution (147 mM sodium chloride, 2.3 mM calcium chloride, 4 mM potassium chloride, pH 6.0) at a rate of 2–5 μl/min. The perfusate also contained 1 μM neostigmine, an acetylcholinesterase inhibitor, to increase the basal release acetylcholine and to improve the recovery of acetylcholine.

When NT was infused into the nucleus basalis, cortical acetylcholine release increased by approximately 47% (Fig. 1). When the rats were injected with the

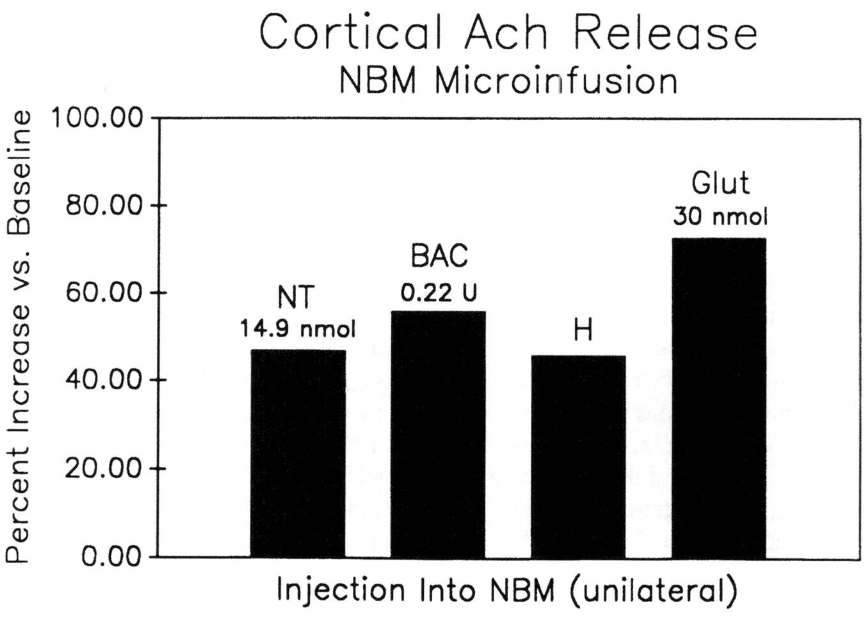

FIGURE 1. Percent increase, as compared to prior baseline levels, in cortical acetylcholine release following microinfusion of neurotensin (NT, nM/μl), vehicle-containing bacitracin (BAC, units of activity), and glutamate (Glut, nmol/μl), or following handling (H) of the rat. The microdialysis probe was placed at a 30° angle within the cortex to avoid touching the caudate nucleus, an area with a high concentration of cholinergic neurons.

vehicle, containing only bacitracin, cortical acetylcholine release increased by slightly more than 50%. A similar increase in cortical acetylcholine release was observed when a rat was simply picked up and connected to the microinjection system but not injected. Therefore, the increase in cortical acetylcholine release following the microinfusion of NT was probably not due to the presence of this neuropeptide in the nucleus basalis, suggesting that NT does not influence cholinergic function directly in this brain region. In contrast, microinfusion of glutamate (30 nM/μl) under similar conditions increased acetylcholine release by 73%.

In order to independently confirm the conclusions obtained from this microdialysis study, another indicator of cholinergic function was investigated following the microinfusion of NT into the nucleus basalis. Hemicholinium-3 (HC-3) binds to the sodium-dependent high-affinity choline uptake site on cholinergic terminals (Swann and Hewitt, 1988). Alterations in the number of HC-3 binding sites correlates directly with the activity of cholinergic neurons immediately prior to sacrifice (Swann and Hewitt, 1988; Wenk et al., 1984). Under similar conditions to those described above, NT was infused unilaterally into the nucleus basalis of awake, freely moving rats. Forty minutes later, each rat was sacrificed by decapitation and samples of frontal cortex were prepared for HC-3 binding according to the method of Swann and Hewitt (1988). The increase in HC-3 binding following either the NT injections or handling seen in this study was similar to the increase in acetylcholine release determined in the microdialysis study described above (Fig. 2). For comparison to the effects of NT, the endogenous opiate neuropeptide enkephalin was microinfused into the nucleus basalis. HC-3 binding decreased by approximately 45% in the frontal cortex. The decrease in HC-3 binding is similar to a previously reported decrease in sodium-dependent high-affinity [^{3}H]choline uptake following enkephalin microinfusion into the nucleus basalis (Wenk, 1984). The results of these preliminary studies also suggest that NT does not directly influence cholinergic function in the nucleus basalis.

Finally, the following study investigated the effects of NT microinfusion into the nucleus basalis on performance in a behavioral task that requires learning and memory. If NT influences cholinergic function in the nucleus basalis, performance in a spatial memory task might be altered. Chronic indwelling cannula were inserted bilaterally into the nucleus basalis of 20 rats. After recovery from surgery, the rats were tested in a spatial alternation task in a T-maze (Wenk et al., 1989). At the beginning of each test session, one piece of food was placed at the end of one arm of the maze. The arm containing the food was varied randomly between sessions. For the first, forced trial, the side of the maze that did not contain the food was blocked with a piece of wood, which forced the rat to go to the opposite (baited) side. After eating the food, the rat was placed back on the starting platform. The wood block was removed from the maze so that rat was able to enter either arm. For all subsequent choice trials, food was placed on the side of the maze not visited during the previous trial. The guillotine door between the starting platform and the stem of the maze was opened and the rat was allowed to choose

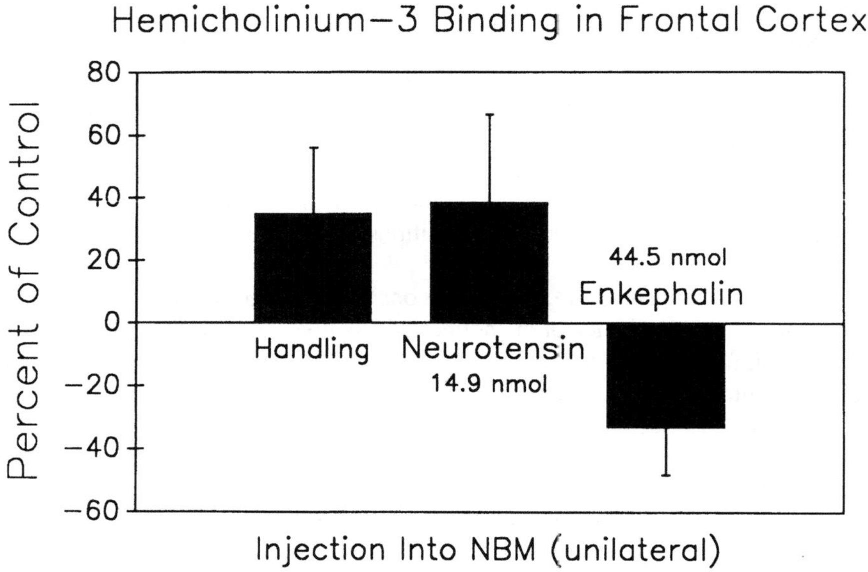

FIGURE 2. Change in the activity of basal nucleus cholinergic cells following unilateral microinfusion of either neurotensin or enkephalin, as determined by binding of [^{3}H]hemicholinium-3 to high-affinity choline uptake sites in the cortex. Data are expressed as the percent change from binding levels following infusion of the vehicle. Four rats were used for each group.

between the arms of the maze. If the rat entered the arm that contained the food, a correct response was recorded and it was allowed to eat the food before the experimenter returned it to the starting platform. If the rat entered the unbaited arm with its forepaws, an incorrect response was recorded and the rat was returned to the starting point without receiving any food. This procedure was repeated until 10 choice trials were completed. Ten choice trials were given daily. The intertrial interval was 5 sec. Approximately 10 min prior to each test session, each rat was bilaterally infused with a single neuropeptide (dissolved in vehicle) into the nucleus basalis. The quantity of each neuropeptide injected (expressed in nmol/μl) is indicated in Figure 3. Four rats were included in each group. Injection of NT into the nucleus basalis had no effect on performance in the T-maze alternation task. Only the rats that received the highest dose (89 nmols) of enkephalin had a significant impairment in performance.

Taken together, these three studies suggest that NT does not influence cholinergic function within the nucleus basalis. The next few sections will outline the results of other studies that are consistent with the hypothesis that NT might influence cholinergic function in the mammillary bodies, neocortex, and amygdala.

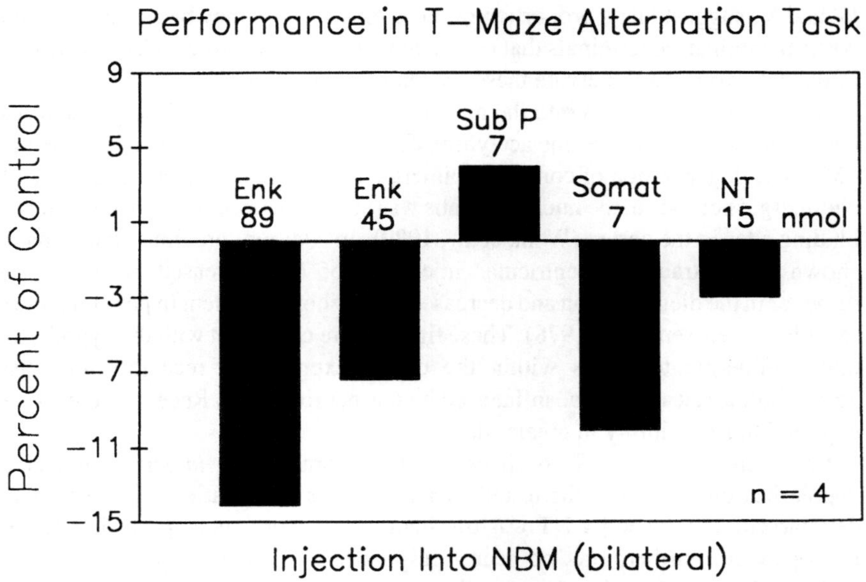

FIGURE 3. Performance of rats in an alternation task in a T-maze following a bilateral microinfusion of four different neuropeptides. Numbers in the figure indicate nanomoles of peptide injected in 1 μl of total volume. Performance is expressed as the percent change from a level following injection of vehicle. Four rats were used for each group. Enk = Leu-Enkephalin; Sub P = substance P; Somat = somatostatin; NT = neurotensin.

Mammillary Bodies

The intense projection of NT-containing neurons into the mammillary bodies from the subiculum may have a role in passive-avoidance behavior (Shibata and Furukawa, 1988). Small doses of NT injected into the mammillary bodies improved performance in a passive-avoidance task. This improvement was not seen when NT was injected into the amygdala, preoptic area, nucleus accumbens, or substantia nigra. However, the alteration in passive-avoidance behavior may have been due to the fact that a rat's responsiveness to electric foot shock is reduced after NT injection into the mammillary bodies (Shibata and Furukawa, 1988). Furthermore, the presence of cholinesterase staining within the supramammillary and lateral mammillary nuclei offers another potential site of interaction for these two neurotransmitter systems and may underlie NT's influence on passive-avoidance behavior.

Neocortex

Cholinergic projections from the nucleus basalis terminate in the frontal and parietal cortex, particularly in layers IV and V (Luiten et al., 1987). [^{125}I]NT binding has also been localized to layers IV and V (Moyse et al., 1987). The coexistence of both [^{125}I]NT binding sites, and NT axons and fibers within layers

IV and V, suggest that a proportion of these binding sites may be associated with cortical cholinergic terminals that originate within the nucleus basalis. The origin of the NT axons that innervate these receptors is unknown.

Ibotenic acid injections into the nucleus basalis destroyed cholinergic neurons and decreased cortical choline acetyltransferase (CAT) activity by approximately 45%. The degeneration of cortical cholinergic terminals subsequent to the loss of cholinergic neurons in the nucleus basalis was also associated with a decline in NT binding sites in the cortex (Wenk et al., 1989). In addition, previous studies have shown that intracerebroventricular injections of NT increased acetylcholine turnover in the diencephalon and decreased acetylcholine content in parietal cortex (Malthe-Sorenssen et al., 1978). These findings are consistent with the hypothesis that cholinergic terminals within the cortex express NT receptors and that acetylcholine release may be influenced by this neuropeptide. Recent studies have explored this possibility in greater detail.

Bath application of NT to slices of rat cerebral cortex *in vitro* induced a depolarization in 88% of the tested pyramidal cells (Audinat et al., 1989). The depolarization induced by NT was transient despite continuous perfusion of the neuropeptide. Furthermore, no neuron responded to a second application of NT within 20 min after the first application. These results suggest a receptor desensitization that develops quickly to the presence of NT. In addition, NT increased the number of spikes elicited in pyramidal cells by a depolarizing current; this increased firing rate was also short lasting. The results of these studies suggest that the effects of NT on medial frontal cortical pyramidal neurons is due primarily to its direct action on tetrodotoxin (TTX)-sensitive sodium currents (Audinat et al., 1989). This study also indicated that the NT-induced depolarization and alteration in firing rate were not due to any presynaptic actions. Furthermore, NT enhanced both inhibitory, GABA-mediated, as well as excitatory inputs onto frontal pyramidal neurons. Indeed, GABAergic interneurons within the cortex may be activated by NT directly to influence cholinergic terminals.

The influence of NT upon cortical acetylcholine release was further investigated in frontal and parietal cortical slices (Lapchak et al., 1990). The results of these studies indicated that NT differentially regulated evoked acetylcholine release from these two cortical areas without altering spontaneous basal release. In the frontal cortex, NT significantly inhibited evoked acetylcholine release by a mechanism that was insensitive to TTX. In contrast, in the parietal cortex NT enhanced the release of acetylcholine by a mechanism that was sensitive to TTX. These data suggest that NT may act on cholinergic terminals in the parietal cortex indirectly via a cortical interneuron, such as GABA.

The rodent cortex contains intrinsic cholinergic interneurons (Eckenstein and Thoenen, 1983). NT may be regulating the release of acetylcholine, either from these intracortical neurons or from the axonal terminals of cells that originate within the nucleus basalis. In order to determine which cholinergic system is controlled by NT, the following investigation was performed. The neurotoxin quinolinic acid was injected unilaterally into the nucleus basalis to destroy cholinergic cells that send afferents to the frontal cortex. Following the production of this lesion, NT microinjections into the frontal cortex no longer influenced cortical acetylcholine release (Lapchak et al., 1990). The results of these studies are consistent with the hypothesis that NT's ability to control acetylcholine release in the frontal cortex is due to its actions at receptors on the terminals of cholinergic afferents from the nucleus basalis.

NT may also influence cholinergic function in the brain via a peripheral mechanism. An i.v. injection of NT in anesthetized rats produced a dramatic increase in the blood glucose level within 2 min. This effect may be partially due to NT's direct effect on liver glycogenolysis. Enhanced availability of glucose for uptake into the brain may alter the function of many different neurotransmitter systems, including cholinergic neurons, and may significantly improve performance of rats in tasks that require learning and memory (Messier et al., 1990; and for a more detailed discussion see Wenk, 1989). The presence of glucose in the brain provides cholinergic neurons with its primary substrate for the production of Coenzyme-A, a precursor to the formation of acetylcholine. In addition to its role in cholinergic biochemistry, glucose also contributes to the production of energy for the neuron (e.g., ATP) and the formation of numerous amino acid neurotransmitters. Some or all of these amino acid neurotransmitter molecules may play an important role in learning and memory. Furthermore, decreased glucose utilization by the brain is associated with a decreased synthesis of specific neurotransmitters, including acetylcholine, and a decline in glucose utilization may underlie age-related changes in memory (Stone et al., 1988). In addition, many studies have documented a significant decline in cholinergic function with aging; this decline is most severe in Alzheimer's disease and may underlie the development of certain symptoms of the dementia (Bartus et al., 1982).

Aging and Dementia

An investigation of age-related changes in endogenous levels of NT in numerous regions of the human and rat central nervous system found a significant decrease in NT levels only in the pars compacta and pars reticulata of the human substantia nigra (Buck et al., 1981). However, the rats used in this study ranged in age from 4 to 28 months and may not have been old enough to demonstrate a significant age-related change; in contrast, the humans ranged in age from 17 to 104 years. Although no changes in endogenous NT levels were identified in aged humans, significant postsynaptic changes may be present in the brains of patients with Alzheimer's disease. An autoradiographic study of the human hippocampus from patients with Alzheimer's disease determined that the number of NT receptor sites was significantly decreased, as compared to age-matched controls. This decrease was most pronounced in the entorhinal cortex and hippocampal area CA3 (Jansen et al., 1990).

Amygdala

In Alzheimer's disease, the amygdala is severely affected by many pathological changes, including senile plaques and neurofibrillary degeneration. In addition, there is a marked reduction in the level of CAT activity, often in the absence of significant changes in noncholinergic neurotransmitter systems. Because the amygdala has been shown to play an important role in memory, affect, and motivation (Mishkin and Aggleton, 1981), the loss of amygdaloid function, in particular, may underlie some of the behavioral manifestations associated with dementia (Kemper, 1978; Vogt et al., 1990). In the amygdala of patients with Alzheimer's disease, NT immunoreactivity was relatively preserved in nuclei with low plaque density, but was significantly decreased in regions of high plaque density (Benzing et al., 1991). This finding is consistent with the hypothesis that plaque formation may influence the degeneration of NT neurons and fibers. This loss in NT immunoreactivity was selective, in that substance P and somatostatin levels were relatively preserved throughout the amygdala, even in areas with elevated plaque counts (Benzing et al., 1991).

Conclusions

Taken together, these anatomical, neurochemical, pharmacological, and behavioral investigations clearly suggest that cholinergic and NT-containing neurons interact within the brain to influence cognitive function. The prominent distribution of NT and cholinergic neurons and fibers within the limbic system is consistent with a potential interaction of these systems in the control of emotional behavior. Furthermore, the loss of both cholinergic and NT innervation of the amygdala in the brains of patients with Alzheimer's disease suggests that both neurotransmitter systems may have an important role in the symptoms of dementia. However, recent research with animals models of the cholinergic hypofunction associated with Alzheimer's disease suggest that the degeneration of cholinergic neurons in the nucleus basalis is not sufficient to impair certain kinds of memory (Dunnett et al., 1987; Etherington et al., 1987; Markowska et al., 1990; Wenk et al., 1989; but see McGurk et al., 1991). Future studies using these animal models might investigate the behavioral consequences of selective loss of forebrain NT neurons, particularly in combination with the loss of forebrain cholinergic neurons. Furthermore, pharmacotherapies might be designed to enhance both the cholinergic and NT neurotransmitter systems, either by supplementation of precursors or application of specific receptor agonists.

Acknowledgments. I wish to thank Irene Malak, Ross Henshaw, Linda Gorman, and Cheryl Harrington for excellent technical assistance with the neurochemistry, behavioral testing, and microdialysis preparations and Luann Snyder for typing the manuscript. The preparation of the chapter was supported by a grant from the National Science Foundation, BNS 89-14941.

References

Alexander MJ, Miller MA, Dorsa DM, Bullock BP, Melloni RH, Dobner PR, Leeman SE (1989): Distribution of neurotensin/neuromedin N mRNA in rat forebrain: Unexpected abundance in hippocampus and subiculum. *Proc Natl Acad Sci USA* 86:5202–5206

Audinat E, Hermel J-M, Crepel F (1989): Neurotensin-induced excitation of neurons of the rat's frontal cortex studied intracellularly in vitro. *Exp Brain Res* 78:358–368

Bartus RT, Dean RL, Beer B, Lippa AS (1982): The cholinergic hypothesis of geriatric memory dysfunction. *Science* 217:408–417

Benzing WC, Mufson EJ, Jennes L, Armstrong DM (1991): Reduction of neurotensin immunoreactivity in the amygdala in Alzheimer's disease. *Brain Res* 537:298–302

Bigl V, Woolf NJ, Butcher LL (1982): Cholinergic projections from the basal forebrain to frontal, parietal, temporal, occipital, and cingulate cortices: A combined fluorescent tracer and acetylcholinesterase analysis. *Brain Res Bull* 8:727–749

Bissette G, Manberg P, Nemeroff CB, Prange AJ (1978): Neurotensin, a biologically active peptide. *Life Sci* 23:2173–2182

Buck SH, Deshmukh PP, Burks TF, Yamamura HI (1981): A survey of substance P, somatostatin, and neurotensin levels in aging in the rat and human central nervous system. *Neurobiol Aging* 2:257–264

Dunnett SB, Whishaw IQ, Jones GH, Bunch ST (1987): Behavioral, biochemical and histochemical effects of different neurotoxic amino acids injected into nucleus basalis magnocellularis of rats. *Neuroscience* 20:653–669

Eckenstein F, Thoenen H (1983): Cholinergic neurons in the rat cerebral cortex demonstrated by immunohistochemical localization of choline acetyltransferase. *Neurosci Lett* 36:211–215

Etherington R, Mittleman G, Robbins TW (1987): Comparative effects of nucleus basalis and fimbria-fornix lesions on delayed matching and alternation tests of memory. *Neurosci Res Comm* 1:135–143

Jansen KLR, Faull RLM, Dragunow M, Synek BL (1990): Alzheimer's disease: Changes in hippocampal N-methyl-d-aspartate, quisqualate, neurotensin, adenosine, benzodiazepine, serotonin and opioid receptors—An autoradiographic study. *Neuroscience* 39:613–627

Kemper TL (1978): Senile dementia: A focal disease in the temporal lobe. In: *Senile Dementia: A Biomedical Approach*, Nandy K, ed. Amsterdam: Elsevier North-Holland

Kiyama H, Shiosaka S, Sakamoto N, Michel JP, Pearson J, Tohyama M (1986): A neurotensin-immunoreactive pathway from the subiculum to the mammillary body in the rat. *Brain Res* 375:357–359

Lapchak PA, Araujo DM, Quirion R, Beaudet A (1990): Neurotensin regulation of endogenous acetylcholine release from rat cerebral cortex: Effect of quinolinic acid lesions of the basal forebrain. *J Neurochem* 55:1397–1403

Luiten PGM, Gaykema RPA, Traber J, Spencer DG (1987): Cortical projection patterns of magnocellular basal nucleus subdivisions as revealed by anterogradely transported *Phaseolus vulgaris* leucoagglutinin. *Brain Res* 413:229–250

Mai JK, Triepel J, Metz J (1987): Neurotensin in the human brain. *Neuroscience* 22:499–524

Malthe-Sorenssen D, Wood PL, Cheney DL, Costa E (1978): Modulation of the turnover rate of acetylcholine in rat brain by intraventricular injections of thyrotropin-releasing hormone, somatostatin, neurotensin and angiotensin II. *J Neurochem* 31:685–691

Markowska AL, Wenk GL, Olton DS (1990): Nucleus basalis magnocellularis and memory: Differential effects of two neurotoxins. *Behav Neural Biol* 54:13–26

Martin LJ, Powers RE, Dellovade TL, Price DL (1991): The bed nucleus-amygdala continuum in human and monkey. *J Comp Neurol* 309:445–485

McGurk SR, Levin ED, Butcher LL (1991): Impairment of radial-arm maze performance in rats following lesions involving the cholinergic medial pathway: Reversal by arecoline and differential effects of muscarinic and nicotinic antagonists. *Neuroscience* 44:137–147

Messier C, Durkin T, Mrabet O, Destrade C (1990): Memory-improving action of glucose: indirect evidence for a facilitation of hippocampal acetylcholine synthesis. *Behav Brain Res* 39:135–143

Mishkin M, Aggleton J (1981): Multiple functional contributions of the amygdala in the monkey. In: *The Amygdala Complex*, Ben-Ari Y, ed. Amsterdam: Elsevier/North Holland Press

Moyse E, Rostene W, Vial M, Leonard K, Mazella J, Kitabgi P, Vincent JP, Beaudet A (1987): Distribution of neurotensin binding sites in rat brain: A light microscopic radioautographic study using monoiodo [^{125}I]Tyr$_3$-neurotensin. *Neuroscience* 22:525–536

Sakamoto N, Michel JP, Kiyama H, Tohyama M, Kopp N, Pearson J (1986): Neurotensin immunoreactivity in the human cingulate gyrus, hippocampal subiculum and mammillary bodies. Its potential role in memory processing. *Brain Res* 375:351–356

Satoh K, Matsumura H (1990): Distribution of neurotensin-containing fibers in the frontal cortex of the macaque monkey. *J Comp Neurol* 298:215–223

Shibata K, Furukawa T (1988): The mammillary body, a potential site of action of neurotensin in passive avoidance behavior in rats. *Brain Res* 443:117–124

Stone WS, Croul CE, Gold PE (1988): Attenuation of scopolamine induced amnesia in mice. *Psychopharmacology* 96:417–420

Swann AC, Hewitt LO (1988): Hemicholinium-3 binding: Correlation with high-affinity choline uptake during changes in cholinergic activity. *Neuropharmacology* 27:611–615

Szigethy E, Wenk GL, Beaudet A (1989): Anatomical substrate for neurotensin-acetylcholine interactions in the rat basal forebrain. *Peptides* 9:1227–1234

Szigethy E, Quirion R, Beaudet A (1990): Distribution of ^{125}I-neurotensin binding sites in human forebrain: Comparison with localization of acetylcholinesterase. *J Comp Neurol* 297:487–498

Tay SSW, Williams TH, Jew JY (1989): Neurotensin immunoreactivity in the central nucleus of the rat amygdala: An ultrastructural approach. *Peptides* 10:113–120

Vogt LJK, Hyman BT, VanHoesen GW, Damasio AR (1990): Pathological alterations in the amygdala in Alzheimer's disease. *Neuroscience* 37:377–385

Wenk GL (1984): Pharmalogical manipulations of the substantia innominata-cortical cholinergic pathway. *Neurosci Lett* 51:99–103

Wenk GL (1989): An hypothesis on the role of glucose in the mechanism of action of cognitive enhancers. *Psychopharmacology* 99:431–438

Wenk GL, Hepler D, Olton D (1984): Behavior alters the uptake of [^{3}H]choline into acetylcholinergic neurons of the nucleus basalis magnocellularis and medial septal area. *Behav Brain Res* 13:129–138

Wenk GL, Markowska AL, Olton DS (1989): Basal forebrain lesions and memory: Alterations in neurotensin, not acetylcholine, may cause amnesia. *Behav Neurosci* 103:765–769

Wenk H, Bigl V, Meyer U (1980): Cholinergic projections from magnocellular nuclei of the basal forebrain to cortical areas in rats. *Brain Res Rev* 2:295–316

Woolf NJ, Eckenstein F, Butcher LL (1984): Cholinergic systems in the rat brain: I. Projections to the limbic telencephalon. *Brain Res Bull* 13:751–784

13

Serotonin Influences on Cholinergic Function: Possible Interactions in Learning and Memory

HOWARD J. NORMILE AND HARVEY J. ALTMAN

Although many neurochemical systems appear to be critically involved in the processes underlying learning and memory, much of the research on the cognitive dysfunction associated with aging and certain age-related neurodegenerative diseases has focused on the cholinergic nervous system. The foundation for this focus, as well as the subsequent cholinergic hypothesis (Bartus et al., 1982), was laid in the 1970s when three independent laboratories reported reduced activity of the acetylcholine-synthesizing enzyme, choline acetyltransferase (CAT), in postmortem brain tissue of Alzheimer's patients (Bowen et al., 1976; Davies and Maloney, 1976; Perry et al., 1977). The loss of CAT activity was especially prominent in the hippocampus and cerebral cortex—brain areas thought to mediate learning and memory processes (see Squire and Zola-Morgan, 1988 for review). Additional support for this view was provided by the demonstration that the cholinergic dysfunction was related to both the density of plaque formation and the severity of the premortem memory deficits (Perry et al., 1978; see Perry, 1986 for review) and the observation that anticholinergic drugs produced memory impairments in young healthy individuals (Drachman and Leavitt, 1974).

Despite these and other data suggesting that a cholinergic deficit may be responsible for the cognitive impairment in degenerative dementia, clinical trials employing agents thought to augment cholinergic activity have produced disappointing results. In addition to the continued search for an effective cholinomimetic agent, the negative results of the clinical trials have encouraged research efforts aimed at identifying the role of noncholinergic neuroregulatory systems in the processes underlying learning and memory, as well as their possible contribution to certain human neural behavioral disorders characterized by cognitive impairment. Moreover, the rapid accumulation of data suggesting that neurotransmitter systems are anatomically and pharmacologically interrelated has recently fostered the study of how multiple systems functionally interact in learning and memory. In this chapter we present evidence that strongly suggests the serotonergic and cholinergic neuroregulatory systems functionally interact in these processes.

Anatomical Sites at which Serotonergic/Cholinergic Interactions Might Occur

Serotonin is known to be widely distributed in the brains of mammals and other vertebrates (see Parent et al., 1984 for review). However, serotonergic cell bodies are restricted to the brainstem, where they are found in nine cell groups, coded B1–B9 by Dahlstrom and Fuxe (1964). These serotonergic cell groups correspond, with some exceptions, to the cell masses of the raphe nuclei. For example, the large B7 cell group, located in the ventral part of the periaqueductal gray matter of the midbrain, along with its caudal section (B6) extending into the rostral pons, comprise the dorsal raphe nucleus. The B8 cell group, the second largest group of serotonergic neurons, lies at the junction between the pons and the midbrain, and corresponds to the median raphe nucleus. The B9 label is used to denote a lateral cluster of serotonergic neurons located in the ventrolateral tegmentum of the pons and midbrain. The dorsal raphe, median raphe, and the B9 cluster are the rostralmost serotonergic cell groups and provide ascending projections to widespread areas of the forebrain, including the cerebral cortex, limbic system, basal ganglia, and diencephalon (Azmitia and Segal, 1978; Ungerstedt, 1971).

While the B9 group is not well characterized, it appears that the serotonergic axons arising from the dorsal and median raphe are morphologically dissimilar (Kosofsky and Molliver, 1987). Serotonergic axons arising from the dorsal raphe are commonly described as "fine" and contain small varicosities that are fusiform in shape. In contrast, serotonergic axons originating from the median raphe contain large spherical varicosities separated by narrow axonal segments, and thus have a "beaded" appearance. These morphological differences have assisted in determining the relative contribution of the dorsal and median raphe to the innervation of specific forebrain areas (Kosofsky and Molliver, 1987; Mamounas and Molliver, 1987). For instance, while both types of fibers appear to coexist in the cerebral cortex, fine fibers predominate, indicating that the dorsal raphe contributes more fibers to the cortex than the median raphe. However, projections to the amygdala are almost exclusively from the dorsal raphe. In contrast, the projections to the hippocampus are derived mostly from the median raphe, with a small input from the dorsal raphe. Functional differences in the serotonergic neurons arising from the dorsal and median raphe are also suggested by ultrastructural studies demonstrating that the individual raphe nuclei may form different types of synaptic contacts (Tork, 1990). The combined data summarized above indicate that forebrain structures, strongly implicated in learning and memory, in particular, the cerebral cortex, hippocampus, and amygdala, are innervated by serotonergic axons arising from three serotonergic cell clusters located in the brainstem. Moreover, it appears that the dorsal and median raphe form two ascending serotonergic systems, which, in addition to origin, differ in terms of axon and synaptic morphology, distribution and pattern of termination, and pharmacological properties (see below).

The cerebral cortex and hippocampus also contain intrinsic cholinergic neurons that provide approximately 30% of their cholinergic innervation (Johnston et al., 1981; Levey et al., 1984). The cortex and hippocampus, in addition to the amygdala, receive extrinsic cholinergic innervation from a discontinuous group of magnocellular neurons located in the basal forebrain (Bigl et al., 1982; Fibiger, 1982; Lehmann et al., 1980; Mesulam et al., 1983; Woolf and Butcher, 1982). The major cholinergic innervation of the hippocampus is provided by the medial septal nucleus and the nucleus of the vertical limb of the diagonal band of Broca, collectively the most rostral portion of the basal forebrain cholinergic system. The nucleus of the horizontal limb of the diagonal band of Broca is the principal source of cholinergic innervation for the medial prefrontal, orbitofrontal, insular, visual, and entorhinal cortices, as well as the olfactory bulb. Cholinergic innervation of the remaining neocortical regions, as well as the amygdala, is provided by the nucleus basalis (NB), the most posterior cells of the basal nuclear complex.

It appears, therefore, that the serotonergic and cholinergic neurotransmitter systems, originating in the raphe nuclei and the basal forebrain, respectively, have overlapping projections to brain structures strongly implicated in the control of learning and memory. Thus, it is possible that these convergent areas represent potential sites at which serotonergic and cholinergic interactions might occur. However, in addition to overlapping terminal projection sites, the nuclei of origin may also provide opportunities for serotonergic/cholinergic interactions. Ipsilateral projections from the brainstem raphe nuclei to the rostral and caudal nuclei of the basal forebrain have been demonstrated using tracing and immunohistochemical techniques (Schober et al., 1989; Semba et al., 1988). For instance, injections into the magnocellular preoptic nucleus and the substantia innominata gave rise to label virtually restricted to the rostral half of the dorsal raphe when compared to other serotonergic perikarya (Vertes, 1988). On the other hand, the median raphe was very heavily labeled following injections into the medial septum and the vertical limb of the diagonal band of Broca. The results suggest a direct raphe influence on basal forebrain nuclei, and, as a consequence, a possible indirect influence on cholinergic input to the cerebral cortex and hippocampus.

Serotonergic Influence on Cholinergic Activity

There is general agreement that serotonin has an inhibitory action on cholinergic neurons. Serotonin or its agonists have been shown to increase acetylcholine levels *in vivo* and to reduce acetylcholine release from certain brain areas *in vitro* (Consolo et al., 1980; Samanin et al., 1978). For example, exogenous serotonin or drugs that mimic the action of serotonin have been shown to decrease K^+-induced or electrically stimulated acetylcholine release from the cortex (Barnes et al., 1989), hippocampus (Maura et al., 1989), and striatum (Gillet et al., 1985; Jackson et al., 1988). In addition, 5,7-dihydroxytryptamine lesions of specific raphe nuclei were shown to increase acetylcholine turnover in the cortex and hippocampus (Quirion and Richard, 1987). However, the view that serotonin has an inhibitory action on cholinergic activity is likely to be an oversimplification, given that the effects of serotonin are often biphasic (Bianchi et al., 1989, 1990) and are mediated by multiple serotonergic receptor subtypes (see below).

Serotonin, Learning, and Memory

Research that focuses on the interaction between neurotransmitter systems in the processes underlying learning and memory would be greatly facilitated if, prior to such efforts, the role of each system in these processes was clearly identified and characterized. Unfortunately, this is not the case for serotonergic/cholinergic interactions, due in large measure to the lack of understanding of the role of serotonin in these processes. While length limitations preclude a thorough review of all serotonin-related learning and memory studies (see Altman and Normile, 1988; McEntee and Crook, 1991; Ogren, 1982 for reviews), the collective results suggest that manipulations designed to augment serotonergic activity often impair the ability of animals to acquire or express previously novel behaviors. Thus, administration of serotonin agonists or serotonin itself (Wetzel et al., 1980; Winter and Petti, 1987), electrical stimulation of the dorsal raphe nucleus (Fibiger et al., 1978), and intrahippocampal implantation of neonatal serotonergic nerve grafts (Ramirez et al., 1989) have all been shown to impair learning and memory in laboratory animals. Similarly, the serotonin agonist m-chlorophenylpiperazine was found to impair memory in Alzheimer's patients (Lawlor et al., 1989a) and elderly controls (Lawlor et al., 1989b). On the other hand, it is evident that the results of studies designed to assess the effects of *impaired* serotonergic function are extremely inconsistent and appear to depend on both the type of experimental intervention used to impair activity and the type of behavioral task employed to test the animal. For example, two different interventions thought to produce similar neurochemical effects (e.g., serotonin depletion) may produce dissimilar behavioral effects in the same task (e.g., Lorens, 1978). Moreover, a single manipulation may produce variable effects across different tasks (Vanderwolf, 1987; Altman et al., 1989). These inconsistencies may be due to a number of factors, including regional variations in serotonin levels, and changes in attention, vigilance, and reward-contingent motivation. The inconsistencies produced by treatments presumed to impede serotonergic activity are unfortunate for they make it very difficult to predict the behavioral consequences, if any, of the apparent regional reduction in certain markers of serotonergic function which appears to occur in aging and in certain age-related diseases, such as Alzheimer's disease (Arai et al., 1984; Jansen et al., 1990; Palmer et al., 1987; see Normile and Altman, 1987 and Cutler, 1989 for reviews). On the other hand, the discrepancies also suggest that serotonin may play a complex role in learning and memory. Accordingly, the continued study of the nature of these discrepancies may provide important insight into the precise role that serotonin plays in these processes.

Serotonergic/Cholinergic Interactions in Learning and Memory

A thorough understanding of the role that a particular neurotransmitter system plays in the processes underlying learning and memory is not a prerequisite for studying how that system may functionally interact with other systems in these processes. However, this lack of knowledge does necessitate the careful design of interactive studies in terms of the experimental intervention employed and the type of task selected for behavioral assessment. For instance, it would be advisable to incorporate dual interventions that, when examined separately, produce a clear and consistent effect on performance.

In this section we will focus on three of our studies, which were designed to examine serotonergic/cholinergic interactions in learning and memory.

Effects of Combined Serotonin Depletion and Cholinergic Lesions on Acquisition of a Complex Discrimination Task in the Rat

The purpose of this first study (Normile et al., 1990) was to determine the effects of NB lesions, alone or in combination with central serotonin depletion, on learning and memory in rats trained in the Stone 14-unit T-maze—a complex, positively reinforced discrimination task. Serotonin depletion was accomplished by the systemic administration of PCA. This particular neurotoxin was selected since we previously demonstrated that PCA consistently facilitated the performance of young (Altman et al., 1989) and aged rats (Normile et al., 1986) in this task (Fig. 1). Destruction of cholinergic neurons within the NB was accomplished by bilateral infusion of ibotenic acid. Each animal was given one trial per day in the maze, 5 days per week, for a total of 25 trials. Performance was assessed by recording the number of errors made during each trial and the total number of trials required to traverse the maze successfully (i.e., making two or less errors on two consecutive days).

The results suggest that serotonin depletion, produced by PCA, facilitated Stone maze learning in rats with an intact cholinergic system (Figs. 2 and 3). However, serotonin depletion had no significant effect on acquisition in NB-lesioned rats. The ability of NB lesions to antagonize the facilitative effects of serotonin depletion occurred despite the fact that, as previously reported (Altman et al., 1984), cholinergic lesions alone had no observable affects on learning in rats trained in this task.

The interactive effects of combined serotonergic and cholinergic manipulations on learning did not appear to result from the attenuation of the neurochemical changes normally produced by each manipulation; that is, the effects of the combined treatment on CAT levels in the frontal cortex (decreased) and hippocampus (no change), as well as [^{3}H]QNB binding in the frontal cortex (no change), were not significantly different from the effects produced by lesions of the NB alone. Similarly, the regional effects of the combined treatment on the levels of serotonin (decreased) and 5-hydroxyindoleacetic acid (decreased), or on the levels

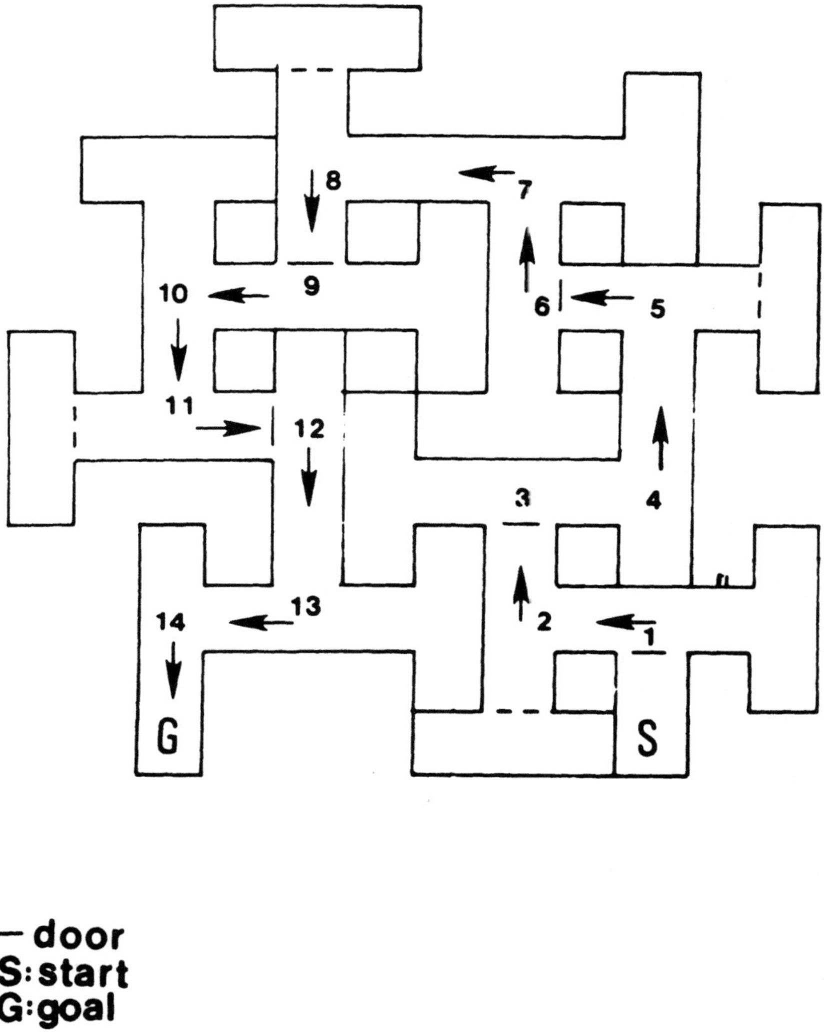

FIGURE 1. Diagram of the Stone 14-unit T-maze. The floor and sides of the maze are constructed of clear acrylic. Sliding doors are located at the start box and at the choice points indicated. Dummy doors (dashed lines) are opposite all true doors. Lights are positioned behind partitions, which completely surround the maze. A cup containing sweetened condensed milk is attached to the end wall of the goal box. The rats (maintained at 80% body weight) are given a total of 25 trials, 1 trial per day.

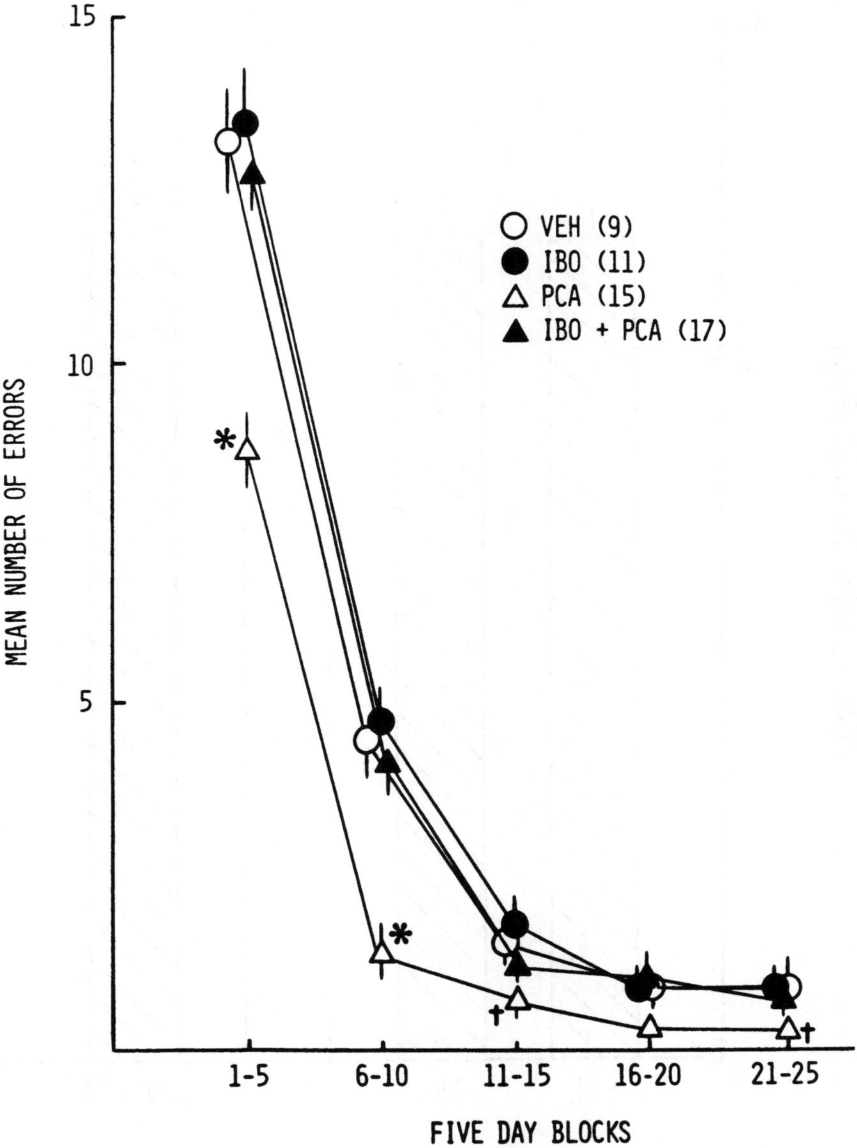

FIGURE 2. Mean (±SEM) number of errors averaged over 5-day blocks for VEH-injected (VEH), NB-lesioned (IBO), PCA-injected (PCA), and combined NB-lesioned and PCA-injected rats (IBO + PCA) trained in the Stone maze. The number in parentheses denotes the number of animals in each group. For blocks 1 and 2, the PCA group made significantly fewer errors compared to all other groups. For blocks 3 and 5, the PCA group differed significantly only from the IBO alone group. *$p < 0.05$ vs. all other groups, †$p < 0.05$ vs. the IBO group.

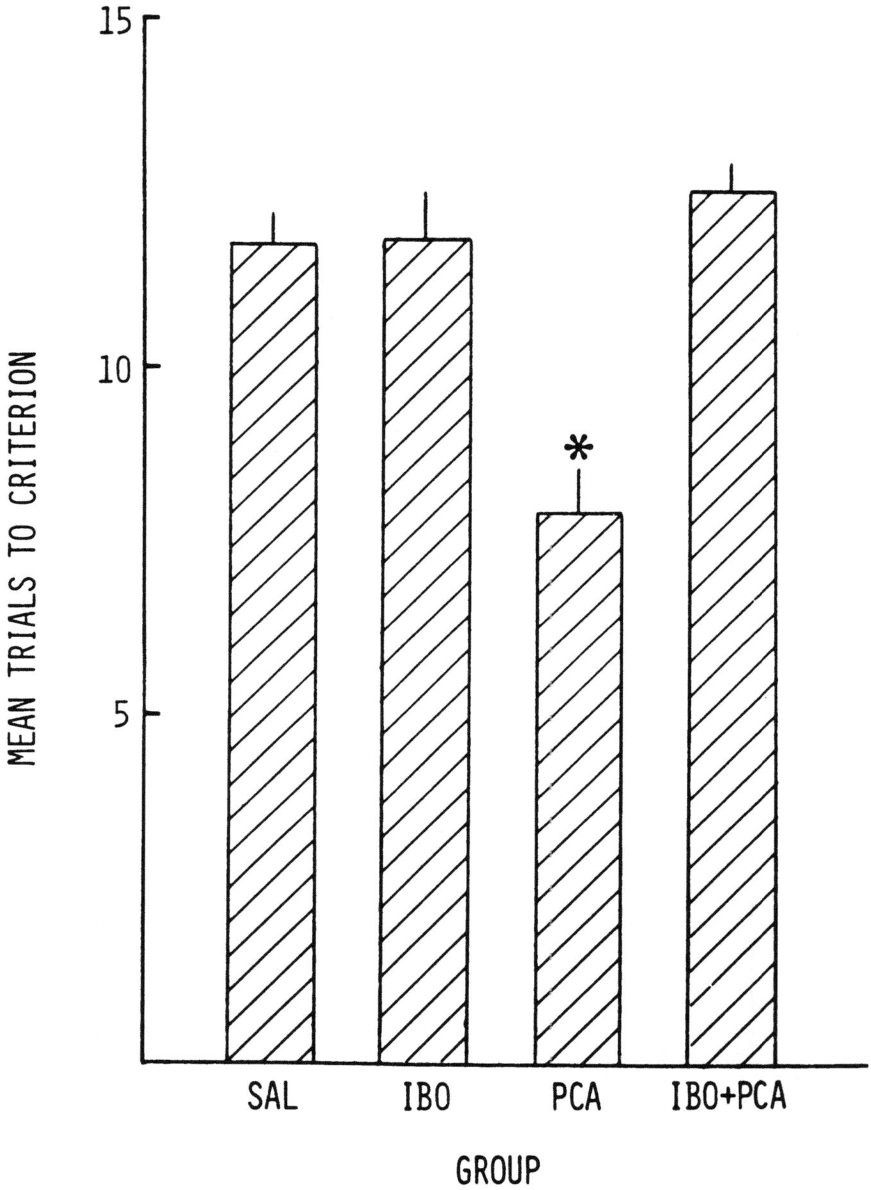

FIGURE 3. Mean (±SEM) number of trials required for rats (same rats represented in Fig. 2) to reach criterion (two or less errors on 2 consecutive days) in the Stone maze. Abbreviations for the four treatment groups are the same as in Figure 2. *$p < 0.05$ vs. all other groups.

of norepinephrine (no change) and dopamine (no change), were not significantly different from that produced by serotonin depletion alone.

It is possible that the neocortex or the amygdala may be sites for the observed serotonergic/cholinergic interaction, since, as described above, a considerable body of evidence indicates that (a) cells within the NB are the primary source of the extrinsic cholinergic input to the neocortex and amygdala, and (b) considerable overlap between serotonergic and cholinergic terminals occurs in these two regions. In addition, the evidence indicating that serotonergic neurons in the raphe nuclei may have an inhibitory influence on cholinergic neurotransmission further suggests that the facilitative effects normally produced by PCA alone may result from a concomitant increase in cholinergic neurotransmission. This interpretation is most likely incomplete since NB lesions alone failed to affect acquisition in this task. Nevertheless, the results of this study support the view that the serotonergic and cholinergic systems may functionally interact in learning and memory.

Following our initial observation that PCA enhanced the performance of rats in the Stone maze, immunocytochemical, electrophysiological, and retrograde axonal transport studies demonstrated that serotonergic projections to the neocortex and hippocampal formation (i.e., dentate gyrus) from the dorsal and medial raphe nuclei were differentially sensitive to PCA (Blier et al., 1990). For example, cortical axons originating from the dorsal raphe (i.e., fine fibers) were shown to be highly vulnerable to PCA, whereas median raphe axons (i.e., beaded fibers) were resistant to PCA's neurotoxic effects (Mamounas and Molliver, 1988). While the neurotoxic effects of PCA on fine serotonergic fibers within other brain regions requires verification, it is likely that PCA and other amphetamine-derived psychotropic drugs affect all fine axons and thus act in all areas that receive serotonergic projections from the dorsal raphe nuclei (e.g., striatum). Accordingly, the effect of PCA on performance observed in this and our previous studies may be mediated by fine serotonergic fibers originating from the dorsal raphe.

The Effects of Serotonergic Receptor Antagonists on NB-Induced Deficits in One-Trial Inhibitory Avoidance Retention

This study represented our initial inquiry into the nature of the serotonergic receptor involved in mediating the possible serotonergic/cholinergic interaction in learning and memory. We previously reported that posttraining administration of serotonergic type-2 receptor antagonists facilitated inhibitory avoidance retention in young mice (Altman and Normile, 1987) and also attenuated the inhibitory avoidance deficits normally exhibited by aged rats (Normile and Altman, 1988). The present study was designed to assess whether posttraining administration of a serotonergic type-2 receptor antagonist could also attenuate an inhibitory avoidance deficit induced by a cholinergic lesion. The cholinergic deficit was produced as above by the bilateral injection of ibotenic acid into the NB. The serotonergic type-2 receptor antagonist selected was the quinazolinedione derivative ketanserin (Awouters, 1985). Consistent with previous results (Berman et al., 1983; Flicker et al., 1983), inhibitory avoidance retention was significantly impaired in rats following cytotoxic destruction of the NB (Fig. 4). In addition, ketanserin failed to attenuate the lesion-induced deficit (Fig. 5), despite the fact that ketanserin alone significantly facilitated performance (Fig. 6). Thus it would appear that, as in the preceding study, an intact cholinergic neurotransmitter system is required to produce the ketanserin-induced effects on performance. It is possible that ketanserin fails to facilitate performance in the combined condition, since the antagonist may normally act on learning and memory processes that are subsequent to cholinergic-mediated events—events that fail to occur as a result of the lesion-induced cholinergic deficit.

It is interesting that recent studies demonstrate that cortical 5-HT_2 receptors may be selectively associated with fine serotonergic axons, thus suggesting that postsynaptic 5-HT_2 receptors mediate the effects of dorsal but not median raphe projections (Blue et al., 1986, 1988). It is, therefore, possible that serotonergic neurons originating in the raphe nuclei may terminate in close association with specific receptor subtypes, thus supporting the view that anatomically and functionally distinct ascending serotonergic systems exist in the brain. Moreover, these data, in addition to the apparent selective action of PCA, indicate that the axons arising from the dorsal raphe and their projection areas may be involved in the mechanisms underlying serotonergic/cholinergic interactions in learning and memory.

The Effects of Combined Acetylcholinesterase Inhibition and Serotonergic Receptor Blockade on Age-Associated Memory Impairments in Rats

Unlike the first two studies, which employed a cytotoxin to produce a cholinergic deficit, this third study incorporated a cholinergic intervention (physostigmine) thought to augment cholinergic activity. In light of one of our previous studies suggesting that posttraining administration of the serotonergic receptor antagonists (e.g., ketanserin) attenuated the memory deficits normally exhibited by aged rats in the inhibitory avoidance task (Normile and Altman, 1988), the present study (Normile and Altman, in press) was designed to determined whether a subeffective dose of ketanserin would augment the facilitative effects produced by the cholinesterase inhibitor physostigmine in this same task. The drugs were injected (i.p.) alone, or in combination, immediately following training. Retention testing occurred 24 hr following training. The results of the study suggest that a dose-dependent enhancement of memory resulted from the dual treatment condition (Fig. 7). The facilitation of memory produced by the combined treatment was observed at doses well below those that produced a similar effect when each drug was administered alone.

The 1.0 mg/kg dose of ketanserin used for the combined-treatment condition is similar to doses used in many previous *in vivo* studies designed to examine the role of 5-HT_2 receptors in serotonin-mediated responses (Eison et al., 1988; Gudelsky et al., 1986; Ashby et al., 1990; Mantz et al., 1990). However, while ketanserin is virtually inactive at 5-HT_1 receptors and muscarinic receptors, the antagonist has

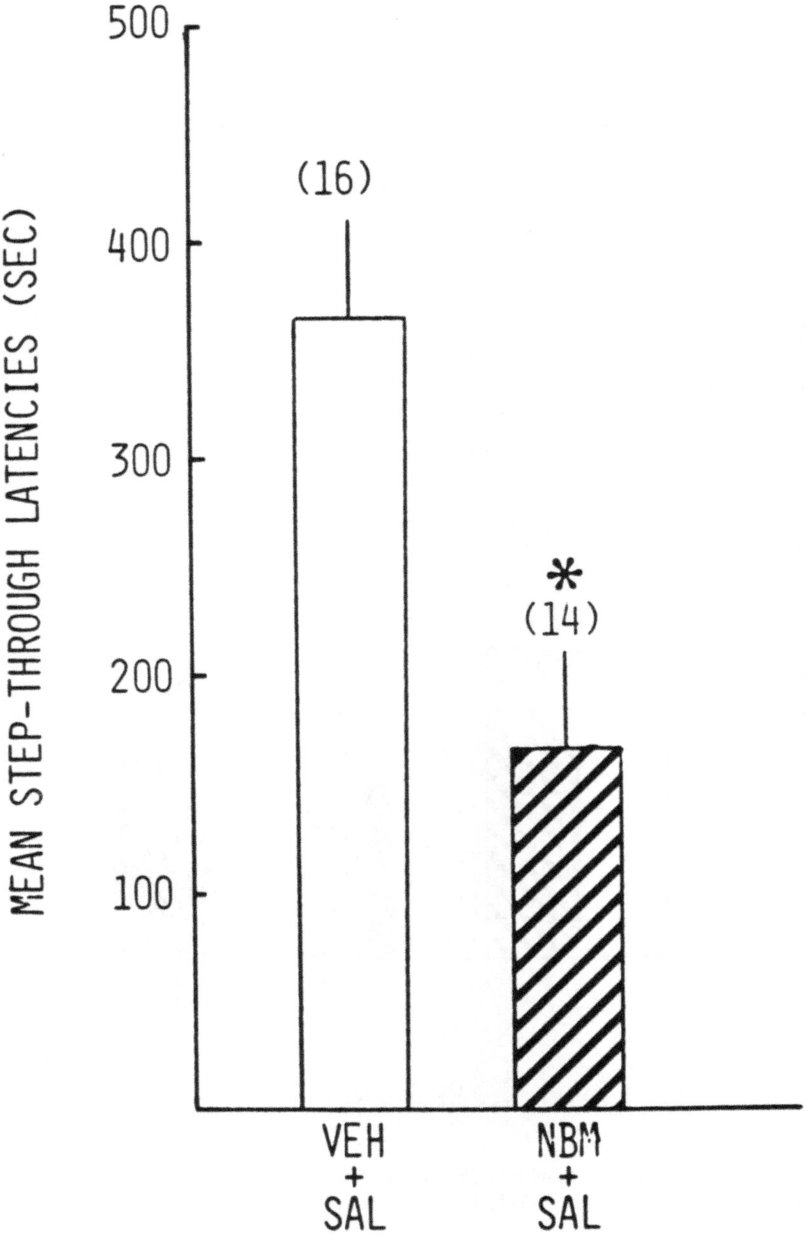

FIGURE 4. The effects of NB lesions (2 weeks prior to behavioral assessment) on retention of an inhibitory avoidance (stepthrough) task. Animals were tested 24 hr following training (shock intensity/duration: 1 mA/3 sec). Bars represent mean (±SEM) test latencies of lesioned and nonlesioned rats. Saline was injected immediately after training in order to provide appropriate control groups for ketanserin-injected animals (data represented in Fig. 5). *$p < 0.02$ compared to VEH+SAL.

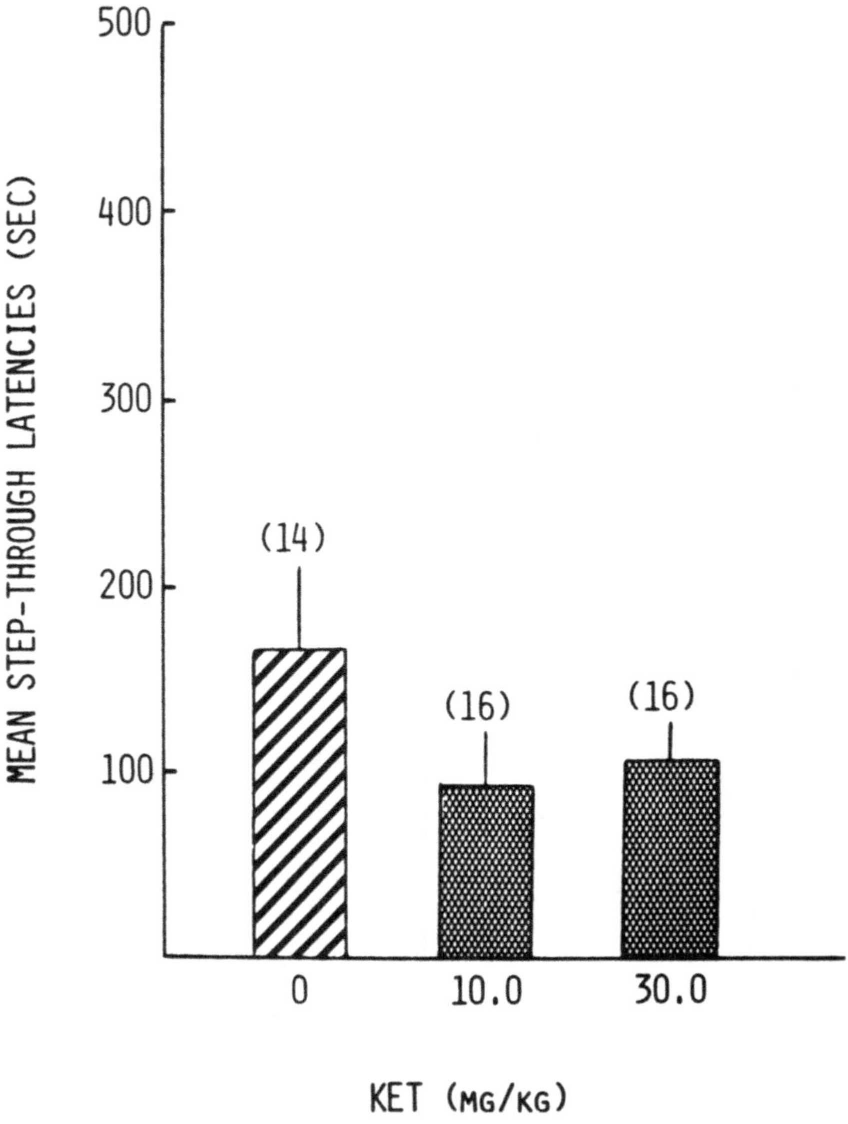

FIGURE 5. The effects of immediate posttraining ketanserin injections on the inhibitory avoidance deficits produced by NB lesions. Ketanserin failed to produce a significant effect on test latencies.

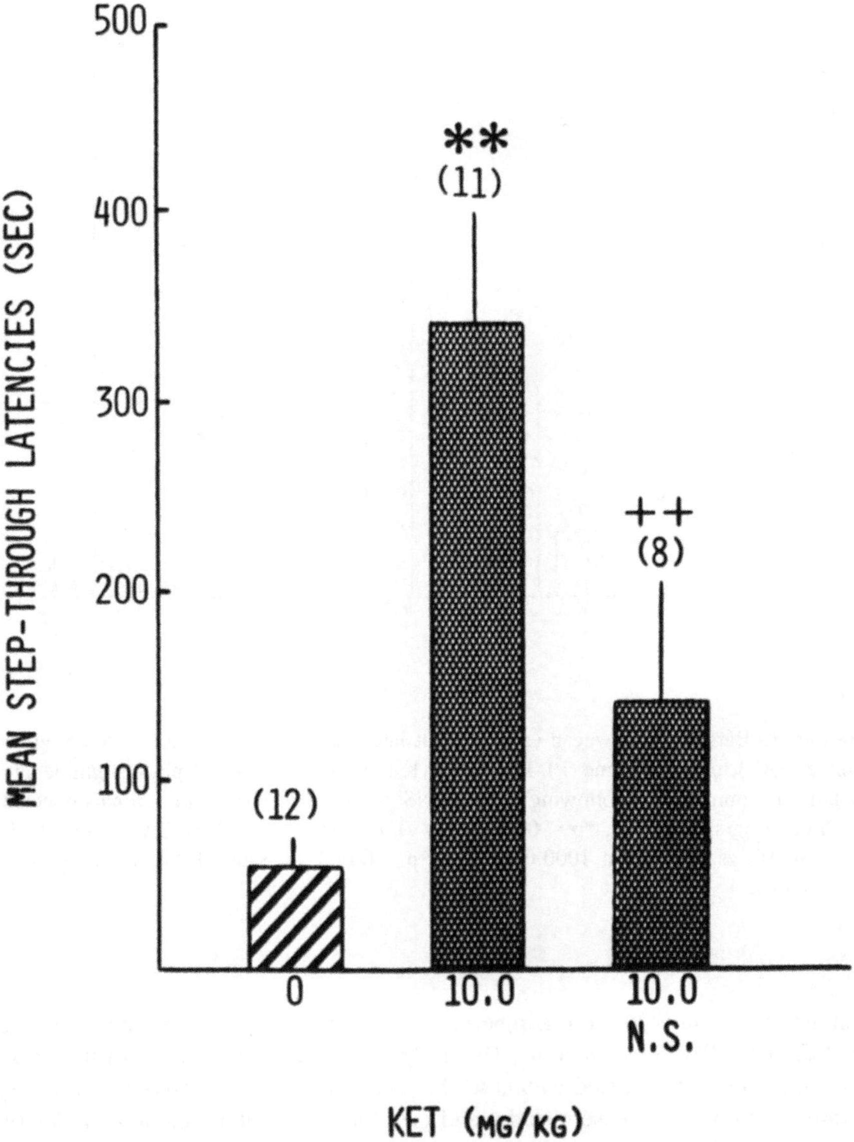

FIGURE 6. Demonstration that posttraining ketanserin administration significantly facilitates inhibitory avoidance retention in young nonlesioned rats. Bars represent mean (±SEM) test latencies. The rats were trained using a low shock strength (0.8 mA/2 sec duration). The N.S. group represents nonshocked (control) animals injected with ketanserin immediately after training. **$p < 0.002$ vs. SAL-injected rats, ‡$p < 0.002$ vs. shocked rats injected with 10.0 mg/kg ketanserin.

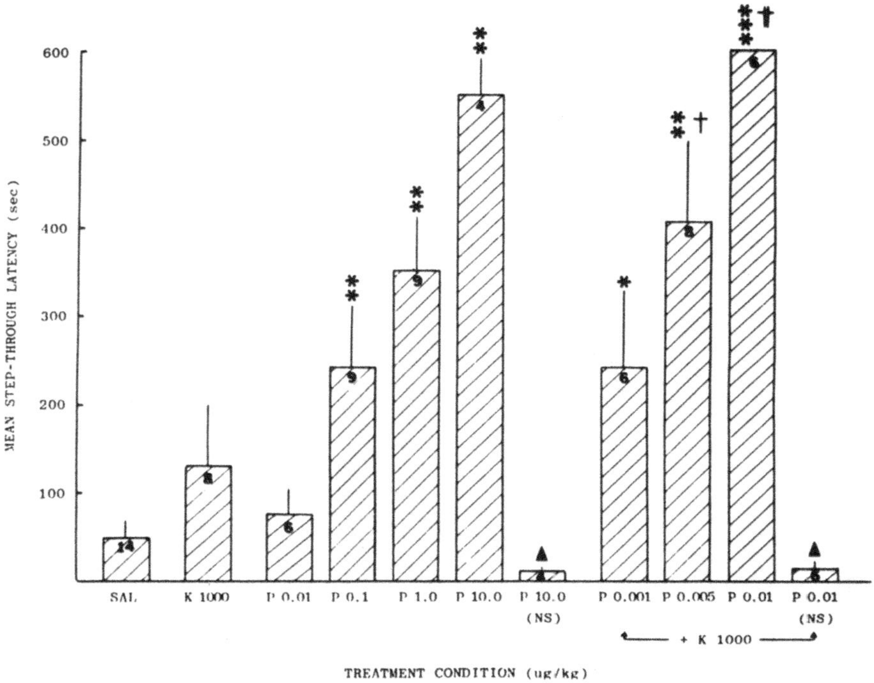

FIGURE 7. Bars represent mean (±SEM) test latencies of aged rats administered either saline (SAL), physostigmine (P), ketanserin (K), or a combination of physostigmine and ketanserin, immediately following training. NS = nonshocked (control) animals injected with the drug(s) indicated. *$p < 0.05$, **$p < 0.02$, ***$p < 0.002$ vs. SAL. †$p < 0.05$, ‡$p < 0.002$ vs. ketanserin 1000.0 μg/kg. ▲$p < 0.002$ vs. same dose administered to trained rats.

nanomolar affinity for α-1 adrenergic and H_1 histaminergic receptors (Lyon and Titeler, 1988; Leysen et al., 1981). The potential contributions of nonserotonergic receptor-mediated events to the ketanserin-induced response were not examined within the present study, and therefore, it is impossible to completely exclude their involvement. However, the potential confounding effects that ketanserin may produce as a result of binding to nonserotonergic receptors (e.g., antagonism of α_1 receptors) would appear to be in opposition to the purported role that these receptors and their putative neurotransmitters play in learning and memory (see McGaugh et al., 1984; Wolkowitz et al., 1985; Martinez, 1986 for reviews).

The design of this study does not allow conclusions regarding the area(s) mediating the effects of combined ketanserin and physostigmine administration on the subsequent expression of the avoidance response. Nevertheless, (a) the high density of 5-HT_2 receptors in the neocortex, (b) the close anatomical association between 5-HT_2 receptors and fine serotonergic terminal projection sites, and (c) the report that the memory-enhancing effects of physostigmine were attenuated by cortical lesions (Haroutunian et al., 1990) suggest that the effects of combined ketanserin and physostigmine administration may be at least partially mediated by the neocortex. However, a recent study showed that a large percentage (68%) of medullary raphe neurons increased spontaneous discharge rate in response to peripheral administration of physostigmine (Futuro-Neto et al., 1989). Physostigmine induced a decrease in the discharge of some raphe cells (12%), whereas no change was observed in the remaining cells. The data suggest a heterogeneous population of cholinergic cells in the medullary raphe, and further illustrate that serotonergic perikarya may be one possible site in which serotonergic/cholinergic interactions occur.

While the data suggest the 5-HT_2 receptor may be involved in the physostigmine-mediated effects on memory, significant complexities and discrepancies exist concerning the serotonin receptor subtypes that mediate the effects of serotonin on acetylcholine release. For example, Barnes et al. (1989), found that serotonin, or the selective 5-HT_3 receptor agonist 2-methyl-serotonin, inhibited potassium-stimulated $[^3H]$ACh efflux from slices of rat entorhinal cortex, but only in the presence of the $5\text{-HT}_2/5\text{-HT}_{1C}$ antagonist ritanserin. In the absence of ritanserin, 2-methyl-serotonin, in the presence of the 5-HT_3 receptor antagonist zacopride, increased potassium-stimulated acetylcholine release. The ritanserin sensitivity of the excitatory effect produced by 2-methyl-serotonin suggests the involvement of 5-HT_2 or 5-HT_{1C} receptor subtypes. On the other hand, Muramatsu et al. (1990) observed that serotonin and 4-bromo-2,5-dimethoxy-phenylisopropylamine (DOB), a 5-HT_2 receptor agonist, inhibited potassium-induced $[^3H]$ACh release from both synaptosomes and slices of rat cerebral cortex. While retanserin and methysergide had no effect on $[^3H]$ACh release alone, both attenuated the inhibition of $[^3H]$ACh release induced by serotonin and DOB. These results contrast with those of Bianchi and collaborators (Bianchi et al., 1986, 1990; Siniscalchi et al., 1990), who failed to observe any influence of serotonin or 8-OH-DPAT on resting or electrically evoked $[^3H]$ACh release from slices of guinea-pig cerebral cortex. However, serotonin and 8-OH-DPAT increased, whereas 2-methyl-serotonin decreased, $[^3H]$ACh efflux from the cerebral cortex of freely moving guinea pigs.

The dissimilar effects of select serotonergic agonists or antagonists on cortical acetylcholine release in cerebral slices vs. freely moving animals suggest a possible complex interaction between postsynaptic serotonergic receptors and inhibition of serotonergic tone in the control of acetylcholine release. In addition, the discrepant results observed in the cortex, as well as those observed in hippocampal tissue, may be related to differences in animal species, site-specific mechanisms, type of stimulation (KCl vs. electrical stimulation), and doses of ligand employed.

Conclusions

Anatomical evidence suggests overlapping serotonergic and cholinergic projections within brain regions strongly implicated in learning and memory. Moreover, electrophysiological and pharmacological studies demonstrate that altered serotonergic activity can influence cholinergic function. It is likely that neuronal activity in these overlapping areas is directly (e.g., through a dual action on postsynaptic neurons) or indirectly (e.g., through presynaptic mediated events) under the influence of both systems. However, the degree of dual influence may vary temporally, resulting in dynamic levels of neuromodulation that could not occur through the input of a single neurotransmitter system. It is also likely that this complexity in neuromodulation afforded by convergence of multiple systems is the major reason why the precise role of the serotonergic "neurotransmitter" or "neuromodulatory" system in learning and memory remains to be characterized. Nevertheless, it appears that experimental manipulations that jointly influence serotonergic and cholinergic function can affect the ability of animals to learn and remember. However, it should be apparent from reading this chapter and Chapters XIV and XVI that there are a number of discrepancies in the results of "interactive" studies, which likely reflect, in part, the inconsistencies often observed when such systems are examined independently. Moreover, whether the serotonergic and cholinergic neurotransmitter systems interact in "normal" learning and memory processes remains to be determined. However, the continued study of the anatomical and pharmacological characteristic of the serotonergic system should greatly assist in these efforts. While the discovery of multiple serotonergic receptor types, dual ascending systems, differential distribution and layering of terminal projections, and the differential effect of drugs on multiple types of serotonergic neurons may initially prove overwhelming, such information will be invaluable in elucidating the role of serotonin in learning and memory. This knowledge will no doubt facilitate our attempts to characterize the functional interrelationship between the serotonergic and cholinergic neurotransmitter systems in these and related processes. Such information should, in turn, assist in the development of more specific pharmacotherapies for the treatment of human memory disorders.

Acknowledgment. Preparation of this paper was supported, in part, by grant AGO7069 from NIA to H.J.N.

References

Altman HJ, Normile HJ (1987): Different temporal effects of serotonergic antagonists on passive avoidance retention. *Pharmacol Biochem Behav* 28:353–359

Altman HJ, Normile HJ (1988): What is the nature of the role of the serotonergic nervous system in learning and memory: Prospects for development of an effective treatment strategy for senile dementia. *Neurobiol Aging* 9:627–638

Altman HJ, Crosland RD, Jenden DJ, Berman RF (1984): Comparison of the effects of nucleus basalis, diagonal band and septal lesions on learning and memory in the Sprague-Dawley rat. *Soc Neurosci Abstr* 9:134

Altman HJ, Ogren SO, Berman RF, Normile HJ (1989): The effects of p-chloroamphetamine, a depletor of brain serotonin, on the performance of rats in two types of positively reinforced complex spatial discrimination tasks. *Behavioral Neural Biol* 52:131–144

Arai H, Kosaka K, Iizuka R (1984): Changes of biogenic amines and their metabolites in postmortem brains from patients with Alzheimer-type dementia. *J Neurochem* 43:388–393

Ashby CR, Jiang LH, Kasser RJ, Wang RY (1990): Electrophysiological characterization of 5-hydroxytryptamine 2 receptors in the rat medial prefrontal cortex. *J Pharmacol Exp Ther* 252:171–178

Awouters F (1985): The pharmacology of ketanserin, the first selective serotonin S_2-antagonist. *Drug Dev Res* 6:263–300

Azmitia EC, Segal M (1978): Autoradiographic analysis of the differential ascending projections of the dorsal and median raphe nuclei in the rat. *J Comp Neurol* 179:641–668

Barnes JM, Barnes WM, Costal B, Naylor RJ, Tyers MB (1989): 5-HT_3 receptors mediate inhibition of acetylcholine release in cortical tissue. *Nature (London)* 338:762–763

Bartus RT, Dean RL, Beer B, Lippa AS (1982): The cholinergic hypothesis of geriatric memory dysfunction. *Science* 217:408–417

Berman RF, Crosland RD, Jenden DJ, Altman HJ (1983): Lesions of the nucleus basalis of Meynert impair memory in Sprague-Dawley rats. *Soc Neurosci Abstr* 8:29

Bianchi C, Siniscalchi A, Beani L (1986): The influence of 5-hydroxytryptamine on the release of acetylcholine from guinea-pig brain ex vivo and in vitro. *Neuropharmacology* 25:1043–1049

Bianchi C, Siniscalchi A, Beani L (1989): The effect of 5-hydroxytryptamine on [^{3}H]-acetylcholine release from guinea-pig striatal slices. *Br J Pharmacol* 25:1043–1049

Bianchi C, Siniscalchi A, Beani L (1990): 5-HT_{1A} agonists increase and 5-HT_3 agonists decrease acetylcholine efflux from the cerebral cortex of freely-moving guinea-pigs. *Br J Pharmacol* 101:448–452

Bigl V, Woolf NJ, Butcher LL (1982): Cholinergic projections from the basal forebrain to frontal, parietal, temporal, occipital, and cingulate cortices: A combine fluorescent tracer and acetylcholinesterase analysis. *Brain Res Bull* 8:727–749

Blier P, Serrano A, Scatton B (1990): Differential responsiveness of the rat dorsal and median raphe 5-HT systems to 5-HT_1 receptor agonists and p-chloroamphetamine. *Synapse* 5:120–133

Blue ME, Yagaloff KA, Mamounas LA, Hartig PR, Molliver ME (1986): Correspondence of 5-HT_2 receptor distribution with serotonin innervation in rat cerebral cortex. *Soc Neurosci Abstr* 12:145

Blue ME, Yagaloff KA, Mamounas LA, Hartig PR, Molliver ME (1988): Correspondence between 5-HT_2 receptor and serotoninergic axons in rat neocortex. *Brain Res* 12:145

Bowen DM, Smith CB, White P, Davison AN (1976): Neurotransmitter related enzymes and indices of hypoxia in senile dementia and abiotrophies. *Brain* 99:459–496

Consolo S, Ladinsky H, Forloni GL, Tirelli AS, Garattini S (1980): Comparison of the effects of the stereoisomers of fenfluramine on the acetylcholine content of rat striatum, hippocampus and nucleus accumbens. *J Pharmacol* 32:201–203

Cutler NR (1989): Neurochemistry of Alzheimer's disease. *Curr Opin Psychiatry* 2:533–536
Dahlstrom A, Fuxe K (1964): Evidence for the existence of monoamine-containing neurons in the central nervous system. I. Demonstration of monoamine cell bodies of brain neurons. *Acta Physiol Scand* 62(Suppl 232):1–55
Davies P, Maloney AJF (1976): Selective loss of cholinergic neurons in Alzheimer's disease. *Lancet* 25:1403
Drachmann DA, Leavitt J (1974): Human memory and the cholinergic system: A relationship to aging. *Arch Neurol* 30:113–121
Eison AG, Yocca FD, Gianutsos G (1988): Noradrenergic denervation alters serotonergic-mediated behavior but not serotonergic receptor number in rats: Modulatory role of beta adrenergic receptors. *J Pharmacol Exper Therap* 246:571–577
Fibiger HC (1982): The organization and some projections of cholinergic neurons of the mammalian forebrain. *Brain Res* 4:327–388
Fibiger HC, Lepiane FB, Phillips AG (1978): Disruption of memory produced by stimulation of the dorsal raphe nucleus: Mediation by serotonin. *Brain Res* 155:380–386
Flicker C, Dean RL, Watkins DL, Fisher SK, Bartus RT (1983): Behavioral and neurochemical effects following neurotoxic lesions of a major cholinergic input to the cerebral cortex in the rat. *Pharmacol Biochem Behav* 18:973–982
Futuro-Neto HA, Dantas MA, Silva SR (1989): Effects of physostigmine on the activity of medullary raphe neurons in anesthetized rats. *Brazil J Med Biol Res* 22:901–904
Gillet G, Ammor S, Fillion G (1985): Serotonin inhibits serotonin release from rat striatal slices: Evidence for a presynaptic receptor-mediated effect. *J Neurochem* 45:1687–1691
Gudelsky GA, Koenig JI, Meltgzer HY (1986): Thermoregulatory responses to serotonin (5-HT) receptor stimulation in the rat: Evidence for opposing roles of 5-HT2 and 5-HT1A receptors, *Neuropharmacology* 25:1307–1313
Haroutunian V, Mantin R, Kanof PD (1990): Frontal cortex as the site of action of physostigmine in nbM-lesioned rats. *Physiol Behav* 47:203–206
Jackson D, Bruno JP, Stachowiak MK, Zigmond MJ (1988): Inhibition of striatal acetylcholine release by serotonin and dopamine and the intracerebral administration of 6-hydroxydopamine to neonatal rats. *Brain Res* 457:267–273
Jansen KLR, Faull RLM, Dragunow M, Synek (1990): Alzheimer's disease: Changes in hippocampal N-methyl-D-aspartate, quisqualate, neurotensin, adenosine, benzodiazepine, serotonin and opiod receptors—an autoradiographic study. *Neuroscience* 39:613–627
Johnston MV, McKinney M, Coyle JT (1981): Neocortical cholinergic innervation: A description of extrinsic and intrinsic components in the rat. *Exp Brain Res* 43:159–172
Kosofsky BE, Molliver ME (1987): The serotonergic innervation of cerebral cortex: Different classes of axon terminals arise from dorsal and median raphe nuclei. *Synapse* 1:153–168
Lawlor BA, Sunderland T, Mellow AM, Hill JL, Molchan SE, Murphy DL (1989a): Hyperresponsivity to the serotonin agonist m-chlorophenylpiperazine in Alzheimer's disease. *Arch Gen Psychiatry* 46:542–549
Lawlor BA, Sunderland T, Mellow AM, Hill JL, Newhouse PA, Murphy DL (1989b): A preliminary study of the effects of intravenous m-chlorophenylpiperazine, a serotonin agonist, in elderly subjects. *Biol Psychiatry* 25:679–686
Lehmann J, Nagy JI, Atmadja S, Fibiger HC (1980): The nucleus basalis magnocellularis:

The origin of a cholinergic projection to the neocortex of the rat. *Neuroscience* 5:1161–1174

Leysen JE, Awouter F, Kennis L, Laduron PM, Vandenberk J, Janssen PAJ (1981): Receptor binding profile of R 41 468, a novel antagonist at 5-HT2 receptors. *Life Sci* 28:1015–1022

Levey RI, Rye DB, Wainer BH, Mufson EJ, Mesulam MM (1984): Choline acetyltransferase-immunoreactive neurons intrinsic to rodent cortex and distinction from acetylcholinesterase-positive neurons. *Neuroscience* 13:341–353

Lorens SA (1978): Some behavioral effects of serotonin depletion depend on method: A comparison of 5,7-dihydroxytrptamine, p-chlorophenylalanine, p-chloroamphetamine and electrolyte raphe lesions. *Ann NY Acad Sci* 305:532–555

Lyon RA, Titeler M (1988): Pharmacology and biochcmistry of the 5-HT2 receptor. In: *The Serotonin Receptors*, Sanders-Bush E, ed. Clifton, NJ: Humana Press, pp 59–88

Mamounas LA, Molliver ME (1987): Dual serotonergic projections to the forebrain have separate origins in the dorsal and median raphe nuclei: Retrograde transport after selective ablation by p-chloroamphetamine (PCA). *Soc Neurosci Abstr* 13:907

Mamounas LA, Molliver ME (1988): Evidence for dual serotonergic projections to neocortex: Axons from the dorsal and median raphe nuclei are differentially vulnerable to the neurotoxin p-chloroamphetamine (PCA). *Exp Neurol* 102:23–36

Mantz J, Godbout R, Tassin JP, Glowinski J, Thierry AM (1990): Inhibition of spontaneous unit activity in the rat medial prefrontal cortex by mesencephalic raphe nuclei. *Brain Res* 524:22–30

Martinez JL (1986): Memory: Drugs and hormones. In: *Learning and Memory: A Biological View*. Martinez JL, Kesner RP, eds. Academic Press: Boston, pp 127–163

Maura G, Fedele E, Raiteri M (1989): Acetylcholine release from rat hippocampal slices in modulated by 5-hydroxytryptamine. *Eur J Pharmacol* 165:173–179

McEntee MJ, Crook TH (1991): Serotonin, memory, and the aging brain. *Psychopharmacology* 103:143–149

McGaugh JL, Liang KC, Bennett C, Sternberg DB (1984): Adrenergic influences on memory storage: Interaction of Peripheral and central systems. In: *Neurobiology of Learning and Memory*, Lynch G, McGaugh J, Weinberger N, eds. Guilford Press: New York, pp 313–332

Mesulam MM, Mufson EJ, Wainer BH, Levey AI (1983): Central cholinergic pathways in the rat: An overview based on an alternative nomenclature (Ch1–Ch6). *Neuroscience* 10:1185–1201

Muramatsu M, Chaki S, Usuki-Ito C, Aihara H (1990): Attenuation of serotonin-induced suppression of [^{3}H]acetylcholine release from rat cerebral cortex by minaprine: Possible involvement of the serotonin-2 receptor and K^+ channel. *Neurochem Int* 16:301–307

Normile HJ, Altman HJ (1987): Serotonin, Alzheimer's disease and learning and memory in animals. In: *Alzheimer's Disease*. Altman HJ, ed. New York: Plenum Press, pp 141–156

Normile HJ, Altman HJ (1988): Enhanced passive avoidance retention following post-train serotonergic receptor antagonist administration in middle-aged and aged rats. *Neurobiol Aging* 9:377–382

Normile HJ, Altman HJ (1992): Effects of combined acetylcholinesterase inhibition and serotonergic receptor blockade on age-associated memory impairments in rats. *Neurobiol Aging*

Normile HJ, Altman HJ, Galloway MP (1986): Facilitation of discrimination learning in aged rats following depletion of brain serotonin. *Am Geriatr Soc Abstr* 43:S49

Normile HJ, Jenden DJ, Kuhn DM, Wolf WA, Altman HJ (1990): Effects of combined serotonin depletion and lesions of the nucleus basalis magnocellularis on acquisition of a complex spatial discrimination task in the rat. *Brain Res* 536:245–250

Ogren SO (1982): Central serotonin neurons and learning in the rat. In: *Biology of Serotonergic Neurotransmission*, Osbourne NN, ed. Chichester: John Wiley & Sons

Palmer AM, Stratmann GC, Procter AW, Bowen DM (1987): Possible neurotransmitter basis of behavioral changes in Alzheimer's disease. *Ann Neurol* 23:616–620

Parent A, Poitras D, Dube L (1984): Comparative anatomy of central monoaminergic systems. In: *Handbook of Chemical Anatomy, Vol 2: Classical Neurotransmitters in the CNS*. Bjorkluna A, Hokfelt T, eds. Amsterdam: Elsevier, pp 409–439

Perry R (1986): Recent advances in neuropathology. *Br Med Bull* 42:34–41

Perry EK, Perry R, Blessed G, Tomlinsson B (1977): Necropsy evidence of central cholinergic deficits in senile dementia. *Lancet* 1:189

Perry EK, Tomlinson BE, Blessed G, Bergman K, Gibson PH, Perry RH (1978): Correlation of cholinergic abnormalities with senile plaques and mental test scores in senile dementia. *Br Med J* 2:1457–1459

Quirion R, Richard J (1987): Differential effects of selective lesions of cholinergic and dopaminergic neurons on serotonin-type 1 receptors in rat brain. *Synapse* 1:124–130

Ramirez TM, Altman HJ, Normile HJ, Kuhn DM, Azmitia EC (1989): The effect of serotonergic intrahippocampal neonatal grafts on learning and memory in the Stone 14-unit T-maze. *Soc Neurosci Abstr* 15:465

Samanin R, Quattrone A, Peri G, Ladinsky H, Consolo S (1978): Evidence of an interaction between serotonergic and cholinergic neurons in the corpus striatum and hippocampus of the rat brain. *Brain Res* 151:73–82

Schober A, Schober W, Luppa H (1989): Serotonergic raphe nuclei projection to the basal forebrain in the rat: A combined HRP- and 5-HT-immunohistochemical investigation. *Acta Histochem Cytochem* 22:199–205

Semba K, Reiner PB, McGeer EG, Fibiger HC (1988): Brainstem afferents to the magnocellular basal forebrain studied by axonal transport, immunohistochemistry, and electrophysiology in the rat. *J Comp Neurol* 267:433–453

Siniscalchi A, Beani L, Bianchi C (1990): Different effects of 8-OH-DPAT, a 5-HT_{1A} receptor agonist, on cortical acetylcholine release, electrocortigram and body temperasture in guinea-pigs and rats. *Eur J Pharmacol* 175:219–223

Squire LR, Zola-Morgan S (1988): Memory: brain systems and behavior. *Topics Neurosci* 11:170–175

Tork I (1990): Anatomy of the serotonergic system. In: *The Neuropharmacology of Serotonin*. Whitaker-Azmitia PM, Peroutka SJ, eds. New York: New York Academy of Sciences, pp 9–35

Ungerstedt U (1971): Stereotaxic mapping of the monoamine pathways in the rat brain. *Acta Physiol Scand* 367(Suppl.):1–48

Vanderwolf CH (1987): Near-total loss of "learning" and "memory" as a result of combined cholinergic and serotonergic blockade in the rat. *Behav Brain Res* 23:43–57

Vertes RP (1988): Brainstem afferents to the basal forebrain in the rat. *Neuroscience* 24:907–935

Wetzel W, Getsova VM, Jork R, Matthies H (1980): Effect of serotonin on Y-maze retention and hippocampal protein synthesis in rats. *Pharmacol Biochem Behav* 12:319–322

Winter JC, Petti DT (1987): The effects of 8-hydroxy-2-(di-n-propylamino) tetralin and other serotonergic agonists on performance in a radial maze: A possible role for 5-HT_{1A} receptors in memory. *Pharmacol Biochem Behav* 27:625–628

Wolkowitz DM, Tinklenberg JR, Weingartner HA (1985): Psychopharmacological perspective of cognitive functions. *Neuropsychobiology* 14:133–156

Woolf NJ, Butcher LL (1982): Cholinergic projections to the basolateral amygdala: A combine evans blue dye and acetylcholinesterase analysis. *Brain Res Bull* 8:751–763

14

The Contribution of the Serotonergic Innervation of the Hippocampus to the Effects of Combined Cholinergic/Serotonergic Deficits

GAL RICHTER-LEVIN AND MENAHEM SEGAL

Introduction

A most disturbing aspect of aging is the reduction in cognitive abilities (Bartus et al., 1982; Weingartner et al., 1987). Although acetylcholine (ACh) is suggested to have a prime role in memory processes (Bartus et al., 1982; Kesner, 1988; Lebrun et al., 1990; Toumane et al., 1988), it is now widely accepted that other neuromodulatory systems participate, probably interactively, in these processes (Baker and Reynolds, 1989; Decker and McGaugh, 1991; Rossor and Iversen, 1986; Segal, 1982) and that a reduction in the functioning of these systems may underlie some aspects of memory deficits associated with aging (Allain et al., 1989; Steinbusch et al., 1989).

The hippocampus is one of several brain regions that are more vulnerable to aging (Barnes, 1988b; Scheibel, 1979). Its involvement in learning and memory is well established (Barnes, 1988a; Morris, 1982; O'Keefe and Nadel, 1978; Scheibel, 1979; Sutherland and Rudy, 1989; Teyler and DiScenna, 1985). The hippocampus, and in particular, the dentate gyrus (DG), are major targets of the serotonergic innervation of the forebrain (Azmitia and Segal, 1978; Moore, 1981). Anatomically, the serotonergic forebrain innervation is affected during aging. Fibers with swollen varicosities and a crumpled and folded form were found in areas such as the frotoparietal cortex, the caudate putamen, and the hippocampus (Steinbusch et al., 1989, 1990; Van Luijtelaar et al., 1988). Similar modifications were found in young mature rats after the injection of a specific neurotoxin, 5,7-dihydroxytryptamine (5,7-DHT) (Van Luijtelaar et al., 1989). Physiologically, there is a reduction in the responses of hippocampal cells of aged rats to the application of serotonin (Baskys et al., 1987; Segal, 1982).

Aged rats exhibit a reduced ability to perform learning tasks (Aggleton et al., 1989; Barnes, 1988b; Bartus et al., 1982; Biegon et al., 1986; Lebrun et al., 1990) and, in particular, spatial learning tasks (Barnes, 1988a; Biegon et al., 1986; Fischer et al., 1991b; Gallagher and Pelleymounter, 1988). These deficits are progressive (Fischer et al., 1991a,b; Richter-Levin and Segal, in preparation) and are noticed before a significant reduction of cholinergic transmission can be detected (Fischer et al., 1991a,b).

Severe damage to the cholinergic innervation of the hippocampus and other regions results in a marked reduction in the ability of such rats to perform spatial memory tasks (Di Patre et al., 1990; Hepler et al., 1985a,b; Ikegami et al., 1989a,b; Kelsey and Landry, 1988; McGurk et al., 1991; Miller et al., 1977; Nilsson et al., 1987; Segal et al., 1989; Srebro et al., 1976). This reduced cognitive ability is associated with a marked reduction of cholinergic functions, which is more profound than that observed in the aged rat brain (Table 1; Decker, 1987). The effects of a partial blockade of the cholinergic system are variable and seem to depend on the extent of the lesion (Nilsson et al., 1988; Richter-Levin and Segal, 1989a, 1991a; Segal et al., 1989; Srebro et al., 1976; Wenk et al., 1989). Depletion of the forebrain serotonergic system by itself has no clear effect on the ability of rats to perform spatial memory tasks (Altman and Normile, 1988; Altman et al., 1990; Asin et al., 1985; Nilsson et al., 1988; Richter-Levin and Segal, 1989a, 1991a). Recently it was suggested that when combined with a partial cholinergic impairment, the result is a marked reduction in the ability of rats to perform such tasks (Nilsson et al., 1988; Richter-Levin and Segal, 1989a, 1991a; Vanderwolf, 1987, 1988).

The following reviews our studies aimed at examining three aspects of serotonin/acetylcholine interactions: (a) the hippocampus as a possible site for integrating the effects of the two neurotransmitter systems, (b) possible correlation between electrophysiological and behavioral effects of such lesions, and (c) a preliminary attempt to compare the effects of combined lesions with the deterioration of functions associated with aging of the brain.

Behavioral Studies

The performance of rats in spatial memory tasks depends on the integrity of the hippocampal formation (Morris et al., 1982; O'Keefe and Nadel, 1978). Thus lesioning the hippocampus adversely affects the ability of rats to perform a spatial memory water-maze task (in which rats have to locate and escape to an invisible platform using extra-maze cues) but does not affect their ability to perform a cued task (in which rats have to locate and escape to a visible platform), using the same maze (Morris et al., 1982). We have used the water maze throughout this study as a hippocampus-dependent spatial memory task.

The spatial representation of the environment in the elderly is typically less effective than in young adults (Barnes, 1988b; Weingartner et al., 1987). Rats have a very efficient spatial memory, as evident by their performance in various memory tasks (O'Keefe and Nadel, 1978). Aged rats exhibit a reduced ability to perform such tasks (Aggleton et al., 1989; Barnes, 1988b; Biegon et al., 1986; Fischer et al., 1991b; Lebrun et al., 1990). These deficits start to be noticed as early as 18 months of age (Fischer et al., 1991b) (Fig. $1A_1$). At this age there are still no significant signs of reduction in the functioning of the cholinergic system (Fischer et al., 1991b). When continuously trained, older rats still show longer escape latencies but 18-month-old rats do acquire the task (Fig. $1A_2$ trial #1). At

TABLE 1. Comparison Between the Effects of Lesions and Aging on the Cholinergic System

	% of control:
1. No. of ChAT positive cells	
MS	
FF transection	−20% (Armstrong et al., 1987)
Colchicine	−20% (Peterson and Mcginty, 1988)
Aged rats	−75% (Fischer et al., 1991b)
	−80% (Nunzi et al., 1988)
2. No. of AChE positive cells	
MS	
FF transection	−33% (Gage et al., 1986)
Aged rats	−57% (Fischer et al., 1989)
VDB	
FF transection	−45% (Gage et al., 1986)
Aged rats	−63% (Fischer et al., 1989)
3. ChAT activity	
Hippocampus	
Electrocoagulation	−15% (Nilsson et al., 1988)
Ibotenic acid	−56% (Hepler et al., 1985b)
AF64A	−35% (Ikegami et al., 1989a)
Aged rats	−ns (Fischer et al., 1989)
	−84% (Luine and Hearns, 1990)
Cortex	
Electrocoagulation	−77% (Dubois et al., 1985)
Ibotenic acid	−67% (Dubois et al., 1985)
	−65% (Hepler et al., 1985b)
	−63% (Murray and Fibiger, 1985)
Aged rats	−ns (Fischer et al., 1989)
	−ns (Luine and Hearns, 1990)
Septum	
Lesion by macroelectrode	−25% (Monmaur et al., 1984)
Aged rats	−ns (Fischer, 1989)
	−95% (Luine and Hearns, 1990)
4. ChAT levels	
Hippocampus	
Electrocoagulation	−63% (Hepler et al., 1985a)
Ibotenic acid	−46% (Wenk et al., 1989)
	−50% (Hepler et al., 1985a)
Aged rats	−ns (Springer et al., 1987)
Cortex	
Electrocoagulation	−75% (Hepler et al., 1985a)
Ibotenic acid	−27% (Wenk et al., 1989)
	−43% (Hepler et al., 1985a)
Aged rats	−ns (Springer et al., 1987)
5. AChE levels	
Hippocampus	
FF transection	−5% (Segal et al., 1989)
Septal lesion	−3% (Segal et al., 1989)
AF64A	−35% (Eva et al., 1987)
Aged rats	−ns (Springer et al., 1987)
	−88% (Biegon et al., 1986)
6. ACh-evoked release	
Hippocampus	
FF transection	−7% (Nilsson et al., 1990)
Aged rats	−ns (Fischer et al., 1991c)

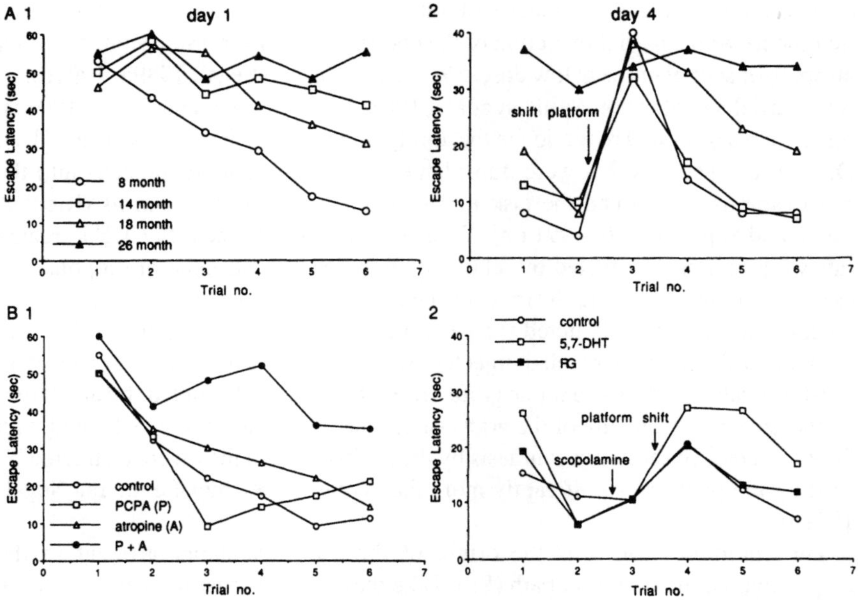

FIGURE 1. A: The effects of age on the performance of rats in a spatial memory water-maze task. Rats were trained for 3 days (six trials per day) to escape to a fixed-position hidden platform. On the fourth day (2), after two retention trials, the platform was shifted to a new location. The acquisition rate (day 1) of younger rats (8 months) was significantly faster than that of the older groups (ANOVA for repeated measurements; $p < 0.01$). By the end of the third day, the 14- and 18-month-old rats performed the maze in a similar manner as the younger rats, whereas the 26-month-old rats had still significantly longer escape latencies ($p < 0.0004$). Likewise, only the 26-month-old rats were significantly slower on the retention trials on day 4 (trials #1–2; $p < 0.002$). Shifting the platform disrupted the performance of the 18-month-old rats ($p < 0.01$). B: The effects of combined serotonergic and cholinergic deficits on the performance of rats in a spatial memory water-maze task. (1) PCPA (300 mg/kg, i.p., 3 days before testing) or atropine (20 mg/kg, i.p., 30 min before the first trial) had no effect on the acquisition rate, whereas the combined treatment significantly increased the escape latencies of such rats ($p < 0.01$). (2) When continuously trained, 5,7-DHT treated rats are no longer affected by scopolamine (2 mg/kg, i.p., 30 min before the first trial). Shifting the platform at this stage disrupted the performance of the 5,7-DHT rats ($p < 0.008$) but not that of controls or of raphe-grafted (RG) rats (all groups were treated with scopolamine). The variability on the first trial is due to a relatively long interval from the previous training day.

this stage, shifting the invisible platform to a new location disrupts the performance of these rats significantly more than that of young adult rats (Fig. $1A_2$ trial #4–6).

Reducing the serotonergic neurotransmission, by either p-chlorophenylalanine (PCPA) (a serotonin synthesis blocker) (Richter-Levin and Segal, 1989a) or by 5,7-DHT (Nilsson et al., 1988; Richter-Levin and Segal, 1989b, 1991a), has no clear effects on the performance of rats in the water maze. Combining such treatments with a partial reduction of the cholinergic neurotransmission, by either atropine or scopolamine at low doses (Richter-Levin and Segal, 1989a, 1991b), or by a partial septal lesion (Nillson et al., 1988; Richter-Levin and Segal, 1991a), resulted in a profound reduction in the ability of rats to acquire this task (Fig. $1B_1$). Double-lesioned rats that were transplanted with embryonic raphe cells into the hippocampus, performed the task in a similar manner to the controls (Richter-Levin and Segal, 1989b, 1991a,b). When continuously trained, 5,7-DHT-treated rats were no longer affected by relatively high doses of atropine or scopolamine (Richter-Levin and Segal, 1991b) (Fig. $1B_2$, trial 3). Shifting the platform to a new location at this stage disrupted the performance of these rats significantly more than controls (that were also injected with scopolamine, Fig. $1B_2$) or than 5,7-DHT rats that received raphe grafts (RG). The effects depended, by and large, on the serotonergic nature of the graft, since when these RG rats were treated with PCPA (to reduce serotonin synthesis by the grafted tissue) they were distracted by shifting the platform significantly more than controls (Richter-Levin and Segal, 1991b).

The entorhinal cortex is the origin of the main excitatory afferent to the hippocampus, the perforant path (PP). Like the hippocampus, it receives a dense serotonergic innervation (Azmitia and Segal, 1978; Moore, 1981). Depleting serotonergic innervation to the entire forebrain may affect the entorhinal cortex as well as the hippocampus and may contribute to the behavioral effects of the lesion. To test for this possibility we compared the effects of raphe grafts in the hippocampus (RG) with those of raphe grafts in the entorhinal cortex (EG). RG rats performed in the maze significantly better than double-lesioned rats, while EG rats did not.

Another experiment designed to focus on the involvement of the serotonergic-hippocampal connection was to compare the effects of intraventricular (IV) injection of 5,7-DHT (which affects the serotonergic innervation of the entire forebrain) with injections of 5,7-DHT into the fornix-fimbria (FF) (which selectively reduces the serotonergic innervation of the hippocampus (Altman et al., 1990)). Both IV and FF rats were significantly affected by atropine at a dose that did not affect the performance of a control group.

These results suggest that a simultaneous reduction in neurotransmission of the serotonergic and cholinergic innervation of the hippocampus underlies the behavioral deficits associated with these lesions.

Electrophysiological Studies

In the behavioral studies only the combined lesions resulted in significant spatial memory deficits. Recently it was reported that a blockade of central cholinergic and serotonergic transmission eliminated all hippocampal rhythmical slow activity as well as neocortical low-voltage fast activity (Vanderwolf, 1988). This was associated with a severe generalized impairment in the ability to perform specific motor acts in coordination with task demands (Vanderwolf, 1988).

A partial blockade of both systems, however, did not produce a total elimination of cortical and hippocampal activation (Vanderwolf, 1988); nor did it produce a generalized behavior-related motor impairment, since such rats, although impaired in the water-maze task, performed well in an active avoidance task (Richter-Levin and Segal, 1991a).

The serotonergic raphe-hippocampal pathway forms multiple synaptic contacts with hippocampal GABAergic interneurons that contain the calcium-binding protein calbindin D_{28k} (Freund et al., 1990). It is possible that this pathway influences hippocampal functions via the modulation of local inhibitory circuits. Looking for local physiological modifications that would correlate with the behavioral deficits, we have studied, following the behavioral tests, feedforward and feedback inhibitory circuits in the hippocampus.

Pulse-paired stimulation of the PP at short interpulse intervals (up to 80 msec in the chloral-hydrate anesthetized rat; Richter-Levin and Segal, 1991a) results in an inhibition of the response to the second stimulus. This type of inhibition is referred to as feedback inhibition. Neither partial septal lesion nor serotonin depletion had any effect on feedback inhibition. Combining the lesions increased the variability among rats but had no significant effect on feedback inhibition (Richter-Levin and Segal, 1991a) (Fig. 2A).

Priming stimulation of the hippocampal commissural pathway results in a blockade of the subsequent response of the dentate gyrus to stimulation of the PP (up to 50 msec in the chloral-hydrate anesthetized rat) (Buzsaki and Czeh, 1981; Douglas et al., 1983). This type of inhibition is referred to as feedforward inhibition. Depletion of serotonin by 5,7-DHT or partial septal lesion had no significant effect on this inhibition, but when combined there was a significant reduction of feedforward inhibition (Richter-Levin and Segal, 1991a) (Fig. 2B). In double-lesioned rats that were grafted with raphe cells in the hippocampus (RG), feedforward inhibition was similar to controls (Richter-Levin and Segal, 1991a) (Fig. 2C).

Aged rats (20 months) exhibited reduced feedforward inhibition that was qualitatively similar to that observed in double-lesioned rats (Fig. 2C).

Discussion

The present results suggest that the behavioral deficits associated with the combined lesions are the result of alterations in hippocampal local circuit activity. There was no difference between the effects of lesioning the serotonergic innervation of the entire forebrain and lesioning the serotonergic innervation of the hippocampus only. Likewise, there was no difference in the effects of blocking cholinergic neurotransmission by atropine or scopolamine, and lesioning the cholinergic innervation of the hippocampus (by lesioning the septum). Furthermore, grafting embryonic raphe cells into the hippocampus but not into the entorhinal cortex could compensate for the effects of lesioning the entire forebrain serotonergic innervation.

The best correlative electrophysiological alteration was the reduction in feed-forward inhibition. This reduction was observed only in the double-lesioned rats and could be reversed by raphe grafts in the hippocampus.

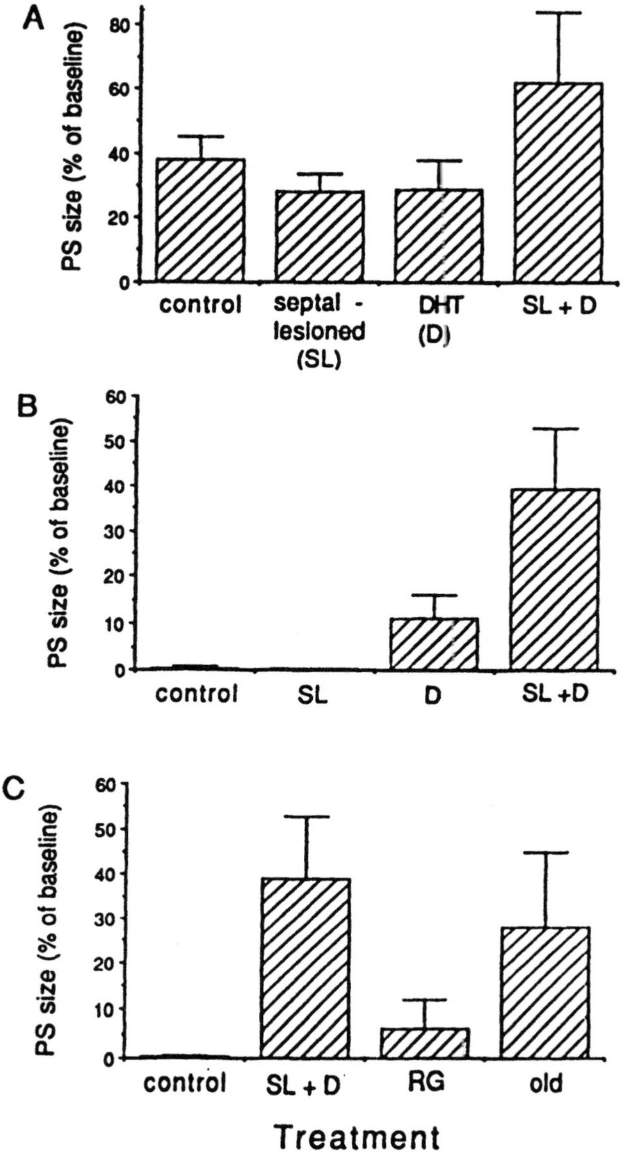

Application of serotonin-releasing drugs (e.g., fenfluramine) resulted in an increase in the responsiveness of the dentate granule cells to stimulation of the PP (Richter-Levin and Segal, 1990). Although these drugs did not directly alter feedforward inhibition, commissurectomy reduced the effects of the drugs (Richter-Levin and Segal, 1990), supporting the assumption that the commissural pathway madiates some effects of the serotonergic innervation of the hippocampus.

A reduction in feedforward inhibition was also observed in the hippocampus of aged rats. This contrasts with the findings of Barnes (1979), who found no differences in either feedforward or feedback inhibition between different age groups. This may result from the differences in methodology, since their study was conducted with freely moving rats, whereas the present study was conducted with anesthetized rats. Feedforward inhibition is much reduced and is highly variable in the awake, compared with the anesthetized, rat (Douglas et al., 1983). It is possible that these may have masked the relatively subtle differences between age groups.

In summary, the hippocampus is assumed to be the behaviorally relevant site of integration of the effects of serotonin and acetylcholine. There is a correlation between feedforward inhibition and the performance of rats in the water maze, but whether there is a causal relation between the two is still not clear. Finally, the effects of the lesions seem to correlate well with the effects of the process of aging of the rat brain. However, additional behavioral and physiological aspects should be studied before a conclusive comparison can be made. The present results encourage such studies.

Acknowledgments. This research was supported by a grant from The Belle S. and Irving E. Meller Center for the Biology of Aging, Rehovot, Israel, and by research grant from the Mario-Negri Institute-Weizmann collaboration fund.

FIGURE 2. A: The effects of combined serotonergic and cholinergic deficits on feedback inhibition in the hippocampus. Baseline is the PS size of the response to the first stimulation of the PP. Interpulse interval = 30 msec. None of the treatments resulted in a significant difference from controls. B: The effects of combined serotonergic and cholinergic deficits on feedforward inhibition. Baseline is the PS size without priming stimulation of the commissural pathway. Interpulse interval = 10 msec. Serotonin depletion reduced feedforward inhibition (control vs. D; $p < 0.03$), but the combined serotonergic/cholinergic deficit reduced it significantly more (D vs. SL + D; $p < 0.05$). C: The effects of raphe grafts and aging on feedforward inhibition. Baseline is the PS size without priming stimulation of the commissural pathway. Interpulse interval = 10 msec. Embryonic raphe cells were grafted into the hippocampus of double-lesioned rats (RG). Feedforward inhibition in these rats was similar to controls and was significantly different from that in the double-lesioned rats ($p < 0.01$). Aged rats (old) were similar to the double-lesioned rats.

References

Aggleton JP, Blint HS, Candy JM (1989): Working memory in aged rats. *Behavioral Neurosci* 103:975–983

Allain H, Moran P, Bentue-Ferrer D, Martinet JP, Lieury A (1989): Pharmacology of the memory process. *Arch Gerontol Geriatr* 1(Suppl):109–120

Altman HJ, Normile HJ (1988): What is the nature of the role of the serotonergic nervous system in learning and memory; prospects for development of effective treatment strategies for senile dementia. *Neurobiol Aging* 9:627–638

Altman HJ, Normile HJ, Galloew MP, Ramirez A, Azmitia EF (1990): Enhanced spatial discrimination learning in rats following 5,7-DHT-induced serotonergic deafferentation of the hippocampus. *Brain Res* 518:61–66

Armstrong DM, Terry RD, Deteresa RM, Bruce G, Hersh LB, Gage FH (1987): Response of septal cholinergic neurons to axotomy. *J Compar Neurol* 264:421–436

Asin KE, Wirtshafter D, Fibiger HC (1985): Electrolytic, but not 5,7-dihydroxytryptamine, lesions of the nucleus medianus raphe impair acquisition of a radial maze task. *Behav Neurol Biol* 44:415–424

Azmitia EC, Segal M (1978): An autoradiographic analysis of the differential ascending projections of the dorsal and median raphe nuclei in the rat. *J Compar Neurol* 179:641–668

Baker GB, Reynolds GP (1989): Biogenic amines and their metabolites in Alzheimer's disease: Noradrenaline, 5-hydroxytryptamine and 5-hydroxyindol-3-acetic acid depleted in hippocampus but not in substantiainnominata. *Neurosci Lett* 100:335–339

Barnes CA (1979): Memory deficits associated with senescence: A neurophysiological and behavioral study in the rat. *J Compar Physiol Psychol* 93:74–109

Barnes CA (1988a): Spatial learning and memory processes: The search for their neurobiological mechanisms in the rat. *Topics Neurosci* 11:163–169

Barnes CA (1988b): Aging and the physiology of spatial memory. *Neurobiol Aging* 9:563–568

Bartus RT, Dean RL III, Beer B, Lippa AS (1982): The cholinergic hypothesis of geriatric memory dysfunction. *Science* 217:408–417

Baskys A, Niesen CE, Carlen PL (1987): Altered modulatory actions of serotonin on dentate granule cells of aged rats. *Brain Res* 419:112–118

Biegon A, Greenberger V, Segal M (1986): Quantitative histochemistry of brain acetylcholinesterase and learning rate in the aged rat. *Neurobiol Aging* 7:215–217

Buzsaki G, Czeh G (1981): Commissural and PP interaction in the rat hippocampus. *Exp Brain Res* 43:429–438

Decker MW (1987): The effects of aging on hippocampal and cortical projections of the forebrain cholinergic system. *Brain Res Rev* 12:423–438

Decker MW, McGaugh JL (1991): The role of interactions between the cholinergic system and other neuromodulatory systems in learning and memory. *Synapse* 7:151–168

Di Patre PL, Oh JD, Simmons JM, Butcher LL (1990): Intrafimbrial colchicine produces transient impairment of radial-arm maze performance correlated with morphologic abnormalities of septohippocampal neurons expressing cholinergic markers and nerve growth factor receptor. *Brain Res* 523:316–320

Douglas RM, McNaughton BL, Goddard GV (1983): Commissural inhibition and facilitation of granule cell discharge in fascia dentata. *J Compar Neurol* 219:285–294

Dubois B, Mayo W, Agid Y, Le Moal M, Simon H (1985): Profound disturbances of

spontaneous and learned behaviors following lesions of the nucleus basalis magnocellularis in the rat. *Brain Res* 338:249–258
Eva C, Fabrazzo M, Costa E (1987): Changes of cholinergic, noradrenergic and serotonergic synaptic transmition indices elicited by ethylcholine aziridinium ion (AF64A) infused intraventricularly. *J Pharmacol Exp Therap* 241:181–186
Fischer W, Gage FH, Bjorklund A (1989): Degenerative changes in forebrain cholinergic nuclei correlate with cognitive impairments in aged rats. *Eur J Neurosci* 1:34–45
Fischer W, Bjorklund A, Chen K, Gage FH (1991a): NGF improves spatial memory in aged rodents as a function of age. *J Neurosci* 11(7):1889–1906
Fischer W, Chen KS, Gage FH, Bjorklund A (1991b): Progressive decline in spatial learning and integrity of forebrain cholinergic neurons in rats during aging. *Neurobiol Aging* 13(1):9–24
Fischer W, Nilsson OG, Bjorklund A (1991c): In vivo acetylcholine release is unaltered in the hippocampus of cognitively impaired aged rats with degenerative changes in the basal forebrain. *Brain Res* 556(1):44–52
Freund TF, Gulyas AI, Acsady L, Gorcs T, Toth K (1990): Serotonergic control of the hippocampus via local inhibitory interneurons. *Proc Natl Acad Sci USA* 87:8501–8505
Gage FH, Wictorin K, Fischer W, Williams LR, Varon S, Bjorklund A (1986): Retrograde cell changes in medial septum and diagonal band following fimbria-fornix transection: Quantitative temporal analysis. *Neuroscience* 19:241–255
Gallagher M, Pelleymounter MA (1988): Spatial learning deficits in old rats: A model for memory decline in the aged. *Neurobiol Aging* 9:549–556
Hepler DJ, Olton DS, Wenk GL, Coyle JT (1985a): Lesions in nucleus basalis magnocellularis and medial septal area of rats produce qualitatively similar memory impairments. *J Neurosci* 5:866–873
Hepler DJ, Wenk GL, Cribbs BL, Olton DS, Coyle JT (1985b): Memory impairments following basal forebrain lesions. *Brain Res* 346:8–14
Ikegami S, Nihonmatsu I, Hatanaka H, Takei N, Kawamura H (1989a): Recovery of hippocampal cholinergic activity by transplantation of septal neurons in AF64A treated rats. *Neurosci Lett* 101:17–22
Ikegami S, Nihonmatsu I, Hatanaka H, Takei N, Kawamura H (1989b): Transplantation of septal cholinergic neurons to the hippocampus improves memory impairments of spatial learning in rats treated with AF64A. *Brain Res* 496:321–326
Kelsey JE, Landry BA (1988): Medial septal lesions disrupt spatial mapping ability in rats. *Behav Neurosci* 102:289–293
Kesner RP (1988): Reevaluation of the contribution of the basal forebrain cholinergic system to memory. *Neurobiol Aging* 9:609–616
Lebrun C, Durkin TP, Marighetto A, Jaffared R (1990): A comparison of the working memory performance of young and aged mice combined with parallel measures of testing and drug-induced activation of septo-hippocampal and nbm-cortical cholinergic neurones. *Neurobiol Aging* 11:515–521
Luine V, Hearns M (1990): Spatial memory deficits in aged rats: Contribution of the cholinergic system assessed by ChAT. *Brain Res* 523:321–324
McGurk SR, Levin ED, Butcher LL (1991): Impairment of radial-arm maze performance in rats following lesions involving the cholinergic medial pathway: Reversal by arecoline and differential effects of muscarinic and nicotinic antagonists. *Neuroscience* 44:137–147
Miller MA, Innes WC, Enloe LJ (1977): Performance on a four-choice search task following septal lesions in the rat. *Physiol Psychol* 5:433–439

Monmaur P, Fage D, M'harzi M, Delacour J, Scatton B (1984): Decrease in both choline acetyltransferase activity and EEG patterns in the hippocampal formation of the rat following septal macroelectrode implantation. *Brain Res* 293:178–183

Moore RY (1981): The anatomy of central serotonin neuron system in the rat brain. In: *Serotonin—Neurotransmission and Behavior*, Jacobs BL, Gelperin A, eds. Cambridge, MA: MIT Press, pp 35–71

Morris RGM, Garrud P, Rawline JNP, O'Keefe J (1982): Place navigation impaired in rats with hippocampal lesions. *Nature* 297:681–683

Murray CL, Fibiger HC (1985): Learning and memory deficits after lesions of the nucleus basalis magnocellularis: Reversal by physostigmine. *Neuroscience* 14:1025–1032

Nilsson OG, Shapiro ML, Gage FH, Olton DS, Bjorklund A (1987): Spatial learning and memory following fimbria-fornix transection and grafting of fetal septal neurons to the hippocampus. *Exp Brain Res* 67:195–215

Nilsson OG, Strecker RE, Daszuta A, Bjorklund A (1988): Combined cholinergic and serotonergic denervation of the forebrain produces severe deficits in a spatial learning task in the rat. *Brain Res* 453:235–246

Nilsson OG, Kalen P, Rosengren E, Bjorklund A (1990): Acetylcholine release from intrahippocampal septal grafts is under control of the host brain. *Proc Natl Acad Sci USA* 87:2647–2651

Nunzi MG, Zanotti A, Milan F, Guidolin D, Polato P, Toffano G (1988): Septal cholinergic neurons and spatial memory in aged rats. Effects of phosphatidylserine administration. In: *Senile Dementias. II International Symposium*, Agnoli A, Cahn J, Lassen N, Mayeux R, eds. Paris: Johan Libbey Eurotex, pp 369–374

O'Keefe J, Nadel L (1978): *The Hippocampus as a Cognitive Map*. Oxford: Clarendon.

Peterson GM, McGinty JF (1988): Direct neurotoxic effects of colchicine on cholinergic neurons in medial septum and striatum. *Neurosci Lett* 94:46–51

Richter-Levin G, Segal M (1989a): Spatial performance is severely impaired in rats with combined reduction of serotonergic and cholinergic transmition. *Brain Res* 477:404–407

Richter-Levin G, Segal M (1989b): Raphe cells grafted into the hippocampus can ameliorate spatial memory deficits in rats with combined serotonergic/cholinergic deficiencies. *Brain Res* 478:184–186

Richter-Levin G, Segal M (1990): Effects of serotonin releasers on dentate granule cell excitability in the rat. *Exp Brain Res* 82:199–207

Richter-Levin G, Segal M (1991a): The effects of serotonin depletion and raphe grafts on hippocampal electrophysiology and behavior. *J Neurosci* 11:1585–1596

Richter-Levin G, Segal M (1991b): Restoration of serotonergic innervation underlies the behavioral effects of raphe grafts. *Brain Res* 556:21–25

Rosser M, Iversen LL (1986) Non-cholinergic neurotransmitter abnormalities in Alzheimer's disease. *Br Med Bull* 42:70–74

Scheibel AB (1979): The hippocampus: Organizational patterns in health and senescence. *Mech Aging Dev* 9:89–102

Segal M (1982): Changes in neurotransmitter actions in the aged rat hippocampus. *Neurobiol Aging* 3:121–132

Segal M, Greenberger V, Pearl E (1989): Septal transplants ameliorate spatial deficits and restore cholinergic functions in rats with damaged septo-hippocampal connection. *Brain Res* 500:139–148

Springer JE, Tayrien MW, Loy R (1987): Regional analysis of age-related changes in the cholinergic system of the hippocampal formation and basal forebrain of the rat. *Brain Res* 407:180–184

Srebro B, Ellertsen B, Ursin H (1976): Deficits in avoidance learning following septal lesions in the albino rat. *Physiol Behav* 16:589–602

Steinbusch HWM, Wolters JG, Van Luijtelaar MPGA, Tonaer JADM (1989): The serotonergic and dopaminergic system in the aged rat brain as studied by immunocytochemistry with antibodies to serotonin and dopamine: Preliminary observations. In: *New Trends in Aging Research*, Pepeu GC, Tomlinson B, Wischik CM, eds. Liviana Press, Springer, pp 193–203

Steinbusch HW, Van Luijtelaar MG, Dijkstra H, Nijssen A, Tonnaer JA (1990): Aging and regenerative capacity of the rat serotonergic system. A morphological, neurochemical and behavioral analysis after transplantation of fetal raphe cells. *Ann NY Acad Sci* 600:384–402

Sutherland RJ, Rudy JW (1989): Configural association theory: The role of the hippocampal formation in learning, memory, and amnesia. *Psychobiology* 17:129–144

Teyler TJ, DiScenna P (1985): The role of the hippocampus in memory: A hypothesis. *Neurosci Biobehav Rev* 9:377–389

Toumane A, Durkin T, Marighetto A, Galey D, Jaffard R (1988): Differential hippocampal and cortical cholinergic activation during the acquisition, retention, reversal and extinction of a spatial discrimination in an 8-arm radial maze by mice. *Behav Brain Res* 30:225–234

Vanderwolf CH (1987): Near total loss of 'learning' and 'memory' as a result of combined cholinergic and serotonergic blockade in the rat. *Behav Brain Res* 23:43–57

Vanderwolf CH (1988): Cerebral activity and behavior: Control by central cholinergic and serotonergic systems. *Int Rev Neurobiol* 30:225–340

Van Luijtelaar MGPA, Steinbusch HWM, Tonnaer JADM (1988): Aberrant morphology of serotonergic fibers in the forebrain of the aged rat. *Neurosci Lett* 95:93–96

Van Luijtelaar MGPA, Steinbusch HWM, Tonnaer JADM (1989): Similarities between aberrant serotonergic fibers in the aged and 5,7-DHT denervated young adult rat brain. *Exp Brain Res* 78:81–89

Weingartner H, Cohen R, Sunderland T, Tariot P, Thompson K, Newhouse P (1987): Diagnosis and assessment of cognitive dysfunctions in the elderly. In: *Psychopharmacology: The Third Generation of Progress*, Meltzer Y, ed. New York: Raven Press, pp 909–919

Wenk GL, Markowaska AL, Olton DS (1989) Basal forebrain lesions and memory: Alterations in neurotensin, not acetylcholine, may cause amnesia. *Behav Neurosci* 103:765–769

15

Overlapping Neural Substrates Underlying Defense Reactions, Aversive Memory, and Convulsive Behavior

CARLOS TOMAZ, MARCUS L. BRANDAO, AND NORBERTO GARCIA-CAIRASCO

Several lines of evidence strongly indicate the existence of an overlapped neural organization that mediates negative affective states and fearful/defensive behaviors. From an evolutionary perspective it seems important for animals to learn, recognize, and integrate information about threatening situations in order to enhance their survival capability. Thus, one can expect an intimate relationship between the neural systems responsible for the elaboration and expression of emotional behavior and those that mediate affective memory. Both terms, *emotion* and *affect*, are here used to describing the reactions of animals to stimuli that produce species-specific behaviors and autonomic responses patterns that can be measured objectively.

The aim of the present chapter is to review some experimental evidence indicating convergence along neural networks underlying defense reactions, convulsive behavior, and affective memories. To accomplish this task, we will review the brain structures involved as well as the neurohumoral substrates subserving these processes.

Anatomical Bases

Some of the main brain regions implicated in defense systems have been identified by electrical stimulation of subcortical areas in different species. Hess and Brugger (1943) first reported that stimulation at specific sites within the medial hypothalamus (MH) elicited defensive behaviors that closely resemble the naturally occurring behavior. Components of the defense reaction have also been elicited by electrical stimulation of the dorsal part of the midbrain periaqueductal gray matter (PAG), and the central, medial, and dorsal nuclei of the amygdaloid complex (AC) in the forebrain (Fernandez de Molina and Hunsperger, 1959; Hilton and Zbrozyna, 1963; Kaada et al., 1954; Skultety, 1963; Ursin and Kaada, 1960). Stimulation of the above structures induces somatic motor responses, such as arching of the back, retraction of the ears, and sometimes, vocalization. In addition, autonomic changes are also observed: increased heart rate and arterial blood pressure, hyperventilation, pupillary dilatation, and piloerection.

In contrast, lesions that destroy the PAG and the AC either eliminate or attenuate defensive patterns of behaviors, including defensive upright postures, escape-related behaviors, avoidance responding following exposure to a learned danger stimulus and immobility (Blanchard and Blanchard, 1972; Blanchard et al., 1981; Liebman et al., 1970). Regarding the MH, ablation or reversible lesion of this area has been reported to produce aggressive behavior and exacerbated reactivity, even to gentle handling (Colpaert and Wiepkema, 1976; Olivier, 1977; Wheatley, 1944). The paradoxical question that emerges from these findings of how both electrical stimulation and lesion of the same region can produce such defense reactions has not yet been clarified. Bandler (1988) has suggested that rather than playing a role as the major integrating center, the MH is involved in the elaboration of the defense reaction, in setting a general level of motivational responsiveness, possibly by modulating the midbrain neurons. In fact, lesions interrupting the pathways between the AC and the MH or electrolytic lesion of the MH have been shown to abolish the defense reactions elicited from intraamygdala and PAG electrical stimulation (Hilton and Zbrozyna, 1963; Olds and Frey, 1971).

Several lines of evidence have suggested that the defense reactions elicited by stimulation of these areas are likely to be the expression of a negative motivational state. For instance, animals are able to learn different behavioral tasks in order to switch off aversive electrical stimulation applied to periventricular structures, such as the MH or PAG (Olds and Olds, 1962; Schmitt et al., 1974). Additionally, electrical stimulation at the PAG, MH, or AC in human patients undergoing neurosurgery has been found to evoke feelings of fear or panic, accompanied by somatic and autonomic changes, including piloerection, increased heart rate, sweating, and vasodilation (Nashold et al., 1969). Therefore, the PAG, MH, and AC seem to constitute the main neural substrates responsible for the elaboration and expression of aversively motivated behaviors, the so-called brain aversive system (BAS; for reviews see Graeff, 1981, 1990).

Anatomical studies show that these brain areas are intimately interconnected by a two-way nerve fiber projections, suggesting that the components of the BAS may work together (Adams, 1979). In addition, it is well known (Bandler et al., 1985; Fuchs et al., 1985; De Olmos et al., 1985; Ricardo and Koh, 1978) that the BAS structures are interconnected to other brain areas, such as the prefrontal cortex, septum, hippocampus, thalamus, and the nucleus of the solitary tract, which have been involved in the modulation of defense reactions and emotional memory. Thus, some brain structures can be capable of modulating emotional behaviors in spite of the fact that they do not seem to be directly implicated in the elaboration of these behaviors. In the following section we will discuss the BAS structures, taking into account the neurotransmitter interactions regulating defensive or emotional behavioral responses triggered by innate or learned aversive stimuli.

Neurotransmitters and Aversion

As was noted above, the view that the dorsal part of the PAG is involved in aversion-motivated behavior is supported by findings that stimulation of this area produces a variety of defensive reactions. Evidence of Pavlovian-conditioned fear following electrical stimulation of the PAG in the rat has been reported by Di Scala and coworkers (1987). Using PAG stimulation as an aversive unconditional stimulus (UCS) and either a light or a tone as a conditional stimulus (CS), these authors showed a suppression of a learned response to lick a saccharine solution in rats in the CS-UCS paired condition. Such a suppression was not observed in the CS-alone group and the CS-UCS random group, indicating that the response suppression observed in the CS-UCS paired group was not due to an experience with the CS itself or to nonassociative mechanisms of pseudoconditioning or sensitization.

In addition, Roberts and Cox (1987) obtained active avoidance conditioning with dorsal PAG electrical stimulation. Employing a simple place preference task that minimized response demands and provided highly salient environmental cues that differentiated "safe" and "unsafe" locations, these authors reported successful avoidance conditioning, indicating that PAG stimulation can support associative learning involving environmental cues and the consequences of aversive PAG stimulation.

Nevertheless, a recent work done by LeDoux and collaborators (1988) shows that neurotoxic lesions of the caudal part of PAG interfere with the behavioral expression of the fear-conditioned response (freezing) but not the autonomic-conditioned responses (increases in arterial pressure). These findings suggest a dissociation of lesion effects, with the PAG involved only in the motoric expression of fear (supposedly by linking the AC to appropriate motor systems).

In order to search for the neurohumoral basis of the defensive reactions observed following PAG and MH electrical stimulation, the behavioral effects of brain drug microinjections have been systematically studied. Sound experimental evidence for a participation of GABA, 5-HT, opioids, and excitatory amino acids in the modulation of aversion and defense in these brain aversive structures has been obtained (Graeff, 1981, 1990). It has been found that GABA agonists inhibit and GABA-A antagonists mimic the effects of the electrical brain stimulation of these regions (Brandão et al., 1982). These results suggest that GABAergic mechanisms exert a tonic inhibitory effect on the neural substrates commanding aversion. As regards to 5-HT, it was found that either 5-HT or the nonselective 5-HT receptor agonist 5-methoxydimethyltryptamine caused dose-dependent increases in the threshold of aversive electrical stimulation of the midbrain tectum (Schütz et al., 1985). Moreover, the antiaversive effect of 5-HT was antagonized by local pretreatment with the 5-HT receptor blockers metergoline and ketanserin, the later showing a considerably higher affinity for 5-HT_2 than for 5-HT_1 receptors. Schütz and coworkers (1985) additionally showed that the 5-HT uptake blocker zimelidine caused similar antiaversive effects, and further experiments by Audi and collaborators (1988) showed that this effect of zimelidine was antagonized by another 5-HT_2 receptor blocker, ritanserin, indicating a mediation by endogenous 5-HT through 5-HT_2 receptors. Support for a 5-HT_2 participation in the modulation of affective behavior in the midbrain tectum have been given by recent electrophysiological studies showing that the majority of 5-HT receptors in this region belongs to the 5-HT_2 type (Brandão et al., 1991). It is also interesting to note that microinjections of 5-HT_2 receptor blockers alone in the midbrain tectum do not induce the aversive behaviors produced by GABA-A receptor blockers. These findings have further suggested that 5-HT inhibitory modulation would be phasic, whereas GABA modulation would be tonic.

An involvement of excitatory amino acids in the generation and elaboration of aversive states in the midbrain tectum has been suggested from studies showing that these compounds applied to this region cause consistent behavioral and autonomic responses characteristic of the defense reaction (Bandler et al., 1985, 1991). On the other hand, opioid mechanisms seem to exert an inhibitory control on the neural substrates of aversive states. In fact, morphine microinjections into PAG attenuated the behavioral and autonomic responses to midbrain stimulation (Jenck et al., 1986; Brandão et al., 1990). Interestingly, behavioral reactivity and autonomic responses, such as an increase in the mean blood pressure and heart rate response to nociceptive stimulation, were not affected by opioid compounds microinjected in the midbrain tectum (Fig. 1). These results suggest that the responses to central aversive and peripheral noxious stimulation are independently controlled in the midbrain tectum (Brandão et al., 1990).

As in the PAG, systemic injections of GABA-A agonists attenuate the defense reaction elicited by electrical stimulation applied to the MH. Liebman (1985) has shown that several benzodiazepine (BZ) anxiolytics selectively increase the latency to switch off electrical brain stimulation by making a shuttling response. Milani and Graeff (1987) demonstrated that microinjections of midazolam and THIP, GABA-BZ receptor agonists, into the MH raised the aversive threshold of electrical brain stimulation of the same area in a dose-dependent way. In contrast, microinjection of bicuculline, a GABA-A receptor blocker, induced the same pattern of behavior observed with electrical stimulation (Graeff 1990). Additionally, the aversive-like effects of bicuculline were antagonized by pretreatment with either THIP or midazolam. These results suggest that in the MH as well as in the PAG, neurons commanding affective states involved in defense reaction seem to be under tonic GABAergic inhibition.

Concerning serotonergic mechanisms, Leroux and Myers (1975) demonstrated that microinjection of 5-HT in the MH of rats increased their tolerance to aversive electrical stimulation, replicating findings observed in PAG studies.

In the area of aversive learning, Di Scala and Sandner (1989) demonstrated that bilateral microinjections of semicarbazide, a GABA synthesis blocker, into the PAG produce conditioned place aversion that is prevented by the concomitant microinjection of the GABA agonist muscimol, which does not produce conditioned place preference or aversion by itself. These results suggest that the

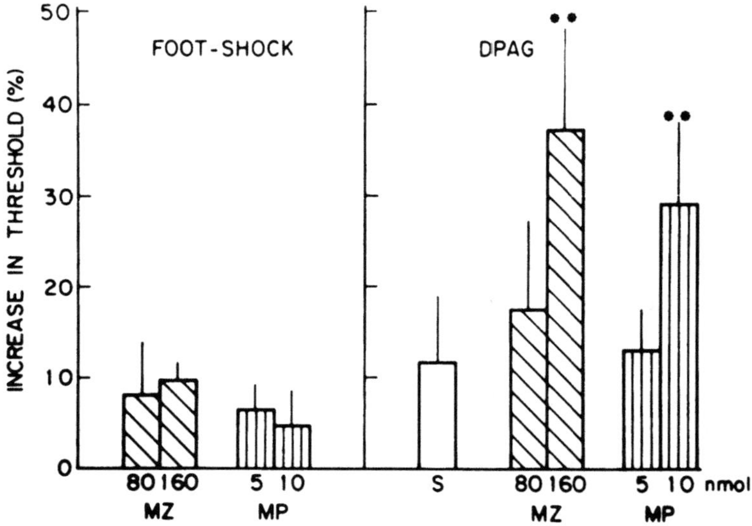

FIGURE 1. Effect of local administration of midazolam (MZ) and morphine (MP) on the aversive thresholds of peripheral noxious stimulation (at left) and of electrical stimulation of PAG of the rat (at right). Increases in threshold are expressed as a percent of pretreatment individual values of the threshold current intensity inducing running or jumps at peripheral or brain stimulations. Columns represent the mean and bars the SEM (N = 8). The behavioral responses were determined 30 min after microinjection of saline (S, control), or of MZ or MP (0.2 μl injected over 30 sec). Vehicle microinjections caused no changes upon the peripheral nociceptive threshold. **$p < 0.01$ compared to control. After Brandão et al., 1990.

blockade of the tonic inhibition exerced by GABAergic terminals in the PAG produces the affective experience of aversion.

Amygdala, Memory, and Emotions

There is a long tradition in associating the AC with a regulatory effect on the states of fear and anger, and defense reactions. Also, several lines of evidence suggest that the AC is involved in the modulation of memory storage, particularly in emotionally based memory (Cahil and McGaugh, 1990; Hitchcock and Davis, 1987; LeDoux et al., 1990). Anatomical studies have shown that the AC receives projections from all sensory modalities, the lateral/basolateral nuclei being the major site of convergence, while efferent fibers depart from the central nucleus to the MH and PAG, as well as to the thalamus, neocortex, hippocampus, striatum, and cells groups of the basal forebrain (Amaral, 1987; Carlsen, 1989; De Olmos et al., 1985). This pattern of connectivity strongly suggests that the AC stands as an interface between hypothalamic and brainstem structures (and the visceral and autonomic functions associated with them), on the one hand, and much of the cerebral cortex (and associated cognitive functions), on the other hand. Thus the internal environment of the body receives equal representation with the outside world in the AC. This anatomical characteristic of the AC has led to the proposal that it is involved in linking information about the sensory aspects of the stimuli processed by the neocortex with the fundamental motivational drive mechanisms and emotions (Mishkin and Aggleton, 1981). Thus, the influence of the AC on learning and memory seems to be inseparable from its role in emotional sphere.

Considering the intraamygdala connections, virtually all of the other amygdaloid nuclei project to the central nucleus. In addition, the central nucleus of the AC receives visceral, taste, and cardiopulmonar information directly from the brainstem (Russchen, 1986). Therefore, the central nucleus provides a way station for the other structures to exert an influence on autonomic and cognitive functions.

The findings that lesions of the AC impair learning of tasks using emotionally arousing training situations suggest the possibility that neurohumoral substances released or activated in response to stress might influence learning by activating the AC. It is well known that stressful stimulation activates neuromodulatory systems, such as opioid peptidergic, GABAergic, and adrenergic systems. Experiments examining the memory-modulating effects of treatments affecting these systems have provided extensive evidence indicating activation of receptors of these neurotransmitters within the AC. Thus, memory retention is enhanced by posttraining intraamygdala injections of norepinephrine (NE) (Liang et al., 1986, 1990). These effects on memory appear to involve activation of the noradrenergic projection to the AC, since they are blocked by injections of the β-adrenergic antagonist propranolol into the AC (Liang et al., 1986). These results are consistent with those reported by Gallagher and coworkers (1981) that found a retention impairment with posttraining intraamygdala injections of the β-adrenergic antagonists propranolol and alprenolol.

Studies examining the memory-modulating effects of treatments affecting GABAergic and opioid peptidergic systems have provided additional evidence suggesting that the effects of these treatments involve the activation of NE receptors in the AC. Findings from different laboratories indicate that systemic as well as intraamygdala injections of opioid antagonists enhance memory retention (Gallagher and Kapp, 1978; Izquierdo, 1979; Tomaz et al., 1990). The view that such effects involve amygdala NE is supported by the finding (McGaugh et al., 1988) that posttraining intraamygdala injections of β-adrenergic antagonists block the memory-enhancing effects of systemically injected naloxone, as the intraamygdala injections of naloxone are blocked when propranolol is injected concurrently with naloxone (Introini-Collison et al., 1989).

There is also extensive evidence indicating that memory can be modulated by systemic as well as intraamygdala GABAergic treatments. When administered shortly after training, GABAergic agonists (muscimol and baclofen) impair retention and GABAergic antagonists (picrotoxin and bicuculline) enhance retention (Breen and McGaugh, 1961; Brioni and McGaugh, 1988; Brioni et al., 1989; Castellano et al., 1989). Recent findings indicate that the memory-modulating effects of systemically administered GABAergic drugs are blocked by lesions of the AC (Ammassari-Teule et al., 1991).

The findings of several experiments suggest that the anxiolytic properties of BZs are due to a facilitation of GABA-mediated neurotransmission in the AC (Nagy et al., 1979; Scheel-Krüger and Petersen, 1982). It is well documented that BZs induce anterograde amnesia in humans and laboratory animals (Lister, 1985; Thiebot, 1985). On the other hand, posttraining intraamygdala, as well as systemic injection of the BZ antagonist flumazenil, causes memory facilitation (Izquierdo et al., 1990). Taken together, these findings suggest that the memory-modulating effects of BZs, like the anxiolytic effects, act via the GABAergic mechanisms in the AC. If this is the case, lesions of the AC should block the amnestic effects of BZs. To examine this possibility one of us (Tomaz et al., 1991) has investigated the effects of systemic-administered diazepam (DZP) on the acquisition and retention of an inhibitory avoidance task in control and AC-lesioned rats. The AC lesions were made with the neurotoxin N-methyl-D-aspartic acid (NMDA) in order to minimize damage to fibers passing through the AC. DZP administered prior to training did not significantly affect the acquisition performance of either the control groups or the AC-lesioned groups. These findings are consistent with other evidence indicating that BZs affect retention without impairing acquisition. However, as is shown in Figure 2, we found DZP-impaired retention in the unoperated and sham control but not in the AC-lesioned animals, showing that the amnestic effects of DZP are blocked by lesions of the AC. These findings suggest that the memory-modulating effects of BZs are mediated, at least in part, through influences involving the AC.

Several studies have documented that specific nuclei of the AC are involved in the anxiolytic properties of BZs. Local infusion of BZs into the basolateral and lateral nuclei produce anxiolytic effects, whereas injections into the medial or central nuclei have little or no effect (Hodges et al., 1987; Petersen et al., 1985; Scheel-Krüger and Petersen, 1982). Thus, going one step further, in an additional study we have addressed this issue by investigating the effects of DZP treatment in control and AC nuclei cell body lesions (Tomaz et al., 1992). Bilateral lesions of the central (CE), lateral (LAT), or basolateral (BL) amygdala were made by infusing ibotenic acid. The retention-test performances are shown in Figure 3. As can be seen, lesions of the BL, but not CE or LAT, nuclei of the AC completely block the retention impairment produced by systemic injections of DZP administered prior to training. The anterograde amnesia produced in sham, as well as CE- and LAT-lesioned, animals may be due to activation of BZ receptors in the BL nucleus of the amygdala. This result indicates that the BL nucleus of the amygdala is an essential structure for DZP influences on memory.

Anatomical literature indicates that the AC has one of the highest densities of BZ receptors in the brain, their density being higher in the BL/LAT nuclei (Nienhoff and Kuhar, 1983; Young and Kuhar, 1980). As mentioned above, specific nuclei of the AC are involved in the anxiolytic action of BZs. For example, local microinjections of low doses of BZs (1 μg of DZP or midazolam) into the BL and LAT nuclei produce anxiolytic effects, whereas infusion into the

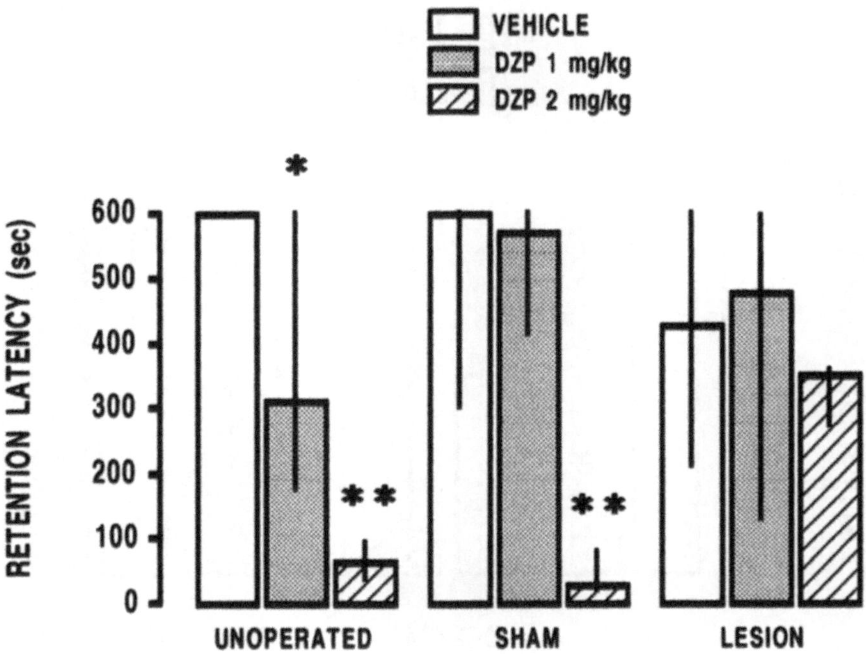

FIGURE 2. Diazepam effect, in NMDA amygdala lesion animals, on the retention test conducted 48 hr after step-through inhibitory avoidance training. Surgical treatment was performed 1 week before behavioral training. Rats were trained 30 min after receiving an injection of either diazepam or vehicle. N = 9–11 per group. Columns and bars represent median step-through latencies and interquartile ranges, respectively. Asterisks indicate the difference from corresponding vehicle control group, $*p < 0.05$, $**p < 0.001$, Mann-Whitney U. After Tomaz et al., 1991.

most anterior part of the AC and in the CE nucleus have little or no effect at all. It is worthy to mention that former studies (Nagy et al., 1979; Shibata et al., 1982) have shown anxiolytic effects of BZs when microinjected into the CE nucleus but not in the BL nucleus. However, it is possible that the higher doses of BZs used in these studies (20 or 25 μg of DZP, 30 μg of midazolam, 50 μg of chlordiazepoxide) led to imprecise localization of the site of anxiolytic effect of BZs in the AC nuclei of rats. Considering everything together, these findings suggest that the memory-modulating effects of BZs as well as the anxiolytic effects of BZs, seem to involve the same or overlapping neuromechanisms in the AC.

Overlapped Substrates of Convulsive Behavior and Aversion

Recently, the inferior colliculus (IC) has been implicated as another area participating in the BAS (Brandão et al., 1988). Interestingly, the IC itself has been considered as one of the most critical areas involved in audiogenic seizures

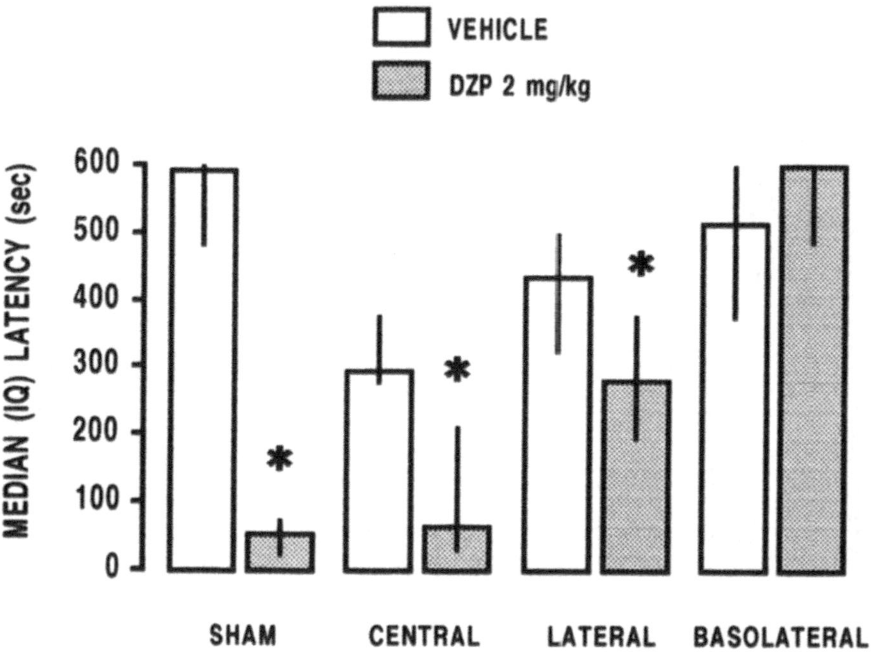

FIGURE 3. Diazepam effect in ibotenic acid amygdala nuclei lesion animals on the retention test conducted 48 hr after step-through avoidance training. Surgical treatment was performed 1 week before behavioral testing. Asterisks indicate the difference from the corresponding vehicle control group, $*p < 0.01$, Mann-Whitney U. After Tomaz et al., 1992.

(AS) (Willott and Lu, 1980; Garcia-Cairasco and Sabbatini, 1991). Then, the question is whether we are referring to some qualitative differences of expression of the same system or of two absolutely different phenomena sharing overlapped substrates. Audiogenic seizures were first described as a model of panic-flight disorders, and experimental data have shown that at least the beginning of AS could be compared to a flight-like pattern (Plotnikoff, 1963). The so-called procursive behaviors denoted an escape reaction through which animals could flee if that option was presented to them (Plotnikoff, 1963). Audiogenic-like seizures are evoked by electrical stimulation (McCown et al., 1987) and by neurochemical activation of the IC by bicuculline and NMDA (Millan et al., 1986). The greatest differences appeared when the seizures so evoked were defined as spontaneous or acoustic facilitated. In our experience higher doses of IC bicuculline evoked overt flight-like behavior, during which the animals are unresponsive to sound stimulation (Garcia-Cairasco et al., 1989). On the other hand, NMDA-evoked audiogenic-like seizures better resemble actual AS, and IC bicuculline produces infrequent spontaneous tonic-clonic seizures (Millan et al., 1986; Garcia-Cairasco et al., 1989).

One of the main questions about the significance of the above-presented data is related to the identification of the neural substrates that underlie these behaviors.

AS are a model of the severe generalized tonic-clonic epileptic syndrome. During the clonic phase vocalization is frequently observed and bizarre motor patterns, such as barrel rolling and clonic spasms are sometimes observed (Garcia-Cairrasco and Sabattini, 1989). Some of these behavioral patterns are the product of activation of neuronal networks widely involved in some affective states, such as flight and aversive responses. The PAG participates in the expression of vocalization, which is typical of fearful situations (Jürgen and Pratt, 1979). Neuroanatomical connections between the IC and PAG are the substrates for that linkage on AS-evoked vocalizations (Herrera et al., 1988). Moreover, acoustically driven emotional memories are processed in a more complex manner through known IC-medial geniculate-amygdala connections (LeDoux et al., 1987, 1990).

Some of the most critical structures that participate in this polymodal integration are the IC cortical nuclei, both the pericentral and external (Aitkin et al., 1975). The IC central nucleus relays primary acoustic information from the lateral lemmniscus to the medial geniculate, and from there to the primary acoustic cortex. On the other hand, the visual extrageniculate pathway relays information from the retina to the superficial layers of the superior colliculus (SC), passing through the pulvinar to the primary visual cortex. However, both IC and SC share common polymodal processing capabilities by receiving multiple inputs from almost the whole brain. Their processing is complex and includes basically the deep layers of the SC and the IC cortical subnuclei. On one hand, the deep layers of the SC participate in a complex network of flight-orientation responses, which could be switched appropriately depending on adaptive needs (Dean et al., 1989). On the other hand, the IC and subcollicular networks are involved in acoustic-evoked startle responses, relevant to the animal's quick awareness or alerting behaviors (Fox, 1979).

In contrast, the SC is part of seizure spreading networks with inputs coming from the substantia nigra reticulata (SNR; Garant and Gale, 1987). Also, the superficial layers or cortical subnuclei of the IC, and not the central nuclei, seem to be the more epileptogenic area in the acoustic midbrain (McCowen et al., 1987). *In vitro* evidence from IC slices indicates that this epileptogenicity is NMDA-dependent (Pierson et al., 1989), in agreement with the NMDA-evoked audiogenic-like responses in freely moving animals (Millan et al., 1986).

A functional interaction between IC and SNR activities has been postulated recently (Garcia-Cairasco and Sabbatini, 1991), indicating that the superior colliculus, reticular formation, and pontine nuclei may be part of sensorimotor interfaces that transform acoustic information coming through the IC into motor responses, the patterns of behavioral AS. IC-SC disconnection by knife cuts blocked the audiogenic-like responses evoked with IC bicuculline (Garcia-Cairasco et al., 1989), indicating that IC-SC connections are necessary parts of this network. Figure 4 illustrates the central position of the cortical subnuclei of the IC as an integrative substrate for midbrain-forebrain interactions in convulsive, aversive, and emotional manifestations.

Some more accurate experiments need to be done to disclose the precise participation of these areas in both aversive and convulsive behavior. The

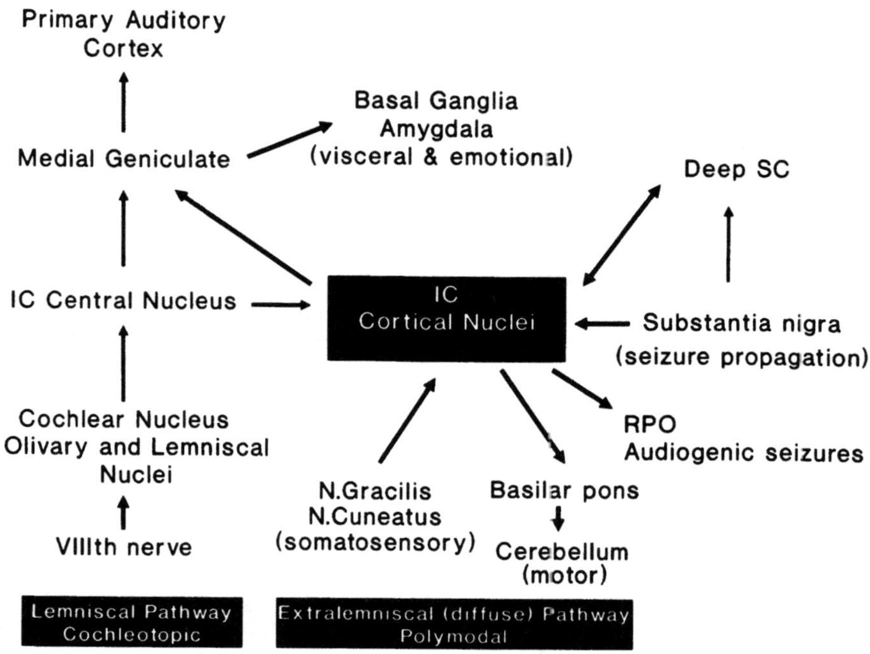

FIGURE 4. Primary acoustic information is processed throughout the lemniscal pathway. Polisensory information is processed through the extralemniscal pathway in which the cortical nuclei of the inferior colliculus play a prominent role. IC = inferior colliculus; RPO = nucleus reticularis pontis oralis; SC = superior colliculus.

interaction between the midbrain and forebrain areas is likely to give us a clue in this regard. For example, when audiogenic-like kindling is evoked by electrical stimulation of the IC cortical nucleus, not only do IC afterdischarges appear (McCown et al., 1987), but so do amygdala and cortical afterdischarges, indicating a recruitment of these areas in the progression of IC kindling. Limbic kindling is a model of temporal lobe epilepsy, and the amygdala has been recognized as one of the areas that is more sensitive to kindling development (Goddard et al., 1969). Once IC kindling is reached, it is more difficult to reach amygdala kindling, indicating some kind of interaction within these two sites (McCown et al., 1987). Moreover, IC afterdischarges are blocked by lidocaine application at the deep prepiriform cortex, one of the most sensitive areas of the forebrain to trigger clonic seizures (Piredda and Gale, 1986). Kindling can be evoked also by chemical means, such as excitatory amino acid microinjections into the amygdala (Croucher and Bradford, 1990), indicating that the electrically induced patterns are probably evoked through these neurochemical systems.

In recent work on cerebral glucose utilization after β-carboline FG 7142 (Pratt et al., 1988), higher doses (10 mg/kg) induced overt convulsions in 30% of the animals. In animals not expressing convulsions, the glucose utilization was increased in limbic structures, such as the mammilary bodies, anterior thalamic nuclei, septal nuclei, and some hippocampal areas. Also, glucose was increased in auditory and visual areas, such as the medial and lateral geniculate bodies and associated cortical areas. In animals expressing overt convulsions, there was an increase of glucose in the cortical amygdala, lateral preoptic area of the hypothalamus, nucleus accumbens, and lateral habenula. Only during seizures did the globus pallidus and substantia nigra present glucose utilization. In conclusion, this work has shown that some limbic structures are functionally involved in both the anxiogenic and convulsive properties of the β-carboline FG 7142. Another related study has shown that kindling induced by FG 7142 allows dissociation of the prolonged changes in seizure threshold from the level of anxiety produced by the drug (Taylor et al., 1988), as demonstrated in elevated social interactions plus Maze and Vogel's conflict tests.

In conclusion, the above-mentioned data indicate a profound interaction between midbrain and forebrain substrates in seizure disorders and aversion. Midbrain-evoked seizures, such as AS, not only share overlapping substrates with aversive states but may in some way affect forebrain substrates and could be affected retrogradely from the same areas. On the other hand, the presence of limbic, particularly amygdala kindling, may share overlapping substrates with aversive phenomena.

Conclusions

Table 1 summarizes a partial list of the neural structures and neurotransmitters involved in the elaboration and expression of defense reactions, affective memories, and convulsive behavior. The experimental evidence reviewed in this chapter suggests a prominent role for GABA as a key neuromodulator in AC, PAG, and MH in the integration of these three processes. The evidence also suggests that none of these brain structures are responsible for the elaboration and expression of

TABLE 1. Partial List of the Most Important Neural Systems Integrating Defense Reactions, Affective Memories, and Convulsive Behaviors

Structure	Function	Neurotransmitter(s)
Adrenal medulla	Internal/visceral information	ADR, NE
Vagus nerve	Visceral/sensory information	GABA, peptides, ACh
Nucleus tractus solitarii	Cardiovascular control	GABA, NE
Periaqueductal gray	Nociception and aversion integration	GABA, opiods, 5-HT
Superior colliculus	Visual reflex signals	GABA, Glu
Inferior colliculus	Acoustic information analyzer	GABA, NMDA
Amygdaloid complex	Integration of sensory information	GABA, NE,ACh
Hippocampus	Context/spatial information	GABA, ACh, Glu
Medial hypothalamus	Neuroendocrine analyser	DA, GABA, NE
Thalamus	Sensory-motor limbic interface	NE, GABA
Prefrontal cortex	Cognitive information processing	ACh

such phenomena. Although these processes have primarily independent neural substrates localized deep in the brain (most likely the midbrain), they interact and overlap in a significant manner at the more rostral level. The evidence accumulated so far strongly supports the view that the AC serves as an interface between the neural systems that produce emotional and convulsive behavior, and those that mediate cognitive functions, especially memory.

Acknowledgments. The researchs from our laboratories reported in this chapter were supported by grants from the Fundação de Amparo a Pesquisa do Estado de São Paulo (FAPESP, Proc no 90/3474-0) and the Conselho Nacional de Desenvolvimento Científico e Tecnológico (CNPq).

References

Adams DB (1979): Brain mechanisms for offense, defense, and submission. *Behav Brain Sci* 2:201–241

Aitkin LM, Webster WR, Veale JL, Crosby DC (1975): Inferior colliculus. I. Comparison of response properties of neurons in central, pericentral, and external nuclei of adult cat. *J Neurophysiol* 38:1196–1207

Amaral D (1987): Memory: Anatomical organization of candidate brain regions. In: *Handbook of Physiology. Sect. 1: Neurophysiology, Vol. 5: Higher Functions of the Brain*, Plum F, ed. Bethesda, MD: American Physiological Society, pp 211–294

Ammassari-Teule M, Pavone F, Castellano C, McGaugh JL (1991): Amygdala and dorsal hippocampus lesions block the effects of GABAergic drugs on memory storage. *Brain Res* 551:104–109

Audi EA, De Aguiar JC, Graeff FG (1988): Mediation by serotonin of the antiaversive effect of zimelidine and propranolol injected into the dorsal midbrain central grey. *J Psychopharmacol* 2:26–32

Bandler R (1988): Brain mechanisms of aggression as revealed by electrical and chemical stimulation: Suggestion of a central role for the midbrain periaqueductal grey region. In: *Progress in Psychobiology and Physiological Psychology*, Vol. 13, Epstein A, Morrison A, eds. New York: Academic Press, pp 67–154

Bandler R, DePaulis A, Vernes M (1985): Identification of midbrain neurons mediating defensive behaviour in the rat by microinjections of excitatory amino acids. *Behav Brain Res* 15:107–119

Bandler R, Carrive P, Zhang SP (1991): Integration of somatic and autonomic reactions within the midbrain periaqueductal grey: Viscerotopic, somatotopic and functional organization. *Prog Brain Res* 87:269–305

Blanchard DC, Blanchard RJ (1972): Innate and conditioned reactions to threat in rats with amygdaloid lesions. *J Comp Physiol Psychol* 81:281–290

Blanchard DC, Williams G, Lee EMC, Blanchard RJ (1981): Taming of wild *Rattus norvegicus* by lesions of the mesencephalic central gray. *Physiol Psychol* 9:157–163

Brandão ML, Aguiar JC, Graeff FG (1982): GABA mediation of the anti-aversive action of the minor tranquilizers. *Pharmacol Biochem Behav* 16:397–402

Brandão ML, Tomaz C, Leao Borges PC, Coimbra NC, Bagri A (1988): Defence reaction induced by microinjections of bicuculline into the inferior colliculus. *Physiol Behav* 44:361–365

Brandão ML, Coimbra NC, Leao Borges PC (1990): Effects of morphine and midazolam on reactivity to peripheral noxious and central aversive stimuli. *Neurosci Biobehav Rev* 14:495–499

Brandão ML, Lopez-Garcia JA, Graeff FG, Roberts MHT (1991): Electrophysiological evidence for excitatory 5-HT_2 and depressant 5-HT_{1A} receptors on neurones of the rat midbrain tectum. *Brain Res* 556:259–266

Breen RA, McGaugh JL (1961): Facilitation of maze learning with posttrial injections of picrotoxin. *J Comp Physiol Psychol* 54:498–501

Brioni JD, McGaugh JL (1988): Posttraining administration of GABAergic antagonists enhance retention of aversively motivated tasks. *Psychopharmacology* 96:505–510

Brioni JD, Nagahara AH, McGaugh JL (1989): Involvement of the amygdala GABAergic system in the modulation of memory storage. *Brain Res* 487:105–112

Cahil L, McGaugh JL (1990): Amygdaloid complex lesions differentially affect retention of tasks using appetitive and aversive reinforcement. *Behav Neurosc* 104:523–543

Carlsen J (1989): New perspectives on the functional anatomical organization of the basolateral amygdala. *Acta Neurol Scand* (Suppl 112)79:4–28

Castellano C, Brioni JD, Nagahara AH, McGaugh JL (1989): Posttraining systemic and intra-amygdala administration of the GABA-B agonist baclofen impair retention. *Behav Neural Biol* 52:170–179

Colpaert FC, Wiepkema PR (1976): Effects of ventromedial hypothalamic lesions on spontaneous intraspecific aggression in male rats. *Behav Biol* 16:117–125

Croucher MJ, Bradford HF (1990): Kindling of full limbic seizures by repeated microinjection of excitatory amino acids into the rat amygdala. *Brain Res* 501:58–65

De Olmos J, Alheid GF, Beltramino CA (1985): Amygdala. In: *The Rat Nervous System, Vol 1, Forebrain and Midbrain*, Paxinos G, ed. Sydney: Academic Press, pp 223–334

Dean P, Redgrave P, Westby GWW (1989): Event or emergency? Two response systems in the mammalian superior colliculus. *Trends Neurosci* 12:137–147

Di Scala G, Sandner G (1989): Conditioned place aversion produced by microinjections of semicarbazide into the periaqueductal gray of the rat. *Brain Res* 483:91–97

Di Scala G, Mana MJ, Jacobs WJ, Phillips AG (1987): Evidence of pavlovian conditioned fear following electrical stimulation of the periaqueductal grey in the rat. *Physiol Behav* 40:55–63

Fernandez de Molina A, Hunsperger RW (1959): Central representation of affective reactions in forebrain and brain stem: Electrical stimulation of amygdala, stria terminalis, and adjacent structures. *J Physiol* (London) 153:251–265

Fox JE (1979): Habituation and prestimulus inhibition of the auditory startle reflex in decerebrate rats. *Physiol Behav* 23:291–297

Fuchs SAG, Edinger H, Siegel A (1985): The organization of hypothalamic pathway mediating affective defense behavior in the cat. *Brain Res* 330:77–92

Gallagher M, Kaap BS (1978): Manipulation of opiate activity in the amygdala alters memory processes. *Life Sciences* 23:1973–1978

Gallagher M, Kapp BS, Pascoe JP, Rapp PR (1981): A neuropharmacology of amygdaloid systems which contribute to learning and memory. In: *The Amygdaloid Complex*, Ben-Ari Y, ed. Amsterdam: Elsevier/North Holland, pp 343–354

Garant D, Gale K (1987): Substantia nigra-mediated anticonvulsant actions: Role of the nigral output pathways. *Exp Neurol* 97:143–159

Garcia-Cairasco N, Sabbatini RME (1989): Neuroethological evaluation of audiogenic seizures in hemidetelencephalated rats. *Behav Brain Res* 33:65–77

Garcia-Cairasco N, Sabbatini RME (1991): Possible interaction between the inferior colliculus and the substantia nigra in audiogenic seizures in rats. *Physiol Behav* 50: 421–427

Garcia-Cairasco N, Terra VC, Tsutsui J, Oliveira JAC (1989): Audiogenic-like and flight behavior after collicular bicuculline are blocked by nigral clobazam and intercollicular transections. Proceedings of the American Epilepsy Society Meeting, *Epilepsia* 30:708

Goddard GV, McIntyre DC, Leech CK (1969): A permanent change in brain resulting from daily electrical stimulation. *Exp Neurol* 25:295–303

Graeff FG (1981): Minor tranquilizers and brain aversive system. *Braz J Med Biol Res* 14:229–265

Graeff FG (1990): Brain defense systems and anxiety. In: *Handbook of Anxiety, Vol. 3: The Neurobiology of Anxiety*, Burrows GD, Roth M, Noyers R Jr, eds. Amsterdam: Elsevier Science Publishers B.V., pp 307–354

Herrera M, Sánchez del Campo F, Ruiz A, Smith Agreda V (1988): Neuronal relationships between the dorsal periaqueductal nucleus and the inferior colliculus (nucleus coomissuralis) in the cat. A Golgi study. *J Anat* 158:137–145

Hess WR, Brugger M (1943): Das subkorticale zentrum der affektiven abwehrreaktion. *Helvetica Physiologica et Pharmacologica Acta* 1:33–52

Hilton SM, Zbrozyna AW (1963): Amygdaloid region for defence reactions and its efferent pathway to the brain stem. *J Physiol* (London) 165:160–173

Hitchock JM, Davis M (1987): Fear-potentiated startle using an auditory conditioned stimulus: Effect of lesions of the amygdala. *Physiol Behav* 39:403–408

Hodges H, Green S, Glenn B (1987): Evidence that the amygdala is involved in benzodiazepine and serotonergic effects on punished responding but not discrimination. *Psychopharmacology* 92:491–504

Introini-Collison IB, Nagahara AH, McGaugh JL (1989): Memory-enhancement with intra-amygdala posttraining naloxone is blocked by concurrent administration of propanolol. *Brain Res* 476:94–101

Izquierdo I (1979): Effect of naloxone and morphine on various forms of memory in the rat: Possible role of endogenous opiate mechanisms in memory consolidation. *Psychopharmacology* 66:199–203

Izquierdo I, Da Cunha C, Huang C, Walz R, Wolfman C, Medina JH (1990): Posttraining down-regulation of memory consolidation by GABA-A mechanisms in the amygdala modulated by endogenous benzodiazepines. *Behav Neural Biol* 54:105–109

Jenck F, Schmitt P, Karli P (1986): Morphine injected into the periaqueductal gray attenuates brain stimulation-induced effects: An intensity discrimination study. *Brain Res* 378:274–284

Jürgen U, Pratt R (1979): Role of the periaqueductal grey in vocal expression of emotion. *Brain Res* 167:367–378

Kaada B, Anderson P, Jansen J (1954): Stimulation of the amygdaloid nuclear complex in unanesthetised cats. *Neurology* 4:48–64

LeDoux JE, Ruggiero DA, Forrest R, Stornetta R, Reis DJ (1987): Topographic organization of convergent projections to the thalamus from the inferior colliculus and spinal cord in the rat. *J Comp Neurol* 264:132–146

LeDoux JE, Iwata J, Cicchetti P, Reis DJ (1988): Different projections of the central amygdaloid nucleus mediate autonomic and behavioral correlates of conditioned fear. *J Neurosci* 8:2517–2529

LeDoux JE, Farb C, Ruggiero DA (1990): Topographic organization of neurons in the acoustic thalamus that project to the amygdala. *J Neurosci* 10:1043–1054

Leroux AG, Myers RD (1975): Action of serotonin microinjected into hypothalamic sites at which electrical stimulation produced aversive responses in the rat. *Physiol Behav* 14:501–505

Liang KC, Juler R, McGaugh JL (1986): Modulating effects of posttraining epinephrine on memory: Involvement of the amygdala noradrenergic system. *Brain Res* 386:125–133

Liang KC, McGaugh JL, Yao HY (1990): Involvement of amygdala pathways in the influence of posttraining amygdala norepinephrine and peripheral epinephrine on memory storage. *Brain Res* 508:225–233

Liebman JC, Mayer DJ, Liebeskind JC (1970): Mesencephalic central gray lesions and fear-motivated behavior in rats. *Brain Res* 23:353–370

Liebman JM (1985): Anxiety, anxiolytics and brain stimulation reinforcement. *Neurosci Biobehav Rev* 9:75–86

Lister RG (1985): The amnestic action of benzodiazepines in man. *Neurosci Biobehav Rev* 9:87–94

McCown TJ, Greenwood RS, Breese GR (1987): Inferior collicular interactions with limbic seizure activity. *Epilepsia* 28:234–241

McGaugh JL, Introini-Collison IB, Nagahara AH (1988): Memory-enhancing effects of posttraining naloxone: Involvement of β-noradrenergic influences in the amigdaloid complex. *Brain Res* 446:37–49

Milani H, Graeff FG (1987): GABA-benzodiazepine modulation of aversion in the medial hypothalamus of the rat. *Pharmacol Biochem Behav* 28:21–27

Millan MH, Meldrum BS, Faingold CL (1986): Induction of audiogenic seizures susceptibility by focal infusion of excitant amino acid or bicuculline into the inferior colliculus of normal rats. *Exp Neurol* 91:634–639

Mishkin M, Aggleton J (1981): Multiple functional contributions of the amygdala in the monkey. In: *The Amygdaloid Complex*, Ben-Ari Y, ed. Amsterdam: Elsevier/North Holland, pp 409–420

Nagy J, Zambo K, Decsi L (1979): Anti-anxiety action of diazepam after intra-amygdaloid application in the rat. *Neuropharmacology* 18:573–576

Nashold BS Jr, Wilson NP, Slaughther GS (1969): Sensations evoked by stimulation in the midbrain of man. *J Neurosurg* 30:14–24

Nienhoff DL, Kuhar MJ (1983): Benzodiazepine receptors: Localization in rat amygdala. *J Neurosci* 10:2091–2097

Olds ME, Olds J (1962): Approach-escape interactions in rat brain. *Am J Physiol* 203:803–810

Olds ME, Frey JH (1971): Effects of hypothalamic lesions on escape behavior produced by midbrain electrical stimulation. *Am J Physiol* 221:8–18

Olivier B (1977): The ventromedial hypothalamus and aggressive behavior in rats. *Aggressive Behav* 3:47–56

Peterson EN, Braestrup C, Schell-Krüger J (1985): Evidence that the anticonflict effect of midazolam in the amygdala is mediated by specific benzodiazepine receptor. *Neurosci Letter* 53:285–288

Pierson MG, Smith KL, Swann JW (1989): A slow NMDA-mediated synaptic potential underlies seizures originating from midbrain. *Brain Res* 486:381–386

Piredda S, Gale K (1986): Role of excitatory amino acid transmission in the genesis of seizures elicited from the deep prepiriform cortex. *Brain Res* 377:205–210

Plotnikoff N (1963): A neuropharmacological study of escape from audiogenic seizures. In: *Psychophysiologie, Neuropharmachologie et Biochimie de la Crise Audiogene*, Paris: Ed. CNRS

Pratt J, Laurie DJ, McCulloch J (1988): The effect of FG7142 upon local cerebral glucose utilization suggest overlap between limbic structures important in anxiety and convulsions. *Brain Res* 475:218–231

Ricardo JA, Koh ET (1978): Anatomical evidence of direct projections from the nucleus of the solitary tract to the hypothalamus, amygdala, and other forebrain structures in the rat. *Brain Res* 153:1–26

Roberts VJ, Cox VC (1987): Active avoidance conditioning with dorsal central gray stimulation in a place preference paradigma. *Psychobiology* 15:167–170

Russchen FT (1986): Cortical and subcortical afferents of the amygdaloid complex. *Adv Exp Med Biol* 1:35–52

Scheel-Krüger J, Petersen E (1982): Anticonflict effect of the benzodiazepines mediated by a GABAergic mechanism in the amygdala. *Eur J Pharmacol* 82:115–116

Schmitt P, Eclancher F, Karli P (1974): Étude des systémes de reforcement negatif et de reforcement positif au niveau de la substance grise centrale chez le rat. *Physiol Behav* 12:271–279

Schütz MTB, De Aguiar JC, Graeff FG (1985): Antiaversive role of serotonin in the dorsal periaqueductal grey matter. *Psychopharmacology* 85:340–345

Shibata K, Kataoka Y, Gomita Y, Ueki S (1982): Localization of the side of the anticonflict action of benzodiazepines in the amygdaloid nucleus of rats. *Brain Res* 234:442–446

Skultety FM (1963): Stimulation of periaqueductal gray and hypothalamus. *Arch Neurol* 8:608–620

Taylor SC, Johnston AL, Wilks LJ, Nocholas JM, File SE, Little HJ (1988): Kindling with the β-carboline FG 7142 suggest separation between changes in seizure threshold and anxiety-related behavior. *Neuropsychobiology* 19:195–201

Thiebot M (1985): Some evidence for amnestic-like effects of benzodiazepines in animals. *Neurosci Biobehav Rev* 9:95–100

Tomaz C, Aguiar MS, Nogueira PJC (1990): Facilitation of memory by peripheral administration of substance P and naloxone using avoidance and habituation learning tasks. *Neurosci Biobehav Res* 14:447–453

Tomaz C, Dickinson-Anson H, McGaugh JL (1991): Amygdala lesions block the amnestic effects of diazepam. *Brain Res* 568:85–91

Tomaz C, Dickinson-Anson H, McGaugh JL (1992): Basolateral amygdala lesions block diazepam-induced anterograde amnesia in an inhibitory avoidance task. *Proc Natl Acad Sci USA* 89:3615–3619

Ursin H, Kaada BR (1960): Functional localization within the amygdaloid complex in the cat. *Electroenceph Clin Neurophysiol* 12:1–20

Wheatley MD (1944): The hypothalamus and affective behavior in cats. A study of the effects of experimental lesions with anatomic correlates. *Arch Neurol* 52:296–316

Willot JF, Lu SM (1980): Midbrain pathway of audiogenic seizures in DBA/2J mice. *Exp Neurol* 70:288–299

Young WS, Kuhar MJ (1980): Radiohistochemical localization of benzodiazepine receptors in rat brain. *J Pharmacol Exp Ther* 212:337–346

16

Potentiation of the Effects of Antimuscarinic Drugs on Behavior by Serotonin Depletion: Specificity and Relation to Learning and Memory

C.H. VANDERWOLF AND D. PENAVA

Introduction

According to a tradition established by the pioneering research of Moruzzi and Magoun (Moruzzi and Magoun, 1949; Moruzzi, 1972; Steriade and Hobson, 1976), an ascending reticulothalamocortical pathway is responsible for the control of cortical activity in relation to consciousness, arousal, and the sleep-waking cycle. More recent research, however, has shown that the Moruzzi-Magoun concept of the ascending reticular activating system is incomplete or inaccurate with respect to both the anatomico-physiological basis of cortical activation and the relation of such activation to behavior (Buzsaki et al., 1988; Riekkinen et al., 1990b; Stewart et al., 1984; Vanderwolf 1988, 1990; Vanderwolf et al., 1990a; Vanderwolf and Robinson, 1981).

This new research has shown that ascending cholinergic projections from the basal forebrain and ascending serotonergic projections from the brainstem play an essential role in the activation of the neocortex and hippocampal formation. The thalamus does not appear to play any essential role in cortical activation. Although ascending cholinergic and serotonergic systems play an important role in the cerebral control of normal waking behavior, they are not essential for the occurrence of the states of sleep and waking as such.

The patterns of cerebral activity that are controlled by ascending cholinergic and serotonergic inputs occur in close correlation with motor activity. During spontaneous waking behaviors, such as walking, swimming, changing posture, or turning or raising the head (Type 1 behavior), cholinergic and serotonergic inputs appear to be active concurrently. Their joint action has a powerful effect on the patterns of unit activity, spontaneous slow waves (the electrocorticogram), and evoked potentials throughout the neocortex and hippocampal formation. Prominent among these effects are the elicitation of rhythmical slow activity (RSA or "theta rhythm") in the hippocampal formation and the elicitation of low-voltage fast activity (LVFA) in the neocortex. Each of these slow wave patterns is accompanied by characteristic local patterns of spontaneous neuronal activity. Hippocampal neurons develop rhythmical synaptic potentials or rhythmical bursts of action potentials in phase with the field potentials of RSA. Neocortical projection cells tend to fire in a continuous irregular pattern with no long-duration synchronized inhibitory pauses during the presence of LVFA.

During such behaviors as standing or lying immobile, face-washing, chewing, gnawing, licking, or shivering (Type 2 behavior), the activity of both ascending cholinergic and serotonergic systems appears to be reduced, but the cholinergic system may be active in isolation, generating RSA or LVFA on its own. The effects are somewhat different in the neocortex as compared to the hippocampus. Thus a normal waking rat, standing immobile, does not display RSA (indicating that neither cholinergic nor serotonergic inputs to the hippocampus are active), but it does often display good LVFA, apparently as a result of a tonic cholinergic input.

If both cholinergic and serotonergic inputs fall to a low level, the hippocampus generates an irregular pattern of slow-wave activity, punctuated by large sharp waves, while the neocortex generates large-amplitude, irregular slow waves or rhythmical spindle activity. Each of these field potential waveforms is associated with distinct patterns of neuronal action potentials.

If either the cholinergic or the serotonergic input is experimentally inactivated, cerebral activity is modified in various ways, but the basic RSA and LVFA patterns remain clearly recognizable in terms of both spontaneous slow-wave activity and unit activity. Therefore, one may speak of an acetylcholine-dependent or cholinergic type of LVFA or RSA, and a serotonin-dependent or serotonergic type of LVFA or RSA. If both cholinergic and serotonergic transmission are inactivated at the same time, LVFA and RSA disappear during spontaneous behavior or during experimental procedures such as the application of noxious stimuli or electrical stimulation of the reticular formation.

The correlation between Type 1 behavior and the occurrence of distinctive patterns of cerebral activity that are dependent on cholinergic and serotonergic neurotransmission suggests a possible causative relation. The occurrence of the neural activity indicated by hippocampal RSA and neocortical LVFA may play a role in the control of Type 1 behavior. One alternative hypothesis, that movement-correlated RSA and LVFA are the result of sensory feedback from movement, can be rejected on the basis of the available evidence.

If cholinergic and serotonergic control of hippocampal RSA and neocortical LVFA plays a role in the control of Type 1 behavior, simultaneous inactivation of central cholinergic and serotonergic activity should produce severe behavioral impairments as a result of a loss of normal cerebral activity. What sort of behavioral impairment should we expect to see? Clues are offered by the vast literature on the effects of surgical lesions of the brain on behavior. Surgical removal of the neocortex, cingulate cortex, and hippocampal formation alters the probability of occurrence and the normal environmental control of behavior but does not produce gross alterations in the topography of most behaviors. Decorticate rats walk, run, climb, swim, wash the face, lick, bite, etc. in nearly normal fashion, but the behaviors often do not occur in normal relation to the environment or to the physiological state, giving the entire performance an appearance of disorganization and aimlessness (Lashley, 1935; Vanderwolf et al., 1978; Whishaw, 1990). In contrast to the effect of cerebral cortical lesions, large lesions of the brainstem, spinal cord, and cerebellum produce gross disturbances of posture and the topography of behavior. These observations suggest that most of the details of the coordination of muscular contractions to produce recognizable motor patterns, such as walking or face-washing, are managed by neurons in the spinal cord, brainstem, and cerebellum. The main role of the cerebral cortex may be to select coordination patterns appropriate to prevailing circumstances.

On the basis of this evidence, one might expect that the loss of cerebral activation and the resulting cerebral cortical dysfunction caused by central cholinergic and serotonergic blockade would produce a syndrome of disorganized, aimless behavior but no gross change in the topography of behavior. Such syndromes as akinesia or coma, which occur as a result of large brainstem or basal diencephalic lesions, would not be expected since blockade of cholinergic and serotonergic transmission can be achieved without gross destruction of these subcortical structures.

In general, the facts are in agreement with these expectations. Behavior has been studied in rats following blockade of central cholinergic and serotonergic transmission by pharmacological or ablation methods. The pharmacological methods include (a) a combination of *p*-cholorophenylalanine (PCPA) with atropine or scopolamine (Richter-Levin and Segal, 1989; Vanderwolf, 1987) and (b) a combination of reserpine and scopolamine or atropine plus amphetamine to reverse the reserpine-induced catalepsy and akinesia (Vanderwolf et al., 1984). Ablation methods have made use of intracerebral injections of 5,7-dihydroxtryptamine (Dickson and Vanderwolf, 1990; Nilsson et al., 1988, 1990; Riekkinen et al., 1990a; Vanderwolf, 1989), combined with scopolamine or with septal lesions that block cholinergic input to the hippocampus. The behavioral effects of combined cholinergic and serotonergic blockade that were observed were widespread and severe, adversely affecting swim-to-platform performance, shock avoidance behavior, hypothalamic self-stimulation behavior, open field behavior, the tendency to walk off the edge of an elevated platform, and the sequential structure of grooming behavior. True potentiation has been observed, since blockade of a single system may have only a small effect or no effect at all, but when combined with blockade of the other system a severe impairment is present. However, as in the case of large cerebral cortical lesions, the animals continue to display normal waking postures and a high level of spontaneous motor activity.

A number of further questions immediately present themselves. Is the potentiating effect of serotonin depletion on the behavioral effect of cholinergic blockade specific to these interventions or is it due to a nonspecific piling up of unrelated neurological impairments? What is the relation between the deficits caused by cholinergic and serotonergic blockade and the amnestic syndromes caused by brain lesions?

Although the anatomical basis of human amnestic syndromes is poorly understood, there is widespread agreement that the temporal lobes and the diencephalon are of particular importance (Milner, 1966; Scoville and Milner, 1957; Mishkin,

1978; Mishkin et al., 1984; Victor et al., 1989). Among temporal lobe structures, the hippocampus is widely suspected of playing a key role in amnesia, but there is strong evidence that projections from the temporal neocortex and pyriform cortex may be even more important (Cirillo et al., 1989; George et al., 1989; Horel, 1978; Horel and Pytko, 1982; Horel et al., 1984; Zola-Morgan et al., 1989).

Consequently, we decided to find out whether depletion of serotonin by PCPA (Koe and Weissman, 1966) would potentiate the behavioral effects of medial thalamic and temporal region destruction in rats in the same way as it does the behavioral effects of cholinergic muscarinic blockade. The effect of mammillary body lesions was also examined, since this structure, too, has been implicated in amnesia (Markowitsch, 1985). In addition, studies were done on the behavioral effects of combinations of classical serotonergic antagonists and scopolamine, an anti-muscarinic drug.

Methods

Behavior

Male hooded or albino rats ranging in age from 2 to 23 months were tested in an apparatus consisting of a plate glass aquarium measuring 43 × 90 cm × 45 cm deep that was filled with clean water at 20°C to a depth of 25 cm. A wire-mesh rat cage with a metal lid was placed in the center of the aquarium so that its exposed upper mesh surface provided a platform (21.5 × 18.5 cm) raised about 1 cm above the surface of the water. On each test trial a rat was placed in the water facing one corner of the tank. The time taken to climb up on the central platform was recorded with a stopwatch; times in excess of 10.0 sec were considered errors. A count was also kept of the number of times each rat swam the length of the aquarium, passing the platform with all four limbs without climbing up. However, since the number of such passes correlates highly with the number of errors, only data on errors will be presented. If a rat failed to climb up on the platform within 60 sec, it was placed on the platform manually. Ten trials were given in a single session, with an intertrial interval of 10–15 sec. Following training, rats were allowed to dry off in a box warmed by a 100 W light bulb. Retention of swim-to-platform behavior was tested by giving 10 additional trials after an interval of 15 min or 24 hr.

Surgery, Histology, and Drug Treatments

Using the Paxinos and Watson (1986) atlas as a guide, a radio frequency heating device was used to make brain lesions in rats anesthetized with pentobarbital (50 mg/kg). Medial thalamic lesions were made by placing the electrode in the midline thalamus and maintaining it at 60–70° for 60 sec. The hippocampus, amygdala, and overlying cortex were damaged or destroyed by maintaining the electrode at 60°C for 60 sec at each of three or five locations in each hemisphere. The white matter and deep layers of the neocortex and pyriform cortex adjacent to the hippocampus were deliberately included in these lesions on the basis of evidence that such damage may also be involved in human temporal lobe amnesia (Horel, 1978). Lesions of the mammillary body were created by maintaining the electrode at 60°C for 60 sec. A sham-operated group was not included, since previous research (Dickson and Vanderwolf, 1990) has shown that a sham neurosurgical operation has no effect on swim-to-platform performance.

At the conclusion of the experiments (3–4 weeks after surgery) all rats that had received brain lesions were killed by an overdose of pentobarbital and perfused through the heart with a 10% formalin and 0.9% saline solution. The brains were removed, stored in the fixative for 24 hr, frozen, and sectioned at 40 μm. Brains containing large lesions (temporal region lesions involving amygdala, hippocampus, neocortex, and pyriform cortex) were embedded in gelatin prior to freezing and sectioning. Sections were mounted on slides, stained with gallocyanin, examined microscopically, and in some cases, photographed.

p-Chlorophenylalanine (PCPA) was prepared as a fine suspension (50 mg/ml) in a 0.5% solution of gum arabic and injected intraperitoneally (500 mg/kg) 3 days prior to behavioral testing. Scopolamine hydrobromide and scopolamine methylbromide were dissolved in saline and injected subcutaneously in the nape of the neck. Methysergide bimaleate, also dissolved in saline, was injected intraperitoneally. Pizotifen and ritanserin were dissolved in a few drops of glacial acetic acid, diluted with saline (final pH 3.5–5.0), and injected intraperitoneally. The relatively low solubility of methysergide, pizotifen, and ritanserin necessitated the use of large injection volumes. Control groups received equivalent volumes of the drug vehicle alone.

Drugs other than PCPA were usually given 15–30 min prior to the first behavioral test session. However, in one series of experiments, scopolamine was given immediately after the first 10-trial test session, and a second 10-trial session was given 15–30 min later.

Statistical Tests

Data were analyzed with the help of nonparametric tests (Siegel, 1956).

Results

Normal rats acquired the habit of swimming to the platform readily and with great consistency. Table 1 shows data obtained from 63 normal rats ranging in age from 2 to 23 months. Of these, 36 rats had been used in a published study (Vanderwolf et al., 1990b) and 27 were used in a hitherto unpublished study of the effects of age on the test. Age was found to have no detectable effect provided that the aged rats were still able to swim (one 23-month-old rat could not be used owing to a IVth ventricle tumor, which prevented normal swimming) and 62 of the 63 rats made 0–3 errors in 10 trials, achieving a mean of 1.6 ± 0.1 (Table 1). Fifty-four of the 63 rats (85.7%) made either one or two errors, six rats made 3 errors, and two rats made no errors, but one normal rat made 5 errors during persistent attempts to

TABLE 1. Number of Errors in 10 Trials in a Swim-to-Platform Test in Rats Following Various Treatments

Group	N	Errors (mean ± S.E.M.)
1. Normal, no treatment	63	1.6 ± 0.1
2. Drug vehicle, one injection	6	1.4 ± 0.2
3. Drug vehicle, two injections	6	1.5 ± 0.2
4. PCPA	8	1.4 ± 0.3
5. Scopolamine (0.5 mg/kg)	16	3.8[a] ± 0.5
6. Scopolamine (5.0 mg/kg)	16	7.3[ab] ± 0.8
7. Methylscopolamine (5.0 mg/kg)	8	1.4 ± 0.2
8. PCPA plus scopolamine (0.5 mg/kg)	9	7.6[ab] ± 0.9
9. PCPA plus scopolamine (5.0 mg/kg)	10	9.4[ab] ± 0.3
10. Medial thalamic lesion plus drug vehicle	21	4.1[a] ± 0.8
11. Medial thalamic lesion plus PCPA	22	5.0[a] ± 0.8
12. Temporal region lesion plus drug vehicle	17	3.4[a] ± 0.4
13. Temporal region lesion plus PCPA	17	3.5[a] ± 0.5
14. Mammillary body lesion plus drug vehicle	9	2.9[a] ± 0.6
15. Mammillary body lesion plus PCPA	8	2.8[a] ± 0.7

PCPA = *p*-chlorophenylalanine, 500 mg/kg. PCPA and the corresponding control injection were given intraperitoneally in a dose of 10.0 ml/kg; scopolamine and the corresponding control injection were given subcutaneously in a dose of 1.0 ml/kg. Thus Groups 5–7 received only a subcutaneous injection; Groups 3, 8, and 9 received an intraperitoneal plus a subcutaneous injection; Groups 2, 4, and 10–15 received only an intraperitoneal injection.
[a]Differs from Group 1, $p < .0001$ or better
[b]Differs from Group 5, $p < .02$ or better; Mann-Whitney test.
Data for Groups 1, 2, 4, 5, 6, and 7 are taken, partly or totally, from work reported in previous publications (Vanderwolf, 1987, 1991; Vanderwolf et al., 1990a,b).

escape from the apparatus. It appears, then, that rats that make 4 or more errors in 10 trials are likely to be abnormal.

Injection of drug vehicles or a single dose of PCPA had no effect on performance, but scopolamine injections resulted in a clear dose-related impairment. Methylscopolamine, an antimuscarinic drug that does not penetrate the blood-brain barrier, had no effect. A combination of PCPA with a moderate dose of scopolamine (0.5 mg/kg) produced a doubling of the number of errors compared to the scopolamine-alone condition, and a combination of PCPA with a large dose of scopolamine (5.0 mg/kg) produced a condition in which errors were made on virtually every trial (Table 1).

Brain Lesions

Medial thalamic lesions (N = 43) were varied in size from small (destruction of the parataenial and periventricular nuclei with some damage to the medial aspect of the nucleus medialis dorsalis and the overlying parts of the hippocampal region CAI and the dentate gyrus) to large (total or near-total destruction of medial, midline, and intralaminar nuclei plus some damage to the nucleus anterior medialis). Figure 1 shows a representative large medial thalamic lesion.

After a recovery period of 2–3 weeks, rats with medial thalamic lesions were given either a single injection of PCPA or an equivalent volume of drug vehicle.

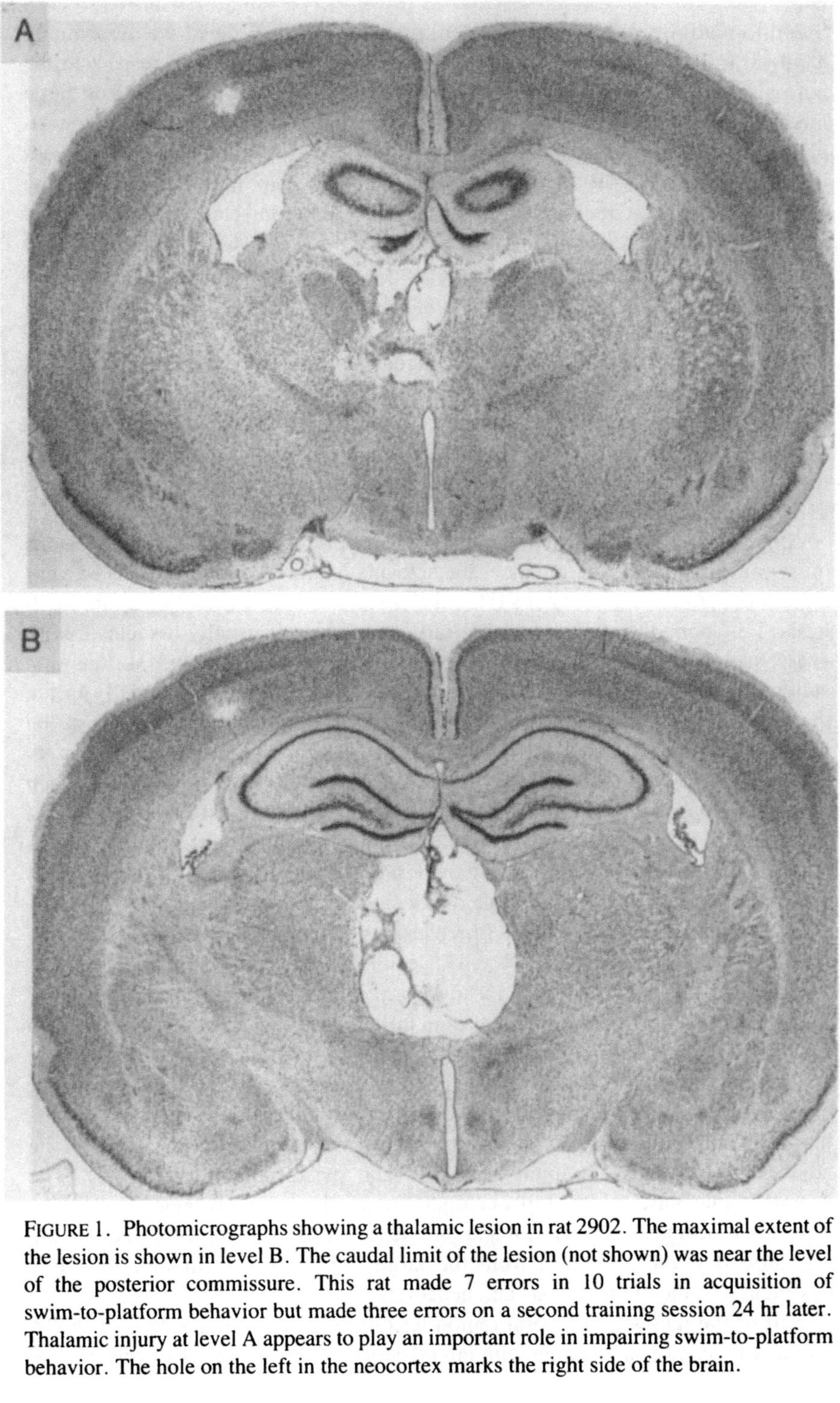

FIGURE 1. Photomicrographs showing a thalamic lesion in rat 2902. The maximal extent of the lesion is shown in level B. The caudal limit of the lesion (not shown) was near the level of the posterior commissure. This rat made 7 errors in 10 trials in acquisition of swim-to-platform behavior but made three errors on a second training session 24 hr later. Thalamic injury at level A appears to play an important role in impairing swim-to-platform behavior. The hole on the left in the neocortex marks the right side of the brain.

Swim-to-platform performance, tested after a delay of 3 days, was essentially identical in the two groups, and both groups were impaired with respect to the normal group (Table 1). Two subgroups of these rats (all these rats had large medial thalamic lesions) were given a second training session in the swim-to-platform task after a delay of 24 hr (data not shown in Table 1). The vehicle-injected group (N = 9) averaged 6.9 ± 1.2 errors in the first session and 4.2 ± 1.2 errors in the second session. The PCPA-injected group (N = 10) averaged 7.0 ± 1.3 errors in the first session and 5.4 ± 1.6 errors in the second session. There is no significant difference between the groups at any time, but there is a significant overall improvement from the first to the second session ($p < .01$, Wilcoxon test).

The remaining medial thalamus-damaged rats, most of which had sustained only small or medium-sized lesions, were given a second training session 15 min after the first session, and (in some cases) a third session 24 hr after the first one. In general, performance was better on the second and third sessions as compared to the first one, but there was no suggestion of any real difference between PCPA-treated and vehicle-treated rats.

Although PCPA treatment did not increase the effect of medial thalamic lesions on swim-to-platform performance, the size and location of the lesion had a very significant effect. Small lesions (N = 15) were associated with 1–3 errors in 10 trials, i.e., normal performance, but large lesions were usually associated with grossly impaired performance (5–10 errors in 10 trials). However, six rats in a subgroup of nine in which a large lesion was placed rather far caudally in the thalamus each made only one error in 10 trials. The remaining three rats of the subgroup made 7–10 errors in 10 trials. Rats in the second subgroup (N = 14) with a large lesion placed more rostrally in the thalamus were severely impaired, each making 5–10 errors in 10 trials. The area that apparently must be destroyed if a large medial thalamic lesion is to impair swim-to-platform performance consistently is illustrated in level A in Figure 1. Precise identification of the essential anatomical structures is not possible on the basis of the data available.

Thirty-four rats sustained small (three lesions per hemisphere, N = 17) or large (five lesions per hemisphere; N = 17) lesions of the hippocampal formation, amygdala, adjacent neocortex, and pyriform cortex (temporal region lesions). These lesions destroyed the amygdala and inflicted varying amounts of damage on the caudal substantia innominata, globus pallidus, internal capsule, optic tract, and ventral geniculate nucleus. Damage to the substantia innominata was never extensive. The large lesions destroyed most of the caudal two thirds of the hippocampal formation and the white matter and deep cellular layers of the adjacent parietal and occipital neocortex and pyriform cortex (Fig. 2). The suprathalamic portion of the hippocampal formation was largely intact in all cases. The small lesions, closely comparable to those described by Dickson and Vanderwolf (1990), differed from the large lesions mainly in involving less destruction of the hippocampus and neocortex.

Half of the rats with temporal region lesions received a single dose of PCPA (eight with small lesions; nine with large lesions), and the other half received an

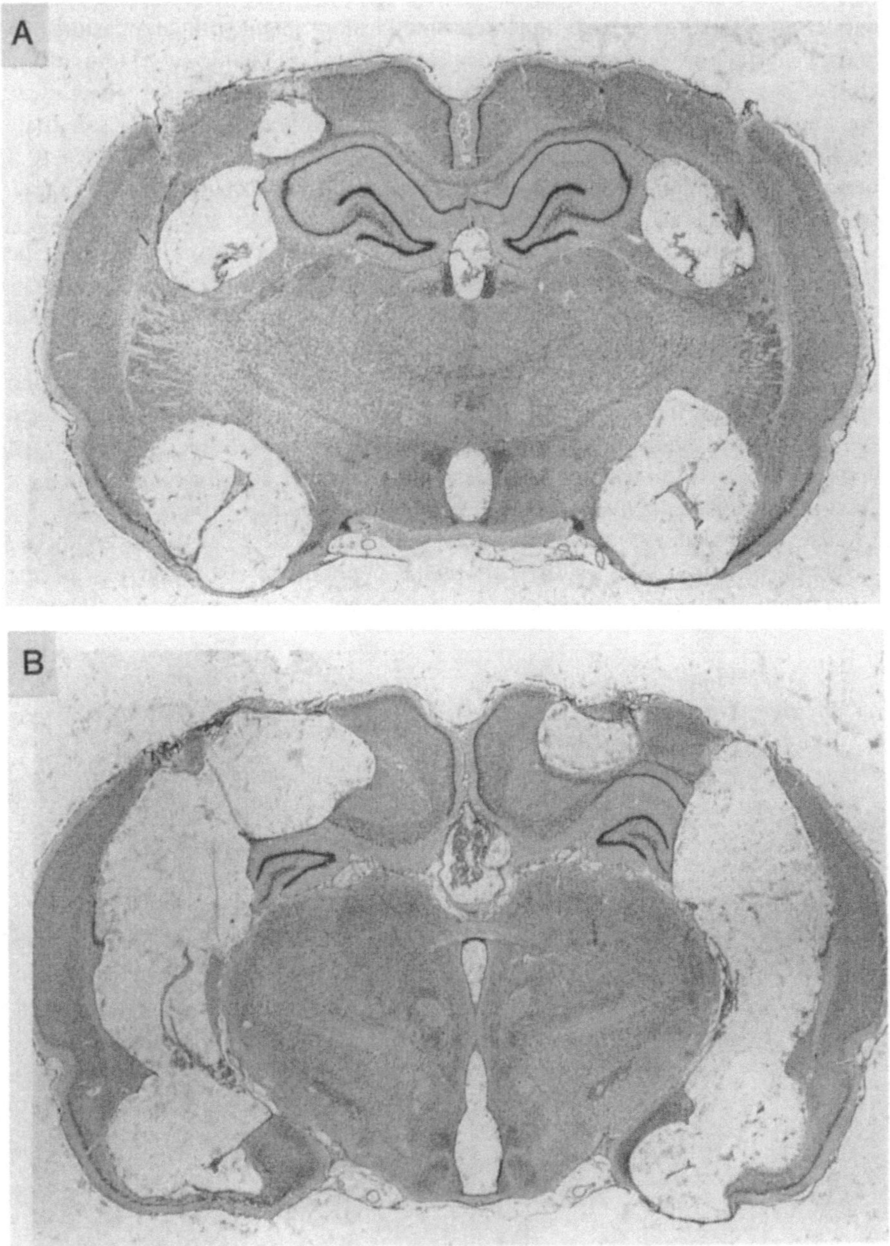

FIGURE 2. Photomicrographs showing damage to the hippocampus, neocortex, pyriform cortex, amygdala, and some adjacent structures in rat 2962. Note ventricular dilatation. Level A shows the maximal extent of the amygdala lesion; level B shows the maximal extent of the hippocampal and neocortical lesion. This rat made two errors in 10 trials in acquisition of swim-to-platform behavior and no errors in a 10-trial retention test 15 min later.

injection of the drug vehicle (nine with small lesions; eight with large lesions). A test 3 days later revealed virtually identical performance in the swim-to-platform test (Table 1). PCPA did not enhance the effect of the lesions. There was, however, a relation between lesion size and swim-to-platform performance (data not shown in Table 1). The rats with large lesions (N = 17) averaged 4.5 ± 0.4 errors, differing significantly from the normal group ($p < .00006$; Mann-Whitney test), while the rats with small lesions (N = 17) averaged 2.4 ± 0.3 errors, a slight but significant ($p < .01$) impairment as compared to the normal group. The large and small temporal region lesion groups also differ significantly ($p < .002$).

Retention of swim-to-platform behavior was tested after an interval of 24 h in the rats with large temporal region lesions (data not shown in Table 1). The vehicle-treated group (N = 8) averaged 4.3 ± 0.5 errors in the first session and 0.6 ± 0.4 errors in the second session. The PCPA-treated rats (N = 9) averaged 4.7 ± 0.6 errors in the first session and 1.1 ± 0.2 errors in the second session. All rats performed better in the second session than in the first session, but there is no hint of a difference between the PCPA-injected and control groups.

The lesions of the mammillary body (N = 17) were less than complete, generally involving the caudal half of the mammillary body and part of the supramammillary area. The largest lesion of the entire series is shown in Figure 3.

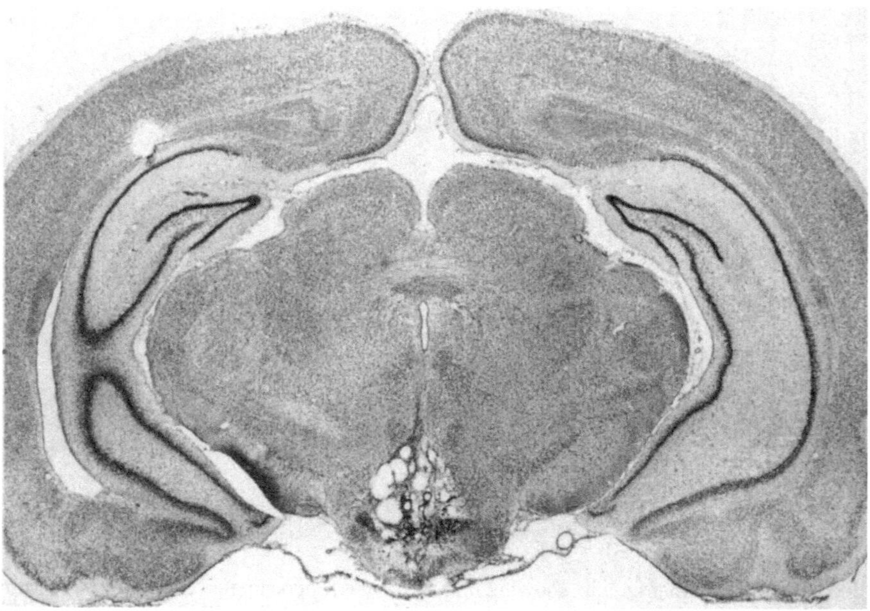

FIGURE 3. Photomicrograph showing damage to the mammillary body and supramammillary region in rat 2862. The maximal extent of the lesion is shown. This rat made three errors in 10 trials in acquisition of swim-to-platform behavior, and no errors in a 10-trial retention test 15 min later. The hole on the left in the neocortex marks the right side of the brain.

One group of these rats (N = 8) received a single dose of PCPA, and the remaining rats (N = 9) received an injection of the drug vehicle. The two groups were virtually identical in swim-to-platform performance, tested 3 days later, but both groups were significantly impaired with respect to normal rats ($p < .0001$, Table 1).

Serotonin Antagonists

Rats injected intraperitoneally with a drug vehicle 15–30 min before training (Table 2; Groups 1, 2, 9a, 10a) or on three successive days prior to training (Group 11a) all performed at approximately the same level, regardless of the volume injected (1–10 ml/kg) or of the nature of the drug vehicle (saline in Groups 1, 2, 9a, 10c; saline plus gum arabic in Group 11a). Immediately after training, scopolamine or saline was injected subcutaneously and the rats were retrained after a delay

TABLE 2. Effects of Some Serotonergic Antagonists, *p*-Chlorophenylalanine, and Scopolamine on the Number of Errors Made by Rats in a Swim-to-Platform Test

Group number	First injection (i.p.)	Training (mean errors ± S.E.M.)	Second injection (s.c.)	Retraining (Mean errors ± S.E.M.)	N
1.	Vehicle (1.0 ml/kg)	0.9 ± 0.1	Vehicle (1.0 ml/kg)	0.1 ± 0.1	8
2.	Vehicle (1.0 ml/kg)	1.1 ± 0.1	Scopolamine (5.0 mg/kg)	1.8 ± 0.4[a]	12
3.	Methysergide (5.0 mg/kg)	1.0 ± 0.2	Scopolamine (5.0 mg/kg)	1.2 ± 0.9[a]	10
4.	Methysergide (10.0 mg/kg)	0.8 ± 0.1	Scopolamine (5.0 mg/kg)	2.1 ± 0.7[a]	12
5.	Pizotifen (5.0 mg/kg)	1.4 ± 0.2	Scopolamine (5.0 mg/kg)	5.5 ± 1.0[ab]	11
6.	Pizotifen (10.0 mg/kg)	2.1 ± 0.4	Scopolamine (5.0 mg/kg)	4.5 ± 1.2[a]	12
7.	Ritanserin (0.1 mg/kg)	1.6 ± 0.3	Scopolamine (5.0 mg/kg)	1.8 ± 0.8[a]	10
8.	Ritanserin (1.0 mg/kg)	1.5 ± 0.3	Scopolamine (5.0 mg/kg)	2.3 ± 0.8[a]	10
9a.	Vehicle (2.0 ml/kg)	1.4 ± 0.2	Scopolamine (5.0 mg/kg)	3.8 ± 1.5[a]	8
9b.	Methysergide (20.0 mg/kg)	1.5 ± 0.2	Scopolamine (5.0 mg/kg)	7.0 ± 1.0[a]	6
10a.	Vehicle (10.0 ml/kg)	1.0 ± 0.4	Scopolamine (5.0 mg/kg)	2.3 ± 0.7[a]	6
10b.	Pizotifen (20.0 mg/kg)	1.8 ± 0.4	Scopolamine (5.0 mg/kg)	2.0 ± 0.9[a]	5
10c.	Ritanserin (10.0 mg/kg)	1.6 ± 0.2	Scopolamine (5.0 mg/kg)	4.1 ± 0.8[a]	8
11a.	Vehicle (5.0 ml/kg × 3)	1.4 ± 0.2	Scopolamine (5.0 mg/kg)	1.3 ± 0.9[a]	11
11b.	PCPA (500 mg/kg)	1.4 ± 0.2	Scopolamine (5.0 mg/kg)	4.3 ± 1.0[ad]	8
11c.	PCPA (500 mg/kg × 3)	2.6 ± 0.7[c]	Scopolamine (5.0 mg/kg)	6.0 ± 1.0[ae]	10

PCPA = *p*-chlorophenylalanine, 500 mg/kg, in a single dose (Group 11b) or three daily doses, (Group 11c). Groups 1–8 received 1.0 ml/kg doses; 9a & 9b received 2.0 ml/kg doses; 10 a–c received 10.0 ml/kg doses; 11a–c received 10.0 ml/kg doses.

Comparisons within columns using the Mann-Whitney test: [a]differs from Group 1, $p < .02$ or better; [b]differs from Group 2, $p < .02$; [c]differs from Group 11a, $p < .05$; [d]differs from Group 11a, $p < .02$; [e]differs from Group 11a, $p < .001$.

During retraining: Groups 2, 9a, and 10a vary significantly, $p < .02$; and Groups 3, 4, and 9b vary significantly, $p < .01$; Kruskal-Wallis one-way analysis of variance.

Data in Groups 11a and 11c are taken from Vanderwolf (1987).

of 15–30 min. Group 1 performed better ($p < .05$, sign test) during retraining (saline vehicle plus saline vehicle) than during training (vehicle, one dose only) but Group 2 (saline vehicle, 1.0 ml/kg, 15–30 min before training) and Group 11a (saline plus gum arabic vehicle, 5.0 ml/kg, on 3 successive days prior to training) performed at about the same level during training (vehicle alone) and during retraining (vehicle plus scopolamine). During retraining, performance in Group 2 (vehicle plus scopolamine) is significantly ($p < .02$) worse than performance in Group 1 (vehicle plus vehicle). Rats given larger injections of saline vehicle (Groups 9a and 10a) showed impaired performance during retraining after scopolamine injection. A Kruskal-Wallis one-way analysis of variance revealed significant ($p < .02$) between-group variation in Groups 2, 9a, 10a during retraining. These data permit two conclusions: (a) Varying volumes of drug vehicle injected alone either a day or more before training or 15–30 min before training have no effect on acquisition of swim-to-platform behavior. (b) Large volumes of drug vehicle given 15–30 min before training tend to potentiate the deleterious effect of scopolamine on performance in a subsequent retraining session.

A single dose of PCPA given 3 days prior to training had no effect by itself but potentiated the effect of scopolamine on subsequent retraining (Table 2, Group 11b). Three daily doses of PCPA tended to impair performance in subsequent training and increased the deleterious effect of scopolamine on subsequent retraining (Table 2, Group 11c).

Treatment with ritanserin and pizotifen had little or no consistent effect on performance in the swim-to-platform test. Kruskal-Wallis tests revealed no significant variation between the three different dosage groups for either drug, either during training (test drug alone) or during retraining (test drug plus scopolamine). Pizotifen did appear to potentiate the effect of scopolamine at a dose of 5.0 mg/kg (Table 2, Group 5), but no comparable significant effect was observed at 10.0 or 20.0 mg/kg (Table 2, Groups 6 and 10b). The Kruskal-Wallis test revealed significant variation ($p < .01$) between the three dosage groups of methysergide during retraining (after scopolamine treatment). However, this effect may be largely due to the increased volume injected at the highest dose of methysergide, since Mann-Whitney tests did not reveal any significant differences between any of the methysergide groups and the appropriate injection volume-equated control groups. Similarly, the poor performance after ritanserin, 10.0 mg/kg, plus scopolamine (Group 10c) does not differ significantly from the appropriate injection volume-equated control performance (Group 10a). Consequently, the data indicate that methysergide, pizotifen, and ritanserin do not have a significant effect on swim-to-platform behavior and do not potentiate the effect of scopolamine.

Discussion

The results show that lesions of varying size in the medial thalamus, temporal region (mainly hippocampal formation, amygdala, and adjacent cortex), or mammillary body impair swim-to-platform behavior but that this impairment is not increased by treatment with a single dose of PCPA. On the other hand, consistent with previous reports (Richter-Levin and Segal, 1989; Vanderwolf, 1987), the impairment produced by a moderate dose of scopolamine (0.5 mg/kg) was markedly increased by PCPA, even when the dose of PCPA used had no effect by itself (potentiation). The effect of a large dose of scopolamine (5.0 mg/kg) on retention of swim-to-platform behavior was also potentiated, as shown in Table 2, but no potentiation was observed in the case of acquisition, probably as a result of a ceiling effect. The potentiation of the effect of a large dose of scopolamine on relatively long-term retention (7 days) of swim-to-platform behavior has been demonstrated previously (Vanderwolf et al., 1990b).

The fact that PCPA potentiates the behavioral impairment produced by scopolamine, but not the behavioral impairment produced by large diencephalic and telencephalic lesions, suggests that the former effect is not due to a multiplication of unrelated neurological deficits. Rather, it is likely that cholinergic and serotonergic neurons participate in a common function in the CNS. A moderate impairment in one of these systems can be compensated if the other one is intact, but if both systems are simultaneously impaired such compensation does not occur and a severe behavioral impairment is the result. These behavioral observations are in agreement with previous electrophysiological work (see Introduction), which has shown that the cholinergic and serotonergic projections are jointly responsible for activation of the cerebral cortex.

The electrophysiological work has also indicated that ascending noradrenergic projections do not play any direct role in the regulation of cortical activation (Vanderwolf, 1988). Consistent with this, depletion of brain noradrenalin has little or no potentiating effect on the behavioral deficits produced by cholinergic blockade (Decker and Gallagher, 1987; Decker and McGaugh, 1989; Prior, 1991; Spangler et al., 1990).

The experiments on the possible potentiating effect of the serotonergic antagonists methysergide, pizotifen, and ritanserin on scopolamine-induced behavioral impairment were largely negative. Such effects as were observed appeared to have been largely due to the necessity of injecting rather large volumes of the drug vehicle. Although no definite explanation can be given for the impairment of behavioral performance following large intraperitoneal doses of drug vehicle plus scopolamine, one of us (CHV) has observed that, in anesthetized rats, intraperitoneal injections of saline tend to result in acute falls in systemic blood pressure. This effect might contribute to an impairment of behavioral performance if it is also present in freely moving rats.

It may seem surprising that methysergide, pizotifen, and ritanserin, in doses that produce strong effects on peripheral serotonergic actions and on the behavioral syndrome elicited by serotonergic agonists (Dixon et al., 1977; Green and Heal, 1985; Green and Backus, 1990), do not appear to potentiate the effects of scopolamine as PCPA does. However, this finding is consistent with the fact that methysergide, pizotifen, ritanserin, and a number of other currently available serotonergic antagonists do not block serotonin-dependent hippocampal RSA and neocortical LVFA (Hargreaves et al., unpublished; Vanderwolf, 1984; Robertson et al., 1991). It is likely, as pointed out originally by Haigler and Aghajanian (1974), that most of the available serotonergic antagonists do not block all of the cerebral effects of serotonin effectively, even though they do antagonize peripheral and spinal serotonergic effects.

Learning and Memory

The brain lesion and drug effects that were investigated in our research have often been discussed in relation to learning and memory. Can the impaired swim-to-platform performance caused by treatment with scopolamine and PCPA, or lesions of the medial thalamus, temporal region, or mammillary body, be attributed to deficits in learning and memory? Such a question is not easily answered since it is not clear what it means.

A common-sense view of memory is that it is a property or part of the mind, having much the same status as other mental entities, such as perception, imagination, emotion, cognition, motivation, or attention. This view appears to be of ancient origin, since it is presented in a well-developed form in Aristotle's writings (Hicks, 1965). The emergence of histology and the cell theory in the 19th century gave rise to an alternative view of memory. Ramon y Cajal (DeFelipe and Jones, 1988) proposed that learning and memory involved changes in the patterns of connectivity between neurons. These two conceptions of learning and memory are often assumed to amount to much the same thing, but in practice they are quite different. For example, the fact that delayed match-to-sample (or non-match-to-sample) performance is impaired by a temporal lobe lesion is often discussed in relation to "memory", but no one supposes that such a finding can tell us anything specific about the location of the synapses that are presumably modified as a result of training in this type of task. Since all of the considerable technical capacity of contemporary neuroscience has, so far, failed to pinpoint the synaptic modifications believed to underlie memory in the mammalian brain, it is inconceivable that such a crude technique as studying the behavioral effects of removing large chunks of brain could accomplish this. In recognition of this, it is widely accepted that it is difficult or impossible to distinguish behaviorally between actual loss of stored information and failure to retrieve stored information after a brain injury. If the amnesia is only temporary, or if apparently lost information can be retrieved by special test procedures (Russell, 1971; Warrington and Weiskrantz, 1970), then an impairment of retrieval is suggested. However, if amnesia is permanent and cannot be overcome by reminder cues and the like, then it is impossible to decide if memory traces are truly absent or present but irretrievable.

Despite these restrictions, experiments involving animal training plus brain lesions attract wide attention among those who profess an interest in learning, memory, or neural plasticity. This may be a result of a widespread implicit acceptance of ancient philosophical doctrines concerning the subdivisions of the psyche. Most people accept the intellectual legitimacy of asking whether a given behavioral deficit is due to an impairment of perception, cognition, attention, motivation, or emotion. If we question this on the grounds that the relevance of these ancient philosophical categories to brain function is an unproven hypothesis (Vanderwolf, 1983), we must ask ourselves: Precisely what conclusions can be justified on the basis of experiments involving behavior plus brain lesions?

A fundamental point, now generally accepted by animal behaviorists, is that performance in behavioral training tasks is dependent partly on learning and partly on the preexisting instinctive propensities of the test animals (Hinde and Stevenson-Hinde, 1973; Seligman and Hager, 1972; Shettleworth, 1972; Tinbergen, 1951). Presumably, this means that good performance in conditioning and learning tasks is partly due to environmentally dependent plastic changes in neural connectivity and partly due to the prior development of genetically controlled sensori-motor linkages that have evolved in the species being tested. It appears that if animals are to peform well, they must have an instinctive ability to deal with the task presented. Training apparently produces only secondary modifications of preexisting behavioral tendencies. In agreement with this, histological studies of the cerebral effects of various training or rearing conditions indicate that development of nerve cell processes and synapses is mainly determined by genetic factors and that variations in individual experience have a smaller effect (Diamond, 1988; Greenough and Volkmar, 1973; Turner and Greenough, 1985).

In terms of brain organization, these considerations suggest that behavioral training will produce synaptic changes in the sensori-motor pathways that are preferentially activated by the particular training task employed. Thus, plastic changes in neuronal connectivity would be expected to occur widely in the central nervous system, the exact loci varying with the tasks required and the species being tested. Brain lesion-behavior studies clearly are of value in delineating the sensori-motor pathways involved in specific behavioral performances but appear to be of little value in locating regions of synaptic plasticity, since destruction of plastic and nonplastic parts of the overall circuitry would be equally effective in disrupting behavioral performance. Similarly, disruption of behavioral performance by pharmacological means does not generally permit conclusions concerning the location or properties of plastic versus nonplastic synapses.

It is likely that there is a close parallel between human amnestic syndromes and loss of conditioned responses in animals following brain lesions or drug treatments. In both cases, it may be difficult or impossible to decide if stored information is actually lost or whether pathways involved in its behavioral expression have been inactivated. The fact that both antimuscarinic treatment and hippocampal lesions impair untrained behavior such as mating or maternal behavior (Kimble et al., 1967; Singer, 1968), as well as specially trained behavior, suggests the importance of innate neural connections in the behavioral impairments produced by these agents.

An alternative theory, based in part on implicit acceptance of the doctrine that traditional subdivisions of the psyche are localized in different parts of the brain, holds that memory can be localized in specific regions, such as the hippocampus or diencephalon. The hippocampal hypothesis of memory, in particular, has been widely supported in the past 30 years. However, there is evidence, both in humans (Gol and Faibisch, 1967; Markowitsch, 1985) and in animals (Horel 1978;

Markowitsch, 1985) that extensive damage to the hippocampal formation is insufficient to produce a severe amnestic syndrome. Horel (1978) and his colleagues (Cirillo et al., 1989; George et al., 1989; Horel and Pytko, 1982; Horel et al., 1984) have presented strong evidence indicating that destruction of neocortical connections is probably involved in the temporal lobe amnestic syndrome, possibly acting in concert with hippocampal damage. It may be that global amnesia cannot be produced by a focal lesion anywhere in the brain (Markowitsch, 1985).

The behavioral effects of combined cholinergic and serotonergic blockade were not restricted to learned behavior in our experiments, since untrained behaviors, such as long sequences of grooming or the tendency to avoid walking off the edge of a vertical decline, were also impaired (Vanderwolf, 1987). Presumably, cholinergic and serotonergic control of cerebral activity plays a role in "instinctive" behavior as well as in "learned" behavior.

Although the brain lesions and drug treatments used in our experiments certainly cannot distinguish between innate and individually acquired synaptic connections (see a more detailed discussion in Vanderwolf, 1987), the results have a certain interest nonetheless. Thus, the fact that serotonergic depletion amplifies the effect of concurrent cholinergic blockade may offer an explanation of the severe behavioral deterioration observed in dementias of the Alzheimer type in which there is concurrent loss of both cholinergic and serotonergic neurons (Bowen and Davison, 1986; Collerton, 1986; Cross, 1990; Davies & Maloney, 1976; Rossor & Iverson, 1986; Whitehouse et al., 1982; Whitford, 1986). Similarly, the finding that PCPA does not enhance the effect of diencephalic or temporal region damage may be of interest in connection with the Wernicke-Korsakoff syndrome. Chronic alcoholism and the resulting thiamine deficiency may depress serotonergic function very severely (Chan-Palay, 1977; Witt and Goldman-Rakic, 1983). However, our data would support clinical evidence that serotonergic depletion does not interact with the thalamic damage, which, it has been suggested, may play a role in the amnestic syndrome in Korsakoff's psychosis (Victor et al., 1989).

In sum, our data are consistent with the view that cholinergic and serotonergic inputs to the cerebral cortex interact in the regulation of cortical activity and behavior but they are insufficient to decide whether these inputs play any role, direct or indirect, in neural plasticity. Whether cholinergic and serotonergic neurons play any role in memory, considered as an Aristotelian subdivision of the psyche, is a very difficult question and may be irrelevant to progress in contemporary neuroscience.

Acknowledgments. The research reported in this paper was supported by an operating grant and a summer studentship award from the Natural Sciences and Engineering Research Council of Canada. We thank R.K. Cooley for expert technical assistance, P. DeShane for testing some of the rats, and D. Gingerich for typing the manuscript.

References

Bowen DM, and Davison AN (1986): Biochemical studies of nerve cells and energy metabolism in Alzheimer's disease. *Br Med Bull* 42:75–80

Buzsaki G, Bickford RG, Ponomareff G, Thal LJ, Mandel R, Gage FH (1988): Nucleus basalis and thalamic control of neocortical activity in the freely moving rat. *J Neurosci* 8:4007–4026

Chan-Palay V (1977): Indoleamine neurons and their processes in the normal rat brain and in chronic diet induced thiamine deficiency demonstrated by uptake of ^{3}H-serotonin. *J Comp Neurol* 176:467–494

Cirillo RA, Horel JA, George PJ (1989): Lesions of the anterior temporal stem and the performance of delayed match-to-sample and visual discrimination in monkeys. *Behav Brain Res* 34:55–69

Collerton D (1986): Cholinergic function and intellectual decline in Alzheimer's disease. *Neuroscience* 19:1–28

Cross AJ (1990): Serotonin in Alzheimer-type dementia and other dementing illnesses. *Ann NY Acad Sci* 600:405–417

Davies P, Maloney AJF (1976): Selective loss of central cholinergic neurons in Alzheimer's disease. *Lancet* 2:1403

Decker MW, Gallagher M (1987): Scopolamine-disruption of radial arm maze performance: Modification by noradrenalin depletion. *Brain Res* 417:59–69

Decker MW, McGaugh JL (1989): Effects of concurrent manipulation of cholinergic and noradrenergic function on learning and retention in mice. *Brain Res* 477:29–37

DeFelipe J, Jones EG (1988): *Cajal on the Cerebral Cortex: an Annotated Translation of the Complete Writings*. New York: Oxford University Press

Diamond MC (1988): *Enriching Heredity: The Impact of the Environment on the Anatomy of the Brain*. New York: The Free Press (Collier Macmillan).

Dickson CT, Vanderwolf CH (1990): Animal models of human amnesia and dementia: Hippocampal and amygdala ablation compared with serotonergic and cholinergic blockade in the rat. *Behav Brain Res* 41:215–227

Dixon AK, Hill RC, Roemer D, Scholtysik G (1977): Pharmacological properties of 4(1-methyl-4-piperidylidine)-9, 10-dihydro-4H-benzo-[4,5]cyclohepta[1,2]-thiophene hydrogen maleate (pizotifen). *Arzneimittelforsch* 27:1968–1979

George PJ, Horel JA, Cirillo RA (1989): Reversible cold lesions of the parahippocampal gyrus in monkeys result in deficits on the delayed match-to-sample and other visual tasks. *Behav Brain Res* 34:163–178

Gol A, Faibish GM (1967): Effects of human hippocampal ablation. *J Neurosurg* 26:390–398

Green AR, Backus LI (1990): Animal models of serotonin behavior. *Ann NY Acad Sci* 600:237–249

Green AR, Heal DJ (1985): The effect of drugs on serotonin-mediated behavioural models. In: *Neuropharmacology of Serotonin*, Green AR, ed. Oxford: Oxford University Press, pp 326–365

Greenough WT, Volkmar FR (1973): Pattern of dendritic branching in occipital cortex of rats reared in complex environment. *Exp Neurol* 40:491–504

Haigler HJ, Aghajanian GK (1974): Peripheral serotonin antagonists: Failure to antagonize serotonin in brain areas receiving a prominent serotonergic input. *J Neur Transmiss* 35:257–273

Hargreaves EL, Watson NV, Penava D, Eckel LA, Vanderwolf CH. Methiothepin alone among nine serotonergic antagonists reduces serotonin-dependent cerebral activation. Unpublished manuscript

Hicks RD (1965): *Aristotle: De Anima*. Amsterdam: AM Hakkert (first published in 1907 by Cambridge University Press).

Hinde RA, Stevenson-Hinde J (1973): *Constraints on Learning: Limitations and Predispositions*. London: Academic Press

Horel JA (1978): The neuroanatomy of amnesia: A critique of the hippocampal memory hypothesis. *Brain* 101:403–445

Horel JA, Pytko DE (1982): Behavioral effects of local cooling in temporal lobe of monkeys. *J Neurophysiol* 47:11–22

Horel JA, Voytko ML, Salsbury KG (1984): Visual learning suppressed by cooling the temporal pole. *Behav Neurosci* 98:310–324

Kimble DP, Rogers L, Hendrickson W (1967): Hippocampal lesions disrupt maternal, not sexual, behavior in the albino rat. *J Comp Physiol Psychol* 63:401–407

Koe BK, Weissman A (1966): *p*-Chlorophenylalanine: A specific depletor of brain serotonin. *J Pharmacol Exp Ther* 154:499–516

Lashley KS (1935): Studies of cerebral function in learning. XI. The behavior of the rat in latch box situations. *Comp Psychol Monog* XI(#2):1–42

Markowitsch HJ (1985): Hypotheses on mnemonic information processing by the brain. *Int J Neurosci* 27:191–227

Milner B (1966): Amnesia following operation on the temporal lobes. In: *Amnesia*, Whitty CWM, Zangwill OL, eds. New York: Appleton-Century Crofts, pp 109–133

Mishkin M (1978): Memory in monkeys severely impaired by combined but not by separate removal of amygdala and hippocampus. *Nature (London)* 273:297–298

Mishkin M, Malamut B, Bachevalier J (1984): Memories and habits: two neural systems. In: *Neurobiology of Learning and Memory*, Lynch G, McGaugh JL, Weinberger NM, eds. New York: Guilford Press, pp 65–77

Moruzzi G (1972): The sleep-waking cycle. *Ergeb Physiol* 64:1–165

Moruzzi G, Magoun HW (1949): Brainstem reticular formation and activation of the EEG. *EEG Clin Neurophysiol* 1:455–473

Nilsson OG, Strecker RE, Daszuta A, Bjorklund A (1988): Combined cholinergic and serotonergic denervation of the forebrain produces severe deficits in spatial learning in the rat. *Brain Res* 453:235–246

Nilsson OG, Brandin P, Björklund A (1990): Amelioration of spatial memory impairment by intrahippocampal grafts of mixed septal and raphe tissue in rats with combined cholinergic and serotonergic denervation of the forebrain. *Brain Res* 515:193–206

Paxinos G, Watson C (1986): *The Rat Brain in Stereotaxic Coordinates, Second Edition*. Sydney: Academic Press

Prior P (1991): *Cholinergic-Noradrenergic and Cholinergic-Serotonergic Interactions in Measures of Working and Reference Memory of the Rat*. Unpublished Ph.D. thesis, University of Western Ontario, London, Ontario, Canada

Richter-Levin G, Segal M (1989): Spatial performance is severely impaired in rats with combined reduction of serotonergic and cholinergic transmission. *Brain Res* 477:404–407

Riekkinen P Jr, Sirvio J, Riekkinen P (1990a): Interaction between raphe dorsalis and nucleus basalis magnocellularis in spatial learning. *Brain Res* 527:342–345

Riekkinen P Jr, Sirvio J, Miettinen R, Riekkinen P (1990b): Interaction between raphe dorsalis and nucleus basalis magnocellularis in the regulation of high-voltage spindle activity in rat neocortex. *Brain Res* 526:31–36

Robertson B, Baker GB, Vanderwolf CH (1991): The effects of serotonergic stimulation on hippocampal and neocortical slow waves and behavior. *Brain Res* 555:265–275

Rossor M, Iverson LL (1986): Non-cholinergic neurotransmitter abnormalities in Alzheimer's disease. *Br Med Bull* 42:70–74

Russell WR (1971): *The Traumatic Amnesias*. London: Oxford University Press

Scoville WB, Milner B (1957): Loss of recent memory after bilateral hippocampal lesions. *J Neurol Neurosurg Psychiat* 20:11–21

Seligman MEP, Hager JL (1972): *Biological Boundaries of Learning*. New York: Appleton-Century Crofts

Shettleworth SJ (1972): Constraints on learning. *Adv Stud Behav* 4:1–68

Siegel S (1956): *Non-Parametric Statistics for the Behavioral Sciences*. McGraw Hill, New York

Singer JJ (1968): The effects of atropine upon the female and male sexual behavior of female rats. *Physiol Behav* 3:377–378

Spangler EL, Wenk GL, Chachich ME, Smith K, Ingram DK (1990): Complex maze performance in rats: Effects of noradrenergic depletion and cholinergic blockade. *Behav Neurosci* 104:410–417

Steriade M, Hobson JA (1976): Neuronal activity during the sleep-waking cycle. In: *Progress in Neurobiology*, Vol 6, Kerkut GA, Phillis JW, eds. Pergamon Press: Oxford, pp 155–376

Stewart DJ, MacFabe DF, Vanderwolf CH (1984): Cholinergic activation of the electrocorticogram: Role of the substantia innominata and effects of atropine and quinuclidinyl benzilate. *Brain Res* 322:219–232

Tinbergen N (1951): *The Study of Instinct*. New York: Oxford University Press

Turner AM, Greenough WT (1985): Differential rearing effects on rat visual cortex synapses. I. Synaptic and neuronal density and synapses per neuron. *Brain Res* 329:195–203

Vanderwolf CH (1983): The influence of psychological concepts on brain-behavior research. In: *Behavioral Approaches to Brain Research*, Robinson TE, ed. New York: Oxford University Press, pp 3–13

Vanderwolf CH (1984): Aminergic control of the electrocorticogram: A progress report. In: *Neurobiology of the Trace Amines*, Boulton AA, Baker GB, Dewhurst W, Sandler M, eds. Clifton, NJ: Humana Press, pp 163–183

Vanderwolf CH (1987): Near-total loss of "learning" and "memory" as a result of combined cholinergic and serotonergic blockade in the rat. *Behav Brain Res* 23:43–57

Vanderwolf CH (1988): Cerebral activity and behavior: Control by cholinergic and serotonergic systems. *Int Rev Neurobiol* 30:225–340

Vanderwolf CH (1989): A general role for serotonin in the control of behavior: Studies with intracerebral 5,7-dihydroxytryptamine. *Brain Res* 504:192–198

Vanderwolf CH (1990): An introduction to the electrical activity of the cerebral cortex: Relations to behavior and control by subcortical inputs. In: *The Cerebral Cortex of the Rat*, Kolb B, Tees RC, eds. Cambridge MA: MIT Press, 151–189

Vanderwolf CH (1991): Anti-muscarinic drug effects in a swim-to-platform test: Dose response relations. *Behav Brain Res* 44:217–219

Vanderwolf CH, Robinson TE (1981): Reticulocortical activity and behavior: A critique of the arousal theory and a new synthesis. *Behav Brain Res* 4:459–514

Vanderwolf CH, Kolb B, Cooley RK (1978): The behavior of the rat following removal of the neocortex and hippocampal formation. *J Comp Physiol Psychol* 92:156–175

Vanderwolf CH, Gutman M, Baker GB (1984): Hypothalamic self-stimulation: The role of dopamine and possible relations to neocortical slow wave activity. *Behav Brain Res* 12:9–19

Vanderwolf CH, Baker GB, Dickson C (1990a): Serotonergic control of cerebral activity and behavior: Models of dementia. *Ann NY Acad Sci* 600:366–383

Vanderwolf CH, Dickson CT, Baker GB (1990b): Effects of *p*-chlorophenylalanine and scopolamine on retention of habits in rats. *Pharmacol Biochem Behav* 35:847–853

Victor M, Adams RD, Collins GH (1989): *The Wernicke-Korsakoff Syndrome and Related Neurologic Disorders Due to Alcoholism and Malnutrition, Second Edition*. Philadelphia: F.A. Davis Co

Warrington EK, Weiskrantz L (1970): Amnesic syndrome: Consolidation or retrieval? *Nature* (London) 228:628–630

Whitehouse PJ, Price DL, Struble RG, Clark AW, Coyle JT, DeLong MR (1982): Alzheimer's disease and senile dementia: Loss of neurons in the basal forebrain. *Science* 215:1237–1239

Whishaw IQ (1990): The decorticate rat. In: *The Cerebral Cortex of the Rat*, Kolb B, Tees RC, eds. Cambridge MA: MIT Press, pp 239–267

Whitford GM (1986): Alzheimer's disease and serotonin: A review. *Neuropsychobiology* 15:133–142

Witt ED, Goldman-Rakic PS (1983): Intermittent thiamine deficiency in the Rhesus monkey. II. Evidence for memory loss. *Ann Neurol* 13:396–401

Zola-Morgan S, Squire LR, Amaral DG, Suzuki WA (1989): Lesions of perirhinal and parahippocampal cortex that spare the amygdala and hippocampal formation produce severe memory impairment. *J Neurosci* 9:4355–4370

17

Interactions of Neurotransmitters and Neuroanatomy: It's Not What You Do, It's The Place that You Do It

DAVID S. OLTON AND KEVIN PANG

Introduction

The subtitle of this manuscript, which is a variation of an old popular song, sets the theme for this paper. Neurotransmitters have their effects on behavior by altering neuronal activity in a particular part of the brain, and the part of the brain in which this activity is altered may substantially influence the effect of the neurotransmitter on behavior. This interaction between neurochemistry and neuroanatomy is illustrated in the context of the basal forebrain cholinergic system (BFCS), which has projections to many different neuroanatomical areas (Bigl et al., 1982; Woolf et al., 1984; Butcher, this volume). For the present purpose, two subdivisions are important: the nucleus basalis magnocellularis (NBM) and its projections to the frontal cortex (FC), and the medial septal area (MSA) and its projections to the hippocampus (HIP). Even though the neurobiological actions of the projections from the NBM and MSA may be similar, the psychological processes affected by these two systems may be different. Dissociations suggest that the NBM-FC system is more involved in attention than memory, whereas the MSA-HIP system is more involved in memory than in attention. This analysis suggests that any functional description of a neurotransmitter system should consider the interaction of both neurochemistry and neuroanatomy. From a functional perspective, it's not just what these neurons do, it's the place that they do it. Much of this material has been extensively reviewed elsewhere (Olton, 1990; Olton et al., 1991b,c; Olton and Wenk, 1987).

Functional descriptions of the hippocampus and of the cholinergic system have both emphasized a role in mnemonic processes, raising the possibility that the cholinergic projections from the MSA to the hippocampus are an important element in the mnemonic system. However, this hypothesis requires direct testing. The experiments investigating hippocampal function were usually not designed to get information about particular transmitters in the hippocampus, and the experiments investigating cholinergic function were usually not designed to obtain information about the role of selective neuroanatomical areas in cholinergic function. The lack of conjoint neuroanatomical and neurochemical specificity means that a strong test of the role of the septohippocampal cholinergic system has not been made.

To achieve this goal, experiments must be designed so that manipulations and measurements can be made of a specific neurotransmitter in a specific neuroanatomical area. The ideal experiment probably does not yet exist, but significant progress has been made in this direction. Neuroanatomically specific measurement of cholinergic activity is possible through *in vivo* microdialysis in behaving animals, and experiments are now pursuing this strategy. Neurochemically specific manipulation of cholinergic neurons is much more difficult. AF64A may be a selective cholinergic neurotoxin, but its selectivity has not been reliable (McGurk et al., 1987). In some cases, the cellular destruction has been confined to cholinergic neurons, whereas in other cases, every cell, regardless of neurotransmitter, has been killed (Chrobak et al., 1987). Intracranial infusions, described in detail elsewhere in this volume (Givens and Olton), offer considerable promise. However, because the receptors on cholinergic neurons are also on other neurons, these intracranial infusions can influence noncholinergic as well as cholinergic function. Nonetheless, this approach provides more specificity than systemic administration of drugs that can alter neuronal function throughout the entire brain (Olton et al., 1991a).

The conceptual framework guiding the research to be described below states that cholinergic projections from the MSA to the HIP mediate working memory but not attention, whereas cholinergic projections from the NBM to the FC mediate certain types of attention, but not recent memory. Some critical dissociations following lesions of the MSA and NBM support the basic hypothesis. Other dissociations are necessary to evaluate the specificity of the effects. Are noncholinergic projections to the HIP and FC also involved in memory and attention, respectively? Are projections from the MSA to areas other than HIP involved in working memory? Are projections from the NBM to targets other than the FC involved in attention? The data providing the basic dissociations are reviewed first. Questions about the specificity of the effects are addressed subsequently.

MSA and Memory

Lesions of the MSA produce mnemonic impairments similar to those produced by lesions of the hippocampus (Gray and McNaughton, 1983). Neurotoxic lesions, usually produced by ibotenic acid, but sometimes by quisqualic acid, impair choice accuracy in tasks that require working memory (see review in Olton et al., 1991b; Olton et al., 1990). These tasks include both spatial and nonspatial delayed conditional discrimination. The magnitude of the impairment is influenced by many parametric variables, including the size of the lesion, the difficulty of the task, and for incomplete lesions, the length of postoperative testing.

Impairments in an operant procedure to assess working memory for a temporal discrimination also followed MSA lesions, and performance in this task shows dissociations between the effects of MSA lesions and NBM lesions (Meck et al.,

1987; Olton et al., 1988). These experiments used variations of a fixed-interval (FI) operant procedure to assess temporal discrimination. A signalled FI schedule was given to each rat. For each FI reinforced trial, the signal was turned on. Following the appropriate FI for the signal, the first lever press turned off the signal and produced reinforcement. The optimal strategy for the rat was to wait until the end of the FI and then press the lever in order to get food. The actual pattern of responding, summed over many trials, was the usual one for a signalled FI reinforced schedule. The rate was low at the beginning of the interval, gradually accelerated as the end of the interval approached, and reached a maximum at the time of reinforcement.

FI probe trials assessed the accuracy of the temporal discrimination. For each FI probe trial, the signal was turned on as in the FI reinforced trial. However, no food was delivered after the FI, and the signal continued. Consequently, no external cue (the delivery of food or the termination of the signal) was available to indicate the end of the FI. During these probe trials, the mean rate of responding increased from the beginning of the interval, reached a peak at the FI, and then decreased. The *peak time*, the time at which the rate reached its maximum, was taken as a measure of the rat's expected time of reinforcement and the accuracy of the temporal discrimination.

The working memory task used gap trials in which the signal was turned on for a period of time less than the associated FI, turned off for the gap, and then turned back on. During these gap trials, the optimal strategy for the rat was to store the duration of the stimulus when the gap began and recall that duration at the end of the gap. Normal rats had successful working memory; the peak time was shifted rightward by an amount of time equal to the duration of the gap. MSA and HIP lesions impaired working memory. When the stimulus was turned on at the end of the gap, these rats showed no evidence of remembering the duration of the stimulus prior to the gap. Timing began from the start so that the peak time was shifted rightward by an amount of time equal to the duration of the stimulus prior to the gap and the gap itself.

Some caution in attributing the effects of these neurotoxic lesions, produced by ibotenic acid, to changes in just the cholinergic system is appropriate. Ibotenic acid (IBO) is not a selective cholinergic neurotoxin, and a comparison of the behavioral effects produced by IBO and by quisqualic acid (QUIS) suggests that noncholinergic systems must contribute to these behavioral changes produced by IBO. QUIS, as compared to IBO, produced a greater decrease in cholinergic markers but a smaller decrease in behavioral performance in some tasks. This lack of correlation between the magnitude of cholinergic pathology and the magnitude of the behavioral impairment suggests that damage to noncholinergic systems may also contribute to these behavioral effects (see reviews in Olton et al., 1991; Olton et al., 1991b). However, the general point about dissociations is still valid. Similar destruction in different parts of the basal forebrain cholinergic system produced different behavioral changes (Marston et al., 1991).

Direct microinfusion of substances into the MSA also indicates the importance of the MSA for working memory. These data are presented in detail in Chapter XIX and will be reviewed only briefly here. In both spatial and nonspatial tasks that require working memory, infusions of compounds into the MSA significantly changed choice accuracy. Muscimol, a GABAergic agonist, and scopolamine, a cholinergic antagonist, both impaired working memory in young rats. Oxotremorine, a cholinergic agonist, reduced the age-associated impairment of working memory in aged rats. The power of hippocampal theta, an electrophysiological indicator of neuronal activity, was highly correlated with choice accuracy. Muscimol and scopolamine reduced the power of hippocampal theta; oxotremorine increased it. Together, the electrophysiological and behavioral data suggest that cholinergic projections from the MSA to the HIP have a significant influence on hippocampal memory processing, which is specially related to both spatial and nonspatial working memory (Olton et al., 1991a).

Behaviorally induced changes in cholinergic activity reflect a mnemonic role of the MSA (Golski et al., 1991; Wenk et al., 1984). Experience in a working memory task increased high-affinity choline uptake and hemicholinium-3 binding, both of which reflect cholinergic activity. This increase in cholinergic activity was due to some mnemonic characteristic of the task because cholinergic activity was not increased by experience walking on a treadmill or being forced to move around a spatial maze without having a mnemonic requirement.

MSA and Attention

The temporal discrimination tasks described previously included a variation to test divided attention. For the divided attention trials, two signals, each associated with a different FI, were presented at the same time. The optimal strategy for the rat was to time both signals simultaneously and to press the lever after the FI for each signal. Probe trials assessed the accuracy of the discrimination, as described above. Normal rats successfully divided attention between both stimuli and had two peak times. One peak was at the FI associated with one stimulus, the other peak was at the FI associated with the other stimulus. MSA and HIP lesions did not impair divided attention.

NBM and Attention

An attentional role for the NBM is indicated by the impairments produced in the divided attention task for timing two signals. As described previously, normal rats successfully divided attention between both stimuli and had two peak times. One peak was at the FI associated with one stimulus; the other peak was at the FI associated with the other stimulus. NBM lesions impaired divided attention. Although rats with these lesions were able to time each stimulus alone, they were unable to divide attention and time both stimuli simultaneously. Consequently, the peak time for the stimulus that was not being timed was shifted rightward for the amount of time that attention was given to the other stimulus.

Infusions of muscimol into the NBM disrupted performance in a two-choice reaction time task in a manner suggesting an impairment of attention. This experiment will be described in detail because it incorporates many new features in the assessment of attentional processes in rats. The general strategy is taken from that used to investigate the neural bases of memory. In both the study of memory and attention, the major theoretical and conceptual frameworks have been developed from experiments with humans. However, the only way to obtain detailed information about the neural mechanisms involved in these cognitive processes is through experiments with animals; only experiments with animals allow sufficient access to the brain to test specific hypotheses about the details of mind/brain relations. The description of the neural bases of memory has proceeded most rapidly when experiments with animals have used the general principles of a comparative cognitive analysis to design experimental procedures that are valid models of those used with humans. These types of models allow the best integration of cognitive and neurobiological analyses.

The experiments described here were designed to take the same strategies that have been successful in the development of behavioral tests for animals to study the neural bases of memory and to apply them to the study of the neural bases of attention. The critical feature incorporated into the present experiments was the use of a two-choice reaction time task to measure the effects of expectancy on reaction time, choice accuracy, response bias, and discriminability (Pang et al., 1992). The experiment for rats was made as similar as possible to those typically used for humans: independent variables, dependent variables, data analyses, and theoretical frameworks. Because some details of the procedure for rats were different from those previously used for people, people were tested with the modified procedures used for rats to be certain that these minor modifications did not alter the basic pattern of results. Although previous experiments with rats and other animals have measured reaction time, these had not used the usual two-choice experimental design and the technical procedures typically used in experiments with humans to assess attentional mechanisms.

The apparatus had a water spout between two levers, which were elevated sufficiently far above the floor of the cage so that the rat stood on his hind legs and placed one front paw on each lever. This arrangement assured that the rat had a consistent physical position with respect to the response mechanisms and the stimuli, and was able to move each forepaw independently to produce two independent responses. Each trial began with presentation of one of two stimuli. If the stimulus was a light, the correct response was to release the left lever. If the stimulus was a tone, the correct response was to release the right lever. Reaction time and choice accuracy were measured, and response bias and discriminability were calculated.

Expectancy was manipulated by varying the proportion of light and tone stimuli in each session. The percentage of trials with each stimulus (light/tone) was 100/0, 80/20, 50/50, 20/80, and 0/100. The data from the first 20 trials of each session were discarded because these trials were intended to provide the information about the ratio of stimuli to be presented in that session. A within-subjects design was used, with each rat being given each stimulus condition for several sessions.

The general pattern of results was clear for both rats and people. As the probability of a stimulus increased, reaction time decreased, choice accuracy increased, response bias shifted to the more probable stimulus, and discriminability was not changed. Although the rats and people differed in a few characteristics, the general pattern of results was the same for both species.

Because the experimental procedures for the rats and the humans were as similar as possible in terms of the independent variables, dependent variables, data analyses, and conceptual frameworks, both species are likely to have used similar cognitive mechanisms to perform the task. This similarity in cognitive mechanisms permits an examination of the neural mechanism of attention in rats.

Considerable evidence suggests that the frontal cortex plays a role in some kinds of attentional processes. The impairment in divided attention as a result of lesions of the NBM and FC is consistent with this view, and other data from both animals and people suggest that frontal systems have some role in attention.

In order to determine the extent to which the NBM-FC system is involved in attention and expectancy, as operationalized in the two-choice reaction time task, infusions of muscimol, 0.5, 1, and 2.5 ng, were made bilaterally into each NBM just prior to behavioral testing. A within-subjects design was used so that each rat was tested both in the control condition and following muscimol in each expectancy condition. At least 2 days occurred between infusions, and no rat received more than 15 infusions.

The infusions had a substantial effect on behavior, especially choice accuracy. Of most significance was an interaction between choice accuracy and the probability of the light stimulus. At 100% or 80% probability, the infusions had no effect. With decreasing probabilities (50% and 20%), muscimol had an increasingly larger effect. This interaction is important because it suggests that the infusion did not alter general perceptual and motoric processes that were common in all stimulus conditions but had a selective effect on attention to an unexpected stimulus (Pang et al., 1991).

NBM and Memory

Some dissociations of NBM and MSA functions occur, but lesions of the NBM can produce impairments similar to those following MSA lesions in many tasks. Consequently, the extent to which the NBM is involved in working memory remains to be determined because no systematic analysis has yet identified the important parameters.

In the working memory tests in the temporal discrimination task described earlier, NBM lesions had no effect. The peak time of rats with NBM lesions was identical to that of control rats. These data suggest that the NBM does not have a role in working memory.

Cholinergic activity in the frontal cortex, a projection site of the NBM, was not increased by performance in a delayed conditional discrimination on a radial arm maze, a task that has often been used to assess working memory (Golski et al., 1991; Wenk et al., 1984). A parametric analysis of task demand is still required to determine if this dissociation is a quantitative or qualitative one (Olton, 1989). However, even if the dissociation is a quantitative one, the results are important because they suggest that mnemonic demands differentially engage the MSA-HIP and NBM-FC systems.

In many delayed conditional discriminations, NBM lesions, produced by ibotenic and quisqualic acid, impaired choice accuracy substantially. The magnitude and duration of these impairments depended on parametric variables, such as the difficulty of the task, the size of the lesion, and the length of postoperative recovery. Although few systematic comparisons of the effects of MSA and NBM lesions have been conducted, the available data suggest that the magnitude of the impairment following both lesions is similar (see reviews in Olton et al., 1990; Olton et al., 1991b).

A comparison of the ways in which NBM and MSA mediate working memory may be helpful to describe functional dissociations between the two regions. The influence of the NBM on mnemonic processes may be substantial, but different from those of the MSA. Because the NBM projects to the FC, an understanding of the role of the FC in the human amnesic syndrome may help to identify important variables responsible for the involvement of the NBM (Schacter, 1987).

Dissociations Among Neurotransmitters

The manipulations and measurements described in the previous sections all indicate that cholinergic activity is a critical component in the NBM and MSA systems. However, the selectivity of this cholinergic contribution remains to be assessed. The lesions destroyed noncholinergic as well as cholinergic cells. The microinfusions stimulated receptors on noncholinergic as well as on cholinergic cells. Assessing the selectivity of the cholinergic component requires manipulations and measurements of individual transmitter systems. Currently, no neurotoxin is available to produce selective cholinergic lesions, and no drug is available to stimulate receptors only on cholinergic cells. The development of appropriate compounds for these uses would provide a significant tool to investigate this question of selectivity. In the meantime, *in vivo* microdialysis of acetylcholine and other transmitter substances is probably the best means to determine the extent to which the behavioral effects should be attributed to just the cholinergic system. A selective cholinergic involvement would be indicated if levels of ACH, but not other neurotransmitters, were correlated with behavioral performance.

Role of FC and HIP

Manipulations of the NBM will alter the activity of cells in the FC because the NBM projects to the FC. Likewise, manipulations of cells in the MSA will alter the activity of cells in the HIP because these MSA cells project to the HIP. Direct comparison of the effects of MSA and HIP lesions, and the effects of NBM and FC lesions, suggests that projections to these target sites are importantly involved in the behavioral effect following these manipulations. For example, in the working memory and divided-attention tasks involving temporal discrimination described previously (Meck et al., 1987; Olton et al., 1988), MSA and HIP lesions both produced one pattern of behavior, while NBM and FC lesions both produced a different pattern of behavior. This pattern of results suggests that the projections to these particular target sites are important for the behavioral effects produced by lesions in the MSA and NBM.

However, both the MSA and the NBM have projections to other target sites. Consequently, these projections may have an important role in mediating the behavioral functions of the MSA and NBM. Experiments investigating the functional importance of projections to these other sites are necessary to determine the extent to which the contributions of the HIP and the FC are unique. Lesions of these areas, and *in vivo* microdialysis in them, can help address this question.

Summary

The general pattern of results obtained from lesions of and infusions into the MSA and NBM is consistent with the functional dissociations between mnemonic and attentional processes presented earlier in this chapter. Obviously, the complementary experiments to test the double dissociation are an important next step. Do infusions into the MSA alter attention? Do infusions into the NBM alter working memory? Are the target sites of the HIP and FC critical for whatever observed dissociations occur? Do noncholinergic as well as cholinergic systems mediate the dissociations? These additional experiments are necessary to test the selectivity of the functional dissociations between the MSA and NBM system. Even without those dissociations, however, these data provide important evidence for a functional relation between the basal forebrain cholinergic system and the cognitive processes of attention and memory depending on the neuroanatomical site to which the BFCS projects.

To return to the title and introduction of this chapter, where you do it in the nervous system significantly influences the behavioral consequences of what you do. Consequently, discussions of the localization of function in the nervous system may proceed most effectively if they consider simultaneously anatomical and neurochemical specificity (Olton et al., 1991a).

Acknowledgments. The research on attention described in this chapter was supported in part by research grant AFOSR-89-0481 from the Air Force Office of Scientific Research. The authors thank A. Dürr for preparation of the manuscript and A. Balogh, A. Caprioli, L. Gorman, A. Markowska, K. Morrione, and R. Wang for helpful comments on the manuscript.

References

Bigl V, Woolf NJ, Butcher LL (1982): Cholinergic projections from the basal forebrain to frontal, parietal, temporal, occipital, and cingulate cortices: A combined fluorescent tracer and acetylcholinesterase analysis. *Brain Res Bull* 8:727–749

Chrobak JJ, Hanin I, Walsh TJ (1987): AF64A (ethylcholine aziridinium ion), a cholinergic neurotoxin, selectively impairs working memory in a multiple component T-maze task. *Brain Res*, 414:15–21

Golski S, Gorman LK, Olton DS (1991): Behaviorally induced cholinergic activity: Role of associative and nonassociative processes. *Soc Neurosci* Abstract 17:137

Gray JA, McNaughton N (1983): Comparison between the behavioral effects of septal and hippocampal lesions: A review. *Neurosci Biobehav Rev* 7:119–188

Marston HM, Everitt BJ, Robbins TW (1991): Evidence for specific relationship between discrimination performance and choline acetyltransferase in the posterior cingulate cortex of the rat. *Soc Neurosci* Abstract 17:479

McGurk SR, Hartgraves SL, Kelly PH, Gordon MN, Butcher LL (1987): Is ethylcholine mustard aziridinium ion (AF64A) a specific cholinergic neurotoxin? *Neuroscience* 22:215–224

Meck WH, Church RM, Wenk GL, Olton DS (1987): Nucleus basalis magnocellularis and medial septal area lesions differentially impair temporal memory. *J Neurosci* 7:3505–3511

Olton DS, Wenk GL (1987): Dementia: Animal models of the cognitive impairments produced by degeneration of the basal forebrain cholinergic system. In: *Psychopharmacology: The Third Generation of Progress*, HY Meltzer, ed. New York: Raven Press pp 941–953

Olton DS, Wenk GL, Church RM, Meck WH (1988): Attention and the frontal cortex as examined by simultaneous temporal processing. *Neuropsychology* 26:307–318

Olton DS (1989): Dimensional mnemonics. In: *The Psychology of Learning and Motivation*, GH Bower, ed. San Diego: Academic Press, pp 1–23

Olton DS (1990): Dementia: Animal models of the cognitive impairments following damage to the basal forebrain cholinergic system. *Brain Res Bull* 25:499–502

Olton DS, Givens B, Markowska AL, Shapiro M, Golski S (1991a): Mnemonic functions of the cholinergic septohippocampal system. In: *Memory: Organization and Locus of Change*, LR Squire, NM Weinberger, JL McGaugh, eds. Oxford: Oxford University Press, pp 250–269

Olton DS, Markowska AL, Voytko ML, Givens B, Gorman L, Wenk GL (1991b): Basal forebrain cholinergic system: A functional analysis. In: *The Basal Forebrain: Anatomy to Function,* TC Napier, PW Kalivas, I Hanin, eds. New York: Plenum Press, pp 353–372

Olton DS, Wenk GL, Markowska AM (1991c): Basal forebrain memory, and attention. In: *Activation to Acquisition: Functional Aspects of the Basal Forebrain*, R. Richardson, ed. Boston: Birkhäuser, pp 246–362

Pang K, Olton DS, Egeth H (1991): Attention and the nucleus basalis magnocellularis. *Soc Neurosci* Abstract 17:137

Pang K, Merkel F, Egeth H, Olton D (1992): Expectancy and stimulus probability: A comparative analysis in rats and humans. *Perception & Psychophysics* 51:607–615

Schacter DL (1987): Memory, amnesia, and frontal lobe dysfunction. *Psychobiology* 15:21–36

Wenk GL, Hepler D, Olton DS (1984): Behavior alters the uptake of (3H)-choline into acetylcholinergic neurons of the nucleus basalis magnocellularis and medial septal area. *Behav Brain Res* 13:129–138

Woolf NJ, Eckenstein F, Butcher LL (1984): Cholinergic systems in the rat brain: I. Projections to the limbic telencephalon. *Brain Res Bull* 13:751–784

18

Intraseptal GABAergic Infusions Disrupt Memory in the Rat: Method and Mechanisms

JAMES J. CHROBAK AND T. CELESTE NAPIER

Introduction

The septum serves as an important focus for coordinating hippocampal neural activity (Petsche et al., 1962). Manipulations within the septum alter hippocampal activity and disrupt the performance of behavioral tasks dependent on the functional integrity of this structure. The present chapter focuses on the effects of GABAergic manipulations within the septum on the rat's performance of radial maze tasks (Fig. 1). We have observed that posttraining intraseptal infusion of the GABA agonist muscimol and the GABA antagonist bicuculline (BIC) can induce acute amnestic deficits in the rat's performance of a delayed non-match-to-sample (DNMTS) radial arm maze (RAM) task (Chrobak et al., 1989a; Chrobak and Napiter, 1991, 1992). These findings suggest that modulation of septal activity can disrupt memory-consolidation processes. Such findings present the challenge of identifying underlying neural events disrupted by these manipulations and determining their relationship to the observed amnestic deficits. We speculate that pretraining intraseptal treatments may alter memory processes by altering neural events operative during hippocampal theta activity, while posttraining intraseptal treatments may modify memory-consolidation processes operative during the hippocampal sharp wave (SPW) state.

While SH activity plays a role in a multiplicity of neuropsychological events, the SH system is a prominent substrate for maintaining representations of recent episodic events, what we refer to as working/episodic (W/E) memory (Chrobak et al., 1991). This aspect of hippocampal function enables "whole (spatial) scenes or episodic memories to be formed, with a snapshot quality . . . such as where, with whom, and what one ate for lunch on the preceding day" (Rolls, 1991, p 258). The absence of this ability characterizes much of the memory impairment observed in many human amnestic syndromes, including the earliest manifestations of dementia, as well as those observed following experimental manipulations of the SH system in rodents and nonhuman primates.

To examine W/E memory in the rat, we have been using a delayed non-match-to-sample (DNMTS) radial arm maze (RAM) task (see Fig. 1 for details). In this

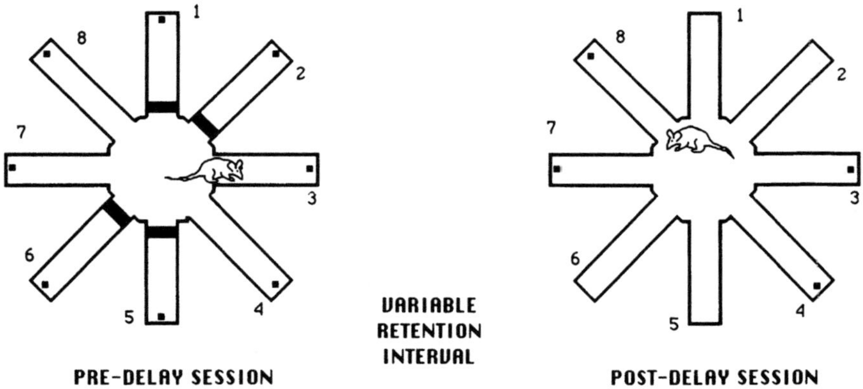

FIGURE 1. In a standard RAM task, the rat is placed into the maze with all eight arms baited. Their task is to enter all eight arms without reentering an arm entered in any daily session. Typically, rats will make seven or eight correct choices in the first eight choices. The task as depicted above (DNMTS-RAM) involves restricting the animal's first four choices to a predetermined set, which varies on a daily basis (left). Following a delay or retention interval, the rat is returned to the maze and is allowed access to all eight arms. The task in the delay condition remains unchanged, and the animal is reinforced for nonmatching to the predelay sample set. Performance indices for the DNMTS-RAM task include the number of correct choices in the first four choices, the number of retroactive errors (repeat entries into predelay chosen arms), and the number of proactive errors (repeated entries into postdelay chosen arms). Repeat entries into predelay chosen arms are counted only once (maximum = 4).

task, the animal is required to remember where they ate breakfast (four of eight maze arms) to efficiently locate their lunch (the location of the four remaining arms not entered during the predelay session). In this respect the task is essentially the same as a standard RAM task in which no delay is imposed, and the rat is free to navigate to the eight available arms as quickly as possible. The insertion of the delay provides (a) for performance based on delayed conditional discrimination, (b) experimental control over the animals' selection strategy, and (c) a time window for manipulating the temporal relation between an acute experimental treatment (i.e., drug administration) and the to-be-remembered event(s).

Important criteria of memory-specific manipulations have been the (a) effectiveness of posttraining treatments, (b) the ineffectiveness of *delayed* posttraining treatments, and (c) an interdependence between treatment strength (i.e., drug dose) and the retention interval over which information must be maintained (a treatment by retention interval interaction). Such criteria are based upon memory consolidation models that posit dynamic alterations in the strength of memory traces (Gold and McGaugh, 1975). Initial traces that may decay spontaneously over time give rise to more consolidated traces, which are then encoded in a semipermanent form that is relatively immune to insult. Treatments imposed prior to, or immediately following, the to-be-remembered events (within seconds or a few minutes) weaken the encoding of an initial trace and subsequent consolidated traces, while delayed treatments (those imposed several minutes or hours after the to-be-remembered event) are typically ineffective (Gold, 1984). Performance after a short retention interval may be relatively unaffected by weak treatments imposed in close temporal contiguity with the to-be-remembered event, yet performance after longer retention intervals may be altered (a treatment by delay interaction may be observed). By examining Figure 1, one can see that the ability to administer treatments at various time points with respect to the to-be-remembered event(s) makes the DNMTS-RAM task particularly advantageous for examining the neurobiology of memory. Further, the ability of the animal to perform this task, to remember where it ate breakfast *each day*, allows for an assessment of treatment effects using repeated-measures designs.

Rationale

A considerable literature demonstrates the involvement of SH neurons in W/E memory processes (Olton, 1986; Walsh and Chrobak, 1990). Investigators have focused mainly on the role of SH cholinergic neurons, given the association of damage to these neurons with dementia (Davies and Maloney, 1976; Henke and Lang, 1983; Rylett et al., 1983), as well as the vast number of studies demonstrating cognitive dysfunction following the administration of cholinergic antagonists (Hagan and Morris, 1988). Our interest was spurred by evidence demonstrating the influence of GABAergic neurons on SH neurons (Allen and Crawford, 1984; Costa et al., 1983; McLennan and Miller, 1974). Acute pharmacological manipulations of intraseptal GABAergic activity seemed a logical extension of examining the behavioral consequences of chronic insult to the SH cholinergic neurons (Chrobak et al., 1989b; Chrobak and Walsh, 1991; Walsh et al., 1984). We hypothesized that intraseptal GABAergic manipulations would disrupt the activity of SH neurons and the presumed feedback regulation of SH neurons from hippocampal pyramidal neurons. This loop is thought to involve an excitatory amino acid pathway from CA3 to the lateral septal nuclei, which in turn sends a GABAergic input to the medial septum. Cholinergic and GABAergic efferents (Freund and Antal, 1988) from the medial septum close the loop by providing an innervation of the HPC (see Jakob and Leranth, 1990 for discussion).

Intraseptal GABAergic manipulations were known to have a disruptive effect on SH electrophysiological activity (Allen and Crawford, 1984; McLennan and Miller, 1974), on neurochemical indices of SH cholinergic activity (Costa et al., 1983), and on behavioral end points (Blaker et al., 1984). Initially, we demonstrated that immediate (within 5 minutes) posttraining intraseptal infusion of the GABA-A agonist muscimol (0.75–3.0 nM; 80–340 ng/0.5 μl) produced a dose-dependent disruption of DNMTS-RAM performance. This deficit was associated with an increase in errors associated with the predelay session (retroactive errors) without a concomitant increase in errors to postdelay entered arms (proactive errors) (Chrobak et al., 1989a, 1991). Such findings indicated that immediate posttraining disruption of intraseptal GABAergic activity was sufficient to disrupt memory processes. The following experiments addressed the question of whether posttraining administration of the GABA-A antagonist BIC would also affect performance, and whether this treatment would disrupt ongoing performance when administered prior to testing.

Methodology

In these experiments, rats were trained to perform the DNMTS or standard RAM task once a day over the course of 6- to 8-week period to a high level of accuracy. At this point, rats were implanted with a chronic indwelling cannula (26 gauge) overlying the MS region. One week later the animals restarted daily testing trials and were tested for five trials without treatment to assess any effect of the cannula implantation. The following week animals began treatments that occurred once a week and involved placing a 33-gauge internal cannula via the guide into the MS (a sham trial) and infusing vehicle or BIC at a rate of 0.125 μl/min. Rats received treatments in their home cage and were unrestrained during the infusion, except to prevent interference with the infusion tubing. It is important to note that, given the extensive training protocol, the rats were well habituated to the handling and restraint required to remove and insert a dummy cannula (which maintained cannulae patency), as well as to insert and remove the internal cannulae for infusion. As standard protocol, rats received at least two control treatments, the initial and final treatment consisting of a sham or vehicle. Given that we have not observed any difference between these treatments within the MS [significant differences have been observed following treatment into the ventral pallidal/ nucleus basalis region (Chrobak and Napier, 1989)], this protocol allows for assessment of any chronic alteration of performance attributable to the repeated infusions.

Effects of Posttraining Treatments

Posttraining intraseptal infusion of BIC (0.2–2.0 nM; 50–500 ng/0.5 μl) produced a dose-dependent disruption of performance as assessed at both 1- and 4-hr retention intervals (Fig. 2). These performance deficits were attributable to an increase in retroactive errors (errors associated with predelay choices), as opposed to an alteration of proactive errors (repeated entries into postdelay choices). Rats were capable of avoiding arms entered within the last minute or several seconds (typically it takes the animal 30–35 sec to complete the task), yet they did not discriminate (by avoiding) arms entered during the predelay session. Significant increases in the number of proactive errors, which are typically minimal, may be interpreted as a memory deficit (this is the most frequent confound of chronic insults or acute pretesting treatment protocols), or may reflect an impairment in sensory discrimination, attention, motivation, or other psychological processes at

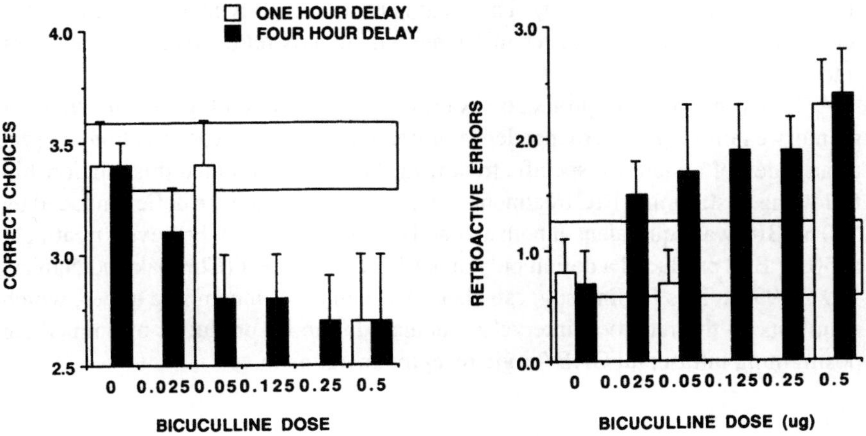

FIGURE 2. Effect of bicuculline (0.0–0.5 μg/0.5 μl) injections into the medial septal nucleus on the number of correct choices and retroactive errors during the postdelay performance of a DNMTS-RAM task. Blocked areas indicates range of mean values for all Tuesdays and Thursdays, the days before and after the weekly treatment day, Wednesday.

the time of testing (Beatty and Rush, 1983; McGaugh, 1989). The multiple choices the animal must make during the DNMTS retention test allows for a discrimination between retrograde deficits (indicative of an effect on memory processes) and proactive deficits (which may reflect a mnemonic impairment or a more generalized performance deficit).

Delayed posttraining infusion of BIC (500 ng), 2 hr into a 4-hr retention interval, had no effect on any performance index. Given that the animal was tested in closer temporal contiguity with the infusion (2 as opposed to 4 hr prior to testing), the deficits do not appear to be related to an attentional, motivational, or motoric deficit at the time of testing or a state-dependent effect. Stackman and Walsh (1992) have more recently observed that immediate posttraining infusions of the benzodiazepine chlordiazepoxide impair postdelay performance: Infusions delayed by 15 min, however, were ineffective. Such findings suggest that a limited temporal window may exist during which these memory processes can be altered by intraseptal treatment, and/or that quantitatively greater treatments (larger drug doses) may be necessary to disrupt memory processes at longer time points after the to-be-remembered events.

Our observations of a selective increase in retroactive, as opposed to proactive, errors would seem to support a disruption of memory consolidation processes. Depending upon the "strength" of the posttraining treatment, other psychological processes at the time of testing could be disrupted. Residual drug effects may persist into the retention test (see Beatty and Rush, 1983, for example). Such effects should wan as the retention interval is increased, while effects on memory specific processes "should" persist. We have observed such residual drug effects up to 1 hr following posttraining intraseptal infusion of the highest dose of muscimol used (3 nM; 252 ng). This treatment significantly increased retroactive and proactive errors at a 1-hr retention interval but only retroactive errors at a 4-hr retention interval.

A larger number of protocols examining the neurobiological substrate of memory examine the interdependence of treatment strength and retention interval as an index of a memory-specific treatment. We have examined this relationship following posttraining BIC treatment and have observed that the deficit induced by 500 ng BIC was equivalent at both a 1- and 4-hr retention test; however, treatment at 50 ng BIC produced a deficit only at a 4-hr retention test (Chrobak and Napier, 1991, 1992). This finding suggests that a weakening of mnemonic traces, which manifests as the retention interval is increased, can be produced by immediate posttraining intraseptal GABAergic receptor antagonism.

Effects of Pretraining Treatments

It is likely that septohippocampal activity contributes to the initial instantiation and the subsequent consolidation of memory traces. Several investigators have shown that pretesting infusion of drugs into the septum can disrupt or enhance (Bostock, et al., 1988; Brioni et al., 1990; Givens, this volume: Givens and Olton, 1990; Mizumori et al., 1990) performance in mnemonic tasks. Based upon findings with posttraining infusion of BIC, we suspected that pretesting infusion of BIC would also disrupt RAM performance. This treatment produced a dose-related alteration of ongoing RAM performance, increasing the number of errors the rat made and decreasing the number of correct choices until an error was made (Fig. 3). Interestingly, rats made several (5–6) correct choices before making an error, and the position of the first error in the previous choice sequence (a repeat of arm #5 in a choice sequence of 5-2-8-4-6-5-1 . . .) was not significantly different from

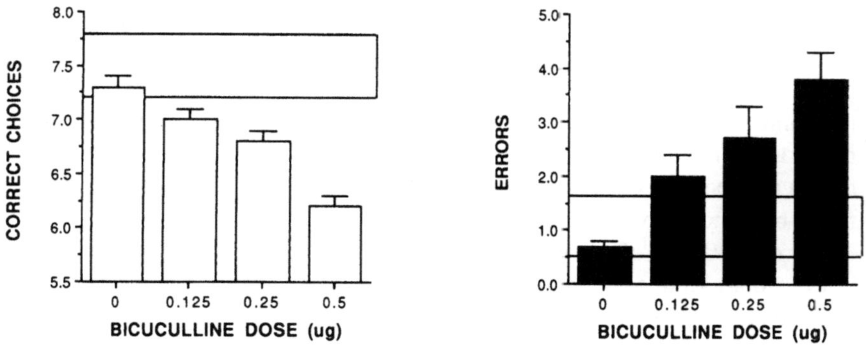

FIGURE 3. Effect of bicuculline (0.0–0.5 μg/0.5 μl) or vehicle infusions into the medial septal nucleus on the number of correct choices and total errors in the performance of a standard RAM task. Blocked areas indicates range of mean values for all Tuesdays and Thursdays, the days before and after the weekly treatment day, Wednesday.

control trials. This finding suggests that rats made errors only as the memory "load" (number of previously entered arms) increased, and that when an error occurred it was a reentry of the first or second arm selection. The later is the typical error the animal will make on control trials in greater than 80% of the first errors in any trial, if an error is made. This suggests that the BIC-induced deficit, occurring when the drug is administered prior to testing, may be related to a disruption of cognitive processes as opposed to an impairment in motivational or motoric processes.

Another more curious observation was an interaction between task performance and drug treatment, in which 3 of the 11 rats tested following 500 ng of BIC manifested seizure activity while performing the maze. In all cases such seizures only occurred after the animals had made more than 10 choices. At this dose rats were significantly slower in making arm selections, that is, they spent more time on the central platform, hesitantly making their selection of the next arm choice. Rats that developed seizures began to enter arms very slowly, circled the center arena, and then displayed head-bobbing behavior. All three were immediately removed from the maze and returned to their home cages. Two ceased head bobbing within seconds of being removed from the maze and appeared normal while being handled. All behaved normally within minutes of the display and consumed their daily food allotments. Maze performance in these three was not notably different on subsequent testing trials. Similar seizures, induced by task performance, have been observed in a few rats performing the maze task following colchicine-induced granule cell lesions (D. Emerich, unpublished observation) and have been reported by Whishaw (1987) following granule and CA3-4 damage.

Mechanisms

Intraseptal Treatments may Affect Memory Processes and Memory Performance by Altering Neural Activity within the HPC

A number of studies have demonstrated that pretraining intraseptal infusions disrupt the performance of W/E memory tasks (Brioni et al., 1990; Chrobak and Napier, 1992; Givens and Olton, 1990; this volume; Mizumori et al., 1990; Staubli and Huston, 1980). Posttraining infusion protocols have demonstrated that memory-consolidation processes can be disrupted by intraseptal infusions (Bostock et al., 1988; Chrobak et al., 1989a, 1991, 1992; Stackman and Walsh, 1992; Staubli and Huston, 1980; Wolfman et al., 1991). Such effects are likely to be manifestations of an alteration in the SH system, as lesions of the septum, fimbria-fornix, HPC, SH cholinergic neurons, and intrinsic HPC neurons are sufficient to induce analogous deficits. It is important to recognize, however, that septal projections to other targets (Raisman, 1966; Swanson and Cowan, 1979), including the thalamus, hypothalamus, amygdala, and other regions, could, and probably do, contribute to the observed deficits. However, manipulation of the SH system alone consistently induces a selective deficit in radial maze performance (Becker et al., 1980).

SH Neurons Regulate Theta Activity within the HPC, and the Alterations of Theta Are Correlated to Deficits in the Performance of Memory Tasks

SH neurons, including SHC neurons, play a predominant role in the manifestation of the hippocampal electroencephalographic (EEG) pattern known as the theta rhythm (Bland, 1986; Buzsaki, 1986). Theta occurs in the rat during locomotion, sensory stimulation, or during paradoxical sleep, and is presumed to represent an attentive or aroused state. Theta is a rhythmic (4–12 Hz) EEG pattern that results from the synchronous fluctuation in the membrane potential of neighboring hippocampal pyramidal and granule cells. Potentials of alternating polarity are generated by the synchronous excitation of the distal dendrites (EPSPs) of pyramidal and granule cells, followed closely in time by an inhibition of the somatic region (IPSPs) (see Bland, 1986; Buzsaki, 1986 for review). Critical to the generation of theta are pacemaker neurons within the MS nucleus (Petsche et al., 1962; Stewart and Fox, 1989; Vinogradova et al., 1980). Medial septal pacemaker neurons and other septal neurons entrained by these pacemakers provide theta rhythmicity to the hippocampus, subiculum, and entorhinal cortex. Theta is thought to contribute to the underlying plasticity of hippocampal circuits and is presumed to maximize the occurrence of synaptic modifications subserving long-term potentiation (LTP) (Larson and Lynch, 1986).

Alterations in theta are associated with behavioral impairments, as observed following SH insult (Mitchell et al., 1982; Winson, 1978) or more recently following pharmacological infusions into the septum (Givens and Olton, 1990; Mizumori et al., 1990). Thus there is reason to suspect that intraseptal infusions of GABAergic agents may disrupt mnemonic performance because they disrupt the activity of hippocampal neurons during theta. In fact, Givens and Olton (1990, this volume) have shown that the working memory deficit induced by pretraining intraseptal infusion of muscimol (10–30 ng), and other agents is correlated with suppression of theta activity. One should note that these doses are considerably less than those used to induce posttraining amnestic deficits (Chrobak et al., 1989a). While many procedural variables may account for such differences, it may be the case that much higher doses are required to induce the amnesia imposed by a posttraining treatment protocol, and perhaps this effect is mediated by another mechanism.

SH Cholinergic Neurons Contribute to the Generation and Maintenance of Theta, and Both May Reflect the Status of Hippocampal Involvement in Memory Processes or Performance

Evidence suggests that SH cholinergic neurons, which represent approximately 50% of the SH neurons, convey theta rhythmicity to the HPC. Activation of hippocampal theta has been correlated to increases in cholinergic neurochemical indices (Dudar et al., 1979), and systemic treatments with anticholinergic agents attenuate theta (Bland, 1986). A number of studies have demonstrated that the activity of SH cholinergic afferents is altered during the performance of learned tasks, and select neurotoxic insult to the these neurons produces memory deficits

(Chrobak et al., 1989b). Both increased and decreased levels of high-affinity choline transport (HAChT, an index of cholinergic neuronal activity) have been observed at distinct time points after performance on tasks dependent upon SH activity, including the Morris water maze (Decker et al., 1988; Gallagher and Pelleymounter, 1988) and radial maze (Galey et al., 1989; Jaffard et al., this volume; Wenk et al., 1984). The picture presented by these studies suggests a significant elevation in cholinergic activity during performance and a subsequent decrease in cholinergic activity. Both groups have observed that the subsequent decrease in cholinergic activity occurring following testing is related to maximal task performance (Decker et al., 1988; Gallagher and Pelleymounter, 1988; Jaffard et al., this volume). These dynamic alterations in cholinergic activity develop over the course of training and are indicative of the rate of acquisition (Jaffard et al., this volume). Such findings support the notion that, in addition to the chronic elevation of the cholinergic system associated with training on learning and memory tasks (Wenk et al., 1984), *acute within-trial dynamics occur*. It may be reasonable to suspect that such changes reflect alterations in theta during task performance and a subsequent transition to a nontheta state. These findings are particularly exciting, given that the level of cholinergic changes are greater than one might presume to be taking place in a rat attentive and aroused, manifesting theta, as the degree of changes reflect accuracy and are not observed in behaviorally active control groups.

Hippocampal and Septal Activity Would Appear to Be Dynamically Regulated During the Performance of Memory Tasks and Intraseptal Manipulations May Modify this Regulation as well as Septohippocampal Interactions

Changes in hippocampal EEG activity reflect changes in behavioral or psychological states and indicate alterations in the sensitivity and efficacy of intrahippocampal neuronal circuits (Winson and Abzug, 1978). Alterations within the SH system reflect not only general behavioral activation, but are also associated with the performance of learned tasks and/or the acquisition of memories. For example, we have pointed to dynamic within-trial changes in SH cholinergic activity and have previously speculated that the appearance of behavioral seizures following intraseptal BIC treatment reflects alterations in hippocampal excitability occurring as a consequence of the animal performing a learned task (Chrobak and Napier, 1992). Thus phasic changes in the excitability of hippocampal neurons, occurring as a consequence of the testing paradigm, may be unregulated following a certain level of disinhibition in the septum. It should be appreciated that subcortical denervation of the HPC, prominently disrupting the SH pathway, promotes the emergence and propagation of epileptiform activity within the HPC (Buzsaki et al., 1989, 1991). Durkin and Koenig (1991) have also reported the induction of seizure activity in mice following intraseptal BIC treatment, and that both muscimol and BIC prevent working memory task-induced elevation of SH cholinergic neurons. These data suggest that intraseptal treatments may produce an acute "functional septal denervation" of the HPC, disrupting the normal regulatory influence of SH neurons on hippocampal excitability. For example, activation of SH GABAergic afferents that innervate the majority of hippocampal interneurons would provide an extremely powerful mechanism for increasing the excitability of hippocampal pyramidal cells (see Freund & Antac, 1988).

Other evidence also supports dynamic intraseptal alterations occurring as a consequence of a trial-unique learning experience. Wolfman and colleagues (1991) have demonstrated that a single trial of passive-avoidance training produces significant alterations in endogenous benzodiazepine-like molecules within the SH system, as well as the amygdala and cortex. Notably, changes in septal levels were elevated threefold by these training trials; such posttraining alterations may result in dramatic alterations in the sensitivity of GABA receptors, which are modulated by benzodiazepines. Thus, a number of changes in the SH system occur as a consequence of presenting the animal with a discrete trial learning event, a unique episodic event, and such changes are likely to give rise to alterations in the activity of hippocampal circuits.

Posttraining Intraseptal Treatments May Disrupt Memory Consolidation by Disrupting the Hippocampal Sharp Wave (SPW) State

The question remains as to how posttraining intraseptal manipulations induce amnestic deficits. The effects of intraseptal infusion of BIC on hippocampal theta have not been examined. It is possible that both activation or blockade of intraseptal receptors in sufficient to disrupt theta by disrupting the pacemaker activity of intraseptal neurons. Thus, pretraining and posttraining disruption of theta may be sufficient to disrupt ongoing W/E memory performance and the maintenance of W/E memories over the protracted retention interval.

Recently, Buzsaki (1989) has suggested that the hippocampal SPW state that occurs during immobility, consummatory behaviors, and slow-wave sleep represent a physiological state wherein information acquired during the exploratory (theta state) is consolidated. Hippocampal sharp waves reflect the synchronous excitation of an ensemble of CA3, CA1, and subicular pyramidal neurons. As opposed to theta waves, which reflect the summation of EPSPs and IPSPs rhythmically influencing pyramidal and granule firing, the high-frequency burst of pyramidal neurons occurring during SPWs is likely to produce a powerful excitation, and perhaps synaptic potentiation, of intrahippocampal and entorhinal targets (Buzsaki et al., 1989). Such activation can result in the LTP of CA3-CA1 synaptic connections (Buzsaki et al., 1987). This hypothesis is intriguing precisely because it supports a two-stage model whereby postacquisition phenomenon serve to consolidate mnemonic traces. It is thus consistent with a large body of experimental data demonstrating the modifiability of memory traces for a limited time period subsequent to acquisition.

We suggest that posttraining intraseptal manipulations are not likely to alter memory consolidation processes by altering hippocampal theta. One may assume that postacquisition consummatory behaviors and the associated nontheta hippocampal EEG accompanying reinforcing behaviors/events are likely to have an important influence on learning and memory processes. Hippocampal theta may represent an exploratory or acquisition stage whereby representations (information) are posited in a temporary form (Buzsaki, 1989). These representations may then be modulated by behaviors/events that reflect the internal milieu and may be disrupted by extended attention to external stimuli (new information, which is a source for retrograde interference). Maintaining hippocampal theta may be, at some time point after the acquisition of information, detrimental to subsequent consolidation of this information. The transition out of the theta (exploratory phase), in this model, would be necessary for the consolidation of information acquired during theta. Such speculation may be indirectly supported by findings that the postperformance reduction in SH cholinergic activity, and presumably the cessation of hippocampal theta observed following the completion of mnemonic tasks, is related to optimal performance. We hypothesize, consistent with dual process models of memory consolidation (Hebb, 1949; Gold and McGaugh, 1975; Buzsaki, 1989), that posttraining intraseptal manipulations disrupt consolidation processes by disrupting the normal regulatory influence of SH neurons during the hippocampal SPW state. Further studies examining the relationship of SH neurons to the generation and character of hippocampal SPW would be a necessary prerequisite for evaluating this hypothesis.

Conclusions

Considerable information suggests that cortical structures support neural activity that allow for higher order cognitive processes. Many investigators have focused on the role of subcortical inputs, predominantly the cortically projecting magnocellular basal nuclear (MBN) neurons, that modulate cortical activity. MBN neurons are part of a diffuse cortical projection system that also includes brainstem monoaminergic neurons. The entire system appears to regulate cortical excitability in relation to an animal's behavioral state (Saper, 1987), and by regulating activity within the entire cortical mantle can influence a broad range of psychological phenomenon (sensory processing, attention, motivation, memory). Our studies have focused on manipulating the septal pole of the MBN, which plays a dominant role in regulating neural activity in the HPC and entorhinal cortex, structures that appear to play a dominant role in mammals ability to form, store, and use information about recent episodic events. Posttraining intraseptal infusions of the GABAergic agonist muscimol and the antagonist bicuculline can induce amnestic deficits in the rat's performance of a DNMTS-RAM task. These deficits can be related to a disruption of memory-consolidation processes. Intraseptal treatments may influence memory processes by altering activity in the HPC and entorhinal cortex. We have speculated that the influence of intraseptal treatments that disrupt ongoing performance (pretraining treatments) may reflect a disruption of processes occurring during hippocampal theta activity. Further we suggest that posttraining intraseptal treatment may disrupt memory consolidation processes operative during SPW activity. Further experiments addressing the effects of these intraseptal treatments on hippocampal theta and the hippocampal SPW state could directly address these hypotheses.

Acknowledgments. The authors would like to thank Drs. H. Read and G. Buzsaki for their comments on this manuscript. This work was supported by USPHS grant #DA05255 to TCN and an Illinois Department of Public Health grant to JJC.

References

Allen CN, Crawford IL (1984): GABAergic agents in the medial septal nucleus affect hippocampal theta rhythm and acetylcholine utilization. *Brain Res* 322:261–267

Beatty WW, Rush J (1983): Retention deficit after d-amphetamine treatment: Memory defect or performance change? *Behav Neural Biol* 37:265–275

Becker JT, Walker JA, Olton DS (1980): Neuroanatomical bases of spatial memory. *Brain Res* 200:307–320

Blaker WD, Peruzzi G, Costa E (1984): Behavioral and neurochemical differentiation of specific projections in the septal-hippocampal cholinergic pathway of the rat. *Proc Natl Acad Sci USA* 81:1880–1882

Bland B (1986): The physiology and pharmacology of hippocampal formation theta rhythms. *Prog Neurobiol* 26:1–54

Bostock E, Gallagher M, King R (1988): Effects of opioid microinjections into the medial septal area on spatial memory in rats. *Behav Neurosci* 102:643–650

Brioni JD, Decker MW, Gamboa LP, Izquierdo I, McGaugh JL (1990): Muscimol injections in the medial septum impair spatial learning. *Brain Res* 522:227–234

Buzsaki G (1986): Generation of hippocampal EEG patterns. In: *The Hippocampus*, Isaacson R, Pribram KH, eds. New York: Plenum Press.

Buzsaki G (1989): A two-stage model of memory trace formation: A role for "noisy" brain states. *Neuroscience* 31:551–570

Buzsaki G, Haas H, Anderson EG (1987): Long-term potentiation induced by physiologically relevant stimulus patterns. *Brain Res* 435:331–333

Buzsaki G, Pnomareff G, Bayardo F, Ruiz R, Gage FH (1989): Neuronal activity in the subcortically denervated hippocampus: A chronic model for epilepsy. *Neuroscience* 28:527–538

Buzsaki G, Hsu M, Slamka C, Gage FH, Horvath Z (1991): Emergence and propagation of interictal spikes in the subcortically denervated hippocampus. *Hippocampus* 1:163–180

Chrobak JJ, Napier TC (1989): Vehicle infusions into the basal forebrain produces task-specific cognitive deficits in the rat. *Soc Neurosci Abstr* 15:1173

Chrobak JJ, Napier TC (1991): Intraseptal administration of bicuculline produces working memory impairments in the rat. *Behav Neural Biol* 55:247–254

Chrobak JJ, Napier TC (1992): Antagonism of GABAergic transmission within the septum disrupts working/episodic memory in the rat. *Neuroscience* 47:833–841

Chrobak JJ, Walsh TJ (1991): Dose- and delay-dependent working/episodic memory impairments following intraventricular administration of ethylcholine aziridinium ion (AF64A). *Behav Neural Biol* 56:200–212

Chrobak JJ, Stackman RR, Walsh TJ (1989a): Intraseptal administration of muscimol produces dose-dependent memory impairments in the rat. *Behav Neural Biol* 52:357–369

Chrobak JJ, Spates MJ, Stackman RW, Walsh TJ (1989b): Hemicholinium-3 prevents the working memory impairments and the cholinergic hypofunction induced by ethylcholine aziridinium ion (AF64A). *Brain Res* 504:269–275

Chrobak JJ, Napier TC, Hanin I, Walsh TJ (1991): The pharmacology of basal forebrain involvement in cognition. In: *Basal Forebrain: Anatomy to Function: Advances in*

Experimental Medicine and Biology, Volume 295, Napier TC, Kalivas P, Hanin I, eds. New York: Plenum Press, pp 383–398

Costa E, Panula P, Thompson HK, Cheney DL (1983): The transynaptic regulation of the septal-hippocampal cholinergic neurons. *Life Sci* 32:165–179

Davies P, Maloney AJF (1976): Selective loss of central cholinergic neurons in Alzheimer's Disease. *Lancet* 2:1403

Decker M, Pelleymounter MA, Gallagher M (1988): Effects of training on a spatial memory task on high affinity choline uptake in hippocampus and cortex in young adult and aged rats. *J Neurosci* 8:90–99

Dudar JD, Whishaw IQ, Szerb JC (1979): Release of acetylcholine from the hippocampus of freely moving rats during memory stimulation and running. *Neuropharmacology* 18:673–678

Durkin T, Koenig J (1991): Septal GABAergic interneurons and the transynaptic control of basal activity and memory test-induce activation of septo-hippocampal cholinergic neurons in mice. *Soc Neurosci Abstr* 17:138

Freund TF, Antal M (1988): GABA-containing neurons in the septum control inhibitory interneurons in the hippocampus. *Nature* 336:170–173

Galey D, Toumane A, Durkin T, Jaffard R (1989): In vivo modulation of septo-hippocampal cholinergic activity in mice: Relationship with spatial reference and working memory performance. *Behav Brain Res* 32:163–172

Gallagher M, Pelleymounter MA (1988): An age-related spatial learning deficit: Choline uptake distinguishes "impaired" and "unimpaired" rats. *Neurobiol Aging* 9:363–369

Givens BS, Olton DS (1990): Cholinergic and GABAergic modulation of medial septal area: Effect on working memory. *Behav Neurosci* 104:849–855

Gold PE (1984): Memory modulation: Roles of peripheral catecholamines. In: *The Neuropsychology of Memory*, Squire L, Butters N, eds. New York: Guilford Press

Gold PE, McGaugh JL (1975): A single-trace, two-process view of memory storage processes. In: *Short-Term memory*, Deutsch D, Deutsch JA, eds. New York: Academic Press

Hagan JJ, Morris RGM (1988): The cholinergic hypothesis of memory: A review of animal experiments. In: *Handbook of Psychopharmacology, Vol 20, Psychopharmacology of the Aging Nervous System*, Iverson LL, Iverson SD, Snyder SH, eds. New York: Plenum Press

Hebb DO (1949): *The Organization of Behavior*. New York: John Wiley & Sons

Henke H, Lang W (1983): Cholinergic enzymes in neocortex, hippocampus and basal forebrain of non-neurological and senile dementia of alzheimer-type patients. *Brain Res* 267:281–291

Jakab RL, Leranth C (1990): Catecholaminergic, GABAergic, and hippocamposeptal innervation of GABAergic somatospiny neurons in the rat lateral septal area. *J Comp Neurol* 302:305–321

Larson J, Lynch G (1986): Induction of synaptic potentiation in hippocampus by pattern stimulation involves two events. *Science* 232:985–988

McGaugh JL (1989): Dissociating learning and performance: Drug and hormone enhancement of memory storage. *Brain Res Bull* 23:339–345

McLennan H, Miller JJ (1974): γ-aminobutyric acid and inhibition in the septal nuclei of the rat. *J Physiol* 237:625–633

Mitchell SJ, Rawlins JNP, Steward O, Olton DS (1982): Medial septal area lesions disrupt theta rhythm and cholinergic staining in medial entorhinal cortex and produce impaired radial arm maze behavior in rats. *J Neurosci* 2:292–302

Mizumori SJY, Perez GM, Alvarado MC, Barnes CA, McNaughton BL (1990): Reversible

inactivation of the medial septum differentially affects two forms of learning in rats. *Brain Res* 528:12–20
Olton DS (1986): Hippocampal function and memory for temporal context. In: *The Hippocampus, Vol 4* Isaacson R, Pribram K, eds. New York: Plenum Press
Petsche H, Stumpf C, Gogolak G (1962): The significance of the rabbit's septum as a relay station between the midbrain and the hippocampus: I. The control of hippocampus arousal activity by the septum cells. *EEG Clin Neurophysiol* 14:202–211
Raisman G (1966): The connections of the septum. *Brain* 9:317–348
Rolls ET (1991): Functions of the primate hippocampus in spatial and nonspatial memory. *Hippocampus* 1:258–261
Rylett RJ, Ball MJ, Colhoun EH (1983): Evidence for high affinity choline transport in synaptosomes prepared from hippocampus and neocortex of patients with Alzheimer's disease. *Brain Res* 289:169–175
Saper CB (1987): Diffuse cortical projection systems: Anatomical organization and role in cortical function. In: *Handbook of Physiology, Section 1, The Nervous System, Volume 5, Part 1*, Mountcastle VB, Plum F, Geiger SR, eds. Bethesda, MD: American Physiology Society
Stackman R, Walsh TJ (1992): Chlordiazepoxide-induced working memory impairment: Site specificity and antagonism with RO15,1788. *Behav Neural Biol*, 57:233–243
Staubli U, Huston JP (1980): Facilitation of learning by post-trial injection of substance P into the medial septal nucleus. *Behav Brain Res* 1:245–255
Stewart M, Fox SE (1989): Two populations of rhythmically bursting neurons in rat medial septum are revealed by atropine. *J Neurophysiol* 61:982–993
Swanson LW, Cowan WM (1979): The connections of the septal region in the rat. *J Comp Neurol* 186:621–656
Vinogradova OS, Brazhnik ES, Karanov AM, Zhadina SD (1980): Neuronal activity of the septum following various types of deafferentiation. *Brain Res* 187:353–368
Walsh TJ, Chrobak JJ (1990): Animal models of Alzheimer's disease: Role of hippocampal cholinergic systems in working memory. In: *Current Topics in Animal Learning: Brain, Emotion and Cognition*, Dachowsky L, Flaherty C, eds. Hillsdale, NJ: Erlbaum
Walsh TJ, Tilson HA, DeHaven DL, Mailman RB, Fisher A, Hanin I (1984): AF64A, a cholinergic neurotoxin, selectively depletes acetylcholine in hippocampus and cortex, and produces long-term passive avoidance and radial-arm maze deficits in the rat. *Brain Res* 321:91–102
Wenk G, Hepler D, Olton DS (1984): Behavior alter the uptake of [^{3}H] choline into acetylcholinergic neurons of the nucleus basalis magnocellularis and medial septal area. *Behav Brain Res* 13:129–138
Whishaw IQ (1987): Hippocampal, granule cell and $CA_{3\text{-}4}$ lesions impair formation of a place learning-set in the rat and induce reflex epilepsy. *Behav Brain Res* 24:59–72
Winson J (1978): Loss of hippocampal theta rhythm results in spatial memory deficit in the rat. *Science* 210:160–163
Winson J, Abzug C (1978): Neuronal transmission through hippocampus pathways dependent on behavior. *J Neurophysiol* 41:463–476
Wolfman C, Da Cunha C, Jerusalinsky D, Levi de Stein M, Viola H, Izquierdo I, Medina JH (1991): Habituation and inhibitory avoidance training alter brain regional levels of benzodiazepine-like molecules and are affected by intracerebral flumazenil microinjection. *Brain Res* 548:74–80

19

Working Memory and Cholinergic Autoregulation in the Medial Septal Area

BENNET GIVENS AND DAVID OLTON

This chapter explores the hypothesis that cholinergic neurotransmission onto GABAergic and cholinergic neurons in the medial septal area (MSA) can elicit rhythmic neural output that coordinates hippocampal information processes and enhances working memory. The MSA is comprised of the medial septal nucleus and the vertical limb of the diagonal band. One of the major efferent projections from the MSA is to the hippocampal, entorhinal, and subicular regions of the temporal lobe. The septohippocampal pathway contains fibers from both cholinergic and GABAergic neurons of the MSA that receive afferent input from GABAergic neurons of the lateral septum, and from cholinergic neurons in the basal forebrain and the brainstem. The GABAergic input to MSA has been well studied anatomically (Leranth and Frotscher, 1989), biochemically (Costa et al., 1983), electrophysiologically (McLennan and Miller, 1974), and behaviorally (Decker and McGaugh, 1991). However, the role of the cholinergic input to MSA neurons is less well characterized and will be the focus of this chapter.

Cholinergic Innervation of MSA

The major source of cholinergic input to MSA neurons is from within the MSA itself. Swanson and Cowan (1979) noted that anterograde tracers injected anywhere within the MSA produced a dense, circumscribed distribution over the rest of the MSA. This "extensive plexus of interconnecting fibers" within the MSA was noted as early as the 1940s (Fox, 1940; Lauer, 1945). Neurons of the vertical limb of the diagonal band drop axon collaterals into the medial septal nucleus as they project to the hippocampus, and these collaterals are immuno-positive for an antibody aginist choline acetyltransferse (ChAT; Ronald Gaykema, personal communication). The other major type of neuron in the MSA contains the GABA-synthesizing enzyme glutamic acid decarboxylase (GAD; Kohler et al., 1984). Unlike other brain regions, these septal GABAergic neurons are projection neurons and do not participate in local interconnectivity. The vast majority of GAD-positive terminals on MSA neurons are lost following transection of fibers from the lateral septum (Leranth and Frotscher, 1989). Thus, cholinergic neurons of the MSA provide collateral innervation to neighboring neurons within the MSA. ChAT-positive (+) terminals are found on both ChAT (+) and GAD (+) neurons in the MSA (Biolawas and Frotscher, 1987). Septal cholinergic neurons appear to innervate cholinergic and GABAergic neurons within the MSA, whereas septal GABAergic neurons apparently do not innervate other septal neurons.

Another potential source for cholinergic innervation of the MSA is from neurons in the brainstem. Two nuclei in the hindbrain, the peripeduncular pontine nucleus and the lateral dorsal tegmental nucleus, contain cholinergic neurons that have ascending projections into forebrain structures. Fibers from these nuclei enter the MSA, where they contact cholinergic neurons (Woolf and Butcher, 1986). This projection from the brainstem to MSA neurons forms the basis for a cholinergic ascending reticular system that can activate the hippocampus via the MSA. The practical consequences of this activation in the MSA is to convert asynchronous volleys into rhythmic output (Brazhnik et al., 1985).

Cholinergic Driving of Rhythmic Septal Output

Cholinergic transmission within the septum activates MSA neurons, i.e., produces an increased firing rate and entrainment into a rhythmic burst pattern of firing. When the muscarinic agonist carbachol is directly applied by iontophoresis, rhythmically bursting MSA neurons respond by increasing their activity (Fig. 1). The rate increases induced by acetylcholine are blocked by the muscarinic antagonist atropine, but not the nicotinic antagonist mecylmylamine (Lamour et al., 1984). Although the vast majority of MSA neurons are activated by muscarinic agonists, MSA neurons respond differentially to muscarinic antagonists. When atropine is applied to the MSA neurons, approximately half of these neurons lose their rhythmic bursting pattern of firing (Stewart and Fox, 1989). A temporal correlation between the loss of rhythmicity of individual MSA neurons and the loss of the hippocampal theta rhythm follows atropine. Thus, the rhythmic firing characteristics of many MSA neurons is controlled by a cholinergic mechanism. Neurons in the MSA are readily inhibited by GABA (Fig. 1). However, application of the GABA-A antagonist bicuculline activates MSA neurons but does not disrupt the rhythmicity or burst frequency, indicating that GABA is unlikely to be involved in the mechanism of oscillatory burst firing (Givens and Breese, 1986). The role of other transmitters in rhythmic burst generation is unknown. Based on careful electrophysiological studies in the rat, Brazhnik and Vinogradova (1986) concluded that anticholinergic drugs block the "intraseptal facilitory cholinergic influences" and decrease the proportion of cells that are rhythmically bursting. Thus, acetylcholine release from within the MSA may drive other MSA neurons to fire rhythmically in bursts.

Acetylcholine may induce MSA neurons to fire in rhythmic bursts by inhibiting calcium-activated potassium channels. Acetylcholine brings about a slow depolarization in cortical and hippocampal neurons by interacting with several

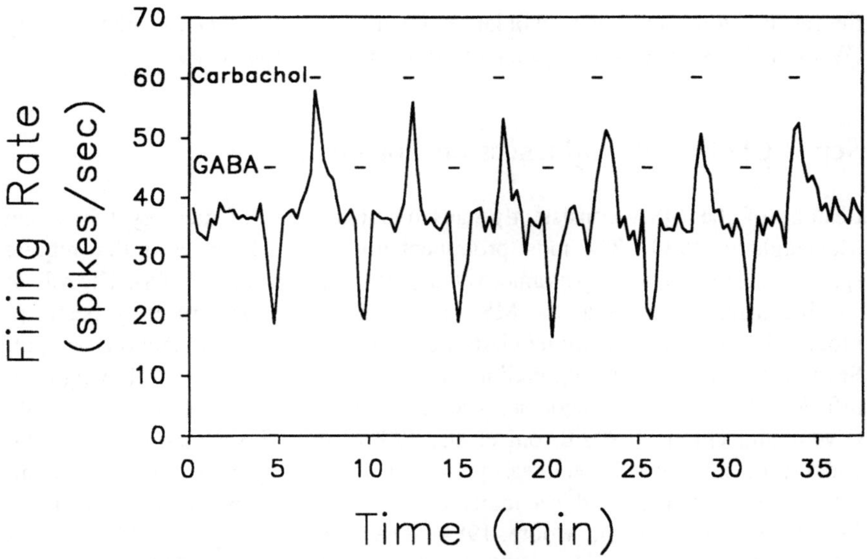

FIGURE 1. Cholinergic activation of a medial septal neuron. Carbachol, a muscarinic receptor agonist, or GABA was iontophoretically applied to a rhythmically bursting neuron in the MSA. Carbachol (15 μA) produced a 45% increase in the firing rate, and GABA (10 μA) produced a 50% decrease in the firing rate. The bars above the line indicate when the current was applied to the pipette. An increase in activity by carbachol was observed in 87% (40/46) of all rhythmic burst firing MSA neurons tested.

potassium-channel conductances: the resting K^+ conductance, the voltage-activated K^+ conductance (m-current), and the calcium-activated K^+ conductance (the afterhyperpolarization current; Madison et al., 1987). The m-current may be brought about by M_1 receptors and the afterhyperpolarization current by M_2 receptors in the hippocampus (Muller and Misgeld, 1986). The afterhyperpolarization is responsible for frequency adaptation. Although devoid of M_1 muscarinic receptors, the septum has many M_2 muscarinic receptors (Spencer et al., 1986). Thus cholinergic transmission in the MSA, if it responds as other forebrain regions, may decrease the afterhyperpolarization and induce burst firing in MSA neurons.

The rhythmically bursting of MSA neurons paces rhythmic slow activity (theta) in the hippocampus (Petsche et al., 1962). Lesions of the MSA eliminate hippocampal theta activity (Green andArduini, 1954). If cholinergic transmission is responsible for the rhythmicity, then disruption of ongoing cholinergic transmission should eliminate theta rhythm in the freely behaving rat. To test this possibility, we have infused the muscarinic antagonist scopolamine directly into the MSA of male Long-Evans rats through indwelling cannulae and have recorded hippocampal theta activity. When scopolamine was infused into the MSA, hippocampal theta rhythm was completely abolished (Givens and Olton, 1990). The hippocampal theta rhythm has previously been linked to memory processes in the rat, including inhibitory avoidance (Landfield et al., 1971), spatial memory (Winson, 1978), and working memory (Givens and Olton, 1990).

Septal Cholinergic Influences on Cognition

Lesions of the MSA consistently caused impairments of memory (Gray and McNaughten, 1983). The most prominent impairment is on tasks that require spatial working memory (Miyamoto et al., 1987; Hagan et al., 1988). The role of specific afferent inputs to the MSA in cognition has only recently been approached. The neurotransmitter GABA clearly regulates MSA functional output. Several laboratories have reported mnemonic impairments in rats following local infusion of the GABA-A agonist muscimol into the MSA (Chrobak et al., 1989; Givens and Olton, 1990; Brioni et al., 1990). Some investigations into beta-endorphin, substance P, and vasopressin have also reported impairments, although the anatomical evidence indicates that the lateral septum may be the locus for these effects (Gall and Moore, 1984). The cholinergic input to MSA neurons has been demonstrated anatomically (Biolawas and Frotscher, 1987) and electrophysiologically (Dutar et al., 1983), but the behavioral consequences of cholinergic activation of MSA have not been investigated. The effects of manipulating cholinergic transmission within the MSA on spatial and nonspatial memory are the focus of the rest of the chapter.

Cognitive Consequences of Septal Muscarinic Antagonism

If cholinergic neurotransmission in the septum is critical for accurate performance of memory tasks, then direct muscarinic antagonism by intraseptal infusion of scopolamine should impair performance in a dose-dependent fashion. Rats were trained on one of three tasks that were designed to test various types of memory: delayed alternation (DA), continuous conditional discrimination (CCD), and two-choice discrimination (2CD). The DA task was spatial working memory, the CCD task was nonspatial working memory, and the 2CD task was nonspatial reference memory.

Each of the three tasks had different stimulus-response contingencies. The DA task used a T-maze with a 20-sec intertrial interval. In order to receive a water reward, the rat was required to visit the arm of the maze opposite to the arm visited on the previous trial. The CCD and 2CD tasks used an operant chamber and required that the rat press one of two levers (left or right) in response to one of two stimuli (tone or light) to receive water. The 2CD task was structured such that a right or left lever press was rewarded following a light or tone stimulus, respectively. The CCD required the rat to remember the previous stimulus in order to respond correctly to the present stimulus. If a light (or tone) stimulus on one trial was followed by another light (or tone) stimulus on the next trial, the rat was rewarded if it responded on the left, or "match" lever. Conversely, if the stimulus changed from one stimulus modality to the other on subsequent trials, then the correct response to the current stimulus was on the right, or "nonmatch" lever.

The procedure for assessing the role of septal cholinergic transmission was nearly identical in all three tasks. After criterion performance was reached, each rat was surgically implanted with a cannula directed towards the MSA, and recording electrodes in the hippocampus. After stable postoperative performance was established, rats were microinfused with scopolamine hydrochloride (0.5 μl over 5 min) at one of five doses: 0, 5, 15, 30, or 60 μg. Electrophysiological recording of theta was done either immediately before and 15 min after microinfusion in the case of the DA task, or immediately after microinfusion and then continuously for 90 min while the rat was performing in the operant chamber in the case of the CCD and 2CD tasks.

The effect of scopolamine on the hippocampal theta rhythm was identical in each of the three tasks. A peak in the spectral power was observed between 7 and 8 Hz in baseline recordings or after intraseptal infusions of saline. Following infusions of scopolamine, this peak was absent. In the CCD and 2CD tasks, where theta was continuously recorded, theta was abolished immediately following infusion, but then slowly returned to baseline levels during the next 30-60 min. By 90 min after the infusion of scopolamine, the power of theta in all three behavioral tasks had returned to baseline levels.

The effects of intraseptal scopolamine on behavior depended on the memory demands required. In the DA task, scopolamine (30 μg) significantly impaired choice accuracy, reducing it from 88% to 64% (Givens and Olton, 1990). Thus, intraseptal scopolamine impaired spatial working memory. In the CCD task, scopolamine (30 μg) had a dramatic effect on choice accuracy at all three delay intervals during the first 30 min after infusion. After 30 min, accuracy at the shortest delay (2.5 sec) recovered rapidly, while accuracy at the longest delay (20 sec) only slowly improved and never fully recovered to baseline levels. The results from the CCD task indicate that nonspatial working memory is also impaired by intraseptal scopolamine (Table 1; Givens and Olton, submitted). In the 2CD task, scopolamine (5, 15, 30, and 60 μg) had no effect on choice accuracy during the 90 min after infusion. Muscarinic blockade in the MSA did not affect nonspatial reference memory. The behavioral results from these three studies suggest that cholinergic transmission in the MSA, and thus activation of the septohippocampal pathway, is involved in working memory processes.

TABLE 1. Mnemonic Effects of Microinfusion of Scopolamine (15 μg) into the MSA

Task	Type of Memory	Effect
Delayed alternation	Spatial working	Impairment
Conditional discrimination	Nonspatial working	Impairment
2 choice discrimination	Nonspatial reference	Unaffected

Modulation of Septal Cholinergic Transmission

If septal cholinergic transmission is functionally important for activity along the septohippocampal pathway and accurate working memory, then modulation of the transmission should affect these physiological and behavioral measures. A candidate neuromodulator of acetylcholine is the 29 amino acid peptide galanin. Galanin is localized in cholinergic neurons of the MSA (Senut et al., 1989) and is believed to be co-released with acetylcholine and to modulate (antagonize) the actions of acetylcholine (Crawley, 1991). Consistent with the idea that the MSA has septo-septal cholinergic fibers, galanin receptors are present in abundance in the septum (Skofitsch et al., 1986). The role of septal galanin in working memory was tested in a series of experiments (Givens et al., 1992).

Galanin was microinfused into the MSA of rats that were trained on a delayed alternation (DA) task. Galanin caused a dose-dependent impairment in choice accuracy in this task and a concomitant suppression of theta activity. However, the effect of galanin on theta was slightly different from that of scopolamine in that high doses of galanin reduced, but did not eliminate, the amplitude of the theta wave. Because these studies did not directly test whether galanin interacted with acetylcholine transmission, an independent action of galanin cannot be eliminated. Nonetheless, the partial antagonism is consistent with a modulatory role of galanin on cholinergic transmission in the MSA, which further supports the hypothesis that cholinergic transmission in the MSA is important for working memory.

Septal Site of Amnesic Action of Peripheral Scopolamine?

Peripheral administration of scopolamine impairs memory in laboratory animals and in humans (Spencer and Lal, 1983). Although not specifically demonstrated, this amnestic effect of scopolamine is assumed to be mediated through muscarinic receptors located in the cortex and hippocampus. However, as described previously, scopolamine microinfused into the MSA also brings about impairments in working memory. Perhaps, the amnestic effect of peripherally administered scopolamine is mediated, in part, by muscarinic antagonism within the MSA. Stewart and Fox (1989) have shown that a subpopulation of neurons in the MSA loses its rhythmic burst pattern of firing when exposed to locally applied atropine. In order to test the idea that the amnestic effects of peripheral scopolamine may involve cholinergic receptors in the MSA, experiments that combined peripheral and central administration of muscarinic agents were carried out.

Activation of septal muscarinic receptor sites can attenuate the behavioral and physiological effects of scopolamine. Rats were trained in delayed alternation (DA) and then received peripheral scopolamine (at 0.2 or 0.4 mg/kg dose) or saline pretreatment, followed 15 min later by direct microinfusion of the muscarinic agonist carbachol (0.5 μg) or saline. Six groups were formed based on the drugs that were administered peripherally/centrally: SAL/SAL, SAL/CARB, SCOP-(0.2)/SAL, SCOP(0.2)/CARB, SCOP(0.4)/SAL, and SCOP(0.4)/CARB. These doses of scopolamine and carbachol that produced consistent effects on memory without affecting general performance. Theta activity was recorded in all rats. Scopolamine caused a predictable dose-dependent decrease in choice accuracy when followed by intraseptal saline. Both doses of scopolamine also shifted the peak theta frequency from 7.5 to 8.5 Hz. The larger dose of scopolamine also slightly decreased the maximal power at the peak frequency. Behavioral performance in the SCOP(0.2)/CARB group was significantly better than the SCOP(0.2)/SAL group and was not different from that of the SAL/SAL group (Fig. 2). Carbachol failed to improve the performance in the SCOP(0.4)/CARB group. Theta in the SCOP(0.2 and 0.4)/CARB groups had peak frequencies intermediate (8 Hz) between those of the SCOP(0.2 and 0.4)/SAL and SAL/SAL groups. These results further underscore the significance of septal cholinergic transmission and suggest that the memory-impairing effects of scopolamine can be overcome by direct muscarinic activation of septal cholinergic receptors.

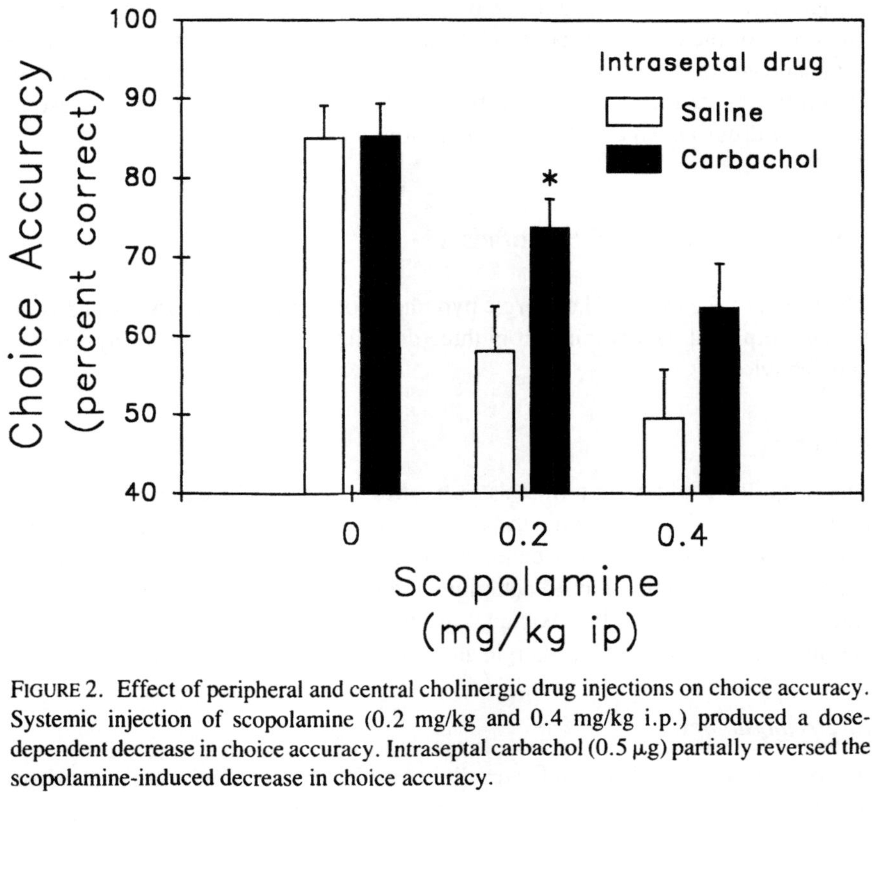

FIGURE 2. Effect of peripheral and central cholinergic drug injections on choice accuracy. Systemic injection of scopolamine (0.2 mg/kg and 0.4 mg/kg i.p.) produced a dose-dependent decrease in choice accuracy. Intraseptal carbachol (0.5 μg) partially reversed the scopolamine-induced decrease in choice accuracy.

Cognitive Enhancement by Intraseptal Muscarinic Agonists

Because pharmacologically induced memory impairments were alleviated by an intraseptal cholinergic agonist, naturally occurring conditions of memory impairment may also benefit by muscarinic activation of the MSA. Declines in memory are characteristic of aged rats (Barnes, 1979; Gallagher and Pellymounter, 1988). In the MSA, the number of neurons (Gilad et al., 1987) but not the synaptic density (Scheff et al., 1991) decreases with increasing age. In order to test the hypothesis that muscarinic activation of the MSA is capable of improving mnemonic function, experiments were designed to measure working memory in aged rats following intraseptal infusion of the muscarinic agonist, oxotremorine (Givens et al., 1991).

Activation of septal muscarinic receptor sites attenuated the behavioral and physiological consequences of aging. Rats, aged 22–24 months old, had a baseline level of choice accuracy in DA of approximately 70%. Intraseptal infusion of oxotremorine produced in a dose-dependent increase in accuracy, with the smallest effective dose being 0.5 μg oxotremorine. When this dose was administered into the lateral ventricles, it was ineffective at improving performance. Moreover, intraseptal oxotremorine shifted the predominant frequency in hippocampal EEG from 7.5 to 6.5 Hz in these aged rats and increased the EPSP amplitude of the evoked response in dentate to perforant path stimulation. These data indicate that muscarinic activation of the MSA can improve memory in memory-impaired aged rats and that the septal muscarinic receptor may be a useful target for drugs aimed at enhancing cognition.

Future Directions for Hypothesis

The future of the "septal cholinergic hypothesis of working memory" depends on further empirical data coming from three domains: anatomy, electrophysiology, and behavior.

Anatomy

The origin of cholinergic terminals in the MSA needs to be determined. These terminals may arise from within the septum or other basal forebrain nuclei, ascend from brainstem cholinergic nuclei, or originate from a combination of these sources. It is also important to determine the precise targets of these cholinergic fibers within the MSA. For example, are there different termination patterns of cholinergic afferents on GABAergic and cholinergic cell populations?

Electrophysiology

Although *in vitro* evidence indicates different electrophysiological characteristic for cholinergic and GABAergic cell populations in the MSA (Griffith and

Matthews, 1987; Markram and Segal, 1990), *in vivo* electrophysiological studies need to confirm the distinctions and to determine how each neuron type responds to cholinergic activation. Moreover, it is critical to correlate single unit firing of the different MSA cell populations to defined behavioral states, especially as the activity relates to sensory processing and motor function in memory tasks (Givens and Olton, 1990).

Behavior

The methods by which the behavioral functions of the MSA have been studied have gradually been refined. Early experiments, using combined MSA and lateral septal lesions, revealed a myriad of behavioral effects (such as "septal rage") that were difficult to organize into a single conceptual framework (Issacson, 1982). Later studies, selective MSA lesions demonstrated consistent impairments in learning and memory (Gray and McNaughton, 1983). In the last 10 years the technique of intraseptal microinfusion has been applied to behavioral studies and has revealed some fairly subtle effects, e.g., different effects on CA1 and CA3 place cell activity (Mizumori et al., 1989). Behavioral data consistently indicate that the MSA is important for spatial working memory, but the specific subsystems within the MSA contributing to these processes are not known. Cholinergic and GABAergic neurons are differentially distributed within the MSA (e.g., one group of cholinergic neurons lies along the midline) and have different termination patterns in the hippocampus. Finer microinfusion techniques with smaller gauge injectors and smaller injection volumes, and more selective pharmacological agents (such as a galanin antagonist), need to be developed to target subpopulations of MSA neurons in order to assess their role in mnemonic processes.

This combination of increasing specificity in anatomy, electrophysiology, and behavior should provide significant advances in our understanding of the neural mechanisms involved in memory.

References

Barnes CA (1979): Memory deficits associated with senescence: A neurophysiological and behavior study in the rat. *J Comp Physiol Psychol* 93:74–104

Biolawas J, Frotscher M (1987): Choline acetyltransferase-immunoreactive neurons and terminals in the rat septal complex: A combined light and electron microscopic study. *J Comp Neurol* 259:298–307

Brazhnik ES, Vinogradova OS, Karonov AM (1985): Frequency modulation of neuronal theta-bursts in rabbit's septum by low-frequency repetitive stimulation of the afferent pathways. *Neuroscience* 14:501–508

Brazhnik ES, Vinogradova OS (1986): Control of neuronal rhythmic bursts in the septal pacemaker of theta-rhythm: Effects of anaesthetic and anticholinergic drugs. *Brain Res* 380:94–106

Brioni JD, Decker MW, Gamboa LP, Izquierdo I, McGaugh JL (1990): Muscimol injections in the medial septum impair spatial learning. *Brain Res* 522:227–234

Chrobak JJ, Stackman RW, Walsh TJ (1989): Intraseptal administration of muscimol produces dose-dependent memory impairments in the rat. *Behav Neural Biol* 52:357–369

Costa E, Panula P, Thompson HK, Cheney DL (1983): The transsynaptic regulation of the septal-hippocampal cholinergic neurons. *Life Sci* 32:165–179

Crawley JN (1991): Coexistence of neuropeptides and "classical" neurotransmitters. Functional interactions between galanin and acetylcholine. *Annals NY Acad Sci* 579:223–245

Decker MW, McGaugh JL (1991): The role of interactions between the cholinergic system and other neuromodulatory systems in learning and memory. *Synapse* 7:151–168

Dutar P, Lamour Y, Jobert A (1983): Acetylcholine excites identified septohippocampal neurons in the rat. *Neurosci Lett* 43:43–47

Fox CA (1940): Certain basal telencephalic centers in the cat. *J Comp Neurol* 72:1–62

Gall C, Moore RY (1984): Distribution of enkephalin, substance P, tyrosine hydroxylase, and 5-hydroxytryptamine immunoreactivity in the septal region of the rat. *J Comp Neurol* 225:212–227

Gallagher M, Pellymounter MA (1988): Spatial learning deficits in old rats: A model for memory decline in the aged. *Neurobiol Aging* 9:549–556

Gilad GM, Rabey JM, Tizabi Y, Gilad VH (1987): Age-dependent loss and compensatory changes of septohippocampal cholinergic neurons in two rat strains differing in longevity and response to stress. *Brain Res* 436:311–322

Givens B, Breese GR (1986): Rhythmically bursting neurons of the medial septum/diagonal band are excited by bicuculline. *Soc Neurosci Abstr* 12:1525

Givens B, Olton DS (1990a): Cholinergic and GABAergic modulation of medial septal area: Effect on working memory. *Behav Neurosci* 104:849–855

Givens B, Olton DS (1990b): Medial septal area codes mnemonic components of a working memory task. *Soc Neurosci Abstr* 16:918

Givens B, Markowska AM, Olton DS (1991): Muscarinic activation of the medial septal area: Improvements in working memory and modulation of hippocampal physiology in aged rats. *Soc Neurosci Abstr* 17:136

Givens B, Olton DS, Crawley JN Galanin in the medial septal area impairs working memory. *Brain Res* 582:71–77

Givens B, Olton DS. Local modulation of basal forebrain activity impairs working and reference memory. (Submitted)

Gray JA, McNaughton N (1983): Comparison of the behavioral effects of septal and hippocampal lesions: A review. *Neurosci Biobehav Rev* 7:119–188

Green JD, Arduini A (1954): Hippocampal electrical activity in arousal. *J Neurophysiol* 17:533–557

Griffith WH, Matthews RT (1986): Electrophysiology of AChE-positive neurons in basal forebrain slices. *Neurosci Lett* 71:169–174

Hagan JJ, Salamone J, Simpson J, Iversen SD, Morris RGM (1988): Place navigation in rats is impaired by lesions of medial septum and diagonal brand but not nucleus basalis magnocellularis. *Behav Brain Res* 27:9–20

Issacson RL (1982): *The Limbic System*. New York: Plenum Press

Kohler C, Chan-Palay V, Wu J-Y (1984): Septal neurons containing glutamic acid decarboxylase immunoreactivity project to the hippocampal region in the rat brain. *Anat Embryol* 169:41–44

Landfield PW, McGaugh JL, Tusa RJ (1971): Theta rhythm: A temporal correlate of memory storage in the rat. *Science* 175:87–89

Lamour Y, Dutar P, Jobert A (1984): Septo-hippocampal and other medial septum-diagonal band neurons: Electrophysiological and pharmacological properties. *Brain Res* 309:227–239

Lauer EW (1945): The nuclear pattern and fiber connections of certain basal telencephalic centers in the macaque. *J Comp Neurol* 82:215–254

Leranth C, Frotscher M (1989): Organization of the septal region in the rat brain: Cholinergic-GABAergic interconnections and the termination of hippocampo-septal fibers. *J Comp Neurol* 289:304–314

Madison DV, Lancaster B, Nicoll RA (1987): Voltage clamp analysis of cholinergic action in the hippocampus. *J Neurosci* 7:733–741

Markram H, Segal M (1990): Electrophysiological characteristics of cholinergic and non-cholinergic neurons in the rat medial septum-diagonal band complex. *Brain Res* 513:171–174

McLennan H, Miller JJ (1974): Gamma-aminobutyric acid and inhibition in the septal nuclei of the rat. *J Physiol* 237:625–633

Miyamoto M, Kata J, Narumi S, Nagaoka A (1987): Characteristics of memory impairment following lesioning of the basal forebrain and medial septal nucleus in rats. *Brain Res* 419:19–31

Mizumori SJY, McNaughton BL, Barnes CA, Fox K (1989): Preserved spatial coding in hippocampal CA1 pyramidal cells during reversible suppression of CA3c output: Evidence for pattern completion in hippocampus. *J Neurosci* 9:3915–3928

Muller W, Misgeld U (1986): Slow cholinergic excitation of guinea pig hippocampal neurons is mediated by two muscarinic receptor subtypes. *Neurosci Lett* 67:107–112

Petsche H, Stumpf C, Golagak G (1962): The significance of the rabbit's septum as a relay station between the midbrain and hippocampus. I. The control of hippocampus arousal activity by the septum cells. *EEG Clin Neurophysiol* 14:202–211

Scheff SW, Scott SA, DeKosky ST (1991): Quantitation of synaptic density in the septal nuclei of young and aged Fischer 344 rats. *Neurobiol Aging* 12:3–12

Senut MC, Menetrey D, Lamour Y (1989): Cholinergic and peptidergic projections from the medial septum and the nucleus of the diagonal band of broca to dorsal hippocampus, cingulate cortex and olfactory bulb: A combined wheatgerm agglutinin-apohorseradish peroxidase-gold immunohistochemical study. *Neuroscience* 30:385–403

Skofitsch G, Sills MA, Jacobowitz DM (1986): Quantitative distribution of galanin-like immunoreactivity in the rat central nervous system. *Peptides* 7:609–613

Spencer DG, Lal H (1983): Effects of anticholinergic drugs on learning and memory. *Drug Devel Res* 3:489–502

Spencer DG, Horvath E, Traber J (1986): Direct autoradiographic determination of M1 and M2 muscarinic acetylcholine receptor distribution in the rat brain: Relation to cholinergic nuclei and projections. *Brain Res* 380:59–68

Stewart M, Fox S (1989): Two populations of rhythmically bursting neurons in rat medial septum are revealed by atropine. *J Neurosci* 61:982–993

Swanson LW, Cowan WM (1979): The connections of the septal region in the rat. *J Comp Neurol* 186:621–656

Winson J (1978): Loss of hippocampal theta rhythm results in spatial memory deficit in the rat. *Science* 201:160–163

Woolf NJ, Butcher LJ (1986); Cholinergic systems in the rat brain: III. Projections from the pontomesencephalic tegmentum to the thalamus, tectum, basal ganglia, and basal forebrain. *Brain Res Bull* 16:603–637

20

Modulation of Memory by Benzodiazepine-Acetylcholine Interactions

THOMAS J. WALSH AND ROBERT W. STACKMAN

Introduction

Memory is one of the most enduring challenges in all of behavioral neuroscience. A memory "system" must be able to engage a diverse array of neural systems that serve perceptual, emotive, and cognitive functions. It needs to (a) access and process sensory information, (b) analyze its biological significance in a specific context, determine (c) whether, (d) how, and (e) where that information should be stored, and (f) the appropriate context in which to later use it. To appreciate the rich complexity of the process we need to examine the interplay of *systems* that support memory and not just focus on a single brain structure or neurotransmitter. The remnants of memory reside in circuits not synapses. To understand the drama of memory we need to listen to the dialogue of all of the characters.

This chapter focuses on the relationships between neurons that utilize acetylcholine (ACh), endogenous benzodiazepines, and GABA. In particular, we address the ways in which these neuronal systems coordinate their activities to modulate memory.

Acetylcholine and Memory

It is well established that brain cholinergic systems participate in learning and memory. In particular, the septohippocampal cholinergic pathway appears to be a critical neural substrate of working memory processes (see Walsh and Chrobak, 1991 for review). Alterations of this system induced by mechanical, electrolytic, or neurotoxic lesions or the administration of pharmacological agents produce impairments in cognitive function. Moreover, this system degenerates in Alzheimer's disease (AD), with the loss of cholinergic parameters correlated with the severity of the prevailing cognitive deficit (Bartus et al., 1982).

A number of experimental strategies have been developed to examine the role of the septohippocampal cholinergic pathway in memory. The most common manipulation has been to destroy either the cholinergic cells of origin that reside in the medial septum or to sever the fimbria-fornix, which conveys the cholinergic afferents from the medial septum to the hippocampal formation. A problem inherent in this strategy is that the medial septum consists of a heterogeneous population of neurons of which only 35% to 60% are likely to be cholinergic (Butcher, 1978; Woolf et al., 1984). Other cells in this region contain GABA, substance P, galanin, and other neuropeptides (Senut et al., 1989; Woodhams et al., 1983; Freund and Antal, 1988). In fact, Peterson and colleagues did observe a significant loss of GABA and substance P-containing neurons in the medial septum after transection of the fimbria-fornix (Petersen et al., 1987). Furthermore, catecholamine and indoleamine projections from the brainstem ascend to the hippocampus (HPC) and forebrain through the fimbria-fornix (Gaspar et al., 1985). Therefore, a variety of transmitter systems are compromised by these common experimental procedures (Chafetz et al., 1981), and the potential contribution of these systems to the behavioral deficits observed following hippocampal damage has often been overlooked. Although the memory impairments induced by these manipulations might depend on the loss of cholinergic input to the HPC, new approaches with greater anatomical and neurochemical specificity will be needed to better define the behavioral properties of the septohippocampal cholinergic pathway.

A number of laboratories have attempted to create models of chronic cholinergic hypofunction with the cholinotoxin AF64A. AF64A (ethylcholine aziridinium ion) is a neurotoxic analog of choline that produces a persistent decrease in all indices of presynaptic cholinergic function in the HPC and selective memory impairments in a variety of tasks. (reviewed in Walsh and Chrobak, 1991; Hanin et al., 1987). Within a narrow range of doses, AF64A produces a permanent disruption of the cholinergic innervation of the HPC. The conceptual advantages of using this approach are threefold. First, it is easier to try to model diseases characterized by a chronic cholinergic hypofunction. Second, it is possible to evaluate strategies to prevent, mitigate, or reverse the neurobiological and/or behavioral deficits resulting from such a compromise. Finally, one can examine the plasticity and synaptic reorganization exhibited by a damaged brain system.

The use of a temporary disruption of the septohippocampal cholinergic pathway might also be a valuable way to explore the functions of this system. A theoretical problem with any lesion technique is that it is not always clear what role the damaged system is playing in the intact state. The ability to produce an acute, reversible model of cholinergic hypofunction might be more useful for investigating the synaptic events that serve to regulate this system and may provide a greater understanding of the neural processes involved in memory. Based upon the synaptic circuitry of the medial septum, we have focused on the interactions of GABA, benzodiazepine (BDZ), and acetylcholine-containing neurons. Precise pharmacological manipulation of these systems in the septum is proving to be a useful strategy for exploring the neurobiology of memory.

Anatomy and Physiology of Acetylcholine-GABA-BDZ Interactions

Cholinergic neurons in the medial septum send fibers to the entire septo-temporal extent of the HPC (Frotscher and Leranth, 1985). These cholinergic fibers directly modulate the physiological responsiveness of the HPC and generate theta activity (Bland, 1986). The theta rhythm coordinates the physiology of the HPC and regulates the inherent plasticity of its synaptic circuitry (Larson et al., 1986). Cholinergic neurons within the medial septum appear to be synaptically regulated by a number of transmitter systems (Costa et al., 1983; Robinson, 1982; Robinson et al., 1979; Wood, 1985). In particular, GABAergic neurons in the lateral septum (Brashear et al., 1986; Onteniente et al., 1987) exert a powerful inhibitory influence on cholinergic cells in the medial septum (McLennan and Miller, 1974). Leranth and Frotscher (1989), using double-labeling techniques, have shown that GABAergic fibers originating from neurons in the lateral septum project to and synapse on cholinergic neurons in the medial septum. Chu and colleagues (1990) have recently demonstrated that the medial septum does contain both types of GABA receptors, but the ratio of GABA-A to GABA-B receptors is approximately 4:1. Furthermore, in all brain areas examined GABA-A receptors have a threefold lower K_D (i.e., greater affinity) and a twofold higher B_{max} (i.e., more binding sites) than GABA-B receptors. The GABA-A receptor is a molecular complex that includes (a) a GABA receptor, (b) a benzodiazepine receptor (BDZ), and (c) a chloride ion channel. Activation of GABA-A/BDZ receptors opens the chloride channel and either hyperpolarizes the postsynaptic membrane or neutralizes the effects of an existing depolarization (Duman et al., 1987). Stimulating GABA-A/BDZ receptors in the medial septum would exert a profound inhibitory influence on cells containing those receptors.

Radiohistochemical studies have found BDZ receptors in highest density in the cerebral cortex, septum, HPC, dentate gyrus, median forebrain bundle, and amygdala of the rat brain (Mohler et al., 1981). The widespread distribution of BDZ receptors in these neuronal areas suggests potential sites for the pharmacological effects of BDZs. BDZ receptors are located in the septum, and the concentration of endogenous benzodiazepine-like compounds (i.e., endozepines) is highest in this structure (Wolfman et al., 1991). There is a 15-fold regional variation in content of endozepines, with limbic structures, such as the septum, amygdala, and HPC, exhibiting the highest concentrations and the cerebellum exhibiting the lowest. GABA, endozepines, and their receptors are located in the septum, and it is likely that they participate in the local modulation of septal neurons.

A variety of neurotransmitters might exert their influence on cholinergic neurons in the septum through a final common GABAergic pathway. Costa and colleagues have demonstrated that the turnover rate (TR) of ACh can be manipulated by intraseptal administration of agonists and antagonists of GABAergic, catecholaminergic, and peptidergic systems (reviewed in Costa et al., 1983).

Intraseptal administration of the GABA-A agonist muscimol produces a 40% to 50% reduction in hippocampal Ach TR, and it also decreases theta activity (Allen and Crawford, 1984). These effects seem to be mediated by the activation of the GABA-A receptor, since they can be blocked by the administration of the GABA-A antagonist, bicuculline, and are not mimicked by intraseptal administration of the GABA-B agonist baclofen (Blaker et al., 1986). There is other pharmacological evidence that afferents to the septum that release opiates, glutamate, or dopamine modulate the activity of cholinergic neurons indirectly through their influence on GABAergic neurons. Thus, inhibitory GABAergic neurons in the septum serve as a final common pathway for the regulation of septal cholinergic neurons that project to the HPC. Pharmacological modulation of GABA-A/BDZ receptors located on cholinergic neurons in the medial septum should lead to behavioral and biochemical effects reflecting a relatively selective compromise of the septohippocampal cholinergic pathway. Systemically administered BDZs should also interact with GABA-A/BDZ receptors in a variety of brain areas.

Behavioral Pharmacology of the Benzodiazepines

The benzodiazepines (BDZ) represent one of the most widely prescribed classes of psychotherapeutic agents. There are more than 25 BDZs used in clinical practice as anxiolytics, anticonvulsants, hypnotics, sedatives, and muscle relaxants (File and Pellow, 1987). In addition to their therapeutic actions, BDZs can also produce undesirable side effects, such as amnesia, ataxia, fatigue, confusion, and depression (Lister, 1985; Thiebot, 1985). While there is a growing appreciation of the pharmacological *mechanism* of action of these drugs, there is a limited understanding of their *sites* of action.

Anxiolytic Sites of Action

A number of studies have attempted to clarify the neuroanatomical site responsible for the anxiolytic effect of BDZs. Nagy and colleagues (1973) have demonstrated a dose-related anxiolytic effect following infusion of diazepam into the anterior nuclei of the amygdala of rats. This antianxiety effect is similar in magnitude to that produced by systemic administration of diazepam. Scheel-Krüger and Petersen (1982) demonstrated that the anticonflict effects of BDZ agonists can also be elicited by injection of midazolam into the medial regions of the amygdala. A recent investigation by Hodges and colleagues (1987) indicated that the infusion of GABA and BDZs into the anterior amygdala produced potent anticonflict effects. The effects of the BDZs were antagonized by coadministration of the BDZ antagonist flumazenil (RO15, 1788). These experiments suggest that components of the amygdala mediate the anxiolytic and anticonflict effects of BDZs. This is further supported by the observation that the anxiolytic efficacy of various BDZs is positively correlated with their affinity for BDZ receptors in the amygdala (Thomas et al., 1985).

Effects of Benzodiazepines on Learning and Memory

There is a large literature describing the dose-related effects of a variety of BDZs on cognitive processes. Systemic administration of BDZs impairs cognition and/or performance in both humans and laboratory animals on a variety of tasks (Lister, 1985; Thiebot, 1985; Cole, 1986). At present, there is no clear conceptual hypothesis concerning the neurobiological or cognitive effects mediating BDZ-induced amnesia. In fact, the amnestic effects of BDZs have been related to either (a) a disruption of the acquisition of new information, (b) altered recall or retrieval of previously acquired information, or (c) a disruption of the process of memory encoding, maintenance, or consolidation.

Peripheral administration of nonsedative doses of BDZs has been shown to induce an amnesia in a number of aversively motivated tasks (see Gamzu, 1988). This amnestic effect is typically observed when the compound is administered prior to, but not following, acquisition. In addition, McNaughton and Morris (1987) demonstrated that administration of CDP prior to training disrupted the acquisition and retention of a spatial memory task. CDP significantly increased the latencies of rats to reach the escape platform in a Morris water-maze task. During a subsequent test of retention, the CDP-treated rats were impaired in their ability to locate the platform. Therefore, the anterograde amnestic property of BDZ treatment has been consistently produced, in both humans and laboratory animals, in a number of aversively motivated paradigms.

In contrast to the well-documented anterograde amnesia of BDZs, Jensen and colleagues (1979) reported that some (flurazepam), but not all (diazepam, lorazepam), BDZs can induce retrograde amnesia for a passive-avoidance task when injected immediately after training. The magnitude and time course of GABA-A/BDZ-induced inhibition produced by each respective BDZ ligand might be responsible for the differential amnestic effects observed (Gamzu, 1988).

Recently, Cole (1988) reported that the systemic injection of CDP impaired successive discrimination performance of rats. Animals who received the BDZs made substantially more operant responses during extinction trials than saline-treated subjects in a go-no go discrimination paradigm. In such a task, a single trial involves the cued alternation between continuous reinforcement and extinction. Therefore it is likely that the BDZs altered the ability to correctly process information regarding the current reinforcement schedule. Furthermore, these impairments were reversed by flumazenil.

Flumazenil is a potent and selective BDZ antagonist that has been widely used to study the pharmacological and behavioral properties of BDZ systems. This compound competitively inhibits the binding of agonists to the BDZ receptor and attenuates the behavioral, neurochemical, and neurophysiological effects of BDZ agonists (Bonetti et al., 1982). The specificity of flumazenil is emphasized by the observation that it does not attenuate the sedative or anticonflict effects of other classes of drugs, including barbiturates, ethanol, or morphine (Bonetti et al., 1982).

A number of studies have shown that the BDZ antagonist flumazenil can attenuate or reverse the amnestic effects of diazepam, lorazepam, triazolam, and chlordiazepoxide in both human subjects (Gentil et al., 1989; Dorow et al., 1987; O'Boyle et al., 1983) and laboratory animals without significantly affecting control performance (Bonetti et al., 1982). O'Boyle and colleagues (1983) reported that flumazenil reversed the amnestic effects of diazepam in a test of delayed word recall. Furthermore, the antagonist did not alter the plasma concentrations or the half-life of diazepam. Therefore, it is clear that flumazenil alters the pharmacodynamic actions of BDZ agonists without altering their pharmacokinetic characteristics. While these studies suggest that BDZs produce their amnestic effects through a specific pharmacological action at the BDZ receptor, they do not address the potential site specificity of these effects.

Amnestic Sites of Action

There have been very few studies that have examined potential sites responsible for the amnestic effects of BDZs. However, since BDZ and GABA-A agonists inhibit the activity of cholinergic neurons, it is reasonable to hypothesize that their amnestic effects are related to these cholinolytic actions. Zsilla and colleagues reported that intravenous injection of diazepam or muscimol decreased the acetylcholine TR in the cortex and midbrain of rats (Zsilla et al., 1976). The intraseptal injection of muscimol is also sufficient to decrease hippocampal theta rhythm and to produce a reduction of ACh turnover (Allen and Crawford, 1984). Neurochemical studies in our laboratory have shown that intraseptal injection of BDZ and GABAergic compounds decreased hippocampal high-affinity choline uptake (HAChU) in a dose-dependent manner (Stackman et al., 1989). HAChU is the rate-limiting step in the synthesis of ACh and is used to determine the in vivo activity of cholinergic neurons. The pharmacological specificity of these effects was shown by the observation that intraseptal bicuculline, a GABA-A antagonist, attenuated the muscimol-induced decrease in HAChU. Thus it is possible to speculate that the amnestic actions of BDZs are related to the ability of these compounds to inhibit the dynamic activity of cholinergic neurons in the medial septum.

Intraseptal administration of muscimol produces a significant decrease in the occurrence of hippocampal theta activity (Allen and Crawford, 1984), which can be attenuated by prior administration of bicuculline. Systemic injection of muscimol or BDZs raises the threshold required to elicit theta activity in the HPC (Quintero et al., 1985). This suggests that activation of GABA receptors, directly or via BDZ binding, produces an inhibition of cholinergic activity in the HPC. Due to the high density of GABA-A and BDZ receptors in the medial septum (Chu et al., 1990; Young and Kuhar, 1980), and the demonstration that GABAergic neurons synapse on and inhibit the activity of septal cholinergic neurons (Leranth and Frotscher, 1989; Brashear et al., 1986; Zaborszky et al., 1986), the medial septum might represent a neural site of action for the amnestic effects of BDZs.

Muscimol-Induced Working Memory Impairments

Preliminary studies from this laboratory have focused on two interrelated questions: (a) the effects of manipulating specific GABA and BDZ receptors in the medial septum on working memory performance in a radial-arm maze (RAM) task, and (b) the dose-related effects of these compounds on HAChU in the HPC. These studies have observed a striking covariation between the cognitive effects of these drugs and their effects on cholinergic parameters in the HPC.

Intraseptal injection of muscimol produced a dose-dependent impairment in a delayed-non-match-to-sample (DNMTS) radial-arm maze task, evidenced by fewer correct choices and more errors (Chrobak et al., 1989). Intraseptal muscimol did not affect latency to perform the task, the ability to locomote through the maze, or the willingness to consume food pellets following a correct arm entry. Furthermore, we found that the amnestic dose of muscimol (3.0 nmoles) produced a significant decrease, while the ineffective dose (0.75 nmoles) produced a nonsignificant decrease in HAChU in the HPC. However, even high doses of the GABA-B agonist baclofen produced only a small nonsignificant decrease of HAChU. Finally, the decrease in HAChU produced by 3 nmoles muscimol could be significantly attenuated by intraseptal injection of 1 nmoles bicuculline, a selective GABA-A antagonist, injected into the medial septum 5 min before muscimol. These data demonstrate that GABA-A/BDZ receptors in the septum exert an inhibitory influence on cholinergic neurons and their activation can disrupt memory processes that depend upon cholinergic function.

Sites and Mechanism of Action of Chlordiazepoxide-Induced Retrograde Amnesia

Based upon our observations with muscimol, we decided to examine the amnestic effects of chlordiazepoxide (CDP), a prototype BDZ that has been widely used in biobehavioral research. We were particularly interested in the temporal characteristics, site specificity, and potential mechanisms of action of CDP-induced working memory impairments. In these studies we used a DNMTS radial-arm maze task. In this task four of eight arms are open and baited with food (94 mg chocolate-flavored pellets). After entering all four open arms, the rat is returned to its home cage for a 1-hr delay. Following the delay rats are returned to the maze and allowed free access to all eight arms, with only those previously blocked arms now baited (DNMTS). Dependent measures during the postdelay session were (a) the number of correct choices in the first four arm selections (C-C), (b) the total number of errors, and (c) the latency per arm choice. The configuration of open and closed arms varied each day so that the rat was required to retain trial-dependent information for only one trial. This repeated acquisition feature promoted the use of within-subject crossover designs in which a single rat can receive different doses of a given drug and serve as its own control (see Chrobak et al., 1989, for a more complete description of the task).

Rats injected systemically with CDP immediately following the predelay session exhibited a dose-related decrease in correct choices and a similar increase in errors during the postdelay session. Doses of 2.5 mg/kg and 5.0 mg/kg, but not 1.25 mg/kg CDP significantly impaired performance (Stackman and Walsh, 1992) (Fig. 1A). None of the treatments affected latency per arm choice. These data illustrate that systemically administered CDP can produce a retrograde amnesia for an appetitive working memory task. These cognitive impairments are evident at doses that are less than those needed to produce an anxiolytic effect in conflict, negative contrast, and punished-responding paradigms (see Flaherty 1990, for review). Consequently, there appears to be a differential dose-response curve for the amnestic and anxiolytic effects of BDZs, with amnestic effects observed at lower doses.

In a second series of studies using the same task, we determined that CDP-induced working memory impairments were site specific. All rats were injected in a random sequence with artifical cerebrospinal fluid (CSF), and 15 nmoles, and 30 nmoles of CDP immediately following the predelay session, and were tested 1 hr later. Implantation of a guide cannula into the medial septum and

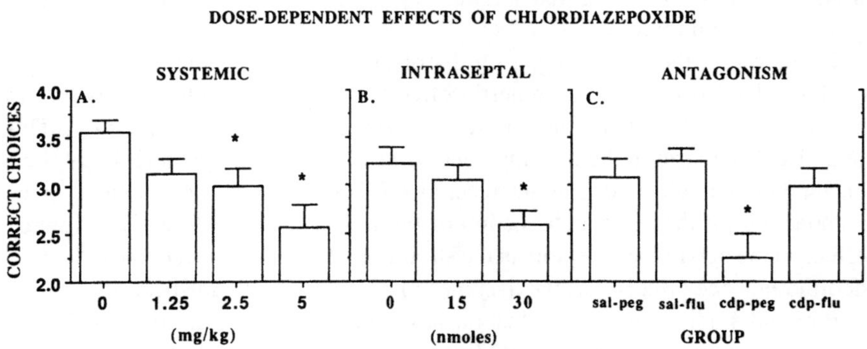

FIGURE 1. This figure illustrates the dose-dependent effects of systemic (A) or intraseptal (B) injection of chlordiazepoxide (CDP) on performance in a delayed-non-match-to-sample radial-arm maze task. Data are presented as the average number of correct choices made following the 1-hr delay. *$p < 0.05$ vs. vehicle-injected controls (0 group). Panel C illustrates that intraseptal injection of flumazenil (1.5 μg) prevented the amnestic effects of systemically injected CDP (5 mg/kg). Sal-peg = systemic saline followed by intraseptal injection of 5% polyethylene glycol (vehicle); sal-flu = systemic saline followed by intraseptal injection of flumazenil; cdp-peg = systemic chlordiazepoxide followed by intraseptal injection of vehicle; and cdp-flu = systemic chlordiazepoxide followed by intraseptal injection of flumazenil. *$p < 0.05$ vs. sal-peg group.

intraseptal injection of CSF had no effect on performance. Injection of 30 nmoles, but not 15 nmoles, CDP into the medial septum immediately following the predelay session produced (a) a dose-related decrease in correct choices, (b) a dose-related increase in errors, and (c) no change in latency per arm choice (Fig. 1B). These deficits were transient, and repeated intraseptal injection has no detrimental effect on baseline performance. The CDP-induced working memory impairments were also time-depenent. Injection of 30 nmoles CDP into the septum immediately, but not 15 min following, the predelay session impaired working memory. These data strongly argue against a state-dependent interpretation. The half-life of CDP is approximately 4 hr following systemic injection (Koechlin et al., 1965). Therefore, subjects injected with CDP either immediately or 15 min following training were most probably experiencing similar "brain states" during the postdelay session.

These data also suggest that CDP disrupts an early phase of working memory encoding or maintenance. There is an apparent time "window" during which memory is susceptible to CDP. Recently, Givens and Olton (1989) reported correlative decreases in choice accuracy in a t-maze task with decreases in hippocampal theta activity following intraseptal injection of amnestic doses of muscimol. It would be interesting to observe the activity of septohippocampal neurons during the performance of the delay RAM task, during the predelay session, and immediately afterward, to determine if the output of those neurons changes dramatically during the different phases of information processing.

Bilateral injection of CDP (10, 20, or 30 nmoles or CSF) into the anterior amygdala had no effect on any index of RAM performance. Higher doses were not used since pilot studies found that doses of CDP greater than 30 nmoles produced gross behavioral toxicity.

In summary, these data suggest a dissociation between neural substrates that mediate the anxiolytic and amnestic effects of CDP. It might be possible to develop regionally targeted anxiolytics that do not impair cognitive function. This logic has been used in the development of the atypical antipsychotic drugs (i.e., clozapine) that block dopamine receptors in mesolimbic sites thought to be responsible for the therapeutic action of these drugs. These drugs have limited effect on the dopamine receptors in the striatum, which are believed to mediate the debilitating side effects (tardive dyskinesia) of antipsychotic medications.

Previous studies have shown that the amnestic effects of a variety of BDZs can be attenuated by flumazenil, a BDZ antagonist, in both humans and laboratory animals (see above). While these studies do support the specific involvement of BDZ receptors in BDZ-induced memory impairments, they do not suggest a potential site of action. If CDP-induced amnesia is mediated through BDZ receptors in the medial septum, then intraseptal injection of flumazenil should prevent these memory impairments.

To avoid the pharmacodynamic problems of coinjecting an agonist and antagonist into the same brain area, we injected an amnestic dose of CDP (5 mg/kg) systemically and then injected flumazenil into the medial septum. Immediately following the predelay session, rats were injected intraperitoneally with saline or 5.0 mg/kg of CDP and then with either flumazenil (10 nmoles; 1.5 μg/0.5 μl) or the polyethylene glycol (5.0%/0.5 μl) vehicle into either the medial septum or anterior nucleus of the amygdala. The doses of flumazenil and PEG were selected based upon literature values and from preliminary neurochemical studies in our laboratory (Hodges et al., 1987; Stackman et al., 1989). Performance was assessed 1 hr following the predelay session.

Systemic injection of CDP impaired performance in the DNMTS task, increasing errors and decreasing the number of correct choices. Intraseptal, but not intraamygdala, injection of flumazenil prevented the amnestic effect of CDP (Fig. 1C).

It appears that CDP must interact with BDZ receptors in the medial septum to disrupt working memory in the appetitive DNMTS task. Pharmacological blockade of these receptors with flumazenil prevented the amnestic effects of CDP. CDP was able to exert all of its biochemical and neurophysiological effects at both its central and peripheral sites of action, except for the medial septum. These data are strong support for the hypothesis that the amnestic effects of CDP are mediated through interactions with BDZ receptors in the medial septum. In addition, infusion of CDP, at doses previously shown to be anxiolytic, into the anterior amygdala had no effect on working memory in this task. This suggests a potential dissociation between the neural sites mediating the therapeutic and debilitating effects of CDP.

The results of these experiments and our preliminary observation that intraseptal CDP (30 nmoles) decreases HAChU in the HPC (Stackman et al., 1989) suggests that the amnestic effects of GABA-A/BDZ receptor agonists may be related to their ability to decrease the activity of the septohippocampal cholinergic pathway during a critical phase of memory. In contrast, BDZ antagonists and inverse agonists might be able to enhance normal cognitive function and to attenuate the deficits associated with cholinergic hypofunction. It has been demonstrated that BDZ antagonists and inverse agonists can (a) enhance memory in a variety of appetitive and aversive paradigms (Duka et al., 1987; Jensen et al., 1987), (b) attenuate scopolamine-induced impairments of passive avoidance retention (Sarter et al., 1988), and (c) attenuate deficits in the performance of a radial maze task induced by ibotenic acid lesions of the nucleus basalis (Sarter and Steckler, 1989). Finally, intraseptal injection of both flumazenil and the β-carboline, β-CCM, increases HAChU in the HPC in a dose-related manner (Stackman et al., 1989). The behavioral significance of this effect is currently under study.

Benzodiazepines, the Amygdala, and Memory

There is a substantial literature supporting a role for the amygdala in memory processes in which a salient affective component is involved, such as in aversive learning. Stimulation and lesions of the amygdala disrupt the acquisition and retention of most aversive tasks, including passive and active avoidance, conditioned taste aversion, shock-motivated discrimination, and others (see McGaugh

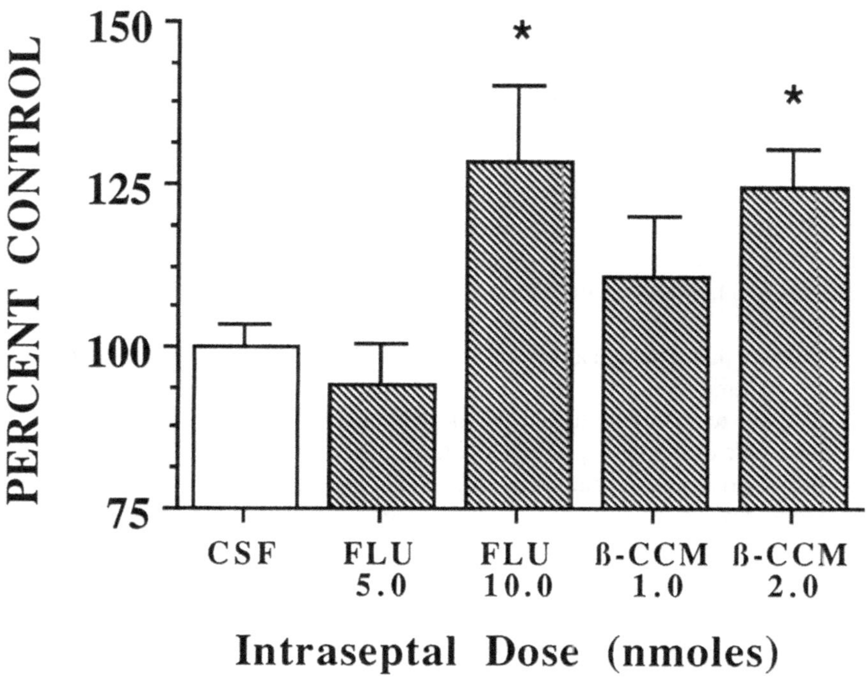

FIGURE 2. This figure illustrates the effects of intraseptal flumazenil and β-CCM on high-affinity choline uptake in the hippocampus. Rats were injected in the medial septum 1 hr prior to sacrifice with either artificial CSF, or 5.0 or 10.0 nmoles flumazenil, or 1.0 or 2.0 nmoles β-CCM. The data are presented as mean (+ SEM) percent of control uptake values. * $p < 0.05$ vs. CSF.

et al., 1990 for review). Manipulation of the amygdala does not typically affect such appetitive tasks as spatial learning in a RAM or delayed-matching or nonmatching discrimination tasks with or without spatial components (Raffaele and Olton, 1988). Therefore, the amygdala seems to participate in memory processes in a task-dependent manner.

A number of laboratories have proposed an important role for GABAergic mechanisms in the amygdala in the modulation of aversive learning (McGaugh et al., 1990; Izquierdo et al., 1990a,b). Systemic or intraamygdala injection of GABA-A antagonists (bicuculline or picrotoxin) enhances retention of a passive-avoidance task, while similar injections of GABA-A (muscimol) or GABA-B (baclofen) agonists impair retention (Brioni et al., 1989; Castellano et al., 1989).

It will be important to examine whether GABA-A/BDZ-mediated processes in different brain areas uniquely contribute to the performance of cognitive tasks with different motivational attributes, degrees of complexity, and cognitive requirements. Are the GABA/BDZ receptors in the amygdala and septum part of a network of structures that play a critical role in different types of memory processes? Such a question becomes even more significant in light of the recent discovery of naturally occurring BDZ-like molecules in the brain (De Robertis et al., 1988).

Endogenous Ligands of the BDZ Receptor and Memory

The discovery of specific BDZ receptors has lead to an extensive search for endogenous ligands for these sites (i.e., endozepines). At present a number of endogenous molecules with either agonist (N-desmethyldiazepam, diazepam) or inverse agonist (diazepam-binding inhibitor (DBI), and its biologically active fragments, ODN and TTN, and n-butyl-β-carboline-3-carboxylate (β-CCB) properties, are candidate ligands for these receptors (for review see Costa and Guidotti, 1991). Diazepam and n-desmethyldiazepam induce anxiolytic effects in animal models, while DBI, ODN, TTN, and β-CCB are anxiogenic (Ferrero et al., 1986). File and Pellow (1986) have proposed that there might be several endogenous BDZ ligands with either agonist or inverse agonist properties that can bidirectionally modulate the GABA-A receptor complex. Furthermore, there is evidence that endogenous BDZ ligands participate in the modulation of aversive memory processes (Izquierdo et al., 1990a,b).

Clinical Implications

The misuse of BDZs is a widespread therapeutic problem, and there are particular patient populations that might be uniquely sensitive to the debilitating effects of these compounds. For example, BDZs might be particularly problematic in geriatric populations, who are predisposed to cognitive decline and who are commonly prescribed BDZs for a host of conditions, despite their limited ability to metabolize and eliminate these drugs. The scope of this problem is highlighted by a study demonstrating that at least 25% of nursing home patients diagnosed with AD were receiving one of the long-acting BDZs: diazepam, flurazepam, or CDP (Beers et al., 1988). The impact of these drugs on cognitive function in these patients needs to be further evaluated. Furthermore, while BDZs might exaggerate the cognitive deficits observed in the normal aged or the AD patient, BDZ antagonists or inverse agonists might be a viable approach to enhancing cognition.

The ability to enhance the activity of cholinergic neurons by disinhibiting them with GABA-A or BDZ antagonists, or inverse agonists, may represent a useful approach to treat cognitive disorders such as AD that are characterized by a prominent cholinergic hypofunction. At the present time there is no effective pharmacological treatment for this disorder. Cholinomimetics have proven to be relatively ineffective due to the prevalence of side effects, their short half-lives, their pharmacodynamic variability between patients, and their nonselective effects on different cholinergic systems and receptor subtypes. Furthermore, the indiscriminate activation of muscarinic receptors by cholinergic agonists might not enhance the functional activity of cholinergic systems that operate in a phasic and rhythmic way (i.e., generation of theta activity). A more useful approach might be to "amplify" the ongoing rhythmic activity of cholinergic neurons by limiting the inhibitory influence of GABA-A/BDZ receptors (Sarter et al., 1990). Furthermore, this might be a necessary first step in the pharmacological treatment of AD, since it has been proposed that surviving cholinergic neurons are "hyperinnervated" by GABAergic terminals in the AD brain (Sarter et al., 1990).

Summary and Conclusions

The literature reviewed here suggests that there is an important neurobiological interaction between neurons that utilize ACh, GABA, and endogenous BDZs. Furthermore, the coordinated activities of these neurotransmitters modulate memory processes. A more complete understanding of these interactions will contribute not only to our knowledge of the biology of memory but could also prove to have important implications for the treatment of age-related memory disorders such as AD.

Acknowledgments. The authors would like to thank Karen Opello for her comments on this manuscript. Some of the original research reported here was supported by NSF grant BNS-9109163 to TJW.

References

Allen CN and Crawford IL (1984): GABAergic agents in the medial septal nucleus affect hippocampal theta rhythm and acetylcholine utilization. *Brain Res* 322:261–267

Bartus RT, Dean RL, Beer B, Lippa AS (1982): The cholinergic hypothesis of geriatric memory dysfunction. *Science* 217:408–417

Beers M, Avorn J, Soumerai SB, Everitt DE, Sherman DS, Salem S (1988): Psychoactive medication use in intermediate-care facility residents. *JAMA* 260:3016–3020

Blaker WD, Cheney DL, Costa E (1986): GABA-A vs. GABA-B modulation of septal-hippocampal interconnections. In: *Dynamics of Cholinergic Function*, Hanin I, ed. New York: Plenum Press, pp 953–961

Bland BH (1986): The physiology and pharmacology of hippocampal formation theta rhythms. *Prog Neurobiol* 26:1–54

Bonetti EP, Pieri L, Cumin R, Schaffner R, Pieri M, Gamzu ER, Muller R, Haefely W (1982): Benzodiazepine antagonist RO 15-1788: Neurological and behavioral effects. *Psychopharmacology* 78:8–18

Brashear HR, Zaborszky L, Heimer L (1986): Distribution of GABAergic and cholinergic neurons in the rat diagonal band. *Neuropharmacology* 17:439–451

Brioni JD, Nagahara AH, McGaugh JL (1989): Involvement of the amygdala GABAergic system in the modulation of memory storage. *Brain Res* 487:105–112

Butcher LL (1978): Recent advances in histochemical techniques for the study of central cholinergic mechanisms. In: *Cholinergic Mechanisms and Psychopharmacology*, Jenden DJ ed. New York: Plenum Press, pp 93–124

Castellano C, Brioni JD, Nagahara AH, McGaugh JL (1989): Post-training systemic and intra-amygdala administration of the GABA-B agonist baclofen impairs retention. *Behav Neural Biol* 52:170–179

Chafetz MD, Thompson RG, Evans SH, Gage FH (1981): Biochemical specificity of septal hyperreactivity: A behavioral discrimination. *Behav Brain Res* 2:409–420

Chrobak JJ, Stackman RW, Walsh TJ (1989): Intraseptal administration of muscimol produces dose-dependent memory impairments in the rat. *Behav Neural Biol* 52:357–369

Chu DCM, Albin RL, Young AB, Penney JB (1990): Distribution and kinetics of GABA-B bindings sites in rat central nervous system: A quantitative autoradiographic study. *Neuroscience* 34:341–357

Cole SO (1986): Effects of benzodiazepines on acquisition and performance: A critical assessment. *Neurosci Biobehav Revs* 10:265–272

Cole SO (1988): Dose-dependent reversal of chlordiazepoxide-induced discrimination impairment by RO 15-1788. *Psychopharmacology* 96:458–461

Costa E, Guidotti A (1991): Diazepam binding inhibitor (DBI): A peptide with multiple biological actions. *Life Sci* 49:325–344

Costa E, Panula P, Thompson HK, Cheney DL (1983): The transynaptic regulation of the septal-hippocampal cholinergic neurons. *Life Sci* 32:165–179

De Robertis E, Pena C, Paladini AC, Medina JH (1988): New developments on the search for the endogenous ligand(s) of central benzodiazepine receptors. *Neurochem Int* 13:1–11

Dorow R, Berenberg D, Duka T, Sauerbrey N (1987): Amnestic effects of lormetazepam and their reversal by the benzodiazepine antagonist RO15-1788. *Psychopharmacology* 93:507–514

Duka T, Stephens DN, Krause W, Dorow R (1987): Human studies on the benzodiazepine receptor antagonist beta-carboline ZK 93426, Preliminary observations on psychotropic activity. *Psychopharmacology* 93:421–427

Duman RS, Sweetnam PM, Gallombardo PA, Tallman JF (1987): Molecular neurobiology of inhibitory amino acid receptors. *Mol Neurobiol* 1:155–189

Ferrero P, Santi MR, Conti-Tronconi B, Costa E, Guidotti A (1986): Study of an octadecaneuropeptide derived from diazepam binding inhibitor (DBI): Biological activity and presence in rat brain. *Proc Natl Acad Sci USA* 83:827–831

File SE, Pellow S (1986): Intrinsic actions of the benzodiazepine receptor antagonist RO15,1788. *Psychopharmacology* 88:1–11

File SE, Pellow S (1987): Behavioral pharmacology of minor tranquilizers. *Pharmacol Ther* 35:265–290

Flaherty CF (1990): Effect of anxiolytics and antidepressants on extinction and negative contrast. *Pharmacol Ther* 46:309–320

Freund TF, Antal M (1988): GABA-containing neurons in the septum control inhibitory interneurons in the hippocampus. *Nature* 336:170–173

Frotscher M, Leranth C (1985): Cholinergic innervation of the rat hippocampus as revealed by choline acetyltransferase immunocytochemistry, a combined light and electron microscopic study. *J Comp Neurol* 239:237–246

Gamzu ER (1988): Animal model studies of benzodiazepine-induced amnesia. In: *Benzodiazepine Receptor Ligands, Memory and Information Processing*, Hindmarch I, Ott H, eds. Berlin: Springer-Verlag, pp 218–229

Gaspar P, Berger B, Alvarez C, Vigny A, Henry JP (1985): Catecholaminergic innervation of the septal area in man: Immunocytochemical study using TH and DBH antibodies. *J Comp Neurol* 241:12–33

Gentil V, Gorenstein C, Camargo C, Singer JM (1989): Effects of flunitrazepam on memory and their reversal by two antagonists. *J Clin Psychopharmacol* 9:191–197
Givens BS, Olton DS (1989): Cholinergic and GABAergic modulation of medial septal area: Effect on working memory. *Behav Neurosci* 104:849–855
Hanin I, Fisher A, Hortnagl H, Leventer SM, Potter PE, Walsh TJ (1987): Ethylcholine mustard aziridinium (AF64A; ECMA) and other potential cholinergic neuron-specific neurotoxins. In: *Psychopharmacology—The Third Generation of Progress*, Meltzer HY, ed. New York: Raven Press, pp 341–349
Hodges H, Green S, Glenn B (1987): Evidence that the amygdala is involved in benzodiazepine and serotonergic effects on punished responding but not discrimination. *Psychopharmacology* 92:491–504
Izquierdo I, Cunha C, Huang CH, Wolfman C, Medina JH (1990a): Post-training down-regulation of memory consolidation by a GABA-A mechanism in the amygdala modulated by endogenous benzodiazepines. *Behav Neural Biol* 54:105–109
Izquierdo I, Cunha C, Medina JH (1990b): Endogenous benzodiazepine modulation of memory processes. *Neurosci Biobehav Rev* 14:419–424
Jensen RA, Martinez JL, Vasquez BJ, McGaugh JL (1979): Benzodiazepines alter acquisition and retention of an inhibitory avoidance response in mice. *Psychopharmacology* 64:125–126
Jensen RA, Stephens DN, Sarter M, Petersen EN (1987): Bidirectional effects of beta-carbolines and benzodiazepines on cognitive processes. *Brain Res Bull* 19:359–364
Koechlin BA, Schwartz MA, Krol G, Oberhansli W (1965): The metabolic fate of C14-labeled chlordiazepoxide in man, in the dog, and in the rat. *J Pharmacol Exper Ther* 148:399–411
Larson J, Wong D, Lynch G (1986): Patterned stimulation at the theta frequency is optimal for the induction of hippocampal long-term potentiation. *Brain Res* 368:347–350
Leranth C, Frotscher M (1989): Organization of the septal region in the rat brain, Cholinergic-GABAergic inter-connections and the termination of hippocampo-septal fibers. *J Comp Neurol* 289:304–314
Lister RG (1985): The amnesic action of benzodiazepines in man. *Neurosci Biobehav Rev* 9:87–94
McGaugh JL, Introini-Collison IB, Nagahara AN, Cahill L, Brioni JD, Castellano C (1990): Involvement of the amygdaloid complex in neuromodulatory influences on memory storage. *Neurosci Biobehav Rev* 14:425–431
McLennan H, Miller JJ (1974): GABA and inhibition in the septal nuclei of the rat. *J Physiol* 237:625–633
McNaughton N, Morris RGM (1987): Chlordiazepoxide, an anxiolytic benzodiazepine, impairs place navigation in rats. *Behav Brain Res* 24:39–46
Mohler H, Wu JY, Richards G (1981): Benzodiazepine receptors: Autoradiographical and immunocytochemical evidence for their localization in regions of GABAergic synaptic contacts. In: *GABA and Benzodiazepine Receptors*, Costa E et al., eds. New York: Raven Press, pp 139–146
Nagy J, Zambo K, Decsi L (1973): Anti-anxiety action of diazepam after intra-amygdaloid application in the rat. *Neuropharmacology* 18:573–576
O'Boyle C, Lambe R, Darragh A, Taffe W, Brick I, Kenny M (1983): RO15-1788 antagonizes the effects of diazepam in man without affecting its bioavailability. *Br J Anaesthes* 55:349–355
Onteniente B, Geffard M, Campistron G, Calas A (1987): An ultrastructural study of

GABA-immunoreactive neurons and terminals in the septum of the rat. *J Neurosci* 7:48–54

Peterson GM, Williams LR, Varon S, Gage FH (1987): Loss of GABAergic neurons in medial septum after fimbria-forniz transection. *Neurosci Lett* 76:140–144

Quintero S, Mellanby J, Thompson MR, Nordeen H, Nutt D, McNaughton N, Gray JA (1985): Septal driving of hippocampal theta rhythm: Role of gamma-aminobutyrate-benzodiazepine receptor complex in mediating effects of anxiolytics. *Neuroscience* 16:875–884

Raffaele KC, Olton DS (1988): Hippocampal and amygdaloid involvement in working memory for nonspatial stimuli. *Behav Neurosci* 102:349–355

Robinson RE (1982): Effect of specific serotonergic lesions on cholinergic neurons of the hippocampus, cortex, and striatum. *Life Sci* 32:345–353

Robinson SE, Malthe-Sorenssen D, Wood PL, Commissiong J (1979): Dopaminergic control of the septal-hippocampal cholinergic pathway. *J Pharmacol Exp Ther* 208:476–479

Sarter M, Steckler T (1989): Spontaneous exploration of a 6-arm radial maze by basal forebrain lesioned rats: Effects of the benzodiazepine receptor antagonist beta-carboline ZK 93,426. *Psychopharmacology* 98:193–202

Sarter M, Bodewitz G, Stephens DN (1988): Attenuation of scopolamine-induced impairment of spontaneous alternation behavior by antagonist but not inverse agonist and agonist beta-carbolines. *Psychopharmacology* 94:491–495

Sarter M, Bruno JP, Dudchenko P (1990): Activating the damaged basal forebrain cholinergic system: Tonic stimulation versus signal amplification. *Psychopharmacology* 101:1–17

Scheel-Krüger J, Petersen EN (1982): Anticonflict effect of the benzodiazepines mediated by a GABAergic mechanism in the amygdala. *Eur J Pharmacol* 82:115–116

Senut MC, Menetrey D, Lamour Y (1989): Cholinergic and peptidergic projections from the medial septum and the nucleus of the diagonal band of broca to dorsal hippocampus, cingulate cortex and olfactory bulb: A combined wheatgerm agglutinin-apohorseradish peroxidase-gold immunohistochemical study. *Neuroscience* 30:385–403

Stackman RW, Walsh TJ (1992): Chlordiazepoxide-induced working memory impairments: Site specificity and reversal by flumazenil (RO 15-1788). *Behav Neural Biol* 57:233–243

Stackman RW, Emerich DF, Taylor LA, Walsh TJ (1989): Intraseptal administration of GABA and benzodiazepine agonists and antagonists: Alterations in hippocampal choline uptake and cognitive behavior. *Soc Neurosci Abst* 15:272.14

Thiebot M-H (1985): Some evidence for amnesic-like effects of benzodiazepines in animals. *Neurosci Biobehav Rev* 9:95–100

Thomas SR, Lewis ME, Iversen SD (1985): Correlation of [^{3}H] diazepam binding density with anxiolytic locus in the amygdaloid complex of the rat. *Brain Res* 342:85–90

Walsh TJ, Chrobak JJ (1991): Animal models of Alzheimer's Disease. Role of hippocampal cholinergic system is working memory. In: *Current Topics in Animal Learning, Brain, Emotion and Cognition*, Dachowsky L, Flaherty C, eds. Hillsdale, NJ: Lawrence Erlbaum, pp 347–379

Wolfman C, Da Cunha C, Jerusalinsky D, Levi de Stein M, Viola H, Izquierdo I, Medina JH (1991): Habituation and inhibitory avoidance training alter brain regional levels of benzodiazepine-like molecules and are affected by intracerebral flumazenil microinjection. *Brain Res* 548:74–80

Wood PL (1985): Pharmacological evaluation of GABAergic and glutaminergic inputs to the nucleus basalis-cortical and the septal-hippocampal cholinergic projections. *Can J Physiol Pharmacol* 64:325–328

Woodhams PL, Roberts GW, Polak JM, Crow TJ (1983): Distribution of neuropeptides in the limbic system of the rat: The bed nucleus of the stria terminalis, septum and preoptic area. *Neuroscience* 8:677–693

Woolf NJ, Eckenstein F, Butcher LL (1984): Cholinergic systems in the rat brain: I. Projections to the limbic telencephalon. *Brain Res Bull* 13:751–784

Young WS, Kuhar MJ (1980): Radiohistochemical localization of benzodiazepine receptors in rat brain. *J Pharmacol Exp Ther* 212:337–346

Zaborszky L, Heimer L, Eckenstein F, Leranth C (1986): GABAergic input to cholinergic forebrain neurons: An ultrastructural study using retrograde tracing of HRP and double immunolabelling. *J Comp Neurol* 250:282–295

Zsilla G, Cheney DL, Costa E (1976): Regional changes in the rate of turnover of acetylcholine in rat brain following diazepam or muscimol. *Naunyn-Schmiedeberg's Arch Pharmacol* 294:251–255

21

Cognition Enhancement Based on GABA-Cholinergic Interactions

MARTIN SARTER, PAUL DUDCHENKO, HOLLY MOORE, LEE ANN HOLLEY, AND JOHN P. BRUNO

Transsynaptic Modulation of Activity in Cholinergic Neurons

Since the early anatomical descriptions of the basal forebrain (Brockhaus, 1942; Gorry, 1963) and the determination of acetylcholine (ACh) as a major transmitter of its cortical projections (for review see Fibiger, 1982; Butcher and Wolf, 1986), research aimed at the determination of the functions of forebrain ACh has flourished. Early theories have proposed that central cholinergic systems are involved in the processing of relatively specific components of behavioral functions (e.g., in the effects of unrewarded behavior; Carlton, 1963); however, evidence suggesting the involvement of cortical cholinergic afferents in the symptoms of dementia (Bowen et al., 1976; Whitehouse et al., 1983; Palmer et al., 1987a,b; Reinikainen et al., 1990) has resulted in an almost exclusive research focus on the cholinergic mediation of learning and memory (Bartus et al., 1982).

This focus has governed a vast number of animal experiments that stressed the correlations between basal forebrain lesion-induced decreases in cortical markers of cholinergic activity and behavioral impairments considered to reflect learning and memory dysfunctions (Smith, 1988; Sarter et al., 1992). In spite of the absence of a selective cholinotoxin, lesion-induced behavioral impairments have been consistently attributed to a decrease in the availability of cortical ACh. However, accumulating evidence suggests that the functional consequences of excitotoxin-induced basal forebrain lesions are not primarily due to disruption of cortical cholinergic afferents (Robbins et al., 1989a; Kesner et al., 1990; Sarter and Dudchenko, 1991; Page et al., 1991).

Pharmacologically induced blockade of muscarinic receptors has become another prime tool for the examination of functions of the cholinergic system. Although the behavioral impairments produced by muscarinic antagonists most likely are a result of postsynaptic receptor blockade (and not of the antagonist-induced, autoreceptor-mediated stimulation of presynaptic ACh release), administration of scopolamine or atropine in animals and humans has been often considered as a pharmacological model of the cholinergic degeneration in dementia (Collerton, 1986). The fact that muscarinic blockade spares cognitive functions that are prominently disrupted in dementia (e.g., the retrieval of information from semantic memory; Beatty et al., 1986; Kopelman and Corn, 1988) may be due in part to the neurological incongruity between the disease and the scopolamine model of cholinergic impairment.

Thus, in spite of intensive research activities, the functions of the basal forebrain cholinergic efferents in general and their involvement in dementia in particular have remained unsettled (Olton et al., 1991; Fibiger, 1991). It also seems doubtful that the behavioral effects of cholinesterase inhibitors and muscarinic agonists, i.e., of drugs that escalate the magnitude of postsynaptic muscarinic receptors stimulation but (further) disrupt the spatial and temporal orchestration of presynaptic ACh release (Sarter et al., 1990), contribute to the understanding of cholinergic functions and the behavioral consequences of cholinergic cell loss.

Alternatively, we assume that the examination of effects of modulation of ACh release on behavioral abilities represents a research approach that will eventually allow more valid conclusions about cholinergic functions. *Ex vivo* measurements of the rate-limiting step of ACh-synthesis, i.e., sodium-dependent high-affinity choline uptake (HACU), in animals that differ in their behavioral performance (Wenk et al., 1984; Gallagher and Pelleymounter, 1988; Lebrun et al., 1990) have supported the notion of forebrain ACh being involved in cognitive functions. However, as it appears difficult to isolate the variables that contribute to changes in HACU post hoc (Lai, 1987), studies that vary ACh release (as an independent variable) appear even more promising.

Blockade of HACU by intracranially administered hemicholinium represents such an approach. The available biochemical evidence indicates that hemicholinium-3 specifically binds to the high-affinity uptake site for choline (Manaker et al., 1986; Quirion, 1987; Bekenstein and Booten, 1989). Several behavioral studies (Hagan et al., 1989; see Table 3 in Sarter et al. 1992) have suggested that hemicholinium, administered intracerebroventricularly, disrupts cognitive processes more selectively than muscarinic blockade.

Unfortunately, a pharmacological tool that would allow direct and specific *stimulation* or presynaptic cholinergic mechanisms *in vivo* appears unavailable. The ability of choline supplementation to induce HACU and ACh synthesis remains a matter of debate (Wurtman et al., 1990), although recent studies have demonstrated that a 30-fold increase in extracellular choline content does not affect ACh release (Westerink and de Boer, 1990) and does not affect various behaviors in rats of different strains and ages (van der Staay, 1989). Thus, choline supplementation is unlikely to boost presynaptic cholinergic activity under physiological conditions.

Below we will summarize the evidence that suggests that cortical ACh release can be bidirectionally modulated by means of benzodiazepine receptor (BZR) ligands that act by facilitating or inhibiting the effects of GABA. The available data indicate that this transsynaptic approach to modulating cortical ACh release represents a most effective tool to study the functions of the basal forebrain cholinergic system. While it appears likely that cholinergic activity can also be stimulated transsynaptically via other neurotransmitter systems (e.g., dopamine, Levin et al., 1990; substance P, Nagel and Huston, 1988; for review see Decker and McGaugh, 1991), the effects of BZR partial or selective inverse agonist-induced dampening of GABAergic transmission on cholinergic activity may represent a favorable mechanism. Such BZR ligands do not fully remove or reinstate a neurotransmitter signal but rather increase the gain of cholinergic excitability as far as it is controlled by GABA (Sarter et al., 1990). Finally, we will discuss the hypotheses that the BZR-agonist-induced cognitive impairments, as well as BZR-selective inverse agonist-induced cognition enhancement, are mediated via decreases and increases of cortical ACh release, respectively.

GABAergic Control of Cortical Acetylcholine

GABAergic Innervation of Cholinergic Neurons

GABAergic synaptic contacts appear to be a general feature of basal forebrain cholinergic neurons (Zaborszky et al., 1986; Ingham et al., 1988; for review see Zaborszky et al., 1991). Autoradiographic examination of the distribution of benzodiazepine binding sites in this area (that are assumed to be distributed in parallel with GABA-binding sites; Schoch et al., 1985; but see Unnerstall et al., 1981) demonstrated that the substantia innominata (nomenclature in accordance to Bigl et al., 1982) shows a considerable density of binding sites for [^{3}H]lormetazepam (LMZ; B_{max}: 277 fmol/mg tissue; for comparison: globus pallidus: 132 fmol/mg; Sarter and Schneider, 1988). Among the BZR ligands tested, the β-carboline ZK 93 426 (see below) showed the highest potency to displace specifically bound LMZ (IC_{50}: 45 nM; diazepam: 100 nM; FG 7142: 540 nM).

There is some evidence suggesting that the GABAergic input to the substantia innominata partly originates in the nucleus accumbens (Mogenson et al., 1983; Zahm and Heimer, 1990). However, it is possible that local GABAergic neurons contribute to the basal forebrain GABA-cholinergic link (Walker et al., 1989; Fisher et al., 1988). Activation of cortical ACh, i.e., amplification of sensory and associational stimulus processing (see below), may be considered a mechanism that contributes to the performance of incentive-motivated responding (Robbins et al., 1989b; Everitt et al., 1989). This cortical activation may result from changes in the level of accumbal GABAergic activity (Mogenson and Yang, 1991).

Effects of $GABA_A$-Agonists and BZR Agonists on Cortical ACh

Neurochemical studies have consistently indicated that GABA inhibits forebrain ACh release or turnover (Consolo et al., 1975). Tables 1 and 2 in Sarter et al. (1990) list studies demonstrating uniformly that $GABA_A$ agonists and benzodiazepine receptor agonists administered systemically, as well as the $GABA_A$ agonist muscimol infused into the substantia innominata, reduce cortical ACh turnover, HACU and ACh release, or increase ACh levels (the latter measure also indicating a decrease in release). These studies suggest that the effects of a systemically administered dose of approximately 5–10 mg/kg muscimol on cortical ACh can be reproduced by the bilateral infusion of about 1 μg muscimol in the substantia innominata. However, Tanganelli et al. (1985) reported conflicting effects of GABA given i.c.v., and muscimol and THIP (both are $GABA_A$ agonists) given i.p. or i.c.v. in guinea pigs. They speculated that $GABA_A$ agonists act via two discrete mechanisms, initially yielding behavioral activation and an increase in cortical ACh release, followed by direct inhibition of cholinergic activity. In contrast, BZR agonists exclusively depressed cortical ACh release (using the epidural cup technique). Tanganelli and coworkers concluded that GABAergic drugs represent a nonhomogeneous group of compounds.

Such a view has been supported by data from a recent study using cortical microdialysis in order to measure cortical ACh release *in vivo* in the freely moving animal (Bruno et al., 1992). Quite unexpectedly, another BZR agonist, chlordiazepoxide (CDP; 3, 5, and 10 mg/kg i.p.), failed to affect cortical ACh release. Figure 1 summarizes the effects of CDP on ACh release in the frontoparietal somatosensory cortex in Fischer-344 rats.

The inability of CDP to reduce ACh was unlikely to be a result of floor effects interacting with the detection limits for ACh, as baseline ACh release was almost completely suppressed by TTX (data not shown). The discrepancy between the effects in previous studies (see Table 1 in Sarter et al., 1990) and CDP in our experiment may partly be related to the fact that previous studies (using CDP) measured either ACh levels or ACh release in anesthetized animals using the cup technique. For example, Phillis et al. (1980), using halothane-anesthetized rats, found that small doses of diazepam (0.05 and 0.25 mg/kg, i.p.) depressed cortical ACh release. It cannot be excluded that the potency of this effect was a result of interactions between diazepam and the depressing effects of halothane on cortical ACh (Schmidt, 1966).

However, Cheney et al. (1973) reported in an abstract that while diazepam (2 mg/kg) increased ACh levels in the mouse brain, CDP (10 mg/kg) failed to alter the steady-state concentrations of ACh. A more recent study by Nabeshima et al. (1990a) presented confusing and contradictory effects of 10 mg CDP on ACh levels in irradiated mice cortex (while their Table II indicates an increase in the cortical ACh level, Fig. 1 illustrates no effect). Thus, while there is a surprising paucity of comparisons between the pharmacological effects of CDP and other BZR agonists, the available data suggest that diazepam and CDP may differentially affect cortical ACh (though this may not be the case for hippocampal ACh; Miller and Richter, 1985). It seems unlikely that the dissociative effects found by

FIGURE 1. Effects of the benzodiazepine receptor agonist chlordiazepoxide on *in vivo* acetylcholine release in frontoparietal somatosensory cortex of Fischer-344 rats. Data from the three doses are expressed as a percent change from baseline values. Mean (±SEM) basal ACh release was 0.40 ± 0.13 and 0.35 ± 0.07 pmol/min in 4- and 18 month-old rats, respectively.

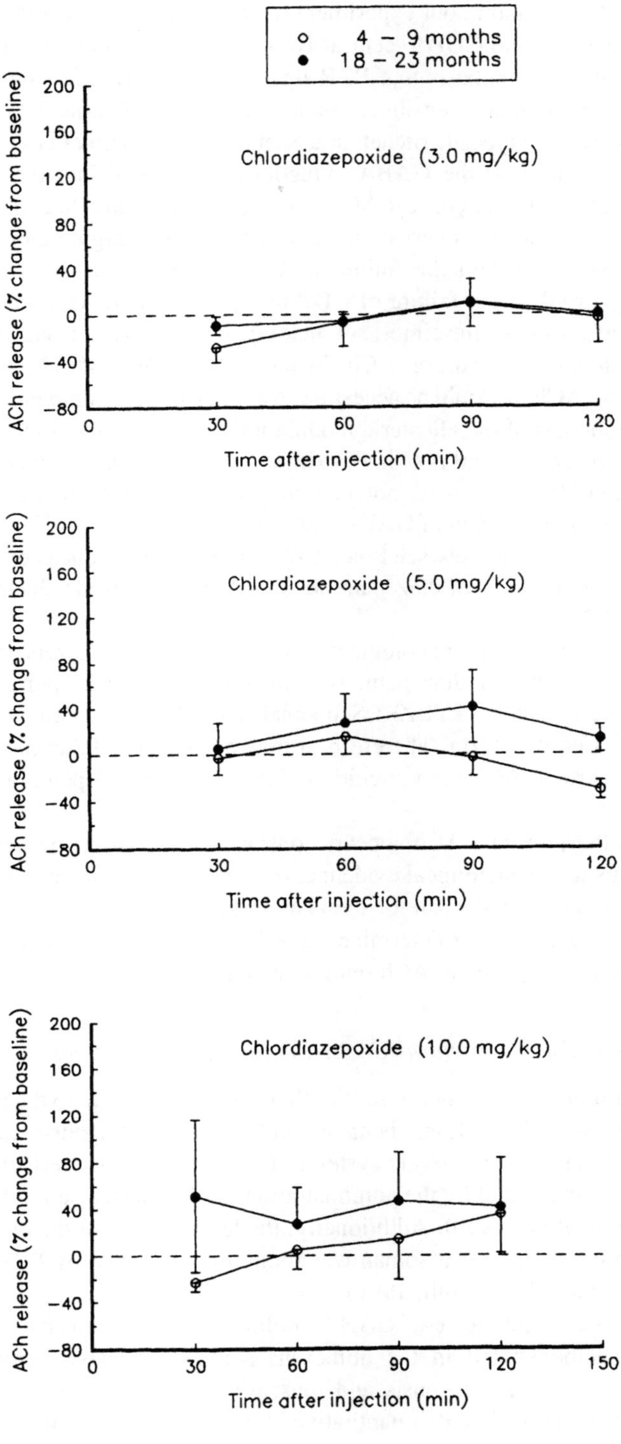

○ 4 – 9 months
● 18 – 23 months
Chlordiazepoxide (3.0 mg/kg)
Chlordiazepoxide (5.0 mg/kg)
Chlordiazepoxide (10.0 mg/kg)
ACh release (% change from baseline)
Time after injection (min)
200
160
120
80
40
0
−40
−80
0
30
60
90
120
150

Cheney et al. (1973) and in our experiment were a result of the differences in the affinity of diazepam and CDP to central BZRs (Richelson et al., 1991). Further examination of the hypothesis that BZR ligands differently affect cortical ACh release appears particularly sensible considering the accumulating data on benzodiazepine receptor subtypes (Pritchett and Seeburg, 1990; Turner et al., 1991).

Blaker (1985) infused the GABA antagonist bicuculline into the substantia innominata and found that cortical ACh remained unaffected. He concluded that GABA does not tonically innervate basal forebrain cholinergic neurons. While this finding would explain the failure of CDP to affect cortical ACh in our experiment (as well as the failure of CDP to replicate the behavioral effects of intrabasalis infusions of muscimol; see below), it is in conflict with the potent effects of systemic diazepam on ACh (Wood, 1985). Thus, if GABA does not tonically inhibit ACh, it would be necessary to assume that, in complete contrast to conventional models of the allosteric modulation of GABAergic transmission by BZR ligands, diazepam, but not CDP, exhibits effects that are independent from GABA. Indeed, BZR agonists, not including CDP, have been found to have effects that were independent of $GABA_A$ receptors (Zhang et al., 1989; Smith and Bierkamper, 1990; for review see Polc, 1991). In addition, Concas et al. (1990) have demonstrated effects of diazepam on GABA-gated chloride channels in the absence of GABA.

Furthermore, a main pharmacological dissociation between diazepam and CDP is given by the fact that diazepam, but not CDP, stimulates peripheral-type benzodiazepine receptors (PTBZR; Schoemaker et al., 1983; Pellow and File, 1984; Blasquez et al., 1991). Therefore, it may be speculated that the effects of diazepam on cortical ACh (and cognition, see below) may depend on PTBZR-mediated effects.

Taken together, cortical ACh appears controlled by GABAergic mechanisms and, analogous to the anatomical evidence, the available data have suggested the basal forebrain as the major site of the anatomical substrate of this interaction. However, the variables that determine the efficacy of benzodiazepine receptor agonists in decreasing cortical ACh release remain unsettled.

GABA/BZR Agonists: Behavioral GABA-Cholinergic Interactions

There are an impressive number of studies that consistently point to the possibility that the toxic, physiological, and behavioral effects of BZR agonists involve the inhibition of forebrain cholinergic systems. For example, these effects of BZR agonists can be attenuated by the administration of cholinesterase inhibitors (see Table 4 in Sarter et al., 1990). Additionally, the lethal effects of the organophosphate anticholinesterase agent soman was found to be blocked by BZR agonists (McDonough et al., 1989; Shih, 1991).

The behavioral significance of GABA-cholinergic interactions has been most impressively demonstrated in the numerous studies that have compared the cognitive effects of BZR agonists and muscarinic blockers in humans. These studies illustrated considerable quantitative and qualitative similarities between the effects of diazepam and scopolamine (Frith et al., 1984; Preston et al., 1989;

Curran et al., 1991; see Table 4 in Sarter et al., 1990; Rusted et al., 1991). The failure of physostigmine to attenuate the memory deficits produced by diazepam or lorazepam (Ghoneim and Mewaldt, 1977; Preston et al., 1989) may be considered as evidence against the hypothesis that BZR-agonist-induced amnesia is a result of inhibition of cholinergic neurons. However, this position may not be supported when one considers the above-mentioned pharmacological properties of cholinesterase inhibitors, i.e., a lack of ability to restore normal cholinergic signal flow (Sarter et al., 1990). In this light, the failure of physostigmine to attenuate BZR-induced amnesia represents an expected result.

Interestingly, studies on the effects of CDP on cognitive abilities in humans appear unavailable (Lister, 1985; Curran, 1986). Thus, it is unknown whether the effects of CDP on human cognitive processes were never tested or whether negative findings were not reported. However, if it is hypothesized that the effects of some BZR agonists (such as diazepam) on cortical ACh and on cognition are correlated, and that these effects are partly based on nonconventional, possibly GABA-independent, mechanisms, these effects would not be expected to be attenuated by the Ro15-1788, because this BZR antagonist does not act via the peripheral-type BZR (Schoemaker et al., 1983). Indeed, Preston et al. (1989) failed to antagonize the cognitive effects of lorazepam by the BZR antagonist. Moreover, Ro15-1788 *exclusively* failed to attenuate the mnemonic effects of a BZR agonist, whereas sedative, subjective, psychophysiological, and motor effects could be antagonized (Curran and Birch, 1991; but see Dorow et al., 1987). It seems likely, therefore, that the mnemonic effects of some BZR agonists are selectively mediated via mechanisms different from the facilitation of GABA-gated chloride channel conductance (see above). These mechanisms may be also responsible for the effects of diazepam on cortical ACh.

Comparable to the effects of BZR agonists on human cognition, the animal literature shows a surprising paucity of data that suggest a potency of CDP in disrupting cognitive functions. File and Mabbutt (1989) failed to find any behavioral consequence of animals treated with CDP for 4 weeks. Several studies demonstrated CDP-induced impairments in discriminative abilities (Hasegawa et al., 1973; Francis and Cooper, 1979; Ksir and Slifer, 1982), but the drug did not affect forgetting in a delayed conditional discrimination task (Tan et al., 1990). Some effects of CDP on spatial learning have been reported (McNaughton and Morris, 1987; Hodges and Green, 1986); however, it can not be excluded that the performance effects of CDP were a result of its noncognitive, particularly anxiolytic, properties. Data on the effects of CDP in delayed response tasks also do not support the idea that CDP affects cognitive abilities (Sahgal and Iversen, 1980).

In contrast to CDP, other BZR agonists, particularly diazepam, have been consistently reported to block acquisition processes (McNamara and Skelton, 1991; for review see Cole, 1986; Thiebot, 1985). Furthermore, the animal literature supports the hypothesis that the disruption of putatively cognitive processes by GABAergic treatments is mediated via inhibition of cholinergic systems (Castellano and McGaugh, 1991; Rupniak et al., 1990), although the available evidence appears less uniform than the findings in humans (see above). For example, Nabeshima et al. (1990b) reported that the disruptive effect of CDP in a passive-avoidance task was attenuated by scopolamine. It seems likely that the heterogeneity of relevant animal data is primarily a result of the very heterogeneous demands of different animal behavioral tests and of the unsettled validity of various tasks in terms of testing learning and memory (see Sarter et al., 1992; Sarter, 1991).

Studies that have dealt with the behavioral effects of infusions of GABAergic drugs into the basal forebrain have supported the hypothesis that the GABAergic innervation of basal forebrain cholinergic neurons represents a major anatomical substrate for mediating cognitive effects of GABAergic drugs. For example, Ridgon and Pirch (1984) demonstrated that infusions of GABA into the nucleus basalis of the basal forebrain blocked cue-elicited firing of frontal cortex units. Dudchenko and Sarter (1991) bilaterally infused muscimol (25 and 50 ng/0.5 µl/hemisphere) into the basal forebrain of animals performing a visual conditional discrimination task. Above-chance performance in this task requires the animals to retrieve propositional rules (e.g., if there is flashing light, go left; if there is constant light, go right) from memory and to select one of the competing response rules. Muscimol dose-dependently decreased correct responding in this task. Furthermore, systemically coadministered physostigmine dose-dependently interacted with the effects of muscimol, augmenting and attenuating the muscimol-induced impairment. These data strongly support the hypothesis that basal forebrain GABA-cholinergic interactions are involved in the cognitive effects of GABAergic drugs.

More recently, however, we found that infusions of CDP into the basal forebrain failed to reproduce the effects of muscimol (Dudchenko and Sarter, 1992). Two different interpretations of this result appear possible. On the one hand, this finding may be considered as a confirmation of Blaker's (1985; see above) conclusion that the GABAergic input to basal forebrain neurons is not tonically active (and that performance in the conditional discrimination task did not activate it). As we observed that well-trained animals failed to learn a reversal of the rules, it appears plausible that well-trained animals performed this task on the basis of a habit-like level of processing (Sarter, 1990a). It seems unlikely, however, that cortical ACh is critical for the performance of habits (Mishkin et al., 1984). On the other hand, the reasons for the failure of CDP to affect performance in this task, when given systemically (Sarter, 1990a) or intracranially (Dudchenko and Sarter, 1992), and its inability to decrease cortical ACh (see above) may be related to CDP's inability to act via those atypical mechanisms that may play a role in the effects of diazepam (as discussed above).

Benzodiazepine Receptor Inverse Agonism and Cognition Enhancement

Full Inverse Agonists

With the discovery of benzodiazepine receptor inverse agonists (Braestrup et al. (1982), it has become evident that benzodiazepine receptor ligands are capable of bidirectionally modulating GABAergic transmission (for review see Haefely, 1989). BZR ligands have been classified as full or partial (inverse) agonists, or antagonists, on the basis of measures of intrinsic activity. Several compounds from the inverse agonist side of the spectrum of BZR ligands are listed in Table 1.

In line with early hypotheses suggesting that inverse agonists exhibit effects that precisely represent the mirror image of the effects of agonists, it was assumed that, as agonists induce amnesia, inverse agonists would attenuate these effects and may even exhibit cognition enhancement. As the proconflict, proconvulsant, and convulsant effects of inverse agonists were assumed to be manifested at larger doses of BZR inverse agonists than the procognitive effects (i.e., at greater BZR occupancy; Potier et al., 1988), cognition enhancement produced by small doses of full inverse agonists was the subject of early studies.

Venault and coworkers (1986) reported that β-CCM enhanced learning in three animal paradigms; (a) β-CCM given to mice prior to a first exposure to an unfamiliar environment increased the animal's food intake in this environment during a second exposure 4 days later; (b) β-CCM, administered before a one-trial passive-avoidance experience increased the percentage of mice that avoided a dark box during a test session following a single-trial passive avoidance experience; (c) β-CCM (2.5 mg/kg), administered to newly hatched chicks during an imprinting session, resulted in an increase in time the chicks followed the decoy 24 hr later. It seems to be a common feature of these three tasks that they did not examine the controlled processing of declarative information but rather tested performance that may be based on the automatic retrieval of information of an unknown (but certainly nondeclarative) quality. Such a qualification of the cognitive processes

TABLE 1. Benzodiazepine Receptor Inverse Agonists: Conventional Classification

Drug	Structure	Classification
ZK 93 426	β-carboline	Weak inverse agonist
ZK 90886	β-carboline	Partial inverse agonist
FG 71 42	β-carboline	Partial inverse agonist
3-ethoxy-β-carboline	β-carboline	Partial inverse agonist
FG 7098	β-carboline	Partial inverse agonist
FG 7338	β-carboline	Partial inverse agonist
Ro15-4513	Diazepine	Partial inverse agonist
Ro15-3505	Diazepine	Partial inverse agonist
Ru 33965	Diazepine	Partial inverse agonist
Ru 34000	Diazepine	Partial inverse agonist
CGS 8216	Pyrazoloquinoline	Partial inverse agonist
Ro19-4603	Diazepine	Partial inverse agonist
S-135	Pyrazoloquinoline	Full inverse agonist
β-CCM	β-carboline	Full inverse agonist
β-CCE	β-carboline	Full inverse agonist
DMCM	β-carboline	Full inverse agonist

Classification of ligands in this table either followed the proposed classification given in relevant publications or is based on information about the drug's effects on GABA binding or chloride conductance, or its (pro-) convulsive efficacy.

addressed by these tasks appears important, as it is likely that different levels of processing and types of information are dependent on distinct neuronal mechanisms (Morris, 1984; Mishkin et al., 1984). Thus, the specific nature of the drug-induced facilitation in Venault's et al. experiments remains unclear. In addition, the putative cognition enhancement of inverse agonists has not been tested in tasks that would permit the examination of the interactions between the effects of parametrically varied demands on cognitive processes (Olton and Markowska, 1987) and of inverse agonists on performance variables (Raffalli-Sebille et al., 1990; Chapouthier et al., 1991). Therefore, the validity of conclusions of inverse-agonist-induced cognition enhancement is presently limited (Sarter et al., 1992; Sarter, 1991).

The putative cognition enhancement produced by BZR-inverse gonists, as well as by partial inverse agonists and antagonists/weak inverse agonists (Kumar et al., 1988; Lal and Forster, 1990; Lal et al., 1988) has been invariably tested in paradigms involving punishment. Different theoretical concepts have been proposed in order to account for the finding that BZR ligands exhibit particularly potent effects during aversive, stressful learning (Izquierdo et al., 1990). However, the stress- and possibly anxiety-inducing effects of inverse agonists (Thiebot et al., 1988) may distort the perceived intensity of the noxious stimulus (Carey, 1987) and thereby facilitate performance. In the only available experiment that dealt with positive reinforcement, Raffalli-Sebille and Chapouthier (1991) trained mice to select a dark alley in order to receive food reward. Mice were treated with one dose of β-CCM (0.3 mg/kg), which was administered during the initial three sessions. β-CCM reduced the number of errors (i.e., entries into the lit alley). However, as all animals may have performed at chance level (according to their Fig. 2, β-CCM-treated mice made 12 errors during the initial 30 trials, which is the chance level when examined with a binomial test; $p = 0.5$ for a correct response) and as an anxiogenic drug may increase the animals' preference for a dark alley, it remains doubtful that β-CCM enhanced cognitive processes. Thus, whether inverse agonists facilitate positively reinforced performance is unclear. However, it appears reasonable to speculate that, based on the proconvulsive, convulsive, stress-like, and possibly anxiogenic properties of full and partial inverse agonists (Pellow, 1985), beneficial effects of such drugs in tasks that require the controlled processing of declarative information are not to be expected, as fear induction or seizures are unlikely to promote cognitive abilities (Holmes et al., 1990; Holmes and Drugan, 1991).

Selective Inverse Agonists

While the terms *full (inverse) agonist* or *partial (inverse) agonist* have been generally thought to indicate the consistent functional correlates of a particular magnitude of a compound's intrinsic activity (Haefely et al., 1990), BZR ligands have been demonstrated to exhibit combinations of pharmacological effets (e.g., anxiogenic and anticonvulsant; Pellow, 1985) that do not fit with the unidimensional classification of a compound as, for example, a partial inverse agonist.

Therefore, it appears likely that cognition enhancement produced by BZR-(partial) inverse agonists does not necessarily have to be accompanied by unwelcome side effects typical of full inverse agonists. The two major examples of compounds that appear to exhibit cognition enhancement via inhibition of GABAergic transmission without inducing full-inverse agonist-like side effects are the β-carboline ZK 93 426 and the triazole MDL 26,479.

The β-carboline ZK 93 426 was originally described as a BZR antagonist (Jensen et al., 1984). However, Jensen et al. found that ZK 93 426 exhibited both potent proconflict effects (an inverse agonist-like property) and weak anticonvulsant effects (a partial agonist-like property). They concluded that "ZK 93 426 exhibits agonist, antagonist, and inverse agonist properties" (p. 255). Furthermore, ZK 93 426 was found to stimulate local cerebral glucose utilization, though the pattern of brain structures involved in this effect neither represented a mirror image of the metabolic effects of BZR agonists nor matched the effects of the partial inverse agonist FG 7142 (Sarter, 1990b). Thus, the term *selective inverse agonist* has been proposed for this drug (Sarter et al., 1990), pointing to the fact that it shows some of the properties of partial inverse agonists (primarily with respect to biochemical measures such as the TBPS shift; Braestrup et al., 1984) but is devoid of the adverse properties of inverse agonists. Furthermore, these compounds were found to be almost exclusively active in tests measuring cognitive abilities in animals and humans.

ZK 93 426 was found to potently antagonize the behavioral impairments of scopolamine in a variety of tasks (Jensen et al., 1987; Sarter et al., 1988a; Stephens and Sarter, 1988). The failure of ZK 93 426 to attenuate the impairments produced by intraventricular hemicholinium-3 (Hagan et al., 1989) suggests that the β-carboline acts via an increase in ACh synthesis to counteract the effects of muscarinic blockade (hemicholinium-3 blocks HACU and therefore the *de novo* synthesis of ACh). These early results have suggested that this β-carboline may exhibit beneficial behavioral effects, particularly in subjects suffering from cholinergic hypofunction. In accordance with this hypothesis, ZK 93 426 was demonstrated to attenuate the effects of basal forebrain lesions and of chronic alcohol administration on memory, the latter treatment also resulting in a decrease in cortical and hippocampal ChAT-activity (Sarter and Steckler, 1989; Hodges et al., 1989). Consequently, a therapeutic potential for drugs such as ZK 93 426 for the treatment of cognitive symptoms associated with cholinergic hypofunction was proposed (Sarter et al., 1988b).

The psychotropic properties of the selective inverse agonist β-carboline ZK 93 426 have been extensively examined in humans (Duka et al., 1987, 1988; Duka, 1991). In these studies, ZK 93 426 appeared to facilitate human performance in a logical reasoning task, a picture differences test, a delayed recall test, and in auditory and visual vigilance tasks. The drug's efficacy to facilitate cognitive processing has been attributed to its "mild inverse agonist" properties (Duka, 1991, p. 453). However, as there is no evidence that inverse agonist-type compounds would exhibit a greater degree of cognition enhancement (see above), but rather interfere with human performance (Gentil et al., 1989), and as ZK 93

426 shows several unique pharmacological properties that do not fit with the notion of a "mild inverse agonist" (see above) or of a BZR ligand (Giorgi et al., 1989), the cognition enhancement induced by this drug may be considered as a major functional equivalent of its selective inverse agonism.

The triazole MDL 26,479 appears to represent an even more striking example of a selective inverse agonist (Table 2). MDL 26,479 acts similarly to the full inverse agonist DMCM in attenuating muscimol-induced ^{36}Cl-influx *in vitro*, but it lacks convulsant, anxiogenic, or depressant effects (Miller et al., 1992). While the characterization of the putative cognition enhancing properties of MDL 26,479 requires further study, the compound was found to potently attenuate the effects of scopolamine on delayed alternation performance in rats (Holley et al., 1992). Similar to the initial behavioral findings on ZK 93 426, this result points to the possibility that the effects of MDL 26,479 are mediated via cholinergic systems.

Inverse Agonists and Selective Inverse Agonists: Effects on Forebrain Acetylcholine

As summarized above, some BZR agonists have been consistently found to decrease forebrain ACh release and to be potent in disrupting cognitive processes. Thus, it is hypothesized that (selective) inverse agonists that exhibit behavioral facilitation may stimulate ACh release. Miller and Chmielewski (1990a) measured cortical and hippocampal high-affinity choline uptake (HACU) *ex vivo* following

TABLE 2. Pharmacological Properties of MDL 26,479

Test	Effect
Displacement of [^{3}H]Ro15-1788 binding	ED_{50} = 0.22 ± 0.05 mg/kg; i.p. *in vivo*, mouse cortex
[^{3}H]hemicholinium-binding measured *ex vivo* (rat cortical membranes; 1 mg/kg)	Cortex: 133% increase Hippocampus: no change
Antagonism of the effect on hemicholinium binding by Ro15-1788 (1 mg/kg)	Antagonism (in terms of B_{max}; none of the treatments affected K_d)
[^{3}H]hemicholinium-binding measured *ex vivo* by quantitative autoradiography (1 mg/kg)	Hippocampus CA1: 224% increase Hippocampus CA3: 216% increase Dentate gyrus: no change
Precipitated withdrawal in mice chronically treated with diazepam	No effects (1–30 mg/kg; ED_{50} for Ro15-1788: 5.83 mg/kg)
Seizure induction in naive mice	No effects (up to 200 mg/kg)
Separation-induced ultrasonic vocalization in rat pups	No effects (up to 30 mg/kg)
Long-term potentiation in hippocampal slices	Increased amplitude of basal population spikes (10–50 μM)
Spatial delayed alternation performance in rats (delays: 2–32 sec)	No effects (0.10–6.25 mg/kg)
Scopolamine (0.03 or 0.1 mg/kg)-induced impairment in spatial delayed alternation performance	Antagonism

Compiled from data in Miller et al. (1992), Miller and Chmielewski (1990a,b), and Holley et al. (1992).

the administration of the partial inverse agonist FG 7142 and several full inverse agonists. FG 7142 resulted in a *de*crease in cortical HACU and did not affect hippocampal HACU. The convulsant effects of DMCM were correlated with a significant increase in hippocampal but not cortical HACU. β-CCE (20 mg/kg) did not produce convulsions and exclusively stimulated cortical HACU. The putative selective inverse agonist MDL 26,479, administered *in vivo*, increased hemicholinium-3 binding to cortical membranes by more than 100% (B_{max}; Miller et al., 1992; calculated from their Table 6) but did not affect hippocampal HACU.

Until recently, data on the effects of ZK 93 426 on measures of cholinergic activity were unavailable. Employing the microdialysis technique for the measurement of cortical acetylcholine release in awake and freely moving animals, we have measured the effects of two doses of ZK 93 426 on ACh release in frontoparietal somatosensory cortex in young and aged Fischer-344 rats (Bruno et al., 1992). As aging has been previously reported to be correlated with a decreased capacity of aged cholinergic neurons to release the neurotransmitter in response to K^+ or muscarinic antagonists (Meyer et al., 1984; Araujo et al., 1990), the experiment was planned to test the hypothesis that ZK 93 426 stimulates cortical ACh and that this effect is reduced in aged animals.

As illustrated in Figure 2, the smaller dose of ZK 93 426 (1.0 mg/kg, i.p.) modestly stimulated ACh release in both groups of animals. There were no systematic differences in the magnitude or pattern of effects between the young and old rats. The larger dose (5 mg/kg) stimulated release in both groups. Interestingly, initial ACh release in the older animals was greater than that seen in the younger rats. The aged animals, however, appeared unable to sustain this release beyond 40 min. As this *in vivo* technique allows the examination of changes in ACh release over time, it seems evident that the effects of age on ACh release are more complex than revealed by previous *in vitro* studies, and that interactions between age and the effects of ZK 93 426 involve temporal dynamics.

Evidence that would directly support our hypothesis that the effects of selective inverse agonists on cortical ACh release involve basal forebrain GABAergic innervation of cholinergic neurons is not yet available. However, as stimulation of the basal forebrain has been generally found to stimulate cortical HACU or ACh release (Wenk, 1984; Kurosawa et al., 1989), and as in the cortex the terminals immunoreactive for ChAT and GAD show distinct laminar patterns (Brady and Vaughn, 1988), it appears likely that the basal forebrain GABA-cholinergic link represents the major anatomical substrate for this effect.

Selective Inverse Agonist-Induced Facilitation of Cognitive Processes and Cortical ACh Release: Atypical Mechanisms

The lack of ability of CDP to decrease cortical ACh release and, possibly, to affect memory, and the lack of ability of Ro15-1788 to attenuate the amnesic effects of diazepam (which decreases cortical ACh) has prompted us to speculate about the involvement of nonclassical GABA-BZR interactions in the effects of diazepam

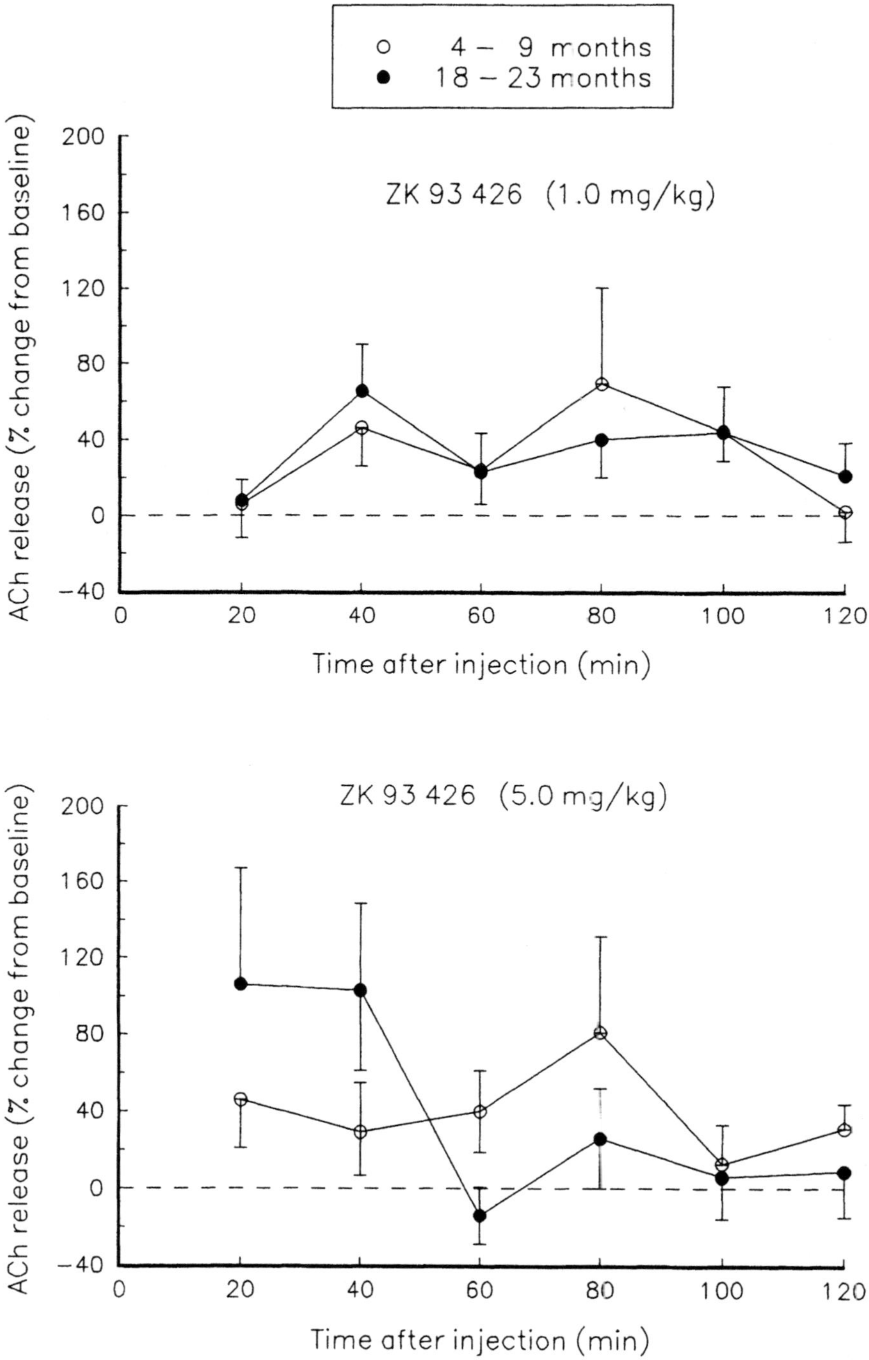

4 – 9 months
18 – 23 months
ZK 93 426 (1.0 mg/kg)
ACh release (% change from baseline)
200
160
120
80
40
0
−40
0
20
40
60
80
100
120
Time after injection (min)
ZK 93 426 (5.0 mg/kg)
ACh release (% change from baseline)
Time after injection (min)

(see above). The available evidence about qualitative pharmacological differences between CDP and diazepam points to a possible role of peripheral-type benzodiazepine receptors (PTBZR).

Consequently, it may also be speculated that the potency of ZK 93 426 and MDL 26,479 to stimulate cortical ACh and to enhance cognitive processes are independent (at least in part) from conventional assumptions about the negative allosteric modulation of GABAergic effects by these compounds. Several findings support such an idea. β-carbolines and some benzodiazepines may not act on isosteric sites (von Blankenfeld et al., 1990). This assertion has gained functional support by the finding that ZK 93 426 and FG 71 42 (but not Ro15-1788, Ro15-1513, and CGS 8216; see Table 2) failed to precipitate withdrawal in cats chronically treated with diazepam (Giorgi et al., 1989). Moreover, noncompetitive interactions between ZK 93 426 and other β-carbolines have been described (Malatynska et al., 1989), making it likely that even different β-carbolines do not act via identical sites and/or identical mechanisms. Likewise, while MDL 26,479 potently inhibits Ro15-1788 binding *in vivo* and $GABA_A$-agonist-induced Cl influx, this compound failed to precipitate withdrawal from chronic diazepam (Miller et al., 1992). Taken together, these data strongly suggest the possibility that selective inverse agonists may preferably be considered as unique compounds, exhibiting cognition enhancement and stimulation of ACh release via mechanisms other than the allosteric modulation of GABAergic transmission.

The extent to which these effects of selective inverse agonists are mediated via PTBZR, or via an interaction between peripheral-type and central BZR (Drugan and Holmes, 1991), remains a matter of speculation. However, the PTBZR-agonist 4′-chlordiazepam (Ro5-4864) has been found to be behaviorally active (Da Cunha et al., 1991) in a way similar to some BZR partial inverse agonists. Future studies will have to determine whether the amnesic and ACh-decreasing effects of diazepam (but not of CDP), and the cognition-enhancing and ACh-increasing effects of selective inverse agonists, are based on interactions with PTBZR mechanisms.

Modulation of Cortical Acetylcholine and of Attentional Abilities

As stressed in the introductory part of this chapter, years of research on the functions of the cholinergic system have failed to result in a generally accepted hypothesis about the behavioral roles of cortical ACh. However, our hypothesis that selective inverse agonist-induced stimulation of cortical ACh release represents the neurobiological substrate of the cognition enhancement exhibited by

---

FIGURE 2. Effects of the benzodiazepine receptor selective inverse agonist ZK 93 426 on *in vivo* acetylcholine release in frontoparietal somatosensory cortex of Fischer-344 rats. Data from the two doses are expressed as a percent change from baseline values. Specific basal ACh release is cited in the legend for Figure 1.

these drugs requires the qualification of the specific cognitive processes facilitated by an increase in cortical ACh release.

We begin such qualification with the assertion that the functions of cortical ACh may best be characterized by its effects on the processing of information, irrespective of its modality (associative, auditive, visual, motor). In other words, changes in cholinergic activity may not be correlated with a particular behavioral function but with modulations in information processing (see also Drachman and Sahakian, 1979).

The potency of ACh to change the general responsiveness of cortical neurons (e.g., Krnjevic et al., 1971) and its effects on "wakefulness and alertness" (Celesia and Jaspers, 1966) have been known for quite some time (see also Buzsaki et al., 1988). More recent results have indicated that ACh-induced excitability changes can be long lasting (Sillito and Kemp, 1983), thus representing a mechanism for neuronal plasticity (see also Juliano and Eslin, 1991; Delacour et al., 1990). Furthermore, it has been assumed that this kind of plasticity may be involved in "the task-dependent selective activation of a cortical area" (Sillito and Murphy, 1987). In accordance with this hypothesis, basal forebrain lesions were found to result in the loss or the weakening of visual stimulation-induced excitability of neurons in the primary visual cortex (Sato et al., 1987).

Metherate et al. (1990) found that ACh affects neuronal responsitivity in the auditory cortex "similar to that resulting from an increase in stimulus intensity" (p. 368). They concluded that auditory information processing is adaptively regulated by ACh. Additional work (Metherate and Weinberger, 1990) suggests that ACh is involved in the reshaping of cortical sensory receptive fields. This process appears to depend on the convergence of auditory input and ACh release "brought about by behaviorally significant stimuli" (Metherate and Weinberger, 1990, p. 144; see also Murphy and Sillito, 1991). Thus, the effects of ACh are bound to the concomitant discharge of afferent inputs, suggesting that "one action of ACh is selective gain control of sensory inputs" (Donoghue and Carroll, 1987, p. 370; see also Rasmussen and Dykes, 1988).

While the significance of these electrophysiological findings for understanding the role of ACh in complex behaviors remains a matter of speculation, the precision and the depth in which stimuli are processed may be a function of cholinergic activity. In this regard, the potency of muscarinic antagonists to disrupt attentional functions is not surprising (Warburton, 1977; Dunne and Hartley, 1986; Callaway et al., 1985). Thus there is good evidence for the idea that cortical ACh, instead of mediating particular behavioral components, generally modulates the efficacy of stimulus perception and early-stage processing. This function may not be restricted to exteroceptive, proprioceptive, and interoceptive information but may also include associative stimuli. Thus, stimulation of the cortical cholinergic system would be assumed to facilitate stimulus evaluation processes that are most impressively revealed in tasks testing attentional abilities. As normal aging appears most prominently correlated with impairments in attentional abilities and with a decrease in the dynamic range of cortical ACh release to respond to stimulation (see above and Bruno et al., 1992), selective inverse agonists are expected to exhibit most potent beneficial effects on the attentional performance of aged subjects.

In normal adult humans, ZK 93 426 indeed has been found to facilitate performance in vigilance tasks (Duka, 1991). Attempts to model the human age-related impairments in simple and choice reaction time tasks (Ferris et al., 1976) in animals have resulted in behavioral paradigms that revealed age-related performance differences (Stephens and Sarter, 1988; Bruno et al., 1992), which, however, have been found relatively insensitive to pharmacological manipulations (Moore et al., 1991). These reaction-time paradigms may not address the specific attentional processing that is most robustly disrupted by aging, i.e., memory-driven changes in signal presentation properties and division of attention (Salthouse, 1985; Rabbit, 1981; McDowd and Craik, 1988). Alternatively, we expect that the potency of selective inverse agonists to facilitate performance will be greatest in paradigms that test selective and divided attention, and that their effects will interact with the demands on cognitive processing. Finally, we predict that these beneficial effects will be correlated with increases in cortical ACh release.

Conclusions

This chapter summarizes the anatomical, biochemical, and behavioral foundations of the hypothesis that benzodiazepine receptor selective inverse agonists are capable of enhancing cognitive processes and that these effects are dependent on the increase of cortical ACh release. However, it appears evident that, though some human data have supported some principal components of this hypothesis, our understanding of the biochemical and behavioral effects of the prototype selective inverse agonists remains extremely poor. Future work will have to devaluate the mere demonstration of effects and focus on the determination of the conditions under which selective inverse agonists stimulate cortical ACh and facilitate behavioral abilities.

Acknowledgments. The authors' research was supported in part by funds from the U.S. Public Health Service (NIA AG10173-01 and MH46869-01A1), The Amercian Federation for Aging Reserch, and the Sandoz Foundation for Gerontological Research. We are grateful to Dr. M. Palfreyman (Marion Merrell Dow Research Institute) for the gift of MDL 26,479 and to Dr. D.N. Stephens (Schering AG, Berlin) for the gift of ZK 93 426.

References

Araujo DM, Lapchak PA, Meaney MJ, Collier B, Quirion R (1990): Effects of aging on nicotinic and muscarinic autoreceptor function in the rat brain: Relationship to presynaptic cholinergic markers and binding sites. *J Neurosci* 10:3069–3078

Bartus RT, Dean RL, Beer B, Lippa AS (1982): The cholinergic hypothesis of geriatric memory dysfunction. *Science* 217:408–417

Beatty WW, Butters N, Janowski DS (1986): Patterns of memory failure after scopolamine treatment: Implications for cholinergic hypotheses of dementia. *Behav Neural Biol* 45:196–211

Bekenstein JW, Booten FG (1989): Hemicholinium-3 binding sites in rat brain: A quantitative autoradiographic study. *Brain Res* 481:97–105

Bigl V, Woolf NJ, Butcher LL (1982): Cholinergic projections from the basal forebrain to frontal, parietal, temporal, occipital, and cingulate cortices. A combined fluorescent tracer and acetylcholinesterase analysis. *Brain Res Bull* 8:722–749

Blaker WD (1985): GABAergic control of the cholinergic projections to the frontal cortex is not tonic. *Brain Res* 325:389–390

Blasquez C, Jegou S, Tranchand Bunel D, Delbende C, Braquet P, Vaudry H (1991): Central-type benzodiazepines inhibit release of α-melanocyte-stimulating hormone from the rat hypothalamus. *Neuroscience* 42:509–516

Bowen DM, Smith CB, White P, Davison AN (1976): Neurotransmitter-related enzymes and indices of hypoxia in senile dementia and other atrophies. *Brain* 99:459–496

Brady DR, Vaughn JE (1988): A comparison of the localization of choline acetyltransferase and glutamate decarboxylase immunoreactivity in rat cerebral cortex. *Neuroscience* 24:1009–1026

Braestrup C, Schmiechen R, Nielsen M, Petersen EN (1982): Interaction of convulsive ligands with benzodiazepine receptors. *Science* 216:1241–1243

Braestrup C, Honore T, Nielsen M, Petersen EN, Jensen LH (1984): Ligands for benzodiazepine receptors with positive and negative efficacy. *Biochem Pharmacol* 33:859–862

Brockhaus H (1942): Vergleichend-anatomische Untersuchungen über den basalkern-komplex. *J Psychol Neurol (Leipzig)* 51:57–95

Bruno JP, Moore H, Dudchenko P, Sarter M (1992): Modulation of frontal cortical acetylcholine release by benzodiazepine receptor ligands: age-dependent effects and behavioral correlates. In: *The Treatment of Dementias: A New Generation of Progress*, Meyer EM, Crews FT, Simpkins JW, eds. New York: Plenum Press

Butcher L, Woolf NJ (1986): Central cholinergic systems: Synopsis of anatomy and overview of physiology and pathology. In: *The Biological Substrates of Alzheimer's Disease*, Scheibel AB, Wechsler AF, eds. New York: Academic Press

Buzsaki G, Bickford RG, Ponomareff G, Thal LJ, Mandel R, Gage FH (1988): Nucleus basalis and thalamic control of neocortical activity in the freely moving rat. *J Neurosci* 8:4007–4026

Callaway E, Halliday R, Naylor H, Schechter G (1985): Effects of scopolamine on human stimulus evaluation. *Psychopharmacology* 85:133–138

Carey RJ (1987): Post-trial hormonal treatment effects: Memory modulation or perceptual distortion? *J Neurosci Meth* 22:27–30

Carlton PL (1963): Cholinergic mechanisms in the control of behavior by the brain. *Psychol Rev* 70:19–39

CastellanoC, McGaugh JL (1991): Oxotremorine attenuates retrograde amnesia induced by post-training administration of the GABAergic agonists muscimol and baclofen. *Behav Neural Biol* 56:25–31

Celesia GG, Jaspers HH (1966): Acetylcholine released from cerebral cortex in relation to state of activation. *Neurology* 16:1053–1063

Chapouthier G, Raffali-Sebille MJ, Venault P, Simiand J, Dodd RH (1991): Comparison between the effects of the benzodiazepine receptor ligands methyl beta-carboline-3-carboxylate and diazepam in two learning situations in mice. *Psychobiology* 19:58–63

Cheney DL, Trabucchci M, Hanin I, Costa E (1973): Effects of several benzodiazepines on concentrations and specific activities of choline and acetylcholine in mouse brain. *Pharmacologist* 15:162 (abstract 043)

Cole SO (1986): Effects of benzodiazepines on acquisition and performance: A critical assessment. *Neurosci Biobehav Rev* 10:265–272

Collerton D (1986): Cholinergic function and intellectual decline in Alzheimer's disease. *Neuroscience* 19:1–28

Concas A, Sanna E, Mascia MP, Serra M, Biggio G (1990): Diazepam enhances bicuculline-induced increase of t[^{35}S]butylbicyclophosphorothionate binding in unwashed membrane preparations from rat cerebral cortex. *Neurosci Lett* 112:87–91

Consolo S, Garattini S, Ladinsky H (1975): Action of the benzodiazepines on the cholinergic system. In: *Mechanisms of Action of Benzodiazepines. Advances in Biochemical Psychopharmacology*, Costa E, Greengard P, eds. New York: Raven Press

Curran HV (1986): Tranquillising memories: A review of the effects of benzodiazepines on human memory. *Biol Psychol* 23:179–213

Curran HV, Birch B (1991): Differentiating the sedative, psychomotor and amnesic effects of benzodiazepines: A study with midazolam and the benzodiazepine antagonist, flumazenil. *Psychopharmacology* 103:519–523

Curran HV, Schifano F, Lader M (1991): Models of memory dysfunction? A comparison of the effects of scopolamine and lorazepam on memory, psychomotor performance and mood. *Psychopharmacology* 103:83–90

Da Cunha C, Huang CH, Walz R, Dias M, Koya R, Bianchin M, Pereira ME, Izquierdo I, Median JH (1991): Memory facilitation by post-training intraperitoneal, intracerebroventricular and intra-amygdala injection of Ro 5-4864. *Brain Res* 544:133–136

Decker MW, McGaugh JL (1991): The role of interactions between the cholinergic system and other neuromodulatory systems in learning and memory. *Synapse* 7:151–168

Delacour J, Houcine O, Costa JC (1990): Evidence for a cholinergic mechanism of "learned" changes in the responses of barrel field neurons of the awake and undrugged rat. *Neuroscience* 34:1–8

Donoghue JP, Carroll KL (1987): Cholinergic modulation of sensory processes in rat primary somatic sensory cortex. *Brain Res* 408:367–371

Dorow R, Berenberg D, Duka T, Sauerbrey N (1987): Amnesic effects of lormetazapm and their reversal by the benzodiazepine antagonist Ro15-1788. *Psychopharmacology* 93:507–514

Drachman DA, Sahakian BJ (1979): Effects of cholinergic agents on human learning and memory. In: *Nutrition and the Brain*, Barbeau A, Growdon JH, Wurtman RJ, eds. New York: Raven Press

Drugan RC, Holmes PV (1991): Central and peripheral benzodiazepine receptors: Involvement in an organism's response to physical and psychological stress. *Neurosci Biobehav Rev* 15:277–298

Dudchenko P, Sarter M (1991): GABAergic control of basal forebrain cholinergic neurons and memory. *Behav Brain Res* 42:33–41

Dudchenko P, Sarter M (1992): Failure of chlordiazepoxide to reproduce the behavioral effects of muscimol into the basal forebrain. *Behav Brain Res* 47:202–205

Duka T (1991): Bidirectional activity of benzodiazepine-receptor ligands in cognitive functions in humans. In: *New Concepts in Anxiety*, Briley M, File SE, eds. Boca Raton, FL: CRC Press

Duka T, Edelmann V, Schütt B, Dorow R (1988): β-carbolines as tools in memory reserch:

Human data with the β-carboline ZK 93 426. In: *Benzodiazepine Receptor Ligands, Memory and Information Processing*, Hindmarch I, Ott H, eds. Berlin: Springer

Duka T, Stephens DN, Krause W, Dorow R (1987): Psychotropic activity of the benzodiazepine receptor antagonist β-carboline, ZK 93 426, in human volunteers. *Psychopharmacology* 93:421–427

Dunne MP, Hartley LR (1985): Scopolamine and the control of attention in humans. *Psychopharmacology* 89:94–97

Everitt BJ, Cador M, Robbins TW (1989): Interactions between the amygdala and ventral striatum in stimulus-reward associations: Studies using a second-order schedule of sexual reinforcement. *Neuroscience* 30:63–75

Ferris S, Crook T, Sathananthan G, Gershon S (1976): Reaction time as a diagnostic measure in senility. *J Am Geriatr Soc* 24:529–533

Fibiger HC (1982): The organization and some projections of cholinergic neurons of the mammalian forebrain. *Brain Res Rev* 4:327–388

Fibiger HC (1991): Cholinergic mechanisms in learning, memory and dementia: A review of recent evidence. *Trends Neurosci* 14:220–223

File SE, Mabbutt PS (1989): Performance in learning tasks after a period of chronic chlordiazepoxide administration in the rat. *Pharmacopsychoecologia* 2:43–48

Fisher RS, Buchwald NA, Hull CD, Levine MS (1988): GABAergic basal forebrain neurons project to neocortex: The localization of glutamic acid decarboxylase and choline acetyltransferase in feline corticopetal neurons. *J Comp Neurol* 272:489–502

Francis RL, Cooper SJ (1979): Chlordiazepoxide-induced disruption of discrimination behavior: a signal detection analysis. *Psychophasmacology* 63:307–310

Frith CD, Richardson JTE, Samuel M, Crow TJ, McKenna PJ (1984): The effects of intravenous diazepam and hyoscine upon human memory. *Q J Exp Psychol* 36A:133–144

Gallagher M, Pelleymounter MA (1988): An age-related sptial learning deficit: Choline uptake distinguishes "impaired" and "unimpaired" rats. *Neurobiol Aging* 9:363–369

Gentil V, Gorenstein C, Camargo CHP, Singer JM (1989): Effects of flunitrazepam on memory and their reversal by two antagonists. *J Clin Psychopharmacol* 9:191–197

Ghoneim MM, Mewaldt SP (1977): Studies on human memory: The interactions of diazepam, scopolamine, and physostigmine. *Psychopharmacology* 52:1–6

Giorgi O, Corda MG, Fernandez A, Biggio G (1989): The β-carboline derivatives ZK 93426 and FG 7142 fail to precipitate abstinence signs in diazepam-dependent cats. *Pharmacol Biochem Behav* 32:671–675

Gorry JD (1963): Studies on the comparative anatomy of the ganglion basale of Meynert. *Acta Anat* 55:51–104

Haefely WE (1989): Pharmacology of the allosteric modulation of $GABA_A$ receptors by benzodiazepine receptor ligands. In: *Allosteric Modulation of Amino Acid Receptors: Therapeutic Implications*. New York: Raven Press

Haefely W, Martin JR, Schoch P (1990): Novel anxiolytics that act as partial agonists at benzodiazepine receptors. *Trends Pharmacol Sci* 11:452–456

Hagan JJ, Jansen JHM, Broekkamp CLE (1989): Hemicholinium-3 impairs spatial learning and the deficit reversed by cholinomimetics. *Psychopharmacology* 98:347– 356

Hasegawa Y, Ibika N, Iwahara S (1973): Effects of chlordiazepoxide upon successive red-green discrimination in Japanese monkeys, macaca fuscata. *Psychopharmacology* 30:89–94

Hodges H, Green S (1986): Effects of chlordiazepoxide on cued radial maze performance in rats. *Psychopharmacology* 88:460–466

Hodges H, Thrasher S, Gray JA (1989): Improved radial maze performance induced by the benzodiazepine antagonist ZK 93 426 in lesioned and alcohol-treated rats. *Behav Pharmacol* 1:45–55

Holley LA, Dudchenko P, Sarter M (1992): Attenuation of muscarinic receptor blockade-induced impairment of spatial delayed alternation performance by the triazole MDL 26,479. *Psychopharmacol,* in press

Holmes GL, Thompson JL, Marchi TA, Gabriel PS, Hogan MA, Carl FG, Feldman DS (1990): Effects of seizures on learning, memory, and behavior in the genetically epilepsy-prone rat. *Ann Neurol* 27:24–32

Holmes PV, Drugan RC (1991): Differential effects of anxiogenic central and peripheral benzodiazepine receptor ligands in tests of learning and memory. *Psychopharmacology* 104:249–254

Ingham CA, Bolam JP, Smith AD (1988): GABA-immunoreactive synaptic boutons in the rat basal forebrain: Comparison of neurons that project to the neocortex with pallidosubthalamic neurons. *J Comp Neurol* 273:263–282

Izquierdo I, Pereira ME, Median JH (1990): Benzodiazepine receptor ligand influences on acquisition: Suggestion of an endogenous modulatory mechanism mediated by benzodiazepine receptors. *Behav Neural Biol* 54:27–41

Jensen LH, Petersen EN, Braestrup C, Honore T, Kehr W, Stephens DN, Schneider H, Seidelmann D, Schmiechen R (1984): Evaluation of the β-carboline ZK 93 426 as a benzodiazepine receptor antagonist. *Psychopharmacology* 83:249–256

Jensen LH, Stephens DN, Sarter M, Petersen EN (1987): Bidirectional effects of β-carbolines and benzodiaepines on memory processes. *Brain Res Bull* 19:359–364

Juliano SL, Ma W, Eslin D (1991): Cholinergic depletion prevents expansion of topographic maps in somatosensory cortex. *Proc Natl Acad Sci USA* 88:780–784

Kesner RP, Crutcher KA, Omana H (1990): Memory deficits following nucleus basalis magnocellularis lesions may be mediated through limbic, but not neocortical targets. *Neuroscience* 38:93–102

Kopelman MD, Corn TH (1988): Cholinergic blockade as a model for cholinergic depletion. *Brain* 111:1079–1110.

Krnjevic K, Pumain R, Renaud L (1971): The mechanism of excitation by acetylcholine in the cerebral cortex. *J Physiol* 215:247–268

Ksir C, Slifer B (1982): Drug effects on discrimination performance at two levels of stimulus control. *Psychopharmacology* 76:286–290

Kumar BA, Forster MJ, Lal H (1988): CGS 8216, a benzodiazepine receptor antagonist, enhances learning and memory in mice. *Brain Res* 460:195–198

Kurosawa M, Sato A, Sato Y (1989): Stimulation of the nucleus basalis of Meynert increases acetylcholine release in the cerebral cortex in rats. *Neurosci Lett* 98:45–50

Lai H (1987): Acute exposure to noise affects sodium-dependent high-affinity choline uptake in the central nervous system of the rat. *Pharmacol Biochem Behav* 28:147–151

Lal H, Foster MJ (1990): Flumazenil improves active avoidance performance in aging NZB/B1NJ and C57BL/6NNia mice. *Pharmacol Biochem Behav* 35:747–750

Lal H, Kumar B, Forster MJ (1988): Enhancement of learning and memory in mice by a benzodiazepine antagonist. *FASEB J* 2:2707–2711

Lebrun C, Durkin TP, Marighetto A, Jaffard R (1990): A comparison of the working memory performances of young and aged mice combined with parallel measures of testing and drug-induced activations of septo-hippocampal and nBm-cortical cholinergic neurones. *Neurobiol Aging* 11:515–521

Levin ED, McGurk SR, Rose JE, Butcher LL (1990): Cholinergic-dopaminergic interactions in cognitive performance. *Behav Neural Biol* 54:271–299

Lister RG (1985): The amnesic action of benzodiazepines in man. *Neurosci Biobehav Rev* 9:87–94

Malatynska E, Knapp R, Keda M, Yamamura HI (1989): β-carboline interactions at the Bz-GABA receptor chloride-ionophore complex in the rat cerebral cortex. *Brain Res Bull* 22:845–848

Manaker S, Wieczorek CM, Rainbow TC (1986): Identification of sodium-dependent, high affinity choline uptake sites in rat brain with [^{3}H]hemicholinium-3. *J Neurochem* 46:483–486

McDonough JH, Jaax NK, Crowley RA, Mays MZ, Modrow HE (1989): Atropine and/or diazepam therapy protects against soman-induced neural and cardiac pathology. *Fundamen Appl Toxicol* 13:256–276

McDowd JM, Craik FIM (1988): Effects of ageing and task difficulty on divided attention performance. *J Exp Psychol Hum Perc Perf* 14:267–280

McNamara RK, Skelton RW (1991): Diazepam impairs acquisition but not performance in the Morris water maze. *Pharmacol Biochem Behav* 38:651–658

McNaughton N, Morris RGM (1987): Chlordiazepoxide, an anxiolytic benzodiazepine impairs place navigation in rats. *Behav Brain Res* 24:39–46

Metherate R, Weinberger NM (1990): Cholinergic modulation of responses to single tones produces tone-specific receptive field alterations in cat auditory cortex. *Synapse* 6:133–145

Metherate R, Ashe JH, Weinberger NM (1990): Acetylcholine modifies neuronal acoustic rate-level functions in guinea pig auditory cortex by an action at muscarinic receptors. *Synapse* 6:364–368

Meyer EM, Onge ES, Crews FT (1984): Effects of aging on rat cortical presynaptic cholinergic processes. *Neurobiol Aging* 5:315–317

Miller JA, Chmielewski PA (1990a): The regulation of high-affinity choline uptake in vitro in rat cortical and hippocampal synaptosomes by β-carbolines administered in vivo. *Neurosci Lett* 114:351–355

Miller JA, Chmielewski PA (1990b): Stimulation of [^{3}H]hemicholinium-3 binding in cortical or hippocampal tissues after treatment with β-carbolines or MDL 26,479. *Soc Neurosci Abs* 16:91.11

Miller JA, Richter JA (1985): Effects of anticonvulsants *in vivo* on high affinity choline uptake *in vitro* in mouse hippocampal synaptosomes. *Br J Pharmacol* 84:19–25

Miller JA, Dudley MW, Kehne JH, Sorenson SM, Kane JM (1992): MDL 26,479: A potential cognition enhancer with benzodiazepine inverse agonist-like properties. *Brit J Pharmacol,* in press

Mishkin M, Malamut B, Bachevalier J (1984): Memory and habits: Two neural systems. In: *Neurobiology of Learning and Memory*, Lynch G, McGaugh JL, Weinberger NM, eds. New York: Guilford Press

Mogenson GJ, Yang CR (1991): The contribution of basal forebrain to limbic-motor integration and the mediation of motivation to action. In: *The Basal Forebrain: Anatomy to Function*, Napier TC, Kalivas PW, Hanin I, eds. New York: Plenum Press

Mogenson GJ, Swanson LW, Wu M (1983): Neural projections from nucleus accumbens to globus pallidus, substantia innominata, and lateral preoptic-lateral hypothalamic area: An anatomical and electrophysiological investigation in the rat. *J Neurosci* 3:189–202

Moore H, Dudchenko P, Bruno JP, Sarter M (1992): Towards the modelling of age-related changes of attentional abilities in rats: Simple and choice reaction time tasks, and sustained and selective attention. *Neurobiol Aging*, in press

Morris RGM (1984): Is the distinction between procedural and declarative memory useful with respect to animal models of amnesia. In: *Neurobiology of Learning and Memory*, Lynch G, McGaugh JL, Weinberger NM, eds. New York: Guilford Press

Murphy PC, Sillito AM (1991): Cholinergic enhancement of direction selectivity in the visual cortex of the cat. *Neuroscience* 40:13–20

Nabeshima T, Tohyama K, Ichihara K, Kameyama T (1990b): Effects of benzodiazepines on passive avoidance response and latent learning in mice: Relationship to benzodiazepine receptors and the cholinergic neuronal system. *J Pharm Exp Ther* 255:789–794

Nabeshima T, Tohyama K, Kameyama T (1990a): Involvement of AChergic systems in chlordiazepoxide induced impairment of passive avoidance response. *Neurosci Lett* 117:200–206

Nagel JA, Huston JP (1988): Enhanced inhibitory learning produced by post-trial injections of substance P into the basal forebrain. *Behav Neural Biol* 49:374–385

Olton DS, Markowska AL (1988): Within-subjects, parametric manipulations to investigate aging. *Neurobiol Aging* 9:469–474

Olton D, Markowska A, Voytko ML (1991): Basal forebrain cholinergic system: A functional analysis: In: *The Basal Forebrain: Anatomy to Function*, Napier TC, Kalivas PW, Hanin I, eds. New York: Plenum Press

Page KJ, Everitt BJ, Robbins TW, Marston HM, Wilkinson LS (1991): Dissociable effects on spatial maze and passive avoidance acquisition and retention following AMPA- and ibotenic acid-induced excitotoxic lesions of the basal forebrain in rats: Differential dependence on cholinergic neuronal loss. *Neuroscience* 43:457–472

Palmer AM, Francis PT, Bowen DM, Benton JS, Neary D, Mann DMA, Snowdon JS (1987a): Catecholaminergic neurones assessed ant-mortem in Alzheimer's disease. *Brain Res* 414:365–375

Palmer AM, Francis PT, Benton JS, Sims NR, Mann DMA, Neary D, Snowdon JS, Bowen DM (1987b): Presynaptic serotonergic dysfunction in patients with Alzheimer's disease. *J Neurochem* 48:8–15

Pellow S (1985): Can drug effects on anxiety and convulsions be separated? *Neurosci Biobehav Rev* 9:55–73

Pellow S, File SE (1984): Behavioral actions of Ro 5-4864; A peripheral-type benzodiazepine. *Life Sci* 35:229–240

Phillis JW, Siemens RK, Wu PH (1980): Effects of diazepam on adenosine and acetylcholine release from rat cerebral cortex: Further evidence for a purinergic mechanism in action of diazepam. *Br J Pharmacol* 70:341–348

Polc P (1991): GABA-independent mechanisms of benzodiazepine action. In: *New Concepts in Anxiety*, Briley M, File SE, eds. Boca Raton, FL: CRC Press

Potier MC, Prado de Carvalho L, Dodd RH, Besselievre R, Rossier J (1988): In vivo binding of β-carbolines in mice: Regional differences and correlation of occupancy to pharmacological effects. *J Pharm Exp Ther* 34:124–128

Preston GC, Ward C, Lines CR, Poppleton P, Haigh JRM, Traub M (1989): Scopolamine and benzodiazepine models of dementia: Cross-reversals by Ro 15-1788 and physostigmine. *Psychopharmacology* 98:487–494

Pritchett DB, Seeburg PH (1990): γ-aminobutyric $acid_A$ receptor α_5-subunit creates novel type II benzodiazepine receptor pharmacology. *J Neurochem* 54:1802–1804

Quirion R (1987): Characterization and autoradiographic distribution of hemicholinium-3 high-affinity choline uptake sites in mammalian brain. *Synapse* 1:293–303

Rabbit P (1981): Cognitive psychology needs models for changes in performance with old age. In: *Attention and Performance IX*, Baddeley A, Long J, eds. Hillsdale, NJ: Erlbaum

Raffali-Sebille MJ, Chapouthier G (1991): Similar effects of a beta-carboline and of flumazenil in negatively and positively reinforced learning tasks in mice. *Life Sci* 48:685–692

Raffali-Sebille MJ, Chapouthier G, Venault P, Dodd RH (1990): Methyl β-carboline-3-carboxylate enhances performance in a multiple-trial learning task in mice. *Pharmacol Biochem Behav* 35:281–284

Rasmussen DD, Dykes RW (1988): Long-term enhancement of evoked potentials in cat somatosensory cortex produced by co-activation of the basal forebrain and cutaneous receptors. *Exp Brain Res* 70:276–286

Reinikainen KJ, Soininen H, Riekkinen PJ (1990): Neurotransmitter changes in Alzheimer's disease: Implications to diagnostics and therapy. *J Neurosci Res* 27:576–586

Richelson E, Nelson A, Neeper R (1991): Binding of benzodiazepines and some major metabolites at their sites in normal human frontal cortex *in vitro*. *J Pharmacol Exp Ther* 256:897–901

Ridgon GC, Pirch JH (1984): Microinjection of procaine or GABA into the nucleus basalis magnocellularis affects cue-elicited unit responses in the rat frontal cortex. *Exp Neurol* 85:283–296

Robbins TW, Everitt BJ, Ryan CN, Marston HM, Jones GH, Page KJ (1989a): Comparative effects of quisqualic and ibotenic acid-induced lesions of the substantia innominata and globus pallidus on the acquisition of a conditional visual discrimination: Differential effects on cholinergic mechanism. *Neuroscience* 28:337–352

Robbins TW, Cador M, Taylor JR, Everitt BJ (1989b): Limbic-striatal interactions in reward-related processes. *Neurosci Biobehav Rev* 13:155–162

Rupniak NMJ, Samson NA, Steventon MJ, Iversen SD (1990): Induction of cognitive impairment by scopolamine and noncholinergic agents in rhesus monkey. *Life Sci* 48:893–899

Rusted JM, Eaton-Williams P, Warburton DM (1991): A comparison of the effects of scopolamine and diazepam on working memory. *Psychopharmacology* 105:442–445

Sahgal A, Iversen SD (1980): Recognition memory, chlordiazepoxide and rhesus monkeys: Some problems and results. *Behav Brain Res* 1:227–243

Salthouse TA (1985): Speed of behavior and its implication for cognition. In: *Handbook of the Psychology of Aging*, Birren JE, Schaie KW, eds. New York: Van Nostrand Reinhold

Sarter M (1990a): Retrieval of well-learned propositional rules: Insensitive to changes in activity of individual neurotransmitter systems? *Psychobiology* 18:451–459

Sarter M (1990b): Elevations of local cerebral glucose utilization by the β-carboline ZK 93 426. *Eur J Pharmacol* 177:155–162

Sarter M (1991): Taking stock of cognition enhancers. *Trends Pharmacol Sci* 12:456–461

Sarter M, Bodewitz G, Stephens DN (1988a): Attenuation of scopolamine-induced impairment of spontaneous alternation behaviour by antagonist but not inverse agonist and agonist β-carbolines. *Psychopharmacology* 94:491–495

Sarter M, Bruno JP, Dudchenko P (1990): Activating the damaged basal forebrain cholinergic system: Tonic stimulation versus signal amplification. *Psychopharmacology* 101:1–17

Sarter M, Bruno JP, Dudchenko P (1991): Cholinergic controversies. *Trends Neurosci* 14:484

Sarter M, Dudchenko P (1991): Dissociative effects of ibotenic acid and quisqualic acid-induced basal forebrain lesions on cortical acetylcholinesterase-positive fiber density and cytochrome oxidase activity. *Neuroscience* 41:729–738

Sarter M, Hagan J, Dudchenko P (1992): Behavioral screening for cognition enhancers: from indiscriminate to valid testing. *Psychopharmacology* 107:144–159

Sarter M, Schneider HH (1988): High density of benzodiazepine binding sites in the substantia innominata of the rat. *Pharmacol Biochem Behav* 30:679–682

Sarter M, Schneider HH, Stephens DN (1988b): Treatment strategies for senile dementia: Antagonist β-carbolines. *Trends Neurosci* 11:13–17

Sarter M, Steckler T (1989): Spontaneous exploration of a 6-arm radial arm radial tunnel maze by basal forebrain lesioned rats: Effects of the benzodiazepine receptor antagonist β-carboline ZK 93 426. *Psychopharmacology* 98:193–202

Sato H, Hata Y, Hagihara K, Tsumoto T (1987): Effects of cholinergic depletion on neuron activities in the cat visual cortex. *J Neurophysiol* 58:781–794

Schmidt KF (1966): Effect of halothane anesthesia on regional acetylcholine levels in rat brain. *Anesthesiology* 27:788–792

Schoch P, Richards JG, Häring P, Takacs B, Stähli C, Staehelin T, Haefely W, Möhler H (1985): Co-localization of $GABA_A$ receptors and benzodiazepine receptors in the brain shown by monoclonal antibodies. *Nature* 314:168–171

Schoemaker H, Boles RG, Horst D, Ymamaura HI (1983): Specific high-affinity binding sites for [^{3}H]Ro5-4864 in rat brain and kidney. *J Pharmacol Exp Ther* 225:61–66

Shih TM (1991): Cholinergic actions of diazepam and atropine sulfate in soman poisoning. *Brain Res Bull* 26:565–573

Sillito AM, Kemp JA (1983): Cholinergic modulation of the functional organization of the cat visual cortex. *Brain Res* 289:143–155

Sillito AM, Murphy PC (1987): The Cholinergic Modulation Of Cortical Function. In: *Cerebral Cortex, Vol. 6*, Jones EG, Peters A, eds. New York: Plenum Press

Smith G (1988): Animal models of Alzheimer's disease: Experimental cholinergic denervation. *Brain Res Rev* 13:103–118

Smith KA, Bierkamper GG (1990): Paradoxical role of GABA in a chronic model of petit mal (absence)-like epilepsy in the rat. *Eur J Pharmacol* 176:45–55

Stephens DN, Sarter M (1988): Bidirectional nature of benzodiazepine receptor ligands extends to effects in vigilance. In: *Benzodiazepine Receptor Ligands, Memory and Information Processing*, Hindmarch I, Ott H, eds. Berlin: Springer

Tan S, Kirk RC, Abraham WC, McNaughton N (1990): Chlordiazepoxide reduces discriminability but not rate of forgetting in delayed conditional discrimination. *Psychopharmacology* 101:550–554

Tanganelli S, Bianchi C, Beani L (1985): The modulation of cortical acetylcholine release by GABA, GABA-like drugs and benzodiazepines in freely moving guinea-pigs. *Neuropharmacology* 24:291–299

Thiebot MH (1985): Some evidence for amnestic-like effects of benzodiazepines in animals. *Neurosci Biobehav Rev* 9:95–100

Thiebot MH, Soubrie P, Sanger D (1988): Anxiogenic properties of beta-CCE and FG 7142: a review of promises and pitfalls. *Psychopharmacology* 94:452–463

Turner DM, Sapp DW, Olson RW (1991): The benzodiazepine/alcohol antagonist Ro 15-4513: Binding to a $GABA_A$ receptor subtype that is insensitive to diazepam. *J Pharmacol Exp Ther* 257:1236–1242

Unnerstall JR, Kuhar MJ, Niehoff DL, Palacios JM (1981): Benzodiazepine receptors are coupled to a subpopulation of γ-aminobutyric acid (GABA) receptors: Evidence from a quantitative study. *J Pharm Exp Ther* 218:797–804 van der Staay FJ (1989): *Behavioral Consequences of Chronic Dietary Choline Enrichment*. Thesis, University of Nijmegen

Venault P, Chapouthier G, Prado de Carvalho L, Simiand J, Morre M, Dodd RH, Rossier J (1986): Benzodiazepine impairs and β-carboline enhances performance in learning and memory tasks. *Nature* 321:864–866

Von Blankenfeld G, Ymer S, Pritchett DB, Sontheimer H, Ewert M, Seeburg PH, Kettenmann H (1990): Differential benzodiazepine pharmacology of mammalian recombinant $GABA_A$ receptors. *Neurosci Lett* 115:269–273

Walker LC, Price DL, Young WS (1989): GABAergic neurons in the primate basal forebrain magnocellular complex. *Brain Res* 499:188–192

Warburton DM (1977): Stimulus selection and behavioral inhibition. In: *Drugs, Neurotransmitters, and Behavior, Handbook of Psychopharmacology, Vol. 8*, Iversen LL, Iversen SD, Solomon SH, eds. New York: Plenum Press

Wenk GL (1984): Pharmacological manipulations of the substantia innominata-cortical cholinergic pathway. *Neurosci Lett* 51:99–103

Wenk GL, Hepler D, Olton D (1984): Behavior alters the uptake of [^{3}H]choline into acetylcholinergic neurons of the nucleus basalis magnocellularis and medal septal area. *Behav Brain Res* 13:129–138

Westerink BHC, de Boer P (1990): Effect of choline administrtion on the release of acetylcholine from the striatum as determined by microdialysis in awake rats. *Neurosci Lett* 112:297–301

Whitehouse P, Price D, Struble R, Clark A, Coyle J, DeLong M (1983): Alzheimer's disease and senile dementia: loss of neurons in the basal forebrain. *Science* 215:1237–1239

Wood PL (1985): Pharmacological evaluation of GABAergic and glutamatergic inputs to the nucleus basalis-cortical and septal-hippocampal cholinergic projections. *Can J Physiol Pharmacol* 64:325–328

Wurtman RJ, Blusztajn JK, Ulus IH, Coviella ILG, Buyukuysal L, Growdon JH, Slack BE (1990): Choline metabolism in cholinergic neurons: Implications for the pathogenesis of neurodegenerative diseases. *Adv Neurol* 51:117–125

Zaborszky L, Heimer L, Eckenstein F, Leranth C (1986): GABAergic input to cholinergic forebrain neurons: An ultrastructural study using retrograde tracing of HRP and double immunolabeling. *J Comp Neurol* 250:282–295

Zaborszky L, Cullinan WE, Braun A (1991): Afferents to basal forebrain cholinergic projection neurons: an update. In: *The Basal Forebrain: Anatomy to Function*, Napier TC, Kalivas PW, Hanin I, eds. New York: Plenum Press

Zahm DS, Heimer L (1990): Two transpallidal pathways originating in the rat nucleus accumbens. *J Comp Neurol* 302:437–446

Zhang H, Rosenberg HC, Tietz EI (1989): Injection of benzodiazepines but not GABA or muscimol into pars reticulata of substantia nigra suppresses pentylenetetrazol seizures. *Brain Res* 488:73–79

22

Transmitter Interactions and Cognitive Function: Appreciation of the Concert

EDWARD D. LEVIN, MICHAEL DECKER, AND LARRY L. BUTCHER

By their nature neurons are interactive. Every set of neurons in the brain interacts closely with other neurons. To comprehend the functioning of the brain we must understand the character of neuronal interactions and how they act together as a system. Even very elementary functions, such as regulation of temperature or accoustic startle response, depend upon the interaction of a variety of different neuronal types. Complex functions, such as cognitive behavior, to even a greater degree depend on interactions of multiple neural systems that use a variety of neurotransmitters. Investigation of the role of neurotransmitters in cognitive function has necessarily focused on understanding the influences of neurotransmitter systems in isolation. For example, interest in the role of the basal forebrain cholinergic system in learning and memory characterizes much of the recent work on this subject. Extensive research of this sort has clearly advanced our understanding of the independent contributions of cholinergic and other transmitter systems to memory processes. More complete understanding of neurotransmitter involvement in cognitive function, however, requires that we also consider interactions between these neurotransmitter systems. The goal of this volume, then, was to examine the relationships among transmitter systems that provide the bases for cognitive function.

In the introductory chapter, Roger Russell provided a compelling rationale for this endeavor by explaining the importance of considering the brain as an integrative organ. Specific attributes of neural interactions involved in cognitive functions are examined from a variety of perspectives in the subsequent chapters, with an emphasis on interactions involving cholinergic mechanisms.

Anatomic Bases for Interactions

Chapters by Butcher and Záborsky detailed anatomic bases for the interactions between different neural systems in the basal forebrain, a critical source of cortical projections important for memory function. Detailed information concerning the anatomic nature of the contacts between transmitter systems is essential for the interpretation of physiological, pharmacological, and behavioral studies of interactions.

Interactions within the Cholinergic System

It has long been known that the cholinergic system is of crucial importance for cognitive function. However, it is probably inaccurate to consider the cholinergic system as a unitary force in isolation. First, there are many components within the cholinergic domain that undoubtably play different roles in cognitive function. Both muscarinic and nicotinic cholinergic receptors are important for cognitive function and, as reviewed in the chapter by Levin and Russell, these two receptor subtypes interact on a number of levels in the basis of cognitive function. The interaction of different cholinergic projection systems, principally the septohippocampal and basalocortical cholinergic systems, is also important. These issues were addressed in the chapter by Olton and Pang.

Interactions among Neurotransmitter Systems

As with any neural system in the brain, the cholinergic system does not function in isolation. It is influenced by some neural systems and, in turn, influences others as a part of the information processing network of the brain. Experimental studies have demonstrated that a variety of transmitter systems, in addition to the cholinergic system, are important for cognitive function; the majority of the chapters in this volume deal with how these systems interact.

The importance of catecholamine-cholinergic interactions was explained in chapters by Napier et al., Jaffard et al., Haroutunian et al., Levin and Rose, and Brito. The influence of neurotensin on cholinergic function was outlined in the chapter by Wenk. Serotonin influences on cholinergic function were addressed in chapters by Normile and Altman, Richter-Levin and Segal, Tomaz et al., and Vanderwolf and Penava. GABA/benzodiazepine-cholinergic interactions were covered in chapters by Chrobak and Napier, Givens and Olton, Walsh and Stackman, and Sarter et al.

Interactions can take many forms. They might be enabling, cooperative interactions between systems working in parallel or may involving a balance of opposing effects. Interactions can also be serial in nature, with one neurotransmitter system regulating the function of another. Furthermore, the form of interaction between even the same two neurotransmitter systems can clearly differ depending on the particular behavior involved or the anatomic site of interaction. Different studies of complex systems can often produce apparently conflicting results. For example, the chapters by Richter-Levin and Segal, and by Vanderwolf and Penava, suggest that the disruption of serotonergic function potentiates the impairments produced by experimentally induced cholinergic dysfunction, whereas the chapter by Normile and Altman suggests the opposite, that serotonergic blockers ameliorate deficits induced by cholinergic dysfunction. We should be neither surprised nor discouraged by inconsistencies such as these because, in all likelihood, discrepancies will prove to be critical for defining the nature of the interactions more precisely.

Conclusions

Study of these interactions gives a more accurate view of how the brain works to provide cognitive function. It also opens new avenues for the development of treatment for cognitive dysfunction. The dementia associated with Alzheimer's disease had been thought to be primarily due to cholinergic hypofunction; however, given that a variety of transmitter systems are affected, it is likely that transmitter interactions are important in the understanding and treatment of the cognitive deficits seen. Even transmitter systems relatively unaffected by the disease may be important to consider for therapy. Rather than directing treatments at a system that is severely compromised by a disease, it may be a better therapeutic strategy to direct treatment at systems that are involved in the same functions as the affected system but are still relatively intact.

Molecular biology has in recent years provided a variety of tools with which to discover the rich detail of biochemical physiology. However, it is also important to determine how individual systems act together to provide the function required of the brain. The tools of molecular biology can also serve to help understand how elementary molecular events are related to the interactions of systems of neurons in the generation and governance of normal behavioral function, how disruption of these interactions may lead to dysfunction, and how treatments can be devised to restore the lost function. For basic understanding of how the brain works, it is essential that one consider how the interactions of systems of neurons relate to behavioral function. The greater challenge than determining the essential components of a system is to understand how those components act together to form the whole.

Index

GPSR Compliance

The European Union's (EU) General Product Safety Regulation (GPSR) is a set of rules that requires consumer products to be safe and our obligations to ensure this.

If you have any concerns about our products, you can contact us on ProductSafety@springernature.com

In case Publisher is established outside the EU, the EU authorized representative is:

Springer Nature Customer Service Center GmbH
Europaplatz 3
69115 Heidelberg, Germany

Batch number: 10577878

Printed by Printforce, the Netherlands